FREQUENTLY USED NOTATION

ℓ_u = unsupported length of a compression member

M_{cr} = cracking moment of concrete

M_1 = smaller end factored moment in a compression member

M_2 = larger end factored moment in a compression member

M_{1b} = smaller factored end moment in a compression member due to loads that result in no appreciable sidesway

M_{2b} = larger factored end moment in a compression member due to loads that result in no appreciable sidesway

M_d = moment due to dead load

n = modular ratio (ratio of modulus of elasticity of steel to that of concrete)

P_c = Euler buckling load of column

P_{no} = pure axial load capacity of column

P_0 = nominal axial load strength of a member with no eccentricity

q_a = allowable soil pressure

q_e = effective soil pressure

R_n = a term used in required percentage of steel expression for flexural members $\left(\dfrac{M_u}{\phi bd^2}\right)$

s = spacing of shear or torsional reinforcing parallel to longitudinal reinforcing

v_{tu} = torsional stress

w_c = unit weight of concrete

y_t = distance from centroidal axis of gross section to extreme fiber in tension

z = a term used to estimate crack sizes and specify distribution of reinforcing

β = ratio of long side to short side of footing

β_c = ratio of long side to short side of concentrated load or reaction area

β_d = ratio of maximum factored dead load moment to maximum total load moment (always positive)

β_d = ratio of maximum factored sustained lateral force to maximum factored total lateral force

β_h = ratio of the distance to the neutral axis from the extreme tension concrete fiber to the distance from the neutral axis to the centroid of the tensile steel

β_1 = a factor to be multiplied by the effective depth d of a member to obtain the depth of the equivalent rectangular stress block

δ = computed deflection

δ_b = moment magnification factor for slender columns in frames braced against sidesway

δ_s = moment magnification factor for slender columns in frames not braced against sidesway

ϵ_c = strain in compression concrete

ϵ_s = strain in tension reinforcement

ϵ_s' = strain in compression reinforcement

λ = a multiplier used in computing long-term deflections

μ = coefficient of friction

ξ = a time-dependent factor for sustained loads used in computing long-term deflections

ρ = ratio of nonprestressed reinforcement in a section

ρ' = ratio of compression reinforcing in a section

ρ_b = ratio of tensile reinforcing producing balanced strain condition

ϕ = capacity reduction factor

ω = crack width

Design of Reinforced Concrete

THIRD EDITION

Jack C. McCormac
Clemson University

HarperCollins*CollegePublishers*

Sponsoring Editor: John Lenchek
Project Editors: Kristin Syverson, Cathy Wacaser
Art Director: Julie Anderson
Cover Design: Lesiak/Crampton Design Inc: Cynthia Crampton
Cover Photo: The Chicago Photographic Company
Production Administrator: Brian Branstetter
Compositor: Better Graphics, Inc.
Printer and Binder: R.R. Donnelley & Sons Company
Cover Printer: The Lehigh Press, Inc.

Design of Reinforced Concrete, Third Edition
Copyright © 1993 by HarperCollins College Publishers

Library of Congress Cataloging-in-Publication Data

McCormac, Jack C.
    Design of reinforced concrete/Jack C. McCormac.—3rd ed.
      p.   cm.
    Includes index.
    ISBN 0-06-500491-4
    1. Reinforced concrete construction.   I. Title.
TA683.2.M39   1992
624.1'8341—dc20   92-35494
                                      CIP

95 9 8 7 6 5

Table of Contents

iii

Preface

The purpose of this book, now in its third edition, remains unchanged: To introduce the fundamentals of reinforced concrete design in such a manner as to stimulate interest in the subject. Although the material was written primarily for students just beginning the study of reinforced concrete, it is hoped that it will be useful to practicing designers as well.

Quite a few changes and additions have been made. The book has been updated to conform to the 1989 ACI Code. A computer disk is enclosed with which many common reinforced concrete problems can be solved. The two chapters relating to short columns subject to bending and slender columns subject to bending have been completely rewritten. Most of the homework problems have been revised and many new ones added. In addition, numerous miscellaneous changes have been made throughout the text.

The author wishes to thank the following persons who reviewed this edition:

Abbas Aminmansour, *Pennsylvania State University*
Mohammed A. Basha, *University of Utah*
Spencer L. Brinkerhoff
Nadim Hassoun, *South Dakota State University*
Jai B. Kim, *Bucknell University*
Steve Leftwich, *West Virginia Institute of Technology*
Jack W. Schwalbe, *Florida Institute of Technology*
Methi Wecharatana, *New Jersey Institute of Technology*

He also thanks the reviewers and users of the previous editions for their suggestions and criticisms. He is always grateful to anyone who takes the time to contact him concerning any part of the book.

Jack C. McCormac
Clemson University

1

Introduction

1.1 CONCRETE AND REINFORCED CONCRETE

Concrete is a mixture of sand, gravel, crushed rock, or other aggregates held together in a rocklike mass with a paste of cement and water. Sometimes one or more admixtures are added to change certain characteristics of the concrete such as its workability, durability, and time of hardening.

As with most rocklike substances, concrete has a high compressive strength and a very low tensile strength. *Reinforced concrete* is a combination of concrete and steel wherein the steel reinforcement provides the tensile strength lacking in the concrete. Steel reinforcing is also capable of resisting compression forces and is used in columns as well as in other situations to be described later.

1.2 ADVANTAGES OF REINFORCED CONCRETE AS A STRUCTURAL MATERIAL

Reinforced concrete may be the most important material available for construction. It is used in one form or another for almost all structures, great or small—buildings, bridges, pavements, dams, retaining walls, tunnels, viaducts, drainage and irrigation facilities, tanks, and so on.

The tremendous success of this universal construction material can be understood quite easily if its numerous advantages are considered. These include the following:

1. It has considerable compressive strength as compared to most other materials.

1

Peachtree Center Hotel, the tallest hotel in the world, Atlanta, Georgia. (Courtesy Symons Corporation.)

2. Reinforced concrete has great resistance to the actions of fire and water and, in fact, is the best structural material available for situations where water is present. During fires of average intensity, members with a satisfactory cover of concrete over the reinforcing bars suffer only surface damage without failure.
3. Reinforced concrete structures are very rigid.
4. It is a low-maintenance material.
5. As compared with other materials, it has a very long service life. Under proper conditions, reinforced concrete structures can be used indefinitely without reduction of their load-carrying abilities. This can be explained by the fact that the strength of concrete does not decrease with time but actually increases over a very long period, measured in years, due to the lengthy process of the solidification of the cement paste.

The 320-foot high Pyramid Sports Arena, Memphis, Tennessee. (Courtesy of Economy Forms Corporation.)

6. It is usually the only economical material available for footings, basement walls, piers, and similar applications.
7. A special feature of concrete is its ability to be cast into an extraordinary variety of shapes from simple slabs, beams, and columns to great arches and shells.
8. In most areas, concrete takes advantage of inexpensive local materials (sand, gravel, and water) and requires relatively small amounts of cement and reinforcing steel.
9. A lower grade of skilled labor is required for erection as compared to other materials such as structural steel.

1.3 DISADVANTAGES OF REINFORCED CONCRETE AS A STRUCTURAL MATERIAL

To use concrete successfully the designer must be completely familiar with its weak points as well as with its strong ones. Among its disadvantages are the following:

1. Concrete has a very low tensile strength, requiring the use of tensile reinforcing.
2. Forms are required to hold the concrete in place until it hardens sufficiently. In addition, falsework or shoring may be necessary to keep the forms in place for roofs, walls, and similar structures until the concrete members gain sufficient strength to support themselves. Formwork is very expensive. Its costs run from one-third to two-thirds of the total cost of a reinforced concrete structure, with average values of about 50%.

3. The low strength per unit of weight of concrete leads to heavy members. This becomes an increasingly important matter for long-span structures where concrete's large dead weight has a great effect on bending moments.
4. Similarly, the low strength per unit of volume of concrete means members will be relatively large, an important consideration for tall buildings and long-span structures.
5. The properties of concrete vary widely due to variations in its proportioning and mixing. Furthermore, the placing and curing of concrete is not as carefully controlled as is the production of other materials such as structural steel and laminated wood.

Two other characteristics that can cause problems are concrete's shrinkage and creep. These characteristics are discussed in Section 1.9 of this chapter.

1.4 HISTORICAL BACKGROUND

The average person thinks that concrete has been in common use for many centuries, but such is not the case. Although the Romans made cement—called *pozzolana*—before the birth of Christ by mixing slaked lime with a volcanic ash from Mount Vesuvius and used it to make concrete for building, the art was lost during the Dark Ages and was not revived until the eighteenth and nineteenth centuries. A deposit of natural cement rock was discovered in England in 1796 and was sold as "Roman cement." Various other deposits of natural cement were discovered in both Europe and America and were used for several decades.

The real breakthrough for concrete occurred in 1824 when an English bricklayer named Joseph Aspdin, after long and laborious experiments, obtained a patent for a cement which he called "portland cement" because its color was quite similar to that of the stone quarried on the Isle of Portland off the English coast. He made his cement by taking certain quantities of clay and limestone, pulverizing them, burning them in his kitchen stove, and grinding the resulting clinker into a fine powder. During the early years after its development, his cement was primarily used in stuccos.[1] This wonderful product was very slowly adopted by the building industry and was not even introduced into the United States until 1868; the first portland cement was not manufactured in the United States until the 1870s.

The first uses of reinforced concrete are not very well known. Much of the early work was done by two Frenchmen, Joseph Lambot and Joseph Monier. In about 1850 Lambot built a concrete boat reinforced with a network of parallel wires or bars. Credit is usually given to Monier, however, for the invention of reinforced concrete. In 1867 he received a patent for the construction of con-

---

[1] Kirby, R. S., and Laurson, P. G., 1932, *The Early Years of Modern Civil Engineering* (New Haven: Yale University Press), p. 266.

crete basins or tubs and reservoirs reinforced with a mesh of iron wire. His stated goal in working with this material was to obtain lightness without sacrificing strength.[2]

From 1867 to 1881 Monier received patents for reinforced concrete railroad ties, floor slabs, arches, footbridges, buildings, and other items in both France and Germany. Another Frenchman, François Coignet, built simple reinforced concrete structures and developed basic methods of design. In 1861 he published a book in which he presented quite a few applications. He was the first person to realize that the addition of too much water in the mix greatly reduced concrete strength. Other Europeans who were early experimenters with reinforced concrete included the Englishmen William Fairbairn and William B. Wilkinson, the German G. A. Wayss, and another Frenchman, François Hennebique.[3,4]

William E. Ward built the first reinforced concrete building in the United States in Port Chester, N.Y., in 1875. In 1883 he presented a paper before the American Society of Mechanical Engineers in which he claimed that he got the idea of reinforced concrete by watching English laborers in 1867 trying to remove hardened cement from their iron tools.[5]

Thaddeus Hyatt, an American, was probably the first person to correctly analyze the stresses in a reinforced concrete beam, and in 1877 he published a 28-page book on the subject, entitled *An Account of Some Experiments with Portland Cement Concrete, Combined with Iron as a Building Material*. In this book he praised the use of reinforced concrete and said that "rolled beams have to be taken largely on faith." Hyatt put a great deal of emphasis on the high fire resistance of concrete.[6]

E. L. Ransome of San Francisco is supposed to have used reinforced concrete in the early 1870s and was the originator of deformed (or twisted) bars, for which he received a patent in 1884. These bars, which were square in cross section, were cold-twisted with one complete turn in a length of not more than 12 times the bar diameter.[7] (The purpose of the twisting was to provide better bonding or sticking together of the concrete and the steel.) In 1890 in San Francisco, Ransome built the Leland Stanford Jr. Museum. It is a reinforced concrete building 312 feet long and two stories high in which discarded wire rope

[2] Kirby, R. S., and Laurson, P. G., 1932, *The Early Years of Modern Civil Engineering* (New Haven: Yale University Press), pp. 273–275.

[3] Straub, H., 1964, *A History of Civil Engineering* (Cambridge: The M.I.T. Press), pp. 205–215. Translated from the German *Die Geschichte der Bauingenieurkunst*, Verlag Birkhauser, Basel, 1949.

[4] Kirby, R. S., and Laurson, P. G., 1932, *The Early Years of Modern Civil Engineering* (New Haven: Yale University Press), pp. 273–275.

[5] Ward, W. E., 1883, "Béton in Combination with Iron as a Building Material," *Transactions ASME*, 4, pp. 388–403.

[6] Kirby, R. S., and Laurson, P. G., 1932, *The Early Years of Modern Civil Engineering* (New Haven: Yale University Press), p. 275.

[7] American Society for Testing Materials, 1911, *Proceedings* (vol. XI), pp. 66–68.

from a cable-car system was used as tensile reinforcing. This building experienced little damage in the 1906 earthquake. Since 1980 the development and use of reinforced concrete in the United States has been very rapid.[8,9]

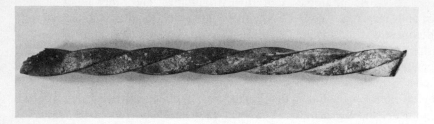

Twisted bar. (Courtesy Clemson University Communications Center.)

1.5 COMPARISON OF REINFORCED CONCRETE AND STRUCTURAL STEEL FOR BUILDINGS AND BRIDGES

When a particular type of structure is being considered, the student may be perplexed by the question, "Should reinforced concrete or structural steel be used?" There is much joking on this point, with the proponents of reinforced concrete referring to steel as that material which rusts, while those favoring structural steel refer to concrete as that material which when overstressed tends to return to its natural state—that is, sand and gravel.

There is no simple answer to this question, inasmuch as both of these materials have many excellent characteristics that can be utilized successfully for so many types of structures. In fact, they are often used together in the same structures with wonderful results.

The selection of the structural material to be used for a particular building depends on the height and span of the structure, the material market, foundation conditions, local building codes, and architectural considerations. For buildings of less than 4 stories, reinforced concrete, structural steel, and wall-bearing construction are competitive. From 4 to about 20 stories, reinforced concrete and structural steel are economically competitive, with steel having taken most of the jobs above 20 stories in the past. Today, however, reinforced concrete is becoming increasingly competitive above 20 stories, and there are a number of reinforced concrete buildings of greater height around the world. The 71-story

[8] Wang, C. K., and Salmon, C. G., 1992, *Reinforced Concrete Design*, 5th ed., (New York: Harper-Collins), pp. 4–5.

[9] "The Story of Cement, Concrete and Reinforced Concrete," *Civil Engineering*, November 1977, pp. 63–65.

969-ft building at 311 South Wacker Drive in Chicago is the tallest reinforced concrete building in the world.

Though we would all like to be involved with the design of tall prestigious reinforced concrete buildings, there are just not enough of them to go around. As a result, nearly all of our work involves much smaller structures. Perhaps 9 out of 10 buildings in the United States are 3 stories or less in height, and more than two-thirds of them contain 15,000 sq ft or less of floor space.

Foundation conditions can very often affect the selection of the material to be used for the structural frame. If foundation conditions are poor, a lighter structural steel frame may be desirable. The building code in a particular city may be favorable to one material over the other. For instance, many cities have fire zones in which only fireproof structures can be erected—a very favorable situation for reinforced concrete. Finally, the time element favors structural steel frames, as they can be erected much more quickly than reinforced concrete ones. The time advantage, however, is not as great as it might seem at first glance because if the structure is to have any type of fire rating, the builder will have to cover the steel with some kind of fireproofing material after it is erected.

To make a decision about using concrete or steel for a bridge will involve several factors, such as span, foundation conditions, loads, architectural considerations, and others. In general, concrete is an excellent compression material and normally will be favored for short-span bridges and for cases where rigidity is required (as, perhaps, for railway bridges).

1.6 COMPATIBILITY OF CONCRETE AND STEEL

Concrete and steel reinforcing work together beautifully in reinforced concrete structures. The advantages of each material seem to compensate for the disadvantages of the other. For instance, the great shortcoming of concrete is its lack of tensile strength; but tensile strength is one of the great advantages of steel. Reinforcing bars have tensile strengths equal to approximately 100 times that of the usual concretes used.

The two materials bond together very well so there is no slippage between the two, and thus they will act together as a unit in resisting forces. The excellent bond obtained is due to the chemical adhesion between the two materials, the natural roughness of the bars, and the closely spaced rib-shaped deformations rolled on the bar surfaces.

Reinforcing bars are subject to corrosion, but the concrete surrounding them provides them with excellent protection. The strength of exposed steel subject to the temperatures reached in fires of ordinary intensity is nil, but the enclosure of the reinforcement in concrete produces very satisfactory fire ratings. Finally, concrete and steel work very well together in relation to temperature changes because their coefficients of thermal expansion are quite close to each other. For steel the coefficient is 0.0000065, while it varies for concrete from about 0.000004 to 0.000007 (average value 0.0000055).

1.7 DESIGN CODES

The most important code in the United States for reinforced concrete design is the American Concrete Institute's *Building Code Requirements for Reinforced Concrete* (ACI 318-89).[10] This code, which is primarily used for the design of buildings, is followed for the majority of the examples given in this text. Frequent references are made to this document and section numbers are provided. Design requirements for various types of reinforced concrete members are presented in the Code along with a "Commentary" on those requirements. The Commentary provides explanations, suggestions, and additional information concerning the design requirements. As a result, users will obtain a better background and understanding of the Code.

The ACI Code is not in itself a legally enforceable document. It is merely a statement of current good practice in reinforced concrete design. It is, however, written in the form of a code or law so that various public bodies such as city councils can easily vote it into their local building codes, and as such it becomes legally enforceable in that area. In this manner the ACI Code has been incorporated into law by countless government organizations throughout the United States. It is also widely accepted in Canada and Mexico and has had a tremendous influence on concrete codes of all countries throughout the world.

As more knowledge is obtained pertaining to the behavior of reinforced concrete, the ACI revises its code. The present objective is to make yearly changes in the code in the form of supplements and to provide major revisions of the entire code every six or seven years.

Other well-known reinforced concrete specifications are those of the American Association of State Highway and Transportation Officials (AASHTO) and the American Railway Engineering Association (AREA).

1.8 TYPES OF PORTLAND CEMENT

Concretes made with normal portland cement require about two weeks to achieve a sufficient strength to permit the removal of forms and the application of moderate loads. Such concretes reach their design strengths after about 28 days and continue to gain strength at a slower rate thereafter.

On many occasions it is desirable to speed up construction by using *high-early-strength cements* which, though more expensive, enable us to obtain desired strengths in 7 to 14 days rather than the normal 28 days. These cements are particularly useful for the fabrication of precast members where the concrete is placed in forms where it quickly gains desired strengths and is then removed from the forms and the forms used to produce more members. Obviously the

---

[10] *Building Code Requirements for Reinforced Concrete* (ACI 318-89). Detroit: American Concrete Institute, 1989.

quicker the desired strength is obtained, the more efficient the operation. A similar discussion can be made for the forming of concrete buildings floor by floor. High-early-strength cements can also be used advantageously for emergency repairs of concrete and for *shotcreting* (where a mortar or concrete is blown through a hose at a high velocity onto a prepared surface).

There are other special types of portland cements available. The chemical process which occurs during the setting or hardening of concrete produces heat. For very massive concrete structures such as dams and piers, the heat will dissipate very slowly and can cause serious problems. It will cause the concrete to expand during hydration. When cooling, the concrete will shrink and severe cracking will often occur.

Concrete may be used where it is exposed to various chlorides and/or sulfates. Such situations occur in seawater construction and for structures exposed to various types of soil. Portland cements are manufactured which have lower heat of hydration, and others are manufactured with greater resistance to attack by chlorides and sulfates.

In the United States the American Society for Testing and Materials (ASTM) recognizes five types of portland cement. These different cements are manufactured from just about the same raw materials, but their properties are changed by using various blends of those materials. Type I cement is the normal cement used for most construction, but there are four other types which are useful for special situations where high early strength or low heat or sulfate resistance are needed. A brief description of these cement types follows.

Type I—the common all-purpose cement used for general construction work.

Type II—a modified cement which has a lower heat of hydration than does Type I cement and which can withstand some exposure to sulfate attack.

Type III—a high-early-strength cement which will produce in the first 24 hours a concrete with a strength about twice that of Type I cement. This cement does have a much higher heat of hydration.

Type IV—a low-heat cement which produces a concrete which dissipates heat very slowly. It is used for very large concrete structures.

Type V—a cement used for concretes which are to be exposed to high concentrations of sulfate.

Should the desired type of cement not be available, various admixtures may be purchased with which the properties of Type I cement can be modified to produce the desired effect.

1.9 PROPERTIES OF REINFORCED CONCRETE

A thorough knowledge of the properties of concrete is necessary for the student before he or she begins to design reinforced concrete structures. An introduction to several of these properties is presented in this section.

Compressive Strength

The compressive strength of concrete (f'_c) is determined by testing to failure 28-day-old 6-in. by 12-in. concrete cylinders at a specified rate of loading. For the 28-day period the cylinders are usually kept under water or in a room with constant temperature and 100% humidity. Although concretes are available with 28-day ultimate strengths from 2500 psi up to as high as 10,000 to 20,000 psi, most of the concretes used fall into the 3000- to 7000-psi range. For ordinary applications, 3000- and 4000-psi concretes are used, whereas for prestressed construction, 5000- and 6000-psi strengths are common. For some applications, such as for the columns of the lower stories of high-rise buildings, concretes with strengths up to 9000 psi have been used and furnished by ready-mix companies. At Two Union Square in Seattle, concrete with strengths up to 19,000 psi was used.

It is quite feasible to move from 3000-psi concrete to 5000-psi concrete without requiring excessive amounts of labor or cement. The approximate increase in cost for such a strength increase is 15% to 20%. To move above 5000- or 6000-psi concrete, however, requires very careful mix designs and considerable attention to such details as mixing, placing, and curing. These requirements cause relatively larger increases in cost.

Several comments are made throughout the text regarding the relative economy of using different strength concretes for different applications, such as for beams, columns, footings, and prestressed members.

It will be noted that field conditions are not the same as those in the curing room, and the 28-day strengths described here cannot be achieved in the field unless almost perfect proportioning, mixture, vibration, and moisture conditions are present. The result is that the same strength probably will not be obtained in the field with the same mixes. As a result, Section 5.3 of the ACI Code requires that the concrete compressive strengths used as a basis for selecting the concrete proportions must exceed the specified 28-day strengths by anywhere from a few hundred psi up to as much as 1400 psi, the actual amount depending on the quality control records of the concrete plant.

The stress–strain curves of Figure 1.1 represent the results obtained from compression tests of sets of 28-day-old standard cylinders of varying strengths. You should carefully study these curves because they bring out several significant points:

(a) The curves are roughly straight while the load is increased from zero to about one-third to one-half the concrete's ultimate strength.

(b) Beyond this range the behavior of concrete is nonlinear. This lack of linearity of concrete stress–strain curves at higher stresses causes some problems in the structural analysis of concrete structures because their behavior is also nonlinear at higher stresses.

(c) Of particular importance is the fact that regardless of strengths, all the concretes reach their ultimate strengths at strains of about 0.002.

(d) Concrete does not have a definite yield strength; rather, the curves run smoothly on to the point of rupture at strains of from 0.003 to 0.004.

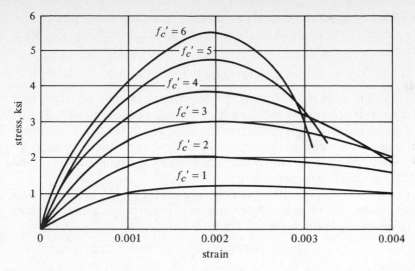

Figure 1.1 Typical concrete stress–strain curve, with short-term loading.

It will be assumed for the purpose of future calculations in this text that concrete fails at 0.003. *The reader should note that this value, which is conservative for normal-strength concretes, may not be conservative for higher-strength concretes in the 8000-psi-and-above range.*

(e) Many tests have clearly shown that stress–strain curves of concrete cylinders are almost identical with those for the compression sides of beams.

(f) It should be further noticed that the weaker grades of concrete are less brittle than the stronger ones—that is, they will take larger strains before breaking.

Modulus of Elasticity

Concrete has no clear-cut modulus of elasticity. Its value varies with different concrete strengths, concrete age, type of loading, and the characteristics of the cement and aggregates. Furthermore, there are several different definitions of the modulus:

(a) The *initial modulus* is the slope of the stress–strain diagram at the origin of the curve.

(b) The *tangent modulus* is the slope of a tangent to the curve at some point along the curve—for instance, at 50% of the ultimate strength of the concrete.

(c) The slope of a line drawn from the origin to a point on the curve somewhere between 25% and 50% of its ultimate compressive strength is referred to as a *secant modulus*.

(d) Another modulus, called the *apparent modulus* or the *long-term modulus*, is determined by using the stresses and strains obtained after the load has been applied for a certain length of time.

Section 8.5 of the ACI Code states that the following expression can be used for calculating the modulus of elasticity of concretes weighing from 90 to 155 lb/ft^3.

$$E_c = w_c^{1.5} 33 \sqrt{f_c'}$$

In this expression, w_c is the weight of the concrete in pounds per cubic foot and f_c' is its 28-day compressive strength in pounds per square inch. This is actually a secant modulus with the line (whose slope equals the modulus) drawn from the origin to a point on the stress–strain curve corresponding approximately to the stress $(0.45 f_c')$ that would occur under the estimated dead and live loads the structure has to support.

For normal-weight concrete weighing approximately 145 lb/ft^3, the ACI Code states that the following simplified version of the previous expression may be used to determine the modulus:

$$E_c = 57,000 \sqrt{f_c'}$$

Table A.1 (see the Appendix at the end of the book) shows values of E_c for different-strength concretes. These values were calculated with the first of the preceding formulas assuming 145 lb/ft^3 concrete.

Poisson's Ratio

As a concrete cylinder is subjected to compressive loads, it not only shortens in length but also expands laterally. The ratio of this lateral expansion to the longitudinal shortening is referred to as *Poisson's ratio*. Its value varies from about 0.11 for the higher-strength concretes to as high as 0.21 for the weaker-grade concretes, with average values of about 0.16.

Shrinkage

When the materials for concrete are mixed together, the cement and water paste fills the voids between the aggregate and bonds the aggregate together. This mixture needs to be sufficiently workable or fluid so that it can be made to flow in between the reinforcing bars and all through the forms. To achieve this desired workability, considerably more water (perhaps twice as much) is used than is necessary for the cement and water to react together (called *hydration*).

After the concrete has been cured and begins to dry, the extra mixing water that was used begins to work its way out of the concrete to the surface, where it evaporates. As a result, the concrete shrinks and cracks. The resulting cracks may reduce the shear strength of the members and be detrimental to the appearance of the structure. In addition, the cracks may permit the reinforcing to be exposed to the atmosphere, thereby increasing the possibility of corrosion. Shrinkage continues for many years, but under ordinary conditions probably about 90% of it occurs during the first year. The amount of moisture that is lost varies with the distance from the surface. Furthermore, the larger the surface area of a member in proportion to its size, the larger the shrinkage; that is, members with small cross sections shrink more proportionately than do those with large ones.

The amount of shrinkage is heavily dependent on the type of exposure. For instance, if concrete is subjected to a considerable amount of wind during curing, its shrinkage will be greater. In a related fashion a humid atmosphere means less shrinkage, whereas a dry one means more.

It should also be realized that it is desirable to use low absorptive aggregates such as those from granite and many limestones. When certain absorptive slates and sandstone aggregates are used, the result may be $1\frac{1}{2}$ or even 2 times the shrinkage with other aggregates.

To minimize shrinkage it is desirable to: (1) keep the amount of mixing water to a minimum; (2) cure the concrete well; (3) place the concrete for walls, floors, and other large items in small sections (thus allowing some of the shrinkage to take place before the next section is placed; (4) use construction joints to control the position of cracks; (5) use shrinkage reinforcement; and (6) use appropriate dense and nonporous aggregates.[11]

Creep

Under sustained compressive loads concrete will continue to deform for long periods of time. This additional deformation is called *creep*, or *plastic flow*. If a compressive load is applied to a concrete member, an immediate or instantaneous or elastic shortening occurs. If the load is left in place for a long time, the member will continue to shorten over a period of several years and the final deformation will usually be 2 to 3 times the initial deformation. We will find in Chapter 5 that this means that long-term deflections may also be as much as 2 or 3 times initial deflections. Perhaps 75% of the total creep will occur during the first year.

Should the long-term load be removed, the member will recover most of its elastic strain and a little of its creep strain. If the load is replaced, both the elastic and creep strains will again be developed.

The amount of creep is very dependent on the amount of stress. It is almost directly proportional to stress as long as the sustained stress is not greater than about one-half of f'_c. Beyond this level, creep will increase rapidly.

Long-term loads not only cause creep but also can adversely affect the strength of the concrete. For loads maintained on concentrically loaded specimens for a year or longer, there may be a strength reduction of perhaps 15% to 25%. *Thus a member loaded with a sustained load of say 85% of its ultimate compression strength f'_c may very well be satisfactory for a while but may fail later.*[12]

---

[11] Leet, K., 1991, *Reinforced Concrete Design*, 2nd ed. (New York: McGraw–Hill), p. 35.

[12] Rüsch, H., 1960, "Researches Toward a General Flexure Theory for Structural Concrete," *Journal ACI*, 57 (1), pp. 1–28.

Several other items affecting the amount of creep are listed below with a comment or two about each.

1. The longer the concrete cures before loads are applied, the smaller will be the creep. Steam curing, which causes quicker strengthening, will also reduce creep.
2. Higher-strength concretes have less creep than do lower-strength concretes stressed at the same values. However, applied stresses for higher-strength concretes are in all probability higher than those for lower-strength concretes, and this fact tends to cause increasing creep.
3. Creep increases with higher temperatures. It is highest when the concrete is at about 150° to 160°F.
4. The higher the humidity, the smaller will be the free pore water which can escape from the concrete. Creep is almost twice as large at 50% humidity than at 100% humidity. It is obviously quite difficult to distinguish between shrinkage and creep.
5. Concretes with the highest percentage of cement–water paste have the highest creep because the paste, not the aggregate, does the creeping. This is particularly true if a good limestone aggregate is used.
6. Obviously the addition of reinforcing to the compression areas of concrete will greatly reduce creep because steel exhibits very little creep at ordinary stresses. As creep tends to occur in the concrete the reinforcing will block it and pick up more and more of the load.

Tensile Strength

The tensile strength of concrete varies from about 8% to 15% of its compressive strength. A major reason for this small strength is the fact that concrete is filled with fine cracks. The cracks have little effect when concrete is subjected to compression loads because the loads cause the cracks to close and permit compression transfer. You can see this is not the case for tensile loads.

Although tensile strength is normally neglected in design calculations, it is nevertheless an important property that affects the sizes and extent of the cracks that occur. Furthermore, the tensile strength of concrete members has a definite reduction effect on their deflections. (Due to the small tensile strength of concrete, very little effort has been made to determine its tensile modulus of elasticity. Based on this limited information, however, it seems that its value is equal to its compression modulus.)

Later on, you might wonder why concrete is not assumed to resist a portion of the tension in a flexural member and the steel the remainder. The reason is that concrete cracks at such small tensile strains that the low stresses in the steel up to that time would make its use uneconomical.

The tensile strength of concrete doesn't vary in direct proportion to its ultimate strength f'_c. It does, however, vary approximately in proportion to the square root of f'_c. This strength is quite difficult to measure with direct axial tension loads because of problems in gripping test specimens so as to avoid stress concentrations and because of difficulties in aligning the loads. As a result of these problems, two rather indirect tests have been developed to measure concrete's tensile strength. These are the *modulus of rupture* and the *split-cylinder tests.*

The tensile strength of concrete in flexure is quite important when considering beam cracks and deflections. For these considerations the tensile strengths obtained with the modulus of rupture test have long been used. The modulus of rupture is usually measured by loading a 6-in. × 6-in. × 30-in. plain (i.e., unreinforced) rectangular beam (with simple supports placed 24 in. on center) to failure with equal concentrated loads at its one-third points as per ASTM C-78.[13] The load is increased until failure occurs by cracking on the tensile face of the beam. The modulus of rupture f_r is then determined from the flexure formula. In the following expressions, b is the beam width, h its depth, and M is the maximum computed moment:

$$f_r = \frac{Mc}{I} = \frac{M(h/2)}{\frac{1}{12}bh^3}$$

$$f_r = \text{modulus of rupture} = \frac{6M}{bh^2}$$

The stress determined in this manner is not too accurate because in using the flexure formula we are assuming the concrete is perfectly elastic with stresses varying in direct proportion to distances from the neutral axis. These assumptions are not too good.

Based on hundreds of tests the Code (Section 9.5.2.3) provides a modulus of rupture f_r equal to $7.5\sqrt{f'_c}$, where f'_c is in psi. This same ACI section provides modifications of f_r for lightweight concretes.

The tensile strength of concrete may also be measured with the split-cylinder test.[14] A cylinder is placed on its side in the testing machine and a compressive load is applied uniformly along the length of the cylinder with support supplied along the bottom for the cylinder's full length (see Figure 1.2). The cylinder will split in half from end to end when its tensile strength is reached. The tensile stress at which splitting occurs is referred to as the *split-cylinder strength* and can be calculated by the following expression, in which P is the maximum compressive force, L is the length, and D is the diameter of the cylinder:

$$f_t = \frac{2P}{\pi L D}$$

Even though pads are used under the loads, some local stress concentrations occur during the tests. In addition, some stresses develop at right angles

---

[13] *Standard Test Method for Flexural Strength of Concrete* (*Using Simple Beam with Third-Point Loading*) (ASTM C78-75) (reapproved 1982). Philadelphia: American Society for Testing and Materials, 1982.

[14] *Standard Method of Test for Splitting Tensile Strength of Cylindrical Concrete Specimens.* (ASTM C496-86). Philadelphia: American Society for Testing and Materials, 1979.

P

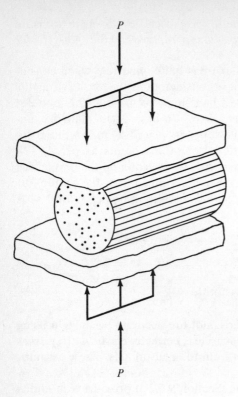

P

Figure 1.2 Split-cylinder test.

to the tension stresses. As a result the tensile strengths obtained are not very accurate.

Shear Strength

It is extremely difficult in testing to obtain pure shear failures unaffected by other stresses. As a result, the tests of concrete shearing strengths through the years have yielded values all the way from one-third to four-fifths of the ultimate compressive strengths. You will learn in Chapter 7 that you do not have to worry about these inconsistent shear strength tests because the pure shear strength of concrete is normally not of importance.

1.10 REINFORCING STEEL

The reinforcing used for concrete structures may be in the form of bars or welded wire fabric. Reinforcing bars are referred to as being *plain* or *deformed*. The deformed bars, which have ribbed projections rolled onto their surfaces (patterns differing with different manufacturers) to provide better bonding between the concrete and the steel, are used for almost all applications. Instead of rolled-on deformations, deformed wire has indentations pressed into it. Plain bars are not used very often except for wrapping around longitudinal bars, primarily in columns. Such bars will be referred to later in this text as *ties* and *spirals*.

Plain round bars are indicated by their diameters in fractions of an inch as $\frac{3}{8}''\phi$, $\frac{1}{2}''\phi$, and $\frac{5}{8}''\phi$. Deformed bars are round and vary in sizes from #3 to #11, with two very large sizes, #14 and #18, also available. For bars up to and including #8, the number of the bar coincides with the bar diameter in eighths of an inch. Bars were formerly manufactured in both round and square cross sections, but today all bars are round.

The #9, #10, and #11 bars have diameters that provide areas equal to the areas of the old 1-in. × 1-in. square bars, $1\frac{1}{8}$-in. × $1\frac{1}{8}$-in. square bars and $1\frac{1}{4}$-in. × $1\frac{1}{4}$-in. square bars, respectively. Similarly the #14 and #18 bars correspond to the old $1\frac{1}{2}$-in. × $1\frac{1}{2}$-in. square bars and 2-in. × 2-in. square bars, respectively. Table A.2 (see Appendix) provides details as to areas, diameters, and weights of reinforcing bars. Although #14 and #18 bars are shown in this table, the designer should check his or her suppliers to see if they have these very large sizes in stock. Reinforcing bars may be purchased in lengths up to 60 ft. Longer bars have to be specially ordered. Normally they are too flexible and difficult to handle.

Welded wire fabric is also frequently used for reinforcing slabs, pavements and shells, and places where this is normally not sufficient room for providing the necessary concrete cover required for regular reinforcing bars. The mesh is made of cold-drawn wires running in both directions and welded together at the points of intersection. The sizes and spacings of the wire may be the same in both directions or may be different, depending on design requirements.

Table A.3(a) in the Appendix provides information concerning certain styles of welded wire fabric that have been recommended by the Wire Reinforcement Institute as common stock styles (normally carried in stock at the mills or at warehousing points and thus usually immediately available). Table A.3(b) in the Appendix provides detailed information as to diameters, areas, weights, and spacings of quite a few wire sizes normally used to manufacture welded wire fabric.

Smooth wire is denoted by the letter "W" followed by a number that equals the cross-sectional area of the wire in hundredths of a square inch. Deformed wire is denoted by the letter "D" followed by a number giving the area. Smooth wire fabric is actually included within the ACI Code's definition of deformed reinforcement because of its mechanical bonding to the concrete caused by the wire intersections. Wire fabric that actually has deformations on the wire surfaces bonds even more to the concrete because of the deformations as well as the wire intersections.

The fabric is usually indicated on drawings by the letters WWF followed by the spacings of the longitudinal wires and the transverse wires and then the total wire areas in hundredths of a square inch per foot of length. For instance, WWF6 × 12–W16 × 8 represents smooth welded wire fabric with a 6-in. longitudinal and a 12-in. transverse spacing with cross-sectional areas of 0.32 in.^{2}/ft and 0.08 in.^{2}/ft, respectively.

1.11 GRADES OF REINFORCING STEEL

Reinforcing bars may consist of billet steel, axle steel, or rail steel. Most bars are made from new or billet steel, but very occasionally they are rolled from old train rails or locomotive axles. These latter steels, having been cold-worked for many years, are not as ductile as the new billet steels.

There are several types of reinforcing bars designated by the ASTM, which are listed at the end of this paragraph. In this listing, Grade 40 means the steel has a specified yield point of 40,000 psi; Grade 50 means 50,000 psi; and so on.

1. ASTM A615, billet steel, Grades 40 and 60.
2. ASTM A615, billet steel, Grade 75 for #11, #14, and #18 bars.
3. ASTM A616, rail steel, Grades 50 and 60.
4. ASTM A617, axle steel, Grades 40 and 60.
5. ASTM A706, low-alloy steel, Grade 60.

Designers in almost all parts of the United States will probably never encounter rail or axle steel bars (A616 and A617) because they are available in such limited areas of the country. Of the several dozen U.S. manufacturers of reinforcing bars listed by the Concrete Reinforcing Steel Institute,[15] only three manufacture rail steel bars and only one manufactures axle bars.

Almost all reinforcing bars conform to the A615 specification and almost all of the material used to make them is not new steel, but is melted reclaimed steel as from old car bodies. Bars conforming to the A706 specification are intended for certain uses when welding and/or bending are of particular importance. Bars conforming to this specification may not always be available from local suppliers.

There is only a small difference between the prices of reinforcing steel with yield strengths of 40 ksi and 60 ksi. As a result, the 60-ksi bars are the most commonly used ones in reinforced concrete design.

When bars are made from steels with yield stresses higher than 60 ksi, the ACI (Section 3.5.3.2) states that the specified yield strength must be the stress corresponding to a strain of 0.35%. The ACI (Section 9.4) has established an upper limit of 80 ksi on yield strengths permitted for reinforced concrete. If they were to permit the use of steels with yield strengths greater than 80 ksi they would have to provide other design restrictions, since the yield strain of 80 ksi steel is almost equal to the ultimate concrete strain in compression. (This last sentence will make sense to the reader after he or she has studied Chapter 2.)

There has been a gradually increasing demand through the years for Grade 75 steel, particularly for use in high-rise buildings where they are used in combination with high-strength concretes. The results are smaller columns, more rentable floor space, and smaller foundations for the lighter buildings that result.

---

[15] *Manual of Standard Practice* (25th ed.), 1990 (Chicago: Concrete Reinforcing Steel Institute), Appendix A, pp. A-1 through A-5.

Grade 75 steel is an appreciably higher-cost steel, and the #14 and #18 bars are often unavailable from stock and will probably have to be specially ordered from the steel mills. This means that there may have to be a special rolling to supply the steel. As a result, its use may not be economically justified unless at least 50 or 60 tons are ordered.

Yield stresses above 60 ksi are also available in welded wire fabric, but the specified stresses must correspond to strains of 0.35%. Smooth fabric must conform to ASTM A185, whereas deformed fabric cannot be smaller than size D4 and must conform to ASTM A496.

The modulus of elasticity for nonprestressed steels is considered to be equal to 29×10^6 psi. For prestressed steels it varies somewhat from manufacturer to manufacturer, with a value of 27×10^6 psi being fairly common.

1.12 CORROSIVE ENVIRONMENTS

When reinforced concrete is subjected to deicing salts, seawater, or spray from these items, it is necessary to provide special corrosion protection for the reinforcing. The structures usually involved are bridge decks, parking garages, wastewater treatment plants, and various coastal structures. We must also consider structures which are subjected to occasional chemical spills which contain chlorides.

Should the reinforcement be insufficiently protected, it will corrode; as it corrodes, the resulting oxides occupy a volume far greater than that of the original metal. The results are large outward pressures which cause severe cracking and spalling of the concrete. This reduces the concrete protection or *cover* for the steel and corrosion accelerates. Also, the *bond* or sticking of the concrete to the steel is reduced. The result of all of these factors is a decided reduction in the life of the structure.

Section 7.7.5 of the Code requires that for corrosive environments, more concrete cover be provided for the reinforcing; it also requires that special concrete proportions or mixes be used.

The lives of such structures can be greatly increased if *epoxycoated reinforcing* bars are used. Such bars need to be handled very carefully so as not to break off any of the coating. Furthermore, they do not bond as well to the concrete, and their lengths will have to be increased somewhat for that reason as we will learn in Chapter 6.

1.13 IDENTIFYING MARKS ON REINFORCING BARS

The ASTM specifications require that identification marks be rolled into the surface of one side of reinforcing bars to provide information as to the producing mill, the bar size, and the type of steel. In addition, for Grade 60 or stronger steels a grade mark is required to indicate the minimum yield point of the steel. The information required is shown in detail in Figure 1.3.

BAR IDENTIFICATION MARKS

Type steel:

S for billet meeting supplemental
   requirements S1 (A 615)

N for new billet (A 615)

R for rail meeting ASTM A 617, Grade 60 bend
   test requirement (A 616) [per ACI 318-83]

I for rail (A 616)

A for axle (A 617)

W for low alloy (A 706)

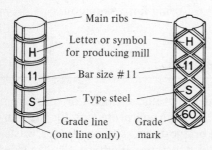

Grade 60

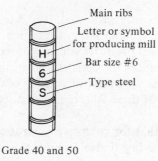

Grade 40 and 50

Figure 1.3 Identification marks for ASTM standard bars. (Courtesy of Concrete Reinforcing Steel Institute.)

1.14 AGGREGATES

The aggregates used in concrete occupy about three-fourths of the concrete volume. Since they are less expensive than the cement, it is desirable to use as much of them as possible. Both fine aggregates (usually sand) and coarse aggregates (usually gravel or crushed stone) are used. Any aggregate that passes a No. 4 sieve (which has wires spaced $\frac{1}{4}$ in. on centers in each direction) is said to be fine aggregate. Material of a larger size is coarse aggregate.

    The maximum-size aggregates that can be used in reinforced concrete are specified in Section 3.3.2 of the ACI Code. These limiting values are as follows: one-fifth of the narrowest dimension between the sides of the forms, one-third of the depth of slabs, or three-quarters of the minimum clear spacing between reinforcing. Larger sizes may be used if in the judgment of the engineer the workability of the concrete and its method of consolidation are such that the aggregate used will not cause the development of honeycomb or voids.

Aggregate must be strong, durable, and clean. Should dust or other particles be present, they may interfere with the bond between the cement paste and the aggregate. The strength of the aggregate has an important effect on the strength of the concrete, and the aggregate properties greatly affect the concrete's durability.

Concretes that have 28-day strengths equal to or greater than 2500 psi and air dry weights equal to or less than 115 lb/ft^3 are said to be structural lightweight concretes. The aggregates used for these concretes are made from expanded shales, clays, or slag. When lightweight aggregates are used for both fine and coarse aggregate, the result is called *all-lightweight* concrete. If sand is used for the fine aggregate and if the coarse aggregate is replaced with lightweight aggregate, the result is referred to as *sand-lightweight* concrete. Concretes made with lightweight aggregates may not be as suitable for heavy wear as those made with normal-weight aggregates.

1.15 LOADS

Perhaps the most important and most difficult task faced by the structural designer is the accurate estimation of the loads that may be applied to a structure during its life. No loads that may reasonably be expected to occur may be overlooked. After loads are estimated, the next problem is to decide the worst possible combinations of these loads that might occur at one time. For instance, would a highway bridge completely covered with ice and snow be simultaneously subjected to fast moving lines of heavily loaded trailer trucks in every lane and to a 90-mile lateral wind, or is some lesser combination of these loads more feasible?

The next two sections of this chapter provide a brief introduction to the types of loads with which the structural designer needs to be familiar. The purpose of these sections is not to discuss loads in great detail but rather to give the reader a "feel" for the subject. As will be seen, loads are classed as being dead loads or live loads.

1.16 DEAD LOADS

Dead loads are loads of constant magnitude that remain in one position. They consist of the structural frame's own weight and other loads that are permanently attached to the frame. For a reinforced concrete building, some dead loads are the frame, walls, floors, roof, plumbing, and fixtures.

To design a structure it is necessary for the weights or dead loads of the various parts to be estimated for use in the analysis. The exact sizes and weights of the parts are not known until the structural analysis is made and the members of the structure selected. The weights, as determined from the actual design, must be compared with the estimated weights. If large discrepancies are present, it will be necessary to repeat the analysis and design using better estimated weights.

Reasonable estimates of structure weights may be obtained by referring to similar-type structures or to various formulas and tables available in most civil engineering handbooks. An experienced designer can estimate very closely the weights of most structures and will spend little time repeating designs because of poor estimates.

1.17 LIVE LOADS

Live loads are loads that may change in position and magnitude. Simply stated, all loads that are not dead loads are live loads. Live loads that move under their own power are said to be moving loads, such as trucks, people, and cranes, whereas those loads that may be moved are movable loads, such as furniture, warehouse materials, and snow. Other live loads include those caused by construction operations, wind, rain, earthquakes, blasts, soils, and temperature changes. A brief description of some of these loads follows:

1. *Snow and ice.* In the colder states, snow and ice loads are often quite important. One inch of snow is equivalent to approximately 0.5 psf (pounds per square foot), but it may be higher at lower elevations where snow is denser. For roof designs, snow loads of from 10 to 40 psf are used, the magnitude depending primarily on the slope of the roof and to a lesser degree on the character of the roof surface. The larger values are used for flat roofs, the smaller ones for sloped roofs. Snow tends to slide off sloped roofs, particularly those with metal or slate surfaces. A load of approximately 10 psf might be used for 45° slopes, and a 40-psf load might be used for flat roofs. Studies of snowfall records in areas with severe winters may indicate the occurrence of snow loads much greater than 40 psf, with values as high as 80 psf in northern Maine.

Snow is a variable load, which may cover an entire roof or only part of it. There may be drifts against walls or buildup in valleys or between parapets. Snow may slide off one roof onto a lower one. The wind may blow it off one side of a sloping roof, or the snow may crust over and remain in position even during very heavy winds.

2. *Rain.* Though snow loads are a more severe problem than rain loads for the usual roof, the situation may be reversed for flat roofs—particularly those in warmer climates. If water on a flat roof accumulates faster than it runs off, the result is called *ponding* because the increased load causes the roof to deflect into a dish shape that can hold more water, which causes greater deflections, and so on. This process continues until equilibrium is reached or until collapse occurs. Ponding is a serious matter as illustrated by the large number of flat-roof failures that occur due to ponding every year in the United States.

3. *Traffic loads for bridges.* Bridges are subjected to series of concentrated loads of varying magnitude caused by groups of truck or train wheels.

4. *Impact loads.* Impact loads are caused by the vibration of moving or movable loads. It is obvious that a crate dropped on the floor of a warehouse or a truck bouncing on uneven pavement of a bridge causes greater forces than would occur if the loads were applied gently and gradually. Impact loads are equal to the difference between the magnitude of the loads actually caused and the magnitude of the loads had they been dead loads.

5. *Lateral loads.* Lateral loads are of two main types: wind and earthquake. A survey of engineering literature for the past 150 years reveals many references to structural failures caused by wind. Perhaps the most infamous of these have been bridge failures such as those of the Tay Bridge in Scotland in 1879 (which caused the deaths of 75 persons) and the Tacoma Narrows Bridge (Tacoma, Washington) in 1940. But there have also been many disastrous building failures due to wind during the same period, such as that of the Union Carbide Building in Toronto in 1958. It is important to realize that a large percentage of building failures due to wind have occurred during their erection.

a. *Wind loads.* The magnitudes of wind loads vary with geographical locations, heights above ground, types of terrain surrounding the buildings including other nearby structures, and other factors. Wind pressures are usually assumed to be uniformly applied to the windward surfaces of buildings and are assumed to be capable of coming from any direction. These assumptions are not altogether correct because wind pressures are not very uniform over large areas, with the pressures near the corners of buildings probably being greater than elsewhere due to wind rushing around the corners, and so on.

Wind forces act as pressures on vertical windward surfaces, pressures or suction on sloping windward surfaces (depending on the slope), and suction on flat surfaces and on leeward vertical and sloping surfaces (due to the creation of negative pressures or vacuums). The student may have noticed this definite suction effect where shingles or other roof coverings have been lifted from the leeward roof surfaces of buildings. Suction or uplift can easily be demonstrated by holding a piece of paper at two of its corners and blowing above it. For some common structures, uplift may be as large as 20 to 30 psf or even more.

During the passing of a tornado or hurricane, a sharp reduction in atmospheric pressure occurs. This decrease in pressure is not reflected inside airtight buildings, and the inside pressures, being greater than the external pressures, cause outward forces against the roofs and walls. Nearly everyone has heard stories of the walls of a building "exploding" outward during a storm.

The average building code in the United States makes no reference to wind velocities in the area or to shapes of the building or to other factors. It just probably requires the use of some specified wind pressure in design such as 20 psf on projected area in elevation up to 300 ft with an increase of 2.5 psf for each additional 100-ft increase in height. The values given in these codes are thought to be rather inaccurate for modern structural design.

The wind pressure on a building can be estimated with the expression to follow in which *p* is the pressure in pounds per square foot acting on vertical surfaces, C_s is a shape coefficient, and V is the basic wind velocity in miles per hour estimated from weather bureau records for that area:

$$p = 0.002558C_sV^2$$

The coefficient C_s depends on the shape of the structure, primarily the roof. For box-type structures C_s is 1.3, of which 0.8 is for the pressure on the windward side and 0.5 is for suction on the leeward side. For such a building the total pressure on the two surfaces equals 20 psf for a wind velocity of 77.8 mph.

As described in the foregoing paragraphs, the accurate determination of the most critical wind loads on a building or bridge is an extremely involved problem; however, sufficient information is available today to permit satisfactory estimates on a reasonably simple basis.

b. *Earthquake loads.* Many areas of the world, including the western part of the United States, fall in earthquake territory, and in these areas it is necessary to consider seismic forces in designs for tall or short buildings. During an earthquake there is an acceleration of the ground surface. This acceleration can be broken down into vertical and horizontal components. Usually the vertical component of the acceleration is assumed to be negligible, but the horizontal component can be severe.

Most buildings can be designed with little extra expense to withstand the forces caused during an earthquake of fairly severe intensity. On the other hand, earthquakes during recent years have clearly shown that the average building that is not designed for earthquake forces can be destroyed by earthquakes that are not particularly severe. The usual practice is to design buildings for additional lateral loads (representing the estimate of earthquake forces) that are equal to some percentage (5% to 10%) of the weight of the building and its contents. An excellent reference on the subject of seismic forces is a publication by the Structural Engineers Association of California (SEAOC), entitled *Recommended Lateral Force Requirements and Commentary*.[16]

It should be noted that many persons look upon the seismic loads to be used in designs as merely being percentage increases of the wind loads. This thought is not really correct, however, because seismic loads are different in their action and because they are not proportional to the exposed area but to the building weight above the level in question.

The effect of the horizontal acceleration increases with the distance above the ground because of the "whipping effect" of the earthquake, and design loads should be increased accordingly. Obviously, towers, water tanks, and penthouses on building roofs occupy precarious positions during an earthquake.

6. *Longitudinal loads.* Longitudinal loads also need to be considered in designing some structures. Stopping a train on a railroad bridge or a truck on a highway bridge causes longitudinal forces to be applied. It is not difficult to

---

[16] *Recommended Lateral Force Requirements and Commentary*, 1987 (San Francisco: Structural Engineers Association of California).

Sewage treatment plant, Redwood City, California. (Courtesy of The Burke Company.)

imagine the tremendous longitudinal force developed when the driver of a 40-ton trailer truck traveling at 60 mph suddenly has to apply the brakes while crossing a highway bridge. There are other longitudinal load situations, such as ships running into docks and the movement of traveling cranes that are supported by building frames.

7. *Other live loads.* Among the other types of live loads with which the structural designer will have to contend are *soil pressures* (such as the exertion of lateral earth pressures on walls or upward pressures on foundations), *hydrostatic pressures* (as water pressure on dams, inertia forces of large bodies of water during earthquakes, and uplift pressures on tanks and basement structures), *blast loads* (caused by explosions, sonic booms, and military weapons), *thermal forces* (due to changes in temperature resulting in structural deformations and resulting structural forces), and *centrifugal forces* (such as those caused on curved bridges by trucks and trains or similar effects on roller coasters).

1.18 SELECTION OF DESIGN LOADS

To assist the designer in estimating the magnitudes of live loads with which he or she should proportion structures, various records have been assembled through the years in the form of building codes and specifications. These publications provide conservative estimates of live-load magnitudes for various situations. One of the most widely used design-load specifications for buildings is that published by the American Society of Civil Engineers (ASCE).[17] Table 1.1 provides a set of floor design loads for use in various types of buildings. These values were taken from the ASCE Code.

The designer is usually fairly well controlled in the design of live loads by the building code requirements in his or her particular area. The values given in these various codes unfortunately vary from city to city, and the designer must be sure to meet the requirements of a particular locality. In the absence of a governing code, the ASCE Code seems to be an excellent one to follow.

Some other commonly used specifications are:

1. For railroad bridges, American Railway Engineering Association (AREA).[18]
2. For highway bridges, American Association of State Highway and Transportation Officials (AASHTO).[19]
3. For buildings, the Uniform Building Code.[20]

These specifications will on many occasions clearly prescribe the loads for which structures are to be designed. Despite the availability of this information, the designer's ingenuity and knowledge of the situation are often needed to predict what loads a particular structure will have to support in years to come. Over the past several decades, insufficient estimates of future traffic loads by bridge designers have resulted in a great amount of replacement with wider and stronger structures.

---

[17] American Society of Civil Engineers, *Minimum Design Loads for Buildings and Other Structures*, ASCE 7-88 (formerly ANSI A58.1), 1988 (New York: American Society of Civil Engineers).

[18] *Manual for Railway Engineering*, 1981 (Washington, D.C.: American Railway Engineering Association).

[19] *Standard Specifications for Highway Bridges* 14th ed., 1989 (Washington. D.C.: American Association of State Highway and Transportation Officials).

[20] *Uniform Building Code*, 1991 edition (Whittier, Calf.: International Conference of Building Officials).

TABLE 1.1 TYPICAL
BUILDING-DESIGN
LIVE LOADS

Type of building	LL (psf)
Apartment houses	
Corridors	80
Apartments	40
Public rooms	100
Office buildings	
Offices	50
Lobbies	100
Restaurants	100
Schools	
Classrooms	40
Corridors	80
Storage warehouses	
Light	125
Heavy	250

1.19 CALCULATION ACCURACY

A most important point, which many students with their amazing computers
and pocket calculators have difficulty in understanding, is that reinforced concrete design is not an exact science for which answers can be confidently calculated to six or eight places. The reasons for this statement should be quite
obvious: The analysis of structures is based on partly true assumptions; the
strengths of materials used vary widely; and maximum loadings can only be
approximated. With respect to this last sentence, how many users of this book
could estimate to within 10% the maximum live load in pounds per square foot
that will ever occur on the building floor they are now occupying? Calculations
to more than two significant figures are obviously of little value and may actually
be dangerous because they may mislead students by giving them a fictitious
sense of accuracy.

2

Flexural Analysis of Beams

2.1 INTRODUCTION

In this section it is assumed that a small transverse load is placed on a concrete beam with tensile reinforcing and that the load is gradually increased in magnitude until the beam fails. As this takes place we will find that the beam will go through three distinct stages before collapse occurs. These are: (1) the uncracked concrete stage, (2) the concrete cracked-elastic stresses range, and (3) the ultimate strength stage. A relatively long beam is considered for this discussion so that shear will not be a factor.

These stages are briefly introduced in this section, and the remainder of the chapter is used to discuss them in detail and provide numerical examples.

Uncracked Concrete Stage

At small loads when the tensile stresses are less than the *modulus of rupture* (the bending tensile stress at which the concrete begins to crack), the entire cross section of the beam resists bending, with compression on one side and tension on the other. Figure 2.1 shows the variation of stresses and strains for these small loads; a numerical example of this type is presented in Section 2.2.

Concrete Cracked—Elastic Stresses Stage

As the load is increased after the modulus of rupture of the beam is exceeded, cracks begin to develop in the bottom of the beam. The moment at which these cracks begin to form—that is, when the tensile stress in the bottom of the beam equals the modulus of rupture—is referred to as the *cracking moment*, M_{cr}. As

Construction of Kingdome, Seattle, Washington. (Courtesy of Economy Forms Corporation.)

the load is further increased, these cracks quickly spread up to the vicinity of the neutral axis, and then the neutral axis begins to move upward. The cracks occur at those places along the beam where the actual moment is greater than the cracking moment, as shown in Figure 2.2(a).

Now that the bottom has cracked, another stage is present because the concrete in the cracked zone obviously cannot resist tensile stresses—the steel must do it. This stage will continue as long as the compression stress in the top fibers is less than about one-half of the concrete's compression strength f'_c and as long as the steel stress is less than its yield point. The stresses and strains for this range are shown in Figure 2.2(b). In this stage the compressive stresses vary linearly with the distance from the neutral axis or as a straight line.

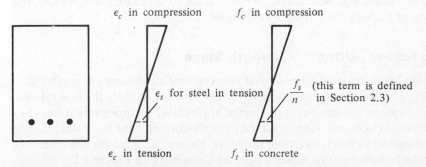

ϵ_c in compression f_c in compression

ϵ_s for steel in tension $\dfrac{f_s}{n}$ (this term is defined in Section 2.3)

ϵ_c in tension f_t in concrete

Figure 2.1 Uncracked concrete stage.

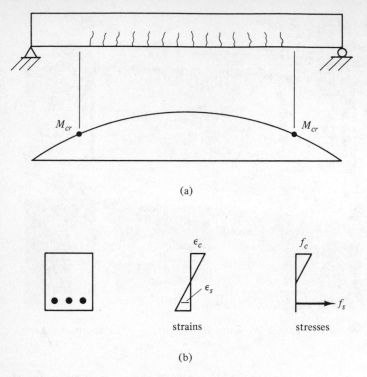

(a)

(b)

Figure 2.2 Concrete cracked—elastic stresses range.

The straight-line stress–strain variation normally occurs in reinforced con-crete beams under normal service-load conditions because at those loads the stresses are generally less than $0.50f'_c$. To compute the concrete and steel stresses in this range, the transformed-area method (to be presented in Section 2.3) is used. The *service* or *working* loads are the loads which are assumed to actually occur when a structure is in use or service. Under these loads, moments develop which are considerably larger than the cracking moments. Obviously the tensile side of the beam will be cracked. We will learn to estimate crack widths and methods of limiting their widths in Chapter 5.

Beam Failure—Ultimate-Strength Stage

As the load is increased further so that the compressive stresses are greater than one-half of f'_c, the tensile cracks move further upward, as does the neutral axis, and the concrete stresses begin to change appreciably from a straight line. For this initial discussion it is assumed that the reinforcing bars have yielded. The stress variation is much like that shown in Figure 2.3. You should relate the information shown in this figure with that given back in Figure 1.1.

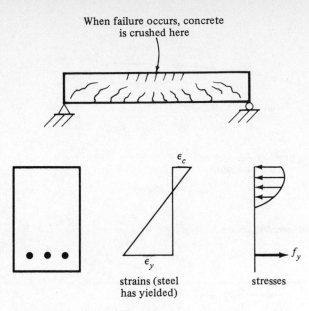

Figure 2.3 Ultimate-strength stage.

To further illustrate the three stages of beam behavior which have just been described, a moment–curvature diagram is shown in Figure 2.4.[1] For this diagram, θ is defined as the angle change of the beam over a certain length and is computed by the following expression in which θ is the strain in a beam fiber at some distance y from the neutral axis of the beam:

$$\theta = \frac{\epsilon}{y}$$

The first stage of the diagram is for small moments less than the cracking moment M_{cr} where the entire beam cross section is available to resist bending. In this range the strains are small and the diagram is nearly vertical and very close to a straight line.

When the moment is increased beyond the cracking moment, the slope of the curve will decrease a little because the beam is not quite as stiff as it was in the initial stage before the concrete cracked. The diagram will follow almost a straight line from M_{cr} to the point where the reinforcing is stressed to its yield point. Up until the steel yields, a fairly large additional load is required to appreciably increase the beam's deflection.

---

[1] MacGregor, J. G., 1992, *Reinforced Concrete Mechanics and Design* (Englewood Cliffs, N.J.: Prentice–Hall), p. 81.

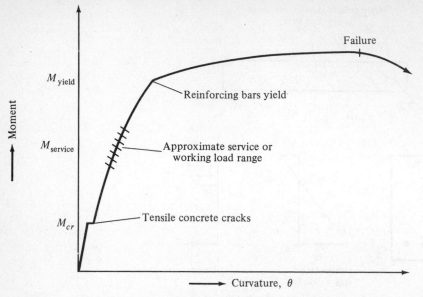

Figure 2.4 Moment–curvature diagram for tensilely reinforced concrete beam.

After the steel yields, the beam has very little additional moment capacity and only a small additional load is required to substantially increase deflections. The slope of the diagram is now very flat.

2.2 CRACKING MOMENT

The area of reinforcing as a percentage of the total cross-sectional area of a beam is quite small (usually 2% or less), and its effect on the beam properties is almost negligible as long as the beam is uncracked. Therefore the calculation of bending stresses in such a beam can be based on the gross properties of the beam's cross section. The stress in the concrete at any point a distance y from the centroid of the cross section can be determined from the following flexure formula in which M is the bending moment, which is equal to or less than the cracking moment of the section, and I_g is the gross moment of inertia of the cross section:

$$f = \frac{My}{I_g}$$

Section 9.5.2.3 of the ACI Code states that the cracking moment of a section may be determined from the expression at the end of this paragraph, in which f_r is the modulus of rupture of the concrete and y_t is the distance from the centroidal axis of the section to its extreme fiber in tension. The Code says that f_r may be taken as $7.5\sqrt{f'_c}$ for normal-weight concrete with f'_c in psi (or as $0.62\sqrt{f'_c}$ when f'_c is in the SI units N/mm²). Other values are provided in the same code section for all-lightweight and sand-lightweight concretes. The cracking moment is as follows:

$$M_{cr} = \frac{f_r I_g}{y_t}$$ (ACI Equation 9-8)

Example 2.1 presents calculations for a reinforced concrete beam where tensile stresses are less than its modulus of rupture. As a result, no tensile cracks are assumed to be present and the stresses are similar to those occurring in a beam constructed with a homogeneous material.

■ EXAMPLE 2.1

(a) Assuming the concrete is uncracked, compute the bending stresses in the extreme fibers of the beam of Figure 2.5 for a bending moment of 25 ft-k. The concrete has an f'_c of 4000 psi and a modulus of rupture of $7.5\sqrt{4000} = 474$ psi.

(b) Determine the cracking moment of the section.

SOLUTION

(a) Bending stresses:

$$I_g = \left(\frac{1}{12}\right)(12)(18)^3 = 5832 \text{ in.}^4$$

$$f = \frac{My}{I_g} = \frac{(12)(25{,}000)(9.00)}{5832} = \underline{463 \text{ psi}}$$

Since this stress is less than the tensile strength or modulus of rupture of the concrete of 474 psi, the section is assumed not to have cracked.

(b) Cracking moment:

$$M_{cr} = \frac{f_r I_g}{y_t} = \frac{(474)(5832)}{9.00} = 307{,}152 \text{ in.-lb} = \underline{\underline{25.6 \text{ ft-k}}}$$ ■

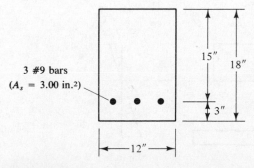

3 #9 bars
($A_s = 3.00$ in.²)

15″

18″

3″

12″

Figure 2.5

2.3 ELASTIC STRESSES—CONCRETE CRACKED

When the bending moment is sufficiently large to cause the tensile stress in the extreme fibers to be greater than the modulus of rupture, it is assumed that all of the concrete on the tensile side of the beam is cracked and must be neglected in the flexure calculations.

The cracking moment of a beam is normally quite small compared to the service load moment. Thus when the service loads are applied, the bottom of the beam cracks. The cracking of the beam does not necessarily mean that the beam is going to fail. The reinforcing bars on the tensile side begin to pick up the tension caused by the applied moment.

On the tensile side of the beam an assumption of perfect bond is made between the reinforcing bars and the concrete. Thus the strain in the concrete and in the steel will be equal at equal distances from the neutral axis. But if the strains in the two materials at a particular point are the same, their stresses cannot be the same since they have different moduli of elasticity. Thus their stresses are in proportion to the ratio of their moduli of elasticity. The ratio of the steel modulus to the concrete modulus is called the *modular ratio n*:

$$n = \frac{E_s}{E_c}$$

If the modular ratio for a particular beam is 10, the stress in the steel will be 10 times the stress in the concrete at the same distance from the neutral axis. Another way of saying this is that when $n = 10$, one sq. inch of steel will carry the same total force as 10 in.^{2} of concrete.

For the beam of Figure 2.6, the steel bars are replaced with an equivalent area of fictitious concrete (nA_s), which supposedly can resist tension. This area is referred to as the *transformed area*. The resulting revised cross section or transformed section is handled by the usual methods for elastic homogeneous beams. Also shown in the figure is a diagram showing the stress variation in the beam. On the tensile side a dotted line is shown because the diagram is discontinuous. There the concrete is assumed to be cracked and unable to resist tension. The value shown opposite the steel is the fictitious stress in the concrete if it could

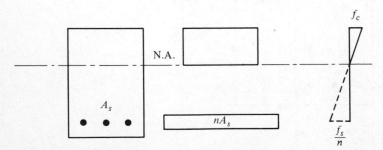

Figure 2.6

Arch bridge, Pittsburgh, Pennsylvania. (Courtesy of Portland Cement Association.)

carry tension. This value is shown as f_s/n because it must be multiplied by n to give the steel stress f_s.

Examples 2.2, 2.3, and 2.4 are transformed-area problems that illustrate the calculations necessary for determining the stresses and resisting moments for reinforced concrete beams. The first step to be taken in each of these problems is to locate the neutral axis, which is assumed to be located a distance x from the compression surface of the beam. The moment of the compression area of the beam cross section about the neutral axis must equal the moment of the tensile area about the neutral axis. The resulting quadratic equation can be solved by completing the squares.

After the neutral axis is located, the moment of inertia of the transformed section is calculated, and the stresses in the concrete and the steel are computed with the flexure formula.

■ EXAMPLE 2.2

Calculate the bending stresses in the beam shown in Figure 2.7 by using the transformed area method: $n = 9$ and $M = 70$ ft-k.

SOLUTION

Taking Moments About Neutral Axis (Referring to Figure 2.8)

$$(12x)\left(\frac{x}{2}\right) = (9)(3.00)(17 - x)$$

$$6x^2 = 459 - 27.00x$$

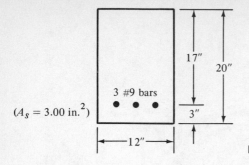

Figure 2.7

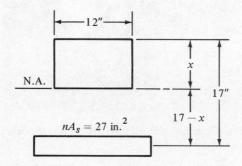

Figure 2.8

Solving by Completing the Square

$$6x^2 + 27.00x = 459$$

$$x^2 + 4.50x = 76.5$$

$$(x + 2.25)(x + 2.25) = 76.5 + (2.25)^2$$

$$x + 2.25 = \sqrt{76.5 + (2.25)^2} = 9.03''$$

$$x = 6.78''$$

Moment of Inertia

$$I = (\tfrac{1}{3})(12)(6.78)^3 + (27.00)(10.22)^2 = 4067 \text{ in.}^4$$

Bending Stresses

$$f_c = \frac{My}{I} = \frac{(12)(70{,}000)(6.78)}{4067} = \underline{\underline{1400 \text{ psi}}}$$

$$f_s = n\frac{My}{I} = (9)\frac{(12)(70{,}000)(10.22)}{4067} = \underline{\underline{18{,}998 \text{ psi}}}$$ ∎

■ EXAMPLE 2.3

Determine the allowable resisting moment of the beam of Example 2.2 if the allowable stresses are $f_c = 1350$ psi and $f_s = 20{,}000$ psi.

SOLUTION

$$M_c = \frac{f_c I}{y} = \frac{(1350)(4067)}{6.78} = 809{,}800'' \text{ lb} = \underline{\underline{67.5 \text{ ft-k}}} \leftarrow$$

$$M_s = \frac{f_s I}{ny} = \frac{(20{,}000)(4067)}{(9)(10.22)} = 884{,}323'' \text{ lb} = 73.7 \text{ ft-k} \qquad \blacksquare$$

Discussion For a given beam the concrete and steel will not usually reach their maximum allowable stresses at exactly the same bending moments. Such is the case for this example beam, where the concrete reaches its maximum permissible stress at 67.5 ft-k while the steel does not reach its maximum value until 73.7 ft-k is applied. The resisting moment of the section is 67.5 ft-k because if that value is exceeded, the concrete becomes overstressed even though the steel stress is less than is allowable.

■ EXAMPLE 2.4

Compute the bending stresses in the beam shown in Figure 2.9 by using the transformed-area method; $n = 8$ and $M = 110$ ft-k.

SOLUTION

Locating Neutral Axis (Assuming Neutral Axis Below Hole)

$$(18x)\left(\frac{x}{2}\right) - (6)(6)(x - 3) = (8)(5.06)(23 - x)$$

$$9x^2 - 36x + 108 = 931 - 40.48x$$

$$9x^2 + 4.48x = 823$$

$$x^2 + 0.50x = 91.44$$

$$(x + 0.25)(x + 0.25) = 91.44 + (0.25)^2 = 91.50$$

$$x + 0.25 = \sqrt{91.50} = 9.57$$

$$x = 9.32'' > 6'' \qquad \therefore \text{N.A. below hole}$$
$$\text{as assumed}$$

Moment of Inertia

$$I = (\tfrac{1}{3})(6)(9.32)^3(2) + (\tfrac{1}{3})(6)(3.32)^3 + (8)(5.06)(13.68)^2 = 10{,}887 \text{ in.}^4$$

Computing Stresses

$$f_c = \frac{(12)(110{,}000)(9.32)}{10{,}887} = \underline{\underline{1130 \text{ psi}}}$$

$$f_s = (8)\frac{(12)(110{,}000)(13.68)}{10{,}887} = \underline{\underline{13{,}269 \text{ psi}}} \qquad \blacksquare$$

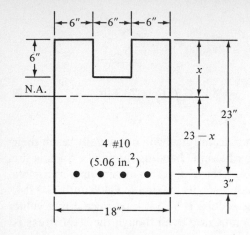

Figure 2.9

Example 2.5 illustrates the analysis of a doubly reinforced concrete beam—that is, one that has compression steel as well as tensile steel. Compression steel is generally thought to be uneconomical, but there are occasional situations where its use is quite advantageous.

Compression steel will permit the use of appreciably smaller beams than those that make use of tensile steel only. Reduced sizes can be very important where space or architectural requirements limit the sizes of beams. Compression steel is quite helpful in reducing long-term deflections, and such steel is useful for positioning stirrups or shear reinforcing, a subject to be discussed in Chapter 7. A detailed discussion of doubly reinforced beams is presented in Chapter 4.

The creep or plastic flow of concrete was described in Section 1.9. Should the compression side of a beam be reinforced, the long-term stresses in that reinforcing will be greatly affected by the creep in the concrete. As time goes by, the compression concrete will compact more tightly, leaving the reinforcing bars (which themselves have negligible creep) to carry more and more of the load.

As a consequence of this creep in the concrete, the stresses in the compression bars computed by the transformed-area method are assumed to double as time goes by. In Example 2.5 the transformed area of the compression bars is assumed to equal $2n$ times their area A_s'.

On the subject of "hair splitting," it will be noted in the example that the compression steel area is really multiplied by $2n - 1$. The transformed area of the compression side equals the gross compression area of the concrete plus $2nA_s'$ minus the area of the holes in the concrete ($1A_s'$), which theoretically should not have been included in the concrete part. This equals the compression concrete area plus $(2n - 1)A_s'$. Similarly, $2n - 1$ is used in the moment of inertia calculations. The stresses in the compression bars are determined by multiplying $2n$ times the stresses in the concrete located at the same distance from the neutral axis.

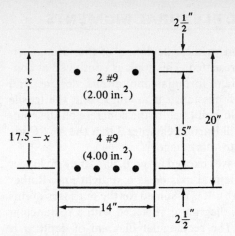

Figure 2.10

■ EXAMPLE 2.5
Compute the bending stresses in the beam shown in Figure 2.10; $n = 10$ and $M = 118$ ft-k.

SOLUTION

Locating Neutral Axis

$$(14x)\left(\frac{x}{2}\right) + (20 - 1)(2.00)(x - 2.5) = (10)(4.00)(17.5 - x)$$

$$7x^2 + 38x - 95 = 700 - 40x$$

$$7x^2 + 78x = 795$$

$$x^2 + 11.14x = 113.57$$

$$x + 5.57 = \sqrt{113.57 + (5.57)^2} = 12.02$$

$$x = 6.45 \text{ in.}$$

Moment of Inertia

$$I = (\tfrac{1}{3})(14)(6.45)^3 + (20 - 1)(2.00)(3.95)^2 + (10)(4.00)(11.05)^2 = 6729 \text{ in.}^4$$

Bending Stresses

$$f_c = \frac{(12)(118,000)(6.45)}{6729} = \underline{\underline{1,357 \text{ psi}}}$$

$$f'_s = 2n\frac{My}{I} = (2)(10)\frac{(12)(118,000)(3.95)}{6729} = \underline{\underline{16,624 \text{ psi}}}$$

$$f_s = (10)\frac{(12)(118,000)(11.05)}{6729} = \underline{\underline{23,253 \text{ psi}}}$$

2.4 ULTIMATE OR NOMINAL FLEXURAL MOMENTS

In this section a very brief introduction to the calculation of the ultimate or nominal flexural strength of beams is presented. This topic is continued at considerable length in the next chapter where formulas, limitations, designs, and other matters are presented. For this discussion it is assumed that the tensile reinforcing bars are stressed to their yield point before the concrete on the compressive side of the beam crushes. We will learn in Chapter 3 that the ACI Code requires all our beam designs to fall into this category.

After the concrete compression stresses exceed about $0.50f_c'$, they no longer vary directly as the distance from the neutral axis or as a straight line. Rather they vary much as shown in Figure 2.11(b). It is assumed for the purposes of this discussion that the curved compression diagram is replaced with a rectangular one with an average stress of $0.85f_c'$. The rectangular diagram of depth a is assumed to have the same c.g. (center of gravity) and total magnitude as the curved diagram. (In Section 3.7 of Chapter 3 we will learn that this distance a is set equal to $\beta_1 c$, where β_1 is a value determined by testing.) These assumptions will enable us to easily calculate the theoretical or nominal flexural strength of reinforced concrete beams. Experimental tests show that with the assumptions used here accurate flexural strengths are determined.

To obtain the nominal or theoretical moment strength of a beam the simple steps to follow are used:

1. Compute total tensile force $T = A_s f_y$.
2. Equate total compression force $C = 0.85f_c'ab$ to $A_s f_y$ and solve for a. In this expression ab is the assumed area in compression at $0.85f_c'$.
3. Calculate the distance between the centers of gravity of T and C. (For a rectangular section it equals $d - a/2$.)
4. Determine M_n which equals T or C times the distance between the centers of gravities.

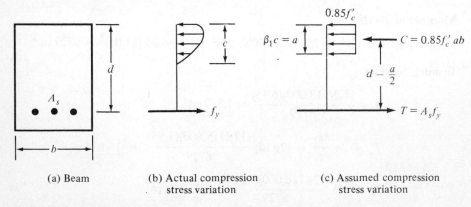

(a) Beam

(b) Actual compression stress variation

(c) Assumed compression stress variation

Figure 2.11 Compression and tension couple at nominal moment.

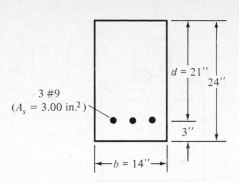
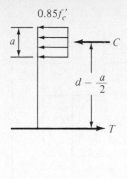

Figure 2.12

■ EXAMPLE 2.6

Determine the nominal or theoretical ultimate moment strength of the beam section shown in Figure 2.12 if $f_y = 60,000$ psi and $f'_c = 3000$ psi.

SOLUTION

$$T = A_s f_y = (3.00)(60) = 180 \text{ k}$$

$$C = 0.85 f'_c ab = (0.85)(3)(a)(14) = 35.7a$$

$$T = C \quad \text{for equilibrium}$$

$$180 = 35.7a$$

$$a = 5.04 \text{ in.}$$

$$d - \frac{a}{2} = 21 - \frac{5.04}{2} = 18.48 \text{ in.}$$

$$M_n = (180)(18.48) = 3326.4 \text{ in.-k} = \underline{\underline{277.2 \text{ ft-k}}} \qquad ■$$

In Example 2.7 the nominal moment capacity of another beam is determined much as it was for Example 2.6. The only difference is that the cross section of the compression area (A_c) stressed at $0.85f'_c$ is not rectangular. As a result, once this area is determined we need to locate its center of gravity. In Figure 2.14 this c.g. is shown as being a distance $\bar{y}$ from the top of the beam. Then the lever arm from C to T is equal to $d - \bar{y}$ (which corresponds to $d - a/2$ in Example 2.6) and M_n equals $A_s f_y (d - \bar{y})$.

With this very simple procedure we can compute M_n for tensilely reinforced beams of any cross section.

■ EXAMPLE 2.7

Calculate the nominal or theoretical ultimate moment strength of the beam section shown in Figure 2.13 if $f_y = 60,000$ psi and $f'_c = 3000$ psi. The 6-in.-wide ledges on top are needed for the support of precast concrete slabs.

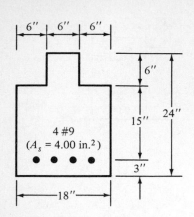

Figure 2.13

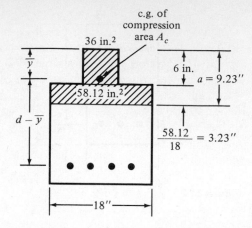

Figure 2.14

SOLUTION

$$T = A_s f_y = (4.00)(60) = 240 \text{ k}$$

$$C = (0.85f_c')(\text{area of concrete } A_c \text{ stressed to } 0.85f_c')$$
$$= 0.85f_c' A_c$$

$$A_c = \frac{T}{0.85f_c'} = \frac{240}{(0.85)(3)} = 94.12 \text{ in.}^2$$

Referring to Figure 2.14, where the top 94.12 in.² is stressed at $0.85f_c'$, we have

$$\bar{y} = \frac{(36)(3) + (58.12)\left(6 + \dfrac{3.23}{2}\right)}{94.12} = 5.85 \text{ in.}$$

$$d - \bar{y} = 21 - 5.85 = 15.15 \text{ in.}$$

$$M_n = (240)(15.15) = 3636 \text{ in.-k} = \underline{\underline{303 \text{ ft-k}}} \quad \blacksquare$$

2.5 COMPUTER PROGRAMS

In this section the reader is introduced to a set of computer programs called CONCRETE, which was prepared for use on IBM and IBM-compatible microcomputers. These programs, which are written in Quick Basic 7, are furnished on the diskette enclosed with this book.

CONCRETE is designed to operate as a "black box"; that is, the user need only supply the specified input data and the computer will perform the calculations and supply the appropriate results automatically. *Even though the reader may have had little or even no previous computer experience he or she can learn to apply these programs in a very short time (minutes, not hours).*

In this section the very simple bits of information needed to use CONCRETE to determine the nominal moment for a reinforced concrete rectangular beam are presented. In later chapters the same diskette is used to solve 11 other types of analysis and design problems. Although CONCRETE is very simple to use and the screen prompts are quite clear as to the steps to be taken, the procedure is described in some detail in the next few paragraphs. Furthermore Appendix D presents an even more detailed description.

The program may be used in one of two ways: run the program from a disk drive or run the program from the hard disk. The procedure for each is as follows:

1. In order to run the program from a disk drive (say the A: drive) insert the diskette in drive A then type A: and press the ENTER key. From the A: prompt, type CONCRETE and press the ENTER key.
2. Create a directory on your hard drive, say CONC, by using the DOS "md conc" command, then change directory to CONC by using the DOS "dc conc" command. Insert the distribution diskette in drive A and copy the contents of the diskette into the new directory by using the DOS "copy a: *.*" command. When all files are copied, type CONCRETE to run the program.

The computer screen shows a list of the programs available. The user moves the cursor to the program desired and presses the RETURN key. In answer to the screen prompts information is inputted as to beam width, effective depth and so on. As each bit of information is inputted the user presses the RETURN key and after the last requested item is supplied the computer shows the value of M_n (the nominal moment for the beam) and a value M_u (the design strength of the member which is described in the next chapter).

With this program you can quickly revise the data for the problem in any way you see fit. The cursor is moved up or down and the input data changed. You then press the RETURN key and the new answers are provided. You may elect to print the results by pressing the F10 key. Finally you can return to the main menu by pressing the ESC key.

A help line is always displayed at the bottom of the screen. On-line help is available for each module by pressing the F2 key. These help messages refer to sections in the textbook and in the ACI Code and help the user understand the method of solution. Users are encouraged to enter the data correctly the first time around. However, the programs can be used to study the effect on the design of changing one or more input values. The program, however, should not be expected to produce reasonable results for unrealistic input.

■ EXAMPLE 2.8

Repeat Example 2.6 using the enclosed computer diskette.

```
STRENGTH ANALYSIS OF RECTANGULAR BEAM WITH TENSILE REINFORCING
-----------------------------------------------------------------
Width                     b (in)     14.0
Depth to Rebars           d (in)     21.0
Area of Steel       As (sq. in.)     3.000
Concrete Strength     f'c (psi)      3000
Steel Yield Strength   fy (psi)      60000


RESULTS
=======
Mn (ft-k) =   277.2
Mu (ft-k) =   249.5
```

■

PROBLEMS

CRACKING MOMENTS

For Problems 2.1 to 2.4 determine the cracking moments for the sections shown if $f'_c =$ 3000 psi and if the modulus of rupture $f_r = 7.5\sqrt{f'_c}$.

Problem 2.1 (*Ans.* 49.3 ft-k)

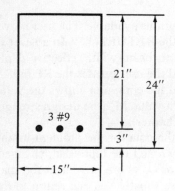

Problem 2.2

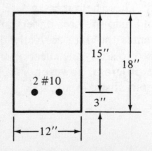

Problem 2.3 (*Ans.* 16.59 ft-k)

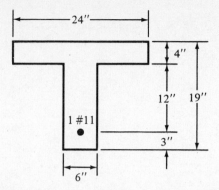

Problem 2.4

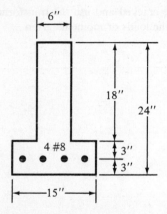

For Problems 2.5 through 2.6 calculate the uniform load (in addition to the beam weight) which will cause the sections to begin to crack if they are used for 25-ft simple spans. $f'_c = 3000$ psi, $f_r = 7.5\sqrt{f'_c}$. Reinforced concrete weight = 150 lb/ft³.

Problem 2.5 (*Ans.* 0.126 k/ft)

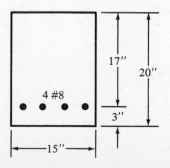

Problem 2.6

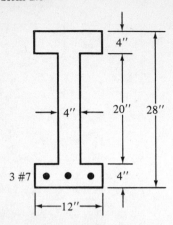

TRANSFORMED-AREA METHOD

For Problems 2.7 to 2.12 assume the sections have cracked and use the transformed-area method to compute their flexural stresses for the loads or moments given.

Problem 2.7 (*Ans.* $f_c = 1163$ psi, $f_s = 13,993$ psi)

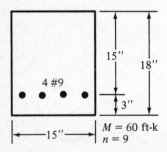

2.8 Repeat Problem 2.7 if 4 #7 bars are used.

Problem 2.9 (*Ans.* $f_c = 898$ psi, $f_s = 11,794$ psi)

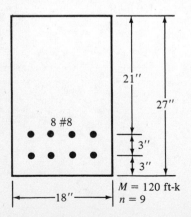

Problem 2.10

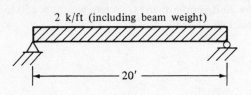

2 k/ft (including beam weight)

20'

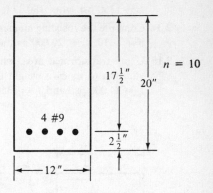

$n = 10$

$17\frac{1}{2}''$ 20''

4 #9

$2\frac{1}{2}''$

12''

Problem 2.11 (*Ans.* $f_c = 1800$ psi, $f_s = 28{,}445$ psi)

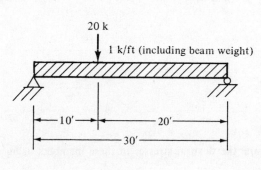

20 k

1 k/ft (including beam weight)

10'   20'

30'

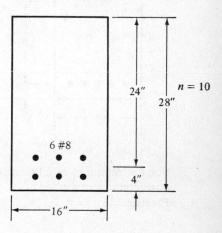

24'' $n = 10$

28''

6 #8

4''

16''

Problem 2.12

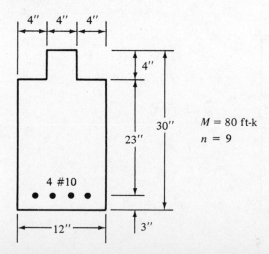

4''   4''   4''

4''

30''

23'' $M = 80$ ft-k

$n = 9$

4 #10

12''   3''

2.13 Compute the resisting moment of the beam of Problem 2.9 if $f_s = 24{,}000$ psi and $f_c = 1350$ psi. (*Ans.* 180.37 ft-k)

2.14 Compute the resisting moment of the beam of Problem 2.11 if 8 #9 bars are used and if $n = 10$, $f_s = 20{,}000$ psi, and $f_c = 1125$ psi. Use the transformed-area method.

2.15 Using transformed area, what allowable uniform load can this beam support in addition to its own weight for a 30-ft simple span? Concrete weight $= 150$ lb/ft^3, $f_s = 24{,}000$ psi, and $f_c = 1350$ psi. (*Ans.* 1.273 k/ft)

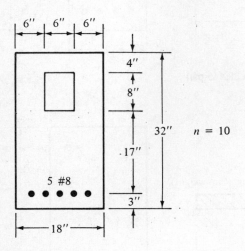

For Problems 2.16 to 2.18 determine the flexural stresses in these members using the transformed-area method.

Problem 2.16

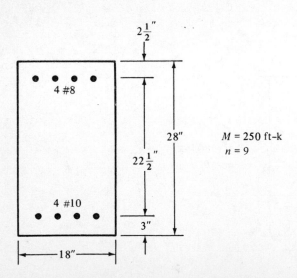

Problem 2.17 (*Ans.* $f_c = 665$ psi, $f_s = 19,196$ psi)

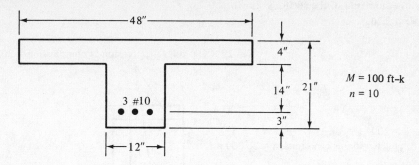

$M = 100$ ft–k
$n = 10$

Problem 2.18

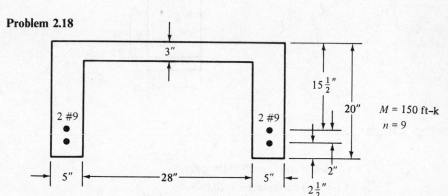

$M = 150$ ft–k
$n = 9$

2.19 Compute the allowable resisting moment of the section shown using transformed area if $f_c = 1800$ psi, $f_s = f'_s = 24,000$ psi, and $n = 8$. (*Ans.* 111.03 ft-k)

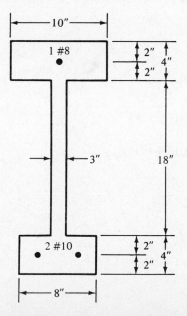

For Problems 2.20 to 2.22, using the transformed-area method determine the allowable resisting moments of the sections shown.
Problem 2.20

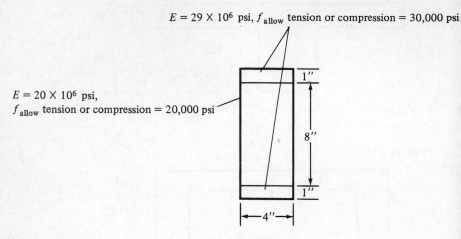

$E = 29 \times 10^6$ psi, f_{allow} tension or compression $= 30,000$ psi

$E = 20 \times 10^6$ psi,
f_{allow} tension or compression $= 20,000$ psi

1″

8″

1″

4″

Problem 2.21 (*Ans.* 19.61 ft-k)

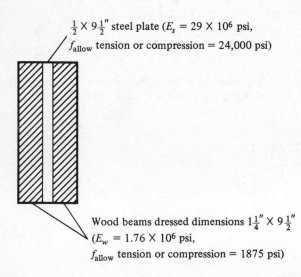

$\frac{1}{2} \times 9\frac{1}{2}''$ steel plate ($E_s = 29 \times 10^6$ psi,
f_{allow} tension or compression $= 24,000$ psi)

Wood beams dressed dimensions $1\frac{1}{4}'' \times 9\frac{1}{2}''$
($E_w = 1.76 \times 10^6$ psi,
f_{allow} tension or compression $= 1875$ psi)

Problem 2.22

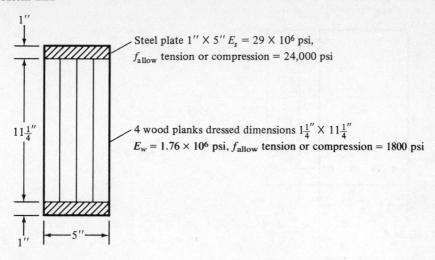

Steel plate $1'' \times 5''$ $E_s = 29 \times 10^6$ psi,
f_{allow} tension or compression = 24,000 psi

4 wood planks dressed dimensions $1\frac{1}{4}'' \times 11\frac{1}{4}''$
$E_w = 1.76 \times 10^6$ psi, f_{allow} tension or compression = 1800 psi

$1''$

$11\frac{1}{4}''$

$1''$

$5''$

NOMINAL STRENGTH ANALYSIS

For Problems 2.23 to 2.30 determine the nominal or theoretical moment capacity of each beam if $f_y = 60,000$ psi and $f'_c = 3000$ psi.

Problem 2.23 (*Ans.* 277.2 ft-k)

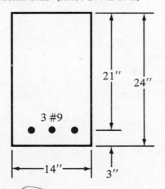

$21''$ $24''$

3 #9

$14''$ $3''$

Problem 2.24

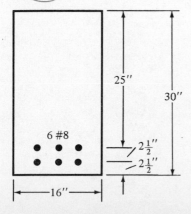

$25''$ $30''$

6 #8

$2\frac{1}{2}''$
$2\frac{1}{2}''$

$16''$

Problem 2.25 (*Ans.* 637.7 ft-k)

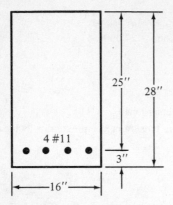

Problem 2.26

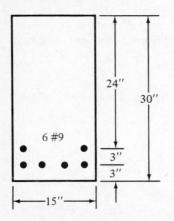

Problem 2.27 (*Ans.* 453.4 ft-k)

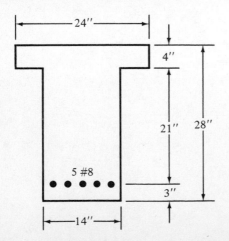

2.28 Repeat Problem 2.27 if 4 #11 bars are used.

2.29 The beam of Problem 2.15 (*Ans*. 518.2 ft-k)

Problem 2.30

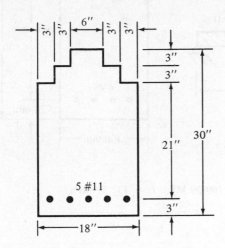

2.31 Using the enclosed computer disk repeat
(a) Problem 2.23 (*Ans*. 277.2 ft-k)
(b) Problem 2.26 (*Ans*. 623.8 ft-k)

2.32 Prepare a flow chart for the determination of M_n for a rectangular beam.

PROBLEMS WITH SI UNITS

In Problems 2.33 to 2.36 compute the flexural stresses in the concrete and steel for the beams shown using the transformed-area method.

Problem 2.33 (*Ans*. $f_c = 8.488$ MPa, $f_s = 112.04$ MPa)

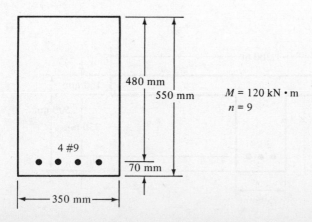

Problem 2.34

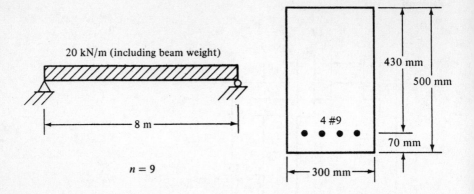

20 kN/m (including beam weight)

8 m

$n = 9$

430 mm

500 mm

4 #9

70 mm

300 mm

Problem 2.35 (*Ans.* f_c = 7.87 MPa, f_s' = 109.39 MPa, f_s = 137.39 MPa)

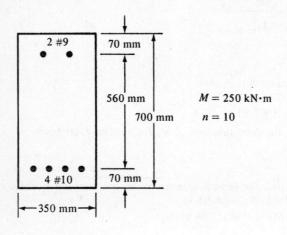

2 #9

70 mm

560 mm

700 mm

70 mm

4 #10

350 mm

$M = 250$ kN·m

$n = 10$

Problem 2.36

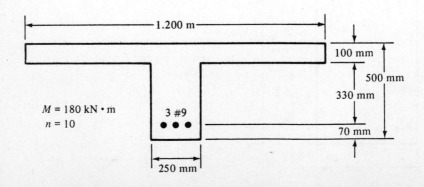

1.200 m

100 mm

500 mm

330 mm

70 mm

$M = 180$ kN · m

$n = 10$

3 #9

250 mm

2.37 Compute the resisting moment of the beam of Problem 2.33 in kilonewton meters if $f_c = 9.3$ MPa and $f_s = 137.9$ MPa. (*Ans.* 131.5 kN·m)

In Problems 2.38 and 2.39 determine the nominal flexural capacity of each of the beams shown if $f_y = 413.7$ MPa and $f'_c = 20.7$ MPa.

Problem 2.38

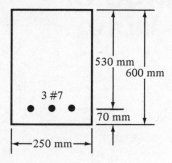

3 #7

530 mm

600 mm

70 mm

250 mm

Problem 2.39 (*Ans.* 360.23 kN·m)

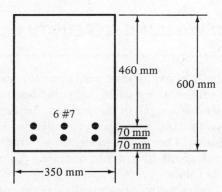

460 mm

600 mm

6 #7

70 mm

70 mm

350 mm

3

Design of Rectangular Beams and One-Way Slabs

3.1 WORKING-STRESS AND WORKING-STRENGTH DESIGN METHODS

From the early 1900s until the early 1960s, nearly all reinforced concrete design in the United States was performed by the working-stress design method (also called *allowable-stress design* or *straight-line design*). In this method, frequently referred to as WSD, the dead and live loads to be supported, called *working loads* or *service loads*, were first estimated. Then the members of the structure were proportioned so that stresses calculated by transformed area did not exceed certain permissible or allowable values.

Since 1963 the ultimate-strength design method has rapidly gained popularity because: (1) it makes use of a more rational approach than does WSD; (2) a more realistic consideration of safety is used; and (3) it provides more economical designs. With this method (now called *strength design*) the working dead and live loads are multiplied by certain load factors (equivalent to safety factors) and the resulting values are called *factored loads*. The members are then selected so they will theoretically just fail under the factored loads.

Even though almost all of the reinforced concrete structures the reader will encounter will be designed by the strength design method, it is still rather desirable to be familiar with WSD for several reasons. These include the following:

1. The design of some highway structures is handled by WSD, although the 1989 AASHTO specifications permit the use of strength design in a manner very similar to that of the ACI Code.
2. Some designers use WSD for proportioning fluid-containing structures (such as water tanks and various sanitary structures). When these structures are designed by WSD, stresses are kept at fairly low levels with

the result that there is appreciably less cracking and less consequent leakage. (If the designer uses strength design and makes use of proper crack control methods as described in Chapter 5 there should be few leakage problems.)

3. The ACI method for calculating the moments of inertia to be used for deflection calculations requires some knowledge of the working-stress procedure.

This textbook is devoted almost entirely to the strength design method. An introduction to WSD, which is called the *Alternate Design Method* in the Code, is presented in Appendix B of this book.

The reader should realize that working-stress design has several disadvantages. When using the method the designer has little knowledge as to the magnitudes of safety factors against collapse; no consideration is given to the fact that different safety factors are desirable for dead and live loads; the method does not account for variations in resistances and loads nor does it account for the possibility that as loads are increased, some increase at different rates than others.

In 1956, as an appendix, the ACI Code for the first time included ultimate-strength design, although the concrete codes of several other countries had been based on such considerations for several decades. In 1963 the Code gave ultimate-strength design equal status with working-stress design, and the 1977 Code made the method now called *strength design* the predominant method and only briefly mentioned the working-stress method. In Section 8.1.2 and Appendix A of the 1989 code the designer is given permission to use the working-stress method for nonprestressed sections, and certain specific provisions for such designs are included.

The general method was called ultimate-strength design for several decades, but the Code now uses the term "strength design." The strength for a particular reinforced concrete member is a value given by the Code and is not necessarily the true ultimate strength of the member. Therefore, the more general term "strength design" is used whether beam strength, column strength, shear strength, or others are being considered.

3.2 ADVANTAGES OF STRENGTH DESIGN

Among the several advantages of the strength design method as compared to the alternate design method are the following:

1. The derivation of the strength design expressions takes into account the nonlinear shape of the stress–strain diagram. When the resulting equations are applied, decidedly better estimates of load-carrying ability are obtained.

2. With strength design a more consistent theory is used throughout the designs of reinforced concrete structures. For instance, with the alternate design method the transformed-area or straight-line method is used for beam design and a strength design procedure is used for columns.

3. A more realistic factor of safety is used in strength design. With working-stress design the same safety factor is used for dead and live loads, whereas such is not the case for strength design. The designer can certainly estimate the magnitudes of the dead loads that a structure will have to support more accurately than he or she can the live loads. For this reason the use of different load or safety factors in strength design for the two types of loads is a definite improvement.

4. A structure designed by the strength method will have a more uniform safety factor against collapse throughout. The strength method takes considerable advantage of higher-strength steels, whereas working-stress design only partly does so. For instance, for the alternate design method the Code (Appendix A) limits the maximum flexural allowable stress in reinforcing bars (in most cases) to 24,000 psi, but in effect much higher values may be used in strength design for higher-strength steels. The result is better economy for strength design.

5. The strength method permits more flexible designs than does the alternate design method. For instance, the percentage of steel may be varied quite a bit. As a result, large sections may be used with small percentages of steel or small sections may be used with large percentages of steel. Such variations are not the case in the relatively fixed alternate design method. If the same amount of steel is used in strength design for a particular beam as would be required by the alternate design method, a

Water Tower Place, Chicago, Illinois. (Courtesy of Symons Corporation.)

smaller section will result. If the same size section is used as required by working-stress design, a smaller amount of steel will be required.

3.3 STRUCTURAL SAFETY

There are two methods by which the structural safety of a reinforced concrete structure might be considered. The first of these methods involves the calculations of the stresses caused by the working or service loads and their comparison with certain allowable stresses. Usually the safety factor against collapse when the working-stress method is used is said to equal the smaller of f'_c/f_c or f_y/f_s.

The second approach to structural safety is the one used in strength design in which the working loads are multiplied by certain load factors that are larger than one. The resulting larger or factored loads are used for designing the structure. The values of the load factors vary depending on the type and combination of the loads.

To accurately estimate the ultimate strength of a structure, it is considered necessary to take into account the uncertainties in material strengths, dimensions, and workmanship. This is done by multiplying the theoretical ultimate strength (called the *nominal strength* herein) of each member by the *capacity reduction factor* ϕ, which is less than one. These values vary from 0.90 for bending down to 0.70 for bearing.

In summary, the strength design approach to safety is to select a member whose computed ultimate load capacity multiplied by its capacity reduction factor will at least equal the sum of the service loads multiplied by their respective load factors.

Member capacities obtained with the strength method are appreciably more accurate than member capacities predicted with the working-stress method.

3.4 LOAD FACTORS

The ACI Code (9.2) states that the required ultimate load-carrying ability of a member U provided to resist the dead load D and the live load L must at least equal

$$U = 1.4D + 1.7L \qquad \text{(ACI Equation 9-1)}$$

The load factors used for live loads logically must be larger than the ones used for dead loads because the designer can estimate the magnitudes of dead loads so much better than the magnitudes of live loads.

Should it be necessary to consider wind load W as well as D and L, the Code states that the structure must be able to support at least the U given at the end of this paragraph. Since wind loads are of lesser duration than dead loads or than the usual gravity live loads, the ACI feels that it is only logical to reduce the value U obtained for this combination of W with D and L. This is done by multiplying the results by 0.75.

$$U = 0.75(1.4D + 1.7L + 1.7W) \qquad \text{(ACI Equation 9-2)}$$

In addition, the following condition, where L is not present, is to be considered:

$$U = 0.9D + 1.3W \qquad \text{(ACI Equation 9-3)}$$

This latter condition is included to cover cases where tension forces develop due to overturning moments. It will govern only for very tall buildings where high wind loads are present. In this expression the dead loads are reduced by 10% to take into account situations where they may have been overestimated.

In no case will the strength of the structure be less than that given by ACI Equation 9-1. Other cases are also given in the Code (9.2) for earthquake forces E, lateral earth pressures H, and so forth.

You will notice that these load factors are not varied in proportion to the seriousness of failure. Although it may seem logical to use a higher load factor for a hospital than for a warehouse, this is not required. It is assumed, instead, that the designer will consider the seriousness of failure while he is specifying the magnitude of the service loads. Another point to remember is that the ACI load factors are minimum values and the designer is free to use larger ones if he or she thinks the consequences of failure so require.

3.5 CAPACITY REDUCTION FACTORS

The purpose of using capacity reduction factors is to take into consideration the uncertainties of material strengths, approximations in analysis, possible variations in dimensions of concrete sections and placement of reinforcement, and other miscellaneous workmanship items. The Code (9.3) prescribes ϕ values or capacity reduction factors for several situations. Among the values given are the following:

0.90 bending in reinforced concrete

0.85 shear and torsion

0.70 bearing on concrete

The sizes of these factors are rather good indications of our knowledge of the subject in question. For instance, calculated ultimate resisting moments in reinforced concrete members seem to be fairly accurate, whereas computed bearing capacities are more questionable.

3.6 UNDERREINFORCED AND OVERREINFORCED BEAMS

Before proceeding with the derivation of beam expressions, it is necessary to define certain terms relating to the amount of tensile steel used in a beam. These terms include *balanced steel ratio*, *underreinforced beams*, and *overreinforced beams*.

A beam that has a balanced steel ratio is one for which the tensile steel will theoretically start to yield and the compression concrete reach its ultimate strain at exactly the same load. Should a beam have less reinforcement than required for a balanced ratio, it is said to be underreinforced; if more, it is said to be overreinforced.

If a beam is underreinforced and the ultimate load is approached, the steel will begin to yield even though the compression concrete is still understressed. If the load is further increased, the steel will continue to elongate, resulting in appreciable deflections and large visible cracks in the tensile concrete. As a result, the users of the structure are given notice that the load must be decreased or else the result will be considerable damage or even failure. If the load is increased further, the tension cracks will become even larger and eventually the compression side of the concrete will become overstressed and fail.

If a beam should be overreinforced, the steel will not yield before failure. As the load is increased, deflections are not noticeable even though the compression concrete is highly stressed, and failure occurs suddenly without warning to the occupants. Rectangular beams will fail in compression when strains are about 0.003 to 0.004 for ordinary grades of concrete.

Obviously, overreinforcing is a situation to be avoided if at all possible, and the Code, by limiting the percentage of tensile steel that may be used in a beam, ensures the design of underreinforced beams and thus the ductile type of failures that provide adequate "running time."

3.7 DERIVATION OF BEAM EXPRESSIONS

Tests of reinforced concrete beams confirm that strains vary in proportion to distances from the neutral axis even on the tension sides and even near ultimate loads. Compression stresses vary approximately in a straight line until the maximum stress equals about $0.50f'_c$. This is not the case, however, after stresses go higher. When the ultimate load is reached, the strain and stress variations are approximately as shown in Figure 3.1.

The compressive stresses vary from zero at the neutral axis to a maximum value at or near the extreme fiber. The actual stress variation and the actual location of the neutral axis vary somewhat from beam to beam depending on

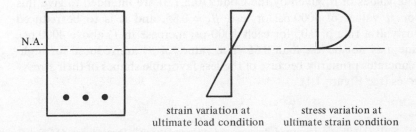

strain variation at   stress variation at
ultimate load condition   ultimate strain condition

Figure 3.1 Ultimate load.

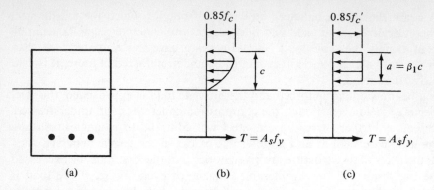

Figure 3.2 Some possible stress distribution shapes.

such items as the magnitude and history of past loadings, shrinkage and creep of the concrete, size and spacing of tension cracks, speed of loading, and so on.

If the shape of the stress diagram was the same for every beam, it would easily be possible to derive a single rational set of expressions for flexural behavior. Because of these stress variations, however, it is necessary to base the strength design method upon a combination of theory and test results.

Although the actual stress distribution given in Figure 3.2(b) may seem to be an important matter, any assumed shape (rectangular, parabolic, trapezoidal, etc.) can be used practically if the resulting equations compare favorably with test results. The most common shapes proposed are the rectangle, parabola, and trapezoid, with the rectangular shape used in this text as shown in Figure 3.2(c) being the most common one.

If the concrete is assumed to crush at a strain of about 0.003 (which is a little conservative for most concretes) and the steel to yield as f_y, it is possible to make a reasonable derivation of beam formulas without knowing the exact stress distribution. However, it is necessary to know the value of the total compression and its centroid.

Whitney[1] replaced the curved stress block with an equivalent rectangular block of intensity $0.85f_c'$ and depth $a = \beta_1 c$, as shown in Figure 3.2(c). The area of this rectangular block should equal that of the curved stress block, and the centroids of the two blocks should coincide. Sufficient test results are available for concrete beams to provide the depths of the equivalent rectangular stress blocks. The values of β_1 given by the Code (10.2.7.3) are intended to give this result. For f_c' values of 4000 psi or less, $\beta_1 = 0.85$, and it is to be reduced continuously at a rate of 0.05 for each 1000-psi increase in f_c' above 4000 psi. Their value may not be less than 0.65. The values of β_1 are reduced for high-strength concretes primarily because of the less favorable shapes of their stress–strain curves (see Figure 1.1).

---

[1] Whitney, C. S., 1942, "Plastic Theory of Reinforced Concrete Design," *Transactions ASCE*, 107, pp. 251–326.

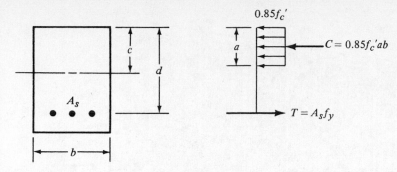

Figure 3.3

Based on these assumptions regarding the stress block, statics equations can easily be written for the sum of the horizontal forces and for the resisting moment produced by the internal couple. These expressions can then be solved separately for a and for the moment M_n.

A very clear statement should be made here regarding the term M_n because it otherwise can be rather confusing to the reader. M_n is defined as the theoretical resisting moment of a section. In Section 3.3 it was stated that the usable strength of a member equals its theoretical strength times the capacity reduction factor, or, in this case, ϕM_n. The usable flexural strength of a beam is defined as M_u:

$$M_u = \phi M_n$$

For writing the beam expressions, reference is made to Figure 3.3. Equating the horizontal forces C and T and solving for a, we obtain

$$0.85 f_c' a b = A_s f_y$$

$$a = \frac{A_s f_y}{0.85 f_c' b} = \frac{\rho f_y d}{0.85 f_c'}, \qquad \text{where } \rho = \frac{A_s}{bd}$$

Because the reinforcing steel is limited to an amount such that it will yield well before the concrete reaches its ultimate strength, the value of the nominal moment M_n can be written as

$$M_n = T\left(d - \frac{a}{2}\right) = A_s f_y\left(d - \frac{a}{2}\right)$$

and the usable flexural strength is

$$M_u = \phi M_n = \phi A_s f_y\left(d - \frac{a}{2}\right)$$

Substituting in this expression the value of a previously obtained gives an alternate form of M_u:

$$M_u = \phi A_s f_y d\left(1 - 0.59 \frac{\rho f_y}{f_c'}\right)$$

Wall construction for sewage treatment plant, Worcester, Massachusetts. (Courtesy of Symons Corporation.)

3.8 MAXIMUM PERMISSIBLE STEEL PERCENTAGE

If a balanced beam (neither underreinforced nor overreinforced) is used, it will theoretically fail suddenly and without warning. Accordingly, the ACI Code (10.3.3) limits the percentage of steel used in singly reinforced concrete beams (without axial loads) to 0.75 times the percentage that would produce a balanced condition.

In this section an expression is derived for ρ_b, the percentage of steel required for a balanced design. At ultimate load for such a beam, the concrete will theoretically fail (at a strain of 0.00300) and the steel will simultaneously yield (see Figure 3.4).

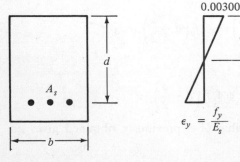

Figure 3.4

The neutral axis is located by the triangular strain relationships that follow, noting that $E_s = 29 \times 10^6$ psi for the reinforcing bars:

$$\frac{c}{d} = \frac{0.00300}{0.00300 + (f_y/E_s)} = \frac{0.00300}{0.003 + (f_y/29 \times 10^6)}$$

$$c = \frac{87,000}{87,000 + f_y} d$$

In Section 3.7 an expression was derived for a by equating the values of C and T. This value can be converted to c by dividing it by β_1:

$$a = \frac{\rho f_y d}{0.85 f'_c}$$

$$c = \frac{a}{\beta_1} = \frac{\rho f_y d}{0.85 \beta_1 f'_c}$$

Two expressions are now available for c, and they are equated to each other and solved for the percentage of steel. This is the balanced percentage ρ_b.

$$\frac{\rho f_y d}{0.85 \beta_1 f'_c} = \frac{87,000}{87,000 + f_y} d$$

$$\rho_b = \left(\frac{0.85 \beta_1 f'_c}{f_y}\right)\left(\frac{87,000}{87,000 + f_y}\right)$$

An overreinforced beam will fail in a brittle manner while an underreinforced one will fail in a ductile manner. To attempt to make sure that only ductile failures can occur the Code (10.3.3) limits the maximum steel percentage as follows

$$\rho_{max} = 0.75 \rho_b$$

Values of ρ_b, ρ_{max}, and so on can easily be calculated for different values of f'_c and f_y and tabulated as shown in Table A.8 (see Appendix). Included in the table are values of ρ_b, $0.75\rho_b$, and $0.50\rho_b$, the latter value being needed in certain continuous members as will be described in Chapter 13.

3.9 MINIMUM PERCENTAGE OF STEEL

A brief discussion of the modes of failure occurring for overreinforced and underreinforced beams was presented in Section 3.6. Sometimes because of architectural or functional requirements, beam dimensions are selected that are much larger than are required for bending alone. Such members theoretically require very small amounts of reinforcing.

There is actually another possible mode of failure that can occur in very lightly reinforced beams. If the ultimate resisting moment of the section is less than its cracking moment, the section will fail immediately when a crack occurs.

This type of failure would occur without warning, and the Code (10.5.1) provides a minimum steel percentage equal to $200/f_y$. The value was obtained by calculating the cracking moment of a plain concrete section and equating it to the strength of a reinforced concrete section of the same size and solving for the steel percentage. *This value, which is not applicable to slabs of uniform thickness, applies to other sections unless the area of reinforcement provided at every section is made at least one-third greater than required by the calculations.* For slabs of uniform thickness, the Code (10.5.3) states the minimum amount of reinforcing in the direction of the span shall not be less than that required for shrinkage and temperature reinforcement. The ACI Commentary (R10.5.3) states that when slabs are overloaded there is a tendency for the loads to be distributed laterally, thus substantially reducing the chances of sudden failure. This explains why a reduction of the minimum reinforcing percentage below $200/f_y$ is permitted in slabs of uniform thickness. Supported slabs such as slabs on grade are not considered to be structural slabs in this section unless they transmit vertical loads from other parts of the structure to the underlying soil.

3.10 FLEXURAL STRENGTH EXAMPLE

In this section the design strength ϕM_n of a beam is computed using the formulas derived in the earlier sections of this chapter. Then this problem is repeated using information given in tables in the Appendix of this text.

■ **EXAMPLE 3.1**
Determine the maximum permissible ultimate-moment capacity of the section shown in Figure 3.5 if $f'_c = 4000$ psi and $f_y = 40,000$ psi.

SOLUTION

$$\rho = \frac{3.00}{(12)(15.5)} = 0.0161$$

ρ_{max} from Table A.8 = $0.0371 > 0.0161$ **OK**

$$\rho_{min} = \frac{200}{f_y} = \frac{200}{40,000} = 0.005 < 0.0161$$ **OK**

Therefore, failure is insured by tensile yielding as desired.

$$a = \frac{A_s f_y}{0.85 f'_c b} = \frac{(3.00)(40)}{(0.85)(4)(12)} = 2.94 \text{ in.}$$

$$M_u = \phi A_s f_y \left(d - \frac{a}{2} \right) = (0.90)(3.00)(40,000)\left(15.5 - \frac{2.94}{2} \right)$$

$$M_u = 1,515,000 \text{ in.-lb} = 126.3 \text{ ft-k}$$ ■

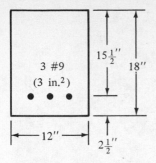

3 #9
(3 in.²)

15½″
18″
12″
2½″

Figure 3.5

Use of Graphs and Tables

In Section 3.7 the following equation was derived:

$$M_u = \phi A_s f_y d \left(1 - 0.59 \frac{\rho f_y}{f_c'} \right)$$

If A_s in this equation is replaced with ρbd, the resulting expression can be solved for $M_u/\phi bd^2$ as follows:

$$M_u = \phi \rho bd f_y d \left(1 - 0.59 \frac{\rho f_y}{f_c'} \right)$$

$$\frac{M_u}{\phi bd^2} = \rho f_y \left(1 - 0.59 \frac{\rho f_y}{f_c'} \right)$$

For a given steel percentage ρ and for a certain concrete f_c' and certain steel f_y, the value of $M_u/\phi bd^2$ can be calculated and plotted in tables, as is illustrated in Tables A.9 through A.14 (see Appendix) or in graphs (see Graph 1 of the Appendix). It is much easier to accurately read the tables than the graphs (at least to the scale to which the graphs are shown in this text). For this reason the tables are used for the examples here. The units for $M_u/\phi bd^2$ in both the tables and the graphs are pounds per square inch.

Once $M_u/\phi bd^2$ is determined for a particular beam, the value of M_u can be calculated as illustrated in the alternate solution for Example 3.1. The same tables and graphs can be used for the design of beams as were used for analysis.

Alternate Solution of Example 3.1 Using Table A.10 (See Appendix)

$$\rho = 0.0161$$

$$\frac{M_u}{\phi bd^2} = 582.8 \text{ psi from table}$$

$$M_u = (0.9)(12)(15.5)^2(582.8) = 1{,}512{,}191 \text{ in.-lb} = 126.0 \text{ ft-k}$$

Barnes Meadow Interchange, Northampton, England. (Courtesy of Cement and Concrete Association.)

Another method of checking to be sure that the strength of the tensile steel does not exceed three-fourths of the compression strength of a beam at balanced conditions is presented here. Though this method does not really add to the $\frac{3}{4}\rho_b$ theory just presented, it provides an alternate procedure that can easily be remembered and that can be utilized without resorting to tables or complicated equations. Furthermore, it is a useful method applicable to beams of other shapes, such as T beams, as will be illustrated in Chapter 4.

For this discussion, the beam of Example 3.1 is used and a balanced strain condition is assumed, as shown in Figure 3.6. The compression concrete is assumed to be at strain 0.00300, and the steel is assumed to be at yield strain f_y/E_s.

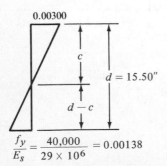

$$\frac{f_y}{E_s} = \frac{40,000}{29 \times 10^6} = 0.00138$$

Figure 3.6

From these strains the value of c can be determined:

$$c = \left(\frac{0.00300}{0.00300 + 0.00138}\right)(15.50) = 10.62 \text{ in.}$$

$$a = \beta_1 c = (0.85)(10.62) = 9.03 \text{ in.}$$

If $a = 9.03$ in., then the area of the beam cross section in compression (A_c) at a stress of $0.85f'_c$ equals ab.

$$A_c = (9.03)(12) = 108.4 \text{ in.}^2$$

$$C = 0.85f'_c A_c = (0.85)(4)(108.4) = 368.6 \text{ k}$$

$$\text{maximum } T = \tfrac{3}{4}C \text{ at balanced condition} = (\tfrac{3}{4})(368.4) = 276.4 \text{ k}$$

$$\text{actual } T = A_s f_y = (3.00)(40) = 120 \text{ k} < 276.4 \text{ k}$$

OK (it's underreinforced)

Overreinforced Beams

Overreinforced beams are almost never encountered in practice because the Code (10.3.3) does not permit their use. They are occasionally found during the review of existing reinforced concrete structures, and their resisting moments may have to be determined. Their analysis is presented in Appendix C of this textbook.

3.11 DESIGN OF RECTANGULAR BEAMS

Before the design of an actual beam is attempted, several miscellaneous topics need to be discussed. These include the following:

1. *Estimated beam weight.* The weight of the beam to be selected must be included in the calculation of the bending moment to be resisted, because the beam must support itself as well as the external loads. The weight estimates for the beams selected in this chapter are very close because the author was able to perform a little preliminary paperwork before making his estimates. You are not expected to be able to glance at a problem and estimate exactly the weight of the beam required. Following the same procedure as did the author, however, you can do a little figuring on the side and make a very reasonable estimate. For instance, you could calculate the moment due to the external loads only, select a beam size, and calculate its weight. From this beam size, you should be able to make a very good estimate of the weight of the final beam section.

2. *Beam proportions.* Unless architectural or other requirements dictate the proportions of reinforced concrete beams, the most economical beam sections are usually obtained for the shorter beams (up to 20 or 25 ft in length) when the ratio of b to d is in the range of $\frac{1}{2}$ to $\frac{2}{3}$. For longer spans, better economy is usually obtained if deep, narrow sections are used. The depths may be as large as 3 or 4 times the widths. In Example 3.2 to follow, the required bd^2 is 4543. Different values of b are assumed and the corresponding values of d computed. A pair of values of b and d are selected that are in the economical ranges given. It will be noticed that the overall beam dimensions are selected to whole inches. This is done for simplicity in constructing forms or for the rental of forms, which are usually available in 1- or 2-in. increments. Furthermore, beam widths are often selected in multiples of 2 or 3 in.

3. *Selection of bars.* After the required reinforcing area is calculated, Table A.4 (see Appendix) can be used to select bars that provide the necessary area. For the usual situations, bars of sizes #11 and smaller are practical. It is usually convenient to use bars of one size only in a beam, although occasionally two sizes will be used.

4. *Cover.* The reinforcing for concrete members must be protected from the surrounding environment; that is, fire and corrosion protection need to be provided. To do this the reinforcing is located at certain minimum distances from the surface of the concrete so that a protective layer of concrete, called *cover*, is provided. In addition, the cover improves the bond between the concrete and the steel. In Section 7.7 of the ACI Code, minimum permissible cover is given for reinforcing bars under different conditions. Values are given for reinforced concrete beams, columns, and slabs, for cast-in-place members, for precast members, for prestressed members, for members exposed to earth and weather, for members not so exposed, and so on. Members that are to be exposed to deicing salts, brackish water, seawater, or spray from these sources must meet the cover requirements of Section 7.7 of the Code as well as being especially proportioned to satisfy the water–cement ratio requirements of Code Section 4.1.2.

The beam designed in Example 3.2 is assumed to be located inside a building and thus protected from the weather. For this case the Code requires a minimum cover of $1\frac{1}{2}$ in. of concrete outside of any reinforcement.

In Chapter 7 you will learn that vertical stirrups are used in most beams for shear reinforcing. Such a stirrup is shown for a beam in Figure 3.7, and $1\frac{1}{2}$-in. cover is assumed to be required outside of the stirrup. Stirrups are commonly #3 bars, and thus a $\frac{3}{8}$-in. diameter is assumed in this figure to determine the minimum edge distance of the bars.

Assuming #9 longitudinal bars, the minimum distance from center of bars to edge of concrete is as follows:

Minimum edge distance = cover + stirrup diameter
$$+ \tfrac{1}{2} \text{ of longitudinal bar diameter}$$

$$= 1.50 + 0.375 + \frac{1.128}{2} = 2.439'' \qquad \text{(say } 2\tfrac{1}{2}'')$$

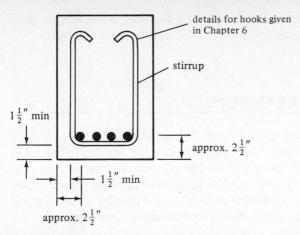

Figure 3.7

The minimum cover required for concrete cast against earth, as in a foot-ing, is 3 in., and for concrete cast not against the earth but later exposed to it, as by backfill, is 2 in. Precast and prestressed concrete or other concrete cast under plant control conditions require less cover, as described in Section 7.7 of the ACI Code.

If concrete members are exposed to very harsh surroundings, such as smoke or acid vapors, the cover should be increased above these minimums.

5. *Minimum spacing of bars.* The Code (7.6) states that the clear distance between parallel bars cannot be less than 1 in. or less than the nominal bar diameter. If the bars are placed in more than one layer, those in the upper layers are required to be placed directly over the ones in the lower layers and the clear distance between the layers must be not less than 1 in.

A major purpose of these requirements is to enable the concrete to pass between the bars. The ACI Code further relates the spacing of the bars to the maximum aggregate sizes for the same purpose. In their Section 3.3.2, maximum permissible aggregate sizes are limited to the smallest of (a) one-fifth of the nar-rowest distance between side forms, (b) one-third of slab depths, (c) three-fourths of the minimum clear spacing between bars.

In selecting the actual bar spacing, the designer will comply with the pre-ceding code requirements and will, in addition, give spacings and other dimen-sions in inches and fractions, not in decimals. The workers in the field are used to working with fractions and would be confused by a spacing of bars such as 3 at 1.45 in. The designer should always strive for simple spacings, as such dimensions will lead to better economy.

Each time a beam is designed, it is necessary to select the spacing and arrangement of the bars. To simplify these calculations, Table A.5 (see Appendix) is given. This table shows the minimum beam widths required for different num-bers of bars. The values given are based on the assumptions that $\frac{3}{8}$-in. stirrups and $1\frac{1}{2}$-in. cover are required. If four #11 bars are required, it can be seen from the table that a minimum beam width of 13.7 in. (say 14 in.) is required.

3.12 BEAM DESIGN EXAMPLES

Example 3.2 illustrates the design of a beam using the maximum permissible steel percentage. This procedure, of course, provides the smallest permissible beam size for the grades of concrete and steel being used. It may yield an economical solution but probably will not. The large amount of steel is expensive and may cause such crowded conditions as to make the placing of the concrete difficult, particularly at places where beams and columns come together.

■ EXAMPLE 3.2

Design a rectangular beam for a 20-ft simple span to support a dead load of 2 k/ft (including the estimated beam weight) and a live load of 3 k/ft. Use ρ_{max}, $f'_c = 4000$ psi, and $f_y = 40,000$ psi.

SOLUTION

$$w_u = 1.4D + 1.7L = (1.4)(2) + (1.7)(3) = 7.9 \text{ k/ft}$$

$$M_u = \frac{(7.9)(20)^2}{8} = 395 \text{ ft-k}$$

From Table A.8, $\rho_{max} = 0.0371$.

$$\rho_{min} = \frac{200}{f_y} = \frac{200}{40,000} = 0.005$$

$$M_u = \phi \rho f_y bd^2 \left(1 - 0.59\rho \frac{f_y}{f'_c} \right)$$

$$(12)(395,000) = (0.90)(0.0371)(40,000)(bd^2)\left[1 - (0.59)(0.0371)\left(\frac{40,000}{4000}\right)\right]$$

$$bd^2 = 4543 \quad \begin{cases} 10 \times 21.31 \\ 12 \times 19.46 \\ 14 \times 18.01 \end{cases}$$

Use 12 × 24 Section ($d = 19.50$)

$$A_s = (0.0371)(12 \times 19.50) = 8.68 \text{ in.}^2 \quad \text{(use 6 #11 bars)}$$

The final section is shown in Figure 3.8.

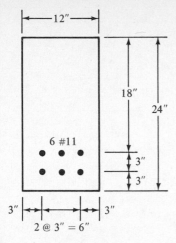

Figure 3.8

Alternate Solution for bd^2 Using Data in the Appendix

$$\rho_{\max} = 0.0371$$

$$\frac{M_u}{\phi bd^2} = 1159.2 \text{ psi (from Table A.10)}$$

$$bd^2 = \frac{(12)(395,000)}{(0.9)(1159.2)} = 4543 \qquad \blacksquare$$

The beam of Example 3.2 is redesigned in Example 3.3 with a smaller steel percentage, equal to $0.18f_c'/f_y$. Through several decades of reinforced concrete experience, it has been found that if the steel percentage is kept down to approximately this value (which is the approximate percentage obtained if the beam were designed by WSD), the beam cross section will be sufficiently large so that deflections will almost never be a problem. It is felt that from the view of both economy and deflection, the use of steel percentages in the range of $0.18f_c'/f_y$ (or perhaps $\frac{1}{2}\rho_{\max} = 0.375\rho_b$) will yield very reasonable results.

If these smaller percentages of steel are used, there will be little difficulty in placing the bars and in getting the concrete between them. Of course, from the standpoint of deflection, higher percentages of steel and thus smaller beams can be used for short spans where deflections present no problem. Whatever steel percentages are used, the resulting members will have to be carefully checked for deflections for long-span beams, cantilever beams, and shallow beams and slabs.

Another reason for using a smaller percentage of steel than $\rho_{\max}$ is given in Section 8.4 of the ACI Code, where a plastic redistribution of moments (a subject to be discussed in Chapter 13) is permitted in continuous members that have a percentage of steel less than $0.5\rho_b$. For the several reasons mentioned

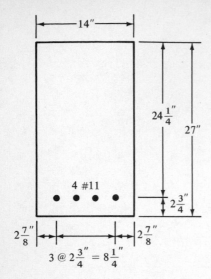

Figure 3.9

in these paragraphs, many designers try to keep their steel percentages some-where in the range between $0.18f'_c/f_y$ and $0.5\rho_{max}$, where they feel they are getting the best economy.

■ **EXAMPLE 3.3**
Repeat Example 3.2, but use $\rho = 0.18f'_c/f_y$.

SOLUTION

$$\rho = \frac{(0.18)(4000)}{40,000} = 0.018$$

$$\frac{M_u}{\phi bd^2} \text{ from Table A.10} = 643.5$$

$$bd^2 = \frac{(12)(395,000)}{(0.90)(643.5)} = 8184 \quad \begin{cases} 14 \times 24.18 \\ 16 \times 22.62 \end{cases}$$

Use 14 × 27 Beam ($d = 24.25$ in.)

$$A_s = (0.018)(14)(24.25) = 6.11 \text{ in.}^2 \qquad \text{(use 4 #11)}$$

The final section is shown in Figure 3.9.   ■

3.13 MISCELLANEOUS BEAM LIMITATIONS

In this section a few general limitations relating to beam design are introduced. Included are such topics as lateral bracing, deflections, and deep beams.

Lateral Support

It is unlikely that laterally unbraced reinforced concrete beams of any normal proportions will buckle laterally, even if they are deep and narrow, unless they are subject to appreciable lateral torsion. As a result the ACI Code (10.4.1) states that lateral bracing for a beam is not required closer than 50 times the least width of the compression flange or face. Should appreciable torsion be present, however, it must be considered in determining the maximum spacing for lateral support.

Deflections

Considerable space is devoted in Chapter 5 to the topic of deflections in reinforced concrete members subjected to bending. Should deflections not be computed and proved thereby to be acceptable, the ACI in its Section 9.5.2 provides minimum permissible beam and slab depths. The purpose of such limitations is to prevent deflections of such magnitudes as would interfere with the use of or cause injury to the structure. If deflections are computed for members of lesser thicknesses than listed in the table and are found to be satisfactory, it is unnecessary to abide by the thickness rules. For simply supported slabs, normal-weight concrete, and grade 60 steel, the minimum depth given when deflections are not computed equals $\ell/20$, where ℓ is the span length of the slab. *For concretes of other weights and for steels of different yield strengths, the minimum depths required by the ACI Code are somewhat revised.* See Table 3.1.

TABLE 3.1 MINIMUM THICKNESS OF NONPRESTRESSED BEAMS OR ONE-WAY SLABS UNLESS DEFLECTIONS ARE COMPUTED[a,b]

	Minimum thickness, h			
	Simply supported	One end continuous	Both ends continuous	Cantilever
Member	Members not supporting or attached to partitions or other construction likely to be damaged by large deflection			
Solid one-way slabs	$\ell/20$	$\ell/24$	$\ell/28$	$\ell/10$
Beams or ribbed one-way slabs	$\ell/16$	$\ell/18.5$	$\ell/21$	$\ell/8$

[a] Span length ℓ is in inches.

[b] Values given shall be used directly for members with normal weight concrete ($w_c = 145$ pcf) and Grade 60 reinforcement. For other conditions, the values shall be modified as follows:

(a) For structural lightweight concrete having unit weights in the range 90–120 lb/ft³, the values shall be multiplied by $(1.65 - 0.005w_c)$ but not less than 1.09, where w_c is the unit weight in lb/ft³.

(b) For f_y other than 60,000 psi, the values shall be multiplied by $(0.4 + f_y/100,000)$.

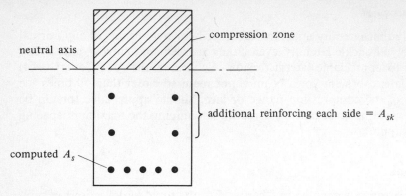

Figure 3.10 A deep beam.

Skin Reinforcement for Deep Beams

Beams with web depths that exceed 3 ft have a tendency to develop excessively wide cracks in the upper parts of their tension zones. To reduce these cracks it is necessary to add some additional longitudinal reinforcing in the zone of flexural tension near the vertical side faces of their webs as shown in Figure 3.10. The Code (10.6.7) states that additional skin reinforcement must be uniformly distributed along both side faces of such members for distances equal to $d/2$ nearest the flexural reinforcing. The area (A_{sk}) of this additional reinforcing per foot of height on each side of the web is given as

$$A_{sk} = 0.012(d - 30)$$

where d is the distance from the outermost compression fiber to the c.g. of the tensile reinforcing. The spacing of this additional reinforcing may not be greater than $d/6$ or 12 in., whichever is smaller. These additional bars may be used in computing the bending strengths of members only if appropriate strains for their positions relative to neutral axes are used to determine bar stresses. There are some special requirements that must be considered relating to shear in deep beams as described in the ACI Code (11.8) and in Section 7.14 of this text.

Other Items

The next four chapters of this book are devoted to several other important items relating to beams. These include different-shaped beams, compression reinforcing, cracks, bar development lengths, and shear.

Further Notes on Beam Sizes

From the standpoints of economy and appearance, only a few different sizes of beams should be used in a particular floor system. Such a practice will save appreciable amounts of money by simplifying the formwork and at the same time will provide a floor system that has a more uniform and attractive appearance.

If a group of college students studying the subject of reinforced concrete were to design a floor system and then compare their work with a design of the same floor system made by an experienced structural designer, the odds are that the major difference between the two designs would be in the number of beam sizes. The practicing designer would probably use only a few different sizes, whereas the average student would probably use a larger number.

The designer would probably examine the building layout to decide where to place the beams and then would make the beam subject to the largest bending moment as small as practically possible (that is, with a high percentage of reinforcing). Then he or she would proportion as many of the other similar beams with the same dimensions. The reinforcing percentages for these latter beams might vary quite a bit because of their different moments.

3.14 DETERMINING STEEL AREA WHEN BEAM DIMENSIONS ARE PREDETERMINED

Sometimes the external dimensions of a beam are predetermined by factors other than moments and shears. The depth of a member may have been selected on the basis of the minimum thickness requirements discussed in Section 3.13 for deflections. The size of a whole group of beams may have been selected to simplify the formwork as discussed in Section 3.13. Finally a specific size may have been chosen for architectural reasons. The paragraphs to follow briefly mention three methods for computing the reinforcing required. Example 3.4 illustrates the application of each of these methods.

Appendix Tables

The value of $M_u/\phi bd^2$ can be computed, and ρ can be selected from the tables. For most situations this is the quickest and most practical method. *The tables given in the Appendix of this text apply only to tensilely reinforced rectangular sections.*

Use of ρ Formula

The equation that follows was previously developed in Section 3.7 for rectangular sections.

$$M_u = \phi A_s f_y d\left(1 - 0.59\frac{\rho f_y}{f'_c}\right)$$

Replacing A_s with $\rho b d$ and letting $R_n = M_u/\phi b d^2$, this expression can be solved for ρ with the following results:

$$\rho = \frac{0.85 f'_c}{f_y}\left(1 - \sqrt{1 - \frac{2R_n}{0.85 f'_c}}\right)$$

This expression is quite useful for cases that do not fall within the tables available. *It also is applicable only to tensilely reinforced rectangular sections.*

Trial-and-Error Method

A value of a can be assumed, the value of A_s computed, the value of a determined for that value of A_s, another value of a assumed, etc. Alternately, a value of the lever arm from C to T (its $d - a/2$ for rectangular sections) can be estimated and used in the trial-and-error procedure. *This method is a general one that will work for all cross sections with tensile reinforcing.* It is particularly useful for T beams, as will be illustrated in the next chapter.

■ EXAMPLE 3.4

The dimensions of the beam shown in Figure 3.11 have been selected for architectural reasons. Determine the reinforcing steel area by each of the methods described in this section.

SOLUTION

Using Appendix Tables

$$\frac{M_u}{\phi b d^2} = \frac{(12)(160,000)}{(0.9)(16)(21)^2} = 302.3$$

$$\rho \text{ from Table A.13} = \sim 0.0054$$

$$A_s = (0.0054)(16)(21) = 1.81 \text{ in.}^2$$

Use 3 #7 bars (1.80 in.^{2})

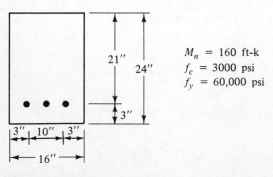

$M_n = 160$ ft-k
$f_c = 3000$ psi
$f_y = 60,000$ psi

Figure 3.11

Using ρ Formula

$$R_n = \frac{M_u}{\phi b d^2} = 302.3$$

$$\rho = \frac{(0.85)(3000)}{60,000}\left(1 - \sqrt{1 - \frac{(2)(302.3)}{(0.85)(3000)}}\right)$$

$$= 0.0054$$

Trial-and-Error Method

Assume $a = 2''$:

$$A_s = \frac{M_u}{\phi f_y\left(d - \dfrac{a}{2}\right)} = \frac{(12)(160,000)}{(0.9)(60,000)\left(21 - \dfrac{2}{2}\right)} = 1.78 \text{ in.}^2$$

$$a = \frac{A_s f_y}{0.85 f'_c b} = \frac{(1.78)(60,000)}{(0.85)(3000)(16)} = 2.61''$$

Assume $a = 2.6''$:

$$A_s = \frac{(12)(160,000)}{(0.9)(60,000)\left(21 - \dfrac{2.61}{2}\right)} = 1.80 \text{ in.}^2$$

$$a = \frac{(1.80)(60,000)}{(0.85)(3000)(16)} = 2.65'' \qquad \underline{\text{(close enough)}} \quad \blacksquare$$

3.15 BUNDLED BARS

Sometimes when large amounts of steel reinforcing are required in a beam or column, it is very difficult to fit all the bars in the cross section. For such situations, groups of parallel bars may be bundled together. Up to four bars can be bundled provided they are enclosed by stirrups or ties. The ACI Code (7.6.6.3) states that bars larger than #11 should not be bundled in beams or girders. This is primarily because of crack control problems, a subject discussed in Chapter 5 of this text. Typical configurations for two-, three-, and four-bar bundles are shown in Figure 3.12.

When spacing limitations and cover requirements are based on bar sizes, the bundled bars may be treated as a single bar for computation purposes; the diameter of the fictitious bar is to be calculated from the total equivalent area of the group. When individual bars in a bundle are cut off within the span of beams or girders, they should terminate at different points. The Code requires that there must be a stagger of at least 40 bar diameters.

Figure 3.12 Bundled-bar arrangements.

3.16 ONE-WAY SLABS

Reinforced concrete slabs are large flat plates that are supported by reinforced concrete beams, walls, or columns, by masonry walls, by structural steel beams or columns, or by the ground. If they are supported on two opposite sides only, they are referred to as *one-way slabs* because the bending is in one direction only—that is, perpendicular to the supported edges. Should the slab be supported by beams on all four edges, it is referred to as a *two-way slab* because the bending is in both directions. Actually, if a rectangular slab is supported on all four sides, but the long side is two or more times as long as the short side, the slab will, for all practical purposes, act as a one-way slab, with bending primarily occurring in the short direction. Such slabs are designed as one-way slabs. You can easily verify these bending moment ideas by supporting a sheet of paper on two opposite sides or on four sides with the support situation described. This section is concerned with one-way slabs; two-way slabs are considered in Chapters 15 and 16. It should be realized that a large percentage of reinforced concrete slabs fall into the one-way class.

A one-way slab is assumed to be a rectangular beam with a large ratio of width to depth. Normally, a 12-in.-wide piece of such a slab is designed as a beam (see Figure 3.13), the slab being assumed to consist of a series of such beams side by side. The method of analysis used is somewhat conservative due to the lateral restraint provided by the adjacent parts of the slab. Normally a beam will tend to expand laterally somewhat as it bends, but this tendency to expand by each of the 12-in. strips is resisted by the adjacent 12-in.-wide strips, which tend to expand also. In other words, Poisson's ratio is assumed to be zero. Actually, the lateral expansion tendency results in a very slight stiffening of the beam strips, which is neglected in the design procedure used here.

The 12-in.-wide beam is quite convenient when thinking of the load calculations, because loads are normally specified as so many pounds per square foot and thus the load carried per foot of the 12-in.-wide beam is the load supported per square foot by the slab. The load supported by the one-way slab

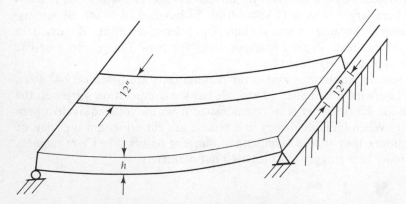

Figure 3.13 A 12-in. strip in a simply supported one-way slab.

including its own weight is transferred to the members supporting the edges of the slab. Obviously, the reinforcing for flexure is placed perpendicular to these supports—that is, parallel to the long direction of the 12-in. beams. This flexural reinforcing may not be spaced farther on center than three times the slab thickness, or 18 in., according to the ACI (7.6.5). Of course, there will be some reinforcing placed in the other direction to resist shrinkage and temperature stresses.

The thickness required for a particular one-way slab depends on the bending, deflection, and shear requirements. As previously described in Section 3.13 of this chapter, the ACI Code (9.5.2.1) provides certain span/depth limitations for concrete flexural members where deflections are not calculated.

Because of the quantities of concrete involved in floor slabs their depths are rounded off to closer values than are used for beam depths. Slab thicknesses are usually rounded off to the nearest $\frac{1}{4}$ in. on the high side for slabs of 6 in. or less in thickness and to the nearest $\frac{1}{2}$ in. on the high side for slabs thicker than 6 in.

As concrete hardens, it shrinks. In addition, temperature changes occur that cause expansion and contraction of the concrete. When cooling occurs, the shrinkage effect and the shortening due to cooling add together. The Code (7.12) states that shrinkage and temperature reinforcement must be provided in a direction perpendicular to the main reinforcement for one-way slabs. (For two-way slabs, reinforcement is provided in both directions for bending.) The Code states that for grades 40 or 50 deformed bars, the minimum percentage of this steel is 0.002 times the gross cross-sectional area of the slab. Notice that the gross cross-sectional area is bh (where h is the slab thickness). The Code (7.12.2.2) states that shrinkage and temperature reinforcement may not be spaced farther apart than five times the slab thickness, or 18 in. When grade 60 deformed bars or welded wire fabric are used, the minimum area is $0.0018bh$.

Shrinkage and temperature steel serves as mat steel in that it is tied perpendicular to the main flexural reinforcing and holds it firmly in place as a mat. This steel also helps to distribute concentrated loads transversely in the slab. (In a similar manner the AASHTO gives minimum permissible amounts of reinforcing in slabs transverse to the main flexural reinforcing for lateral distribution of wheel loads.)

Areas of steel are often determined for 1-ft widths of reinforced concrete slabs, footings, and walls. A table of "areas of bars in slabs" such as Table A.6 in the Appendix is very useful in such cases for the selection of the specific bars to be used. A brief explanation of the preparation of this table is provided here.

For a 1-ft width of concrete the total steel area obviously equals the total or average number of bars in a 1-ft width times the cross sectional area of one bar. This can be expressed as (12 in./bar spacing c to c) (area of one bar). Some examples follow, and the values obtained can be checked in the table. Understanding these calculations enables one to expand the table as desired.

1. #9 bars, 6-in. o.c. Total area in 1-ft width $= (\frac{12}{6})(1.00) = 2.00$ in.^{2}.
2. #9 bars, 5-in. o.c. Total area in 1-ft width $= (\frac{12}{5})(1.00) = 2.40$ in.^{2}.

Example 3.5 illustrates the design of a one-way slab. It will be noted that the Code (7.7.1.c) cover requirement for reinforcement in slabs (#11 and smaller bars) is $\frac{3}{4}$ in. clear when the slab is not exposed directly to the ground or the weather.

■ EXAMPLE 3.5

Design a one-way slab for the inside of a building using the span, loads, and other data given in Figure 3.14.

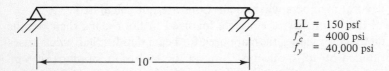

$$LL = 150 \text{ psf}$$
$$f'_c = 4000 \text{ psi}$$
$$f_y = 40,000 \text{ psi}$$

$\longleftarrow$ 10' $\longrightarrow$

Figure 3.14

SOLUTION

Minimum Total Slab Thickness h if Deflections Are Not Computed from Table 3.1

$$h = \left(0.4 + \frac{40,000}{100,000}\right)\left(\frac{\ell}{20}\right) = \frac{\ell}{25} = \frac{(12)(10)}{25} = 4.8'' \qquad \underline{\underline{say\ 5''}}$$

Assume 5-in. Slab (with $d =$ Approximately $5'' - \frac{3}{4}''$ Cover $-\frac{3}{8}''$ for Half-Diameter of Estimated Bar Size = $3\frac{7}{8}''$)

Design a 12-in.-wide strip of the slab:

$$DL = \text{slab wt} = \left(\frac{5}{12}\right)(150) = 62.5 \text{ psf}$$

$$LL = 150 \text{ psf}$$

$$w_u = (1.4)(6.25) + (1.7)(150) = 342.5 \text{ psf}$$

$$M_u = \frac{(0.3425)(10)^2}{8} = 4.28 \text{ ft-k}$$

$$\frac{M_u}{\phi bd^2} = \frac{(12)(4280)}{(0.9)(12)(3.875)^2} = 316.7$$

$$\rho = 0.00833 \text{ (from tables)}$$

$$A_s = (0.00833)(12)(3.875) = 0.387 \text{ in.}^2/\text{ft}$$

Use #4 @ 6'' from Table A.6 ($A_s = 0.39 \text{ in.}^2/\text{ft}$) ■

Placing concrete slab. (Courtesy Bethlehem Steel Corporation.)

Shrinkage and Temperature Steel

$$A_s = 0.002bh = (0.002)(12)(5) = 0.120 \text{ in.}^2/\text{ft}$$

Use #3 @ 11″ (0.12 in.²/ft) as selected from Table A.6

The designers of reinforced concrete structures must be very careful to comply with building code requirements for fire resistance. If the applicable code requires a certain fire resistance rating for floor systems, that requirement may very well cause the designer to use thicker slabs than might otherwise be required to meet the ACI strength design requirements. *In other words, the designer of a building should study carefully the fire resistance provisions of the governing building code before proceeding with the design.*

3.17 CANTILEVER BEAMS AND CONTINUOUS BEAMS

Cantilever beams supporting gravity loads are subject to negative moments throughout their lengths. As a result, their reinforcement is placed on their top or tensile sides as shown in Figure 3.15. The reader will note that for such members the maximum moments occur at the faces of the fixed supports. As a result, the largest amounts of reinforcing are required at those points. You

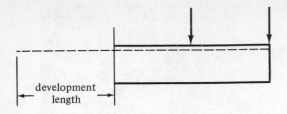

Figure 3.15 Cantilever beam.

should also note that the bars cannot be stopped at the support faces. They must be extended or anchored in the concrete beyond the support face. We will later call this *development length*. The development length does not have to be straight as shown in the figure, because the bars may be hooked at 90° or 180°.

Up to this point, only statically determinate members have been considered by the author. The very common situation, however, is for beams and slabs to be continuous over several supports as shown in Figure 3.16. Because reinforcing is needed on the tensile sides of the beams, we will place it on the bottoms when we have positive moments and in the top when we have negative moments. There are several ways in which the reinforcing bars can be arranged to resist the positive and negative moments in continuous members. One possible arrangement is shown in Figure 3.16(a). These members, including bar arrangements, are discussed in detail in Chapter 13.

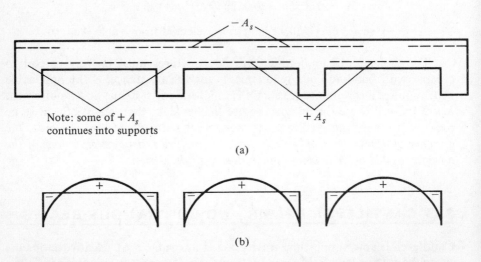

Figure 3.16 Continuous slab showing theoretical placement of bars for given moment diagram.

3.18 COMPUTER EXAMPLES

■ EXAMPLE 3.6
Repeat Example 3.3 using the enclosed computer disk.

SOLUTION

```
DESIGN OF SINGLY REINFORCED RECTANGULAR BEAMS
------------------------------------------------
Ultimate Moment          Mu (ft-k) = 395.0
Concrete Strength        f'c (psi) = 4000
Steel Yield Strength      fy (psi) = 40000
                           Rho =0.018


RESULTS
=======
USE 14.0 in. wide beam with
24.25 in. effective depth
and 4 #11 Bars.
```

■ EXAMPLE 3.7
Repeat Example 3.4 using the enclosed computer disk.

SOLUTION

```
DESIGN OF SINGLY REINFORCED RECTANGULAR BEAMS WITH GIVEN DIMENSIONS
-------------------------------------------------------------------
Ultimate Moment       Mu (ft-k)   160.0
Beam Width               b (in)   16.0
Depth to Rebars          d (in)   21.0
Concrete Strength     f'c (psi)   3000
Steel Yield Strength   fy (psi)   60000

RESULTS
=======
Use  3 # 7 bars.
```

3.19 SI EXAMPLE

Example 3.7 illustrates the design of a beam using SI units. It will be noted that in Tables A.9 through A.14 and in Graph 1 (see Appendix), values of $M_u/\phi bd^2$ are given in pounds per square inch. These may be converted to SI units in newtons per square millimeter by multiplying the table values by 0.006895.

■ **EXAMPLE 3.7**

Design a rectangular beam for a 10-m simple span to support a dead load of 15 kN/m (not including beam weight) and a live load of 20 kN/m. Use $\rho = 0.5\rho_b$, $f'_c = 20.7$ MPa, and $f_y = 275.8$ MPa, and assume the concrete weight is 23.5 kN/m³.

SOLUTION

Assume that the beam weight is 8.5 kN/m.

$$w_u = (1.4)(23.5) + (1.7)(20) = 66.9 \text{ kN/m}$$

$$M_u = \frac{(66.9)(10)^2}{8} = 836.2 \text{ kN·m}$$

$$\rho = 0.0186 \text{ (from Table A.8)}$$

$$M_u = \phi\rho f_y bd^2\left(1 - 0.59\rho\frac{f_y}{f'_c}\right)$$

$$(10^6)(836.2) = (0.9)(0.0186)(275.8)(bd^2)\left[1 - (0.59)(0.0186)\left(\frac{275.8}{20.7}\right)\right]$$

$$bd^2 = 212{,}134{,}550 \text{ mm}^3 \begin{cases} 400 \times 728 \\ 450 \times 687 \\ 500 \times 651 \end{cases}$$

<u>USE 450 mm × 800 mm section ($d = 695$ mm)</u>

$$\text{Beam wt} = \frac{(450)(800)}{(10)^6}(23.5) = 8.46 \text{ kN/m} < 8.5 \text{ kN/m} \qquad \underline{\underline{\text{OK}}}$$

$$A_s = (0.0186)(450)(695) = 5817 \text{ mm}^2$$

<u>USE 6 #11 bars in 2 rows (6036 mm²)</u>

PROBLEMS

In Problems 3.1 to 3.13 design rectangular sections for the beams, loads, and ρ values shown in the accompanying illustrations. Beam weights are not included in the loads shown. Show sketches of cross sections including bar sizes, arrangement, and spacing. Assume concrete weighs 150 lb/ft³, $f_y = 60{,}000$ psi, and $f'_c = 3000$ psi.

Problem 3.1 (*One ans.* 18 in. × 36 in. with 5 #9 bars)

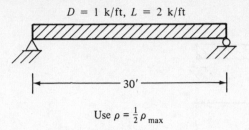

$$D = 1 \text{ k/ft}, \ L = 2 \text{ k/ft}$$

30'

Use $\rho = \frac{1}{2}\rho_{\text{max}}$

3.2 Repeat Problem 3.1 if $L = 3$ k/ft and if the span is 28 ft.

Problem 3.3 (*One ans.* 16 in. × 31 in. with 4 #10 bars)

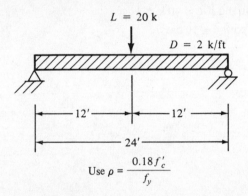

$$L = 20 \text{ k}$$
$$D = 2 \text{ k/ft}$$

12' — 12'

24'

Use $\rho = \dfrac{0.18 f'_c}{f_y}$

3.4 Repeat Problem 3.3 if $D = 1$ k/ft and if $L = 36$ k.

Problem 3.5 (*One ans.* 24 in. × 42 in. with 7 #10 bars plus deep beam reinforcing)

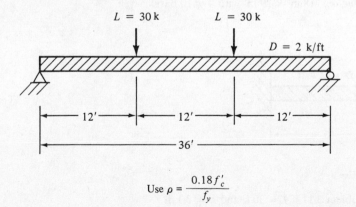

$$L = 30 \text{ k} \qquad L = 30 \text{ k}$$
$$D = 2 \text{ k/ft}$$

12' — 12' — 12'

36'

Use $\rho = \dfrac{0.18 f'_c}{f_y}$

3.6 Repeat Problem 3.5 if $D = 3$ k/ft and if the L concentrated loads are changed to 40 k.

Problem 3.7 (*One ans.* 16 in. × 32 in. with 3 #10)

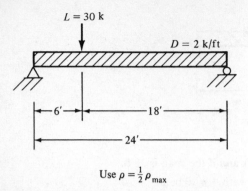

Use $\rho = \frac{1}{2}\rho_{max}$

3.8   Repeat Problem 3.7 if $D = 3$ k/ft and if $L = 40$ k.

Problem 3.9 (*One ans.* 18 in. × 34 in. with 4 #10 bars)

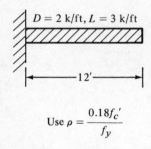

Use $\rho = \dfrac{0.18 f_c'}{f_y}$

3.10   Repeat Problem 3.9 if $L = 4$ k/ft and if the beam span = 15 ft.

Problem 3.11 (*One ans.* 18 in. × 29 in. with 4 #10 bars)

$L = 20$ k

$D = 2$ k/ft

10′

Use $\rho = \frac{1}{2}\rho_b$

3.12   Repeat Problem 3.11 if $L = 30$ k and $D = 3$ k/ft.

Problem 3.13 (*One ans.* 20 in. × 35 in. with 6 #11 bars in 2 layers)

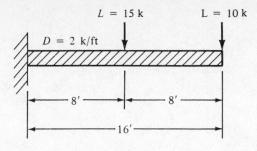

Use $\rho = \rho_{max}$

In Problems 3.14 to 3.16 determine whether the beams are overreinforced or under-reinforced without using tables or formulas if $f_y = 60,000$ psi and $f'_c = 3000$ psi.

Problem 3.14

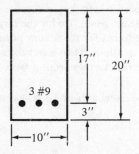

Problem 3.15 (*Ans.* It is overreinforced and can't be used)

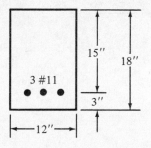

Problem 3.16

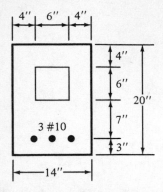

In Problems 3.17 and 3.18 design rectangular sections for the beams and loads shown in the accompanying illustrations. Beam weights are not included in the given loads. $f_y = 60,000$ psi and $f'_c = 3000$ psi. Live loads are to be placed where they will cause the most severe conditions at the sections being considered. Select beam size for the largest moment (positive or negative) and then select the steel required for maximum positive moment and sketch the beam cross section including bar locations. Then select the steel required for maximum negative moment and sketch the beam cross section with bar locations.

Problem 3.17 (*Ans.* 10 in. × 22 in. beam with 3 #7 negative steel bars and 3 #7 positive steel bars)

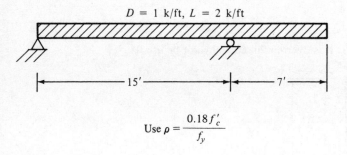

$$\text{Use } \rho = \frac{0.18 f'_c}{f_y}$$

Problem 3.18

$$D = 1 \text{ k/ft}, L = 3 \text{ k/ft}$$

$$\text{Use } \rho = \frac{0.18 f'_c}{f_y}$$

In Problems 3.19 and 3.20 design interior one-way slabs for the situations shown. Concrete weight = 150 lb/ft³, f_y = 60,000 psi, and f'_c = 3000 psi. Do not use the ACI Code's minimum thickness for deflections. Steel percentages are given in the figures.

Problem 3.19 (*One ans.* 6-in. slab with #6 @ 10-in. main steel and #3 @ 10-in. shrinkage and temperature steel)

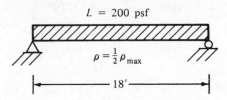

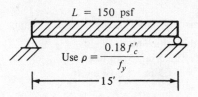

Problem 3.20

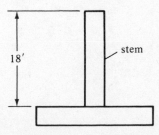

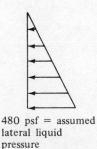

3.21 Repeat Problem 3.19 using the ACI Code's minimum thickness requirement for cases where deflections are not computed. Do not use the *e* given in Prob. 3.19. (*One ans.* 9-in. slab with #5 @ 10-in. main steel and #4 @ 12-in. shrinkage and temperature steel)

3.22 Using f'_c = 3000 psi, f_y = 60,000 psi, and ρ_{max}, determine the depth of a simple beam to support itself for a 200-ft simple span. Neglect cover in calculations.

3.23 Determine the depth required for a beam to support itself only for a 500-ft span. Neglect cover in weight calculations. Given f'_c = 4000 psi, f_y = 60,000 psi and ρ_{max}. (*Ans.* 48.65 ft)

3.24 Determine the stem thickness for maximum moment for the retaining wall shown in the accompanying illustration. Also, determine the steel area required at the bottom and middepth of the stem if f'_c = 3000 psi, and f_y = 60,000 psi. Assume that #8 bars are to be used and that the stem thickness is constant for the 18-ft height. Also, assume that the clear cover required is 2 in. $\rho = 0.18f'_c/f_y$.

480 psf = assumed lateral liquid pressure

3.25 (a) Design a 24-in.-wide precast concrete slab to support a 40-psf live load for a simple span of 15 ft. Assume minimum concrete cover required is $\frac{5}{8}$ in. as per Section 7.7.2 of the Code. Use welded wire fabric for reinforcing, $f_y = 60,000$ psi and $f'_c = 3000$ psi. $\rho = 0.18 f'_c / f_y$. (*One ans*. WWF3 × 8 − W7 × 7 in a 4-in. slab)
   (b) Can a 250-lb man walk across the center of the span when the other live load is not present? Assume 100% impact. (*Ans*. yes. M_u furnished $= 7.81$ ft-k > actual $M_u = 7.12$ ft-k)

3.26 Prepare a flow chart for the design of tensilely reinforced rectangular beams.

3.27 Using the enclosed computer disk solve the following problems.
   (a) Problem 3.1 (*One ans*. $b = 18$ in., $d = 33.5$ in., 5 #9 bars)
   (b) Problem 3.9 (*One ans*. $b = 18$ in., $d = 31.5$ in., 4 #10 bars)

PROBLEMS WITH SI UNITS
In Problems 3.28 to 3.31, design rectangular sections for the beams, loads, and ρ values shown in the accompanying illustrations. Beam weights are not included in the loads given. Show sketches of cross sections including bar sizes, arrangement, and spacing. Assume concrete weighs 23.5 kN/m³, $f_y = 413.7$ MPa, and $f'_c = 20.7$ MPa.
Problem 3.28

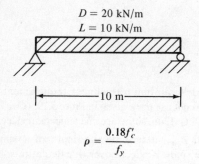

$$D = 20 \text{ kN/m}$$
$$L = 10 \text{ kN/m}$$

|← 10 m →|

$$\rho = \frac{0.18 f'_c}{f_y}$$

Problem 3.29 (*One ans*. 350 mm × 810 mm with 5 #10 bars)

$$D = 16 \text{ kN/m}$$
$$L = 20 \text{ kN/m}$$

|← 5 m →|

$$\rho = \rho_{max}$$

Problem 3.30

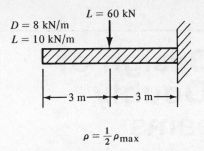

$$\rho = \frac{1}{2}\rho_{max}$$

3.31 Design the one-way slab shown in the accompanying figure to support a live load of 9 kN/m². Do not use the ACI thickness limitation for deflections and assume concrete weighs 23.5 kN/m³. Assume that $f'_c = 20.7$ MPa and $f_y = 413.7$ MPa. Use $\rho = \frac{1}{2}\rho_{max}$. (*One ans.* 270-mm slab with #7 @ 200-mm main steel and #4 @ 250-mm shrinkage and temperature steel)

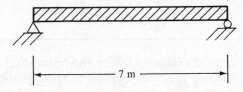

4

Analysis and Design of T Beams and Doubly Reinforced Beams

4.1 T BEAMS

Reinforced concrete floor systems normally consist of slabs and beams that are placed monolithically. As a result, the two parts act together to resist loads. In effect the beams have extra widths at their tops, called *flanges*, and the resulting T-shaped beams are called *T beams*. The part of a T beam below the slab is referred to as the *web* or *stem*. (The beams may be L-shaped if the stem is at the end of a slab. The stirrups (described in Chapter 7) in the webs extend up into the slabs, as perhaps do bent-up bars, with the result that they further make the beams and slabs act together.

There is a problem involved in estimating how much of the slab acts as part of the beam. Should the flanges of a T beam be rather stocky and compact in cross section, bending stresses will be fairly uniformly distributed across the compression zone. If, however, the flanges are wide and thin, bending stresses will vary quite a bit across the flange due to shear deformations. The further a particular part of the slab or flange is away from the stem, the smaller will be its bending stress.

Instead of considering a varying stress situation across the full width of the flange, the ACI Code (8.10.2) calls for a smaller width with an assumed uniform stress distribution for design purposes. The objective is to have the same total compression force in the reduced width that actually occurs in the full width with its varying stresses.

The hatched area in Figure 4.1 shows the effective size of a T beam. For symmetrical T beams, the Code states that the effective flange width may not exceed one-fourth of the beam span, and the overhanging width on each side may not exceed eight times the slab thickness or one-half the clear distance to the adjacent T beam. Isolated T beams must have a flange thickness no less

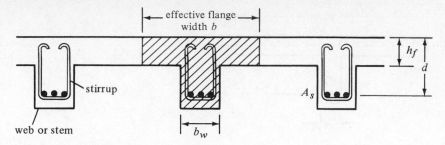

Figure 4.1 Effective width of T beams.

than $\frac{1}{2}$ the web width and its effective flange width may not be larger than 4 times the web width. (ACI 8.10.4).

The analysis of T beams is handled quite similarly to the method used for rectangular beams, and the tensile steel percentage is once again limited to 0.75 times the percentage required for a balanced design. It should be noticed, however, that limiting the steel percentage to a maximum of $0.75\rho_b$ very seldom

Natural History Museum, Kensington, London, England.
(Courtesy of Cement and Concrete Association.)

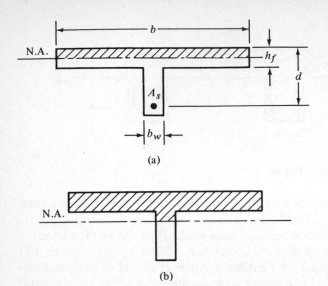

Figure 4.2

presents any difficulty (at least in simple-span T's) because the compression side of the beam is so large that compression stresses are normally quite low, and hence the designer would almost never use an amount of steel greater than $0.75\rho_b$.

The neutral axis for T beams can fall either in the flange or in the stem, depending on the proportions of the slabs and stems. If it falls in the flange, and it almost always does, the rectangular beam formulas apply, as can be seen in Figure 4.2(a). The concrete below the neutral axis is assumed to be cracked and its shape has no effect on the flexure calculations (other than weight). The section above the neutral axis is rectangular. If the neutral axis is below the flange, however, as shown for the beam of Figure 4.2(b), the compression concrete above the neutral axis no longer consists of a single rectangle and thus the normal rectangular beam expressions do not apply.

If the neutral axis is assumed to fall within the flange, the value of a can be computed as it was for rectangular beams:

$$a = \frac{A_s f_y}{0.85 f'_c b} = \frac{\rho f_y d}{0.85 f'_c}$$

The distance to the neutral axis c equals a/β_1. If the computed value of a is equal to or less than the flange thickness, the section for all practical purposes can be assumed to be rectangular even though the computed value of $c = a/\beta_1$ is actually greater than the flange thickness.

A beam does not really have to look like a T beam to be one. This fact is shown by the beam cross sections shown in Figure 4.3. For these cases the compression concrete is T-shaped and the shape or size of the concrete on the tension side, which is assumed to be cracked, has no effect on the theoretical resisting moments. It is true, however, that the shapes, sizes, and weights of

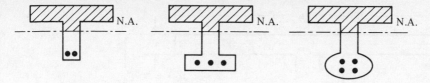

Figure 4.3 Various cross sections of T beams.

the tensile concrete do affect the deflections that occur (as is described in Chapter 5), and their dead weights affect the magnitudes of the moments to be resisted.

4.2 ANALYSIS OF T BEAMS

The calculation of the design strengths of T beams is illustrated in Examples 4.1 and 4.2. In the first of these problems, the neutral axis falls in the flange, while for the second it is in the web. The procedure used for both examples involves the following steps:

1. The calculation of $T = A_s f_y$.
2. The calculation of the area of the concrete in compression (A_c) stressed to $0.85 f'_c$

$$C = T = 0.85 f'_c A_c$$

$$A_c = \frac{T}{0.85 f'_c}$$

3. The location of the center of gravity of the concrete area A_c.
4. The computation of $M_n = T$ times the lever arm from the center of gravity of the steel to the center of gravity of A_c.
5. $M_u = \phi M_n$.

For Example 4.1, where the neutral axis falls in the flange, it would be logical to apply the normal rectangular equations of Sections 3.7 and 3.8 of this book, but the author has used the couple method as a background for the solution of Example 4.2, where the neutral axis falls in the web. This same procedure can be used for determining the design strengths of tensily reinforced concrete beams of any shape ($\top, \ulcorner, \sqcap$, triangular, circular, and so on).

■ **EXAMPLE 4.1**
Determine the permissible design strength of the T beam shown in Figure 4.4, with $f'_c = 3000$ psi and $f_y = 50,000$ psi.

SOLUTION

Computing T

$$T = A_s f_y = (6.00)(50) = 300 \text{ k}$$

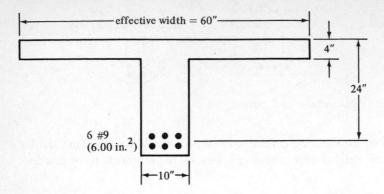

Figure 4.4

Computing A_c

$$A_c = \frac{T}{0.85f'_c} = \frac{300}{(0.85)(3)} = 118 \text{ in.}^2 < 4 \times 60 = 240 \text{ in.}^2$$

Obviously the stress block is entirely within the flange and the rectangular formulas apply. However, using the couple method as follows:

$$a = \frac{118}{60} = 1.97''$$

$$\text{Lever arm} = d - \frac{a}{2} = 24 - \frac{1.97}{2} = 23.02''$$

Calculating the Moment Capacity

$$M_n = T\left(d - \frac{a}{2}\right) = (300)(23.02) = 6906 \text{ in.-k} = 576 \text{ ft-k}$$

$$M_u = \phi M_n = (0.90)(576) = \underline{518 \text{ ft-k}}$$

Checking ρ_{max} (As Previously Described in Section 3.10) Assuming a balanced condition with the strain in the extreme compression fibers of the concrete at 0.00300 and the steel strain equal to f_y/E_s, as shown in Figure 4.5,

$$c = \frac{0.00300}{0.00300 + 0.00172}(24.00) = 15.25''$$

$$a = (0.85)(15.25) = 12.96''$$

$$A_c = (60)(4) + (8.96)(10) = 329.6 \text{ in.}^2$$

$$C = (0.85)(3)(329.6) = 840.5 \text{ k}$$

Maximum $T = (\tfrac{3}{4})(840.5) = 630.4 \text{ k} > $ actual $T = (6)(50) = 300 \text{ k}$ $\underline{\text{OK}}$ ∎

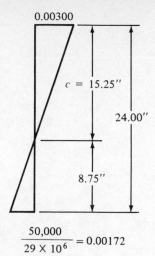

0.00300

$c = 15.25''$

24.00''

8.75''

$$\frac{50,000}{29 \times 10^6} = 0.00172$$

Figure 4.5

■ EXAMPLE 4.2

Compute the design strength for the T beam shown in Figure 4.6, in which $f'_c =$ 3000 psi and $f_y = 50,000$ psi.

SOLUTION

Computing T

$$T = A_s f_y = (10.12)(50) = 506 \text{ k}$$

Computing A_c

$$A_c = \frac{T}{0.85 f'_c} = \frac{506}{(0.85)(3)} = 198.4 \text{ in.}^2 > 30 \times 4 = 120 \text{ in.}^2$$

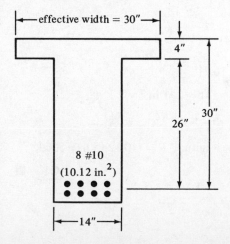

←effective width = 30''→

4''

30''

26''

8 #10
(10.12 in.^{2})

←14''→

Figure 4.6

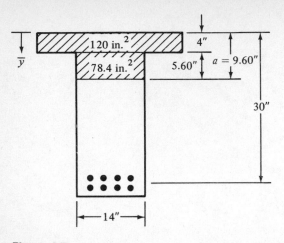

Figure 4.7

Obviously the stress block must extend below the flange to provide the neces-sary compression area $= 198.4 - 120 = 78.4 \text{ in.}^2$, *as shown in Figure 4.7.*

Computing the Distance $\bar{y}$ from the Top of the Flange to the Center of Gravity of A_c

$$\bar{y} = \frac{(120)(2) + (78.4)(6.80)}{198.4} = 3.90''$$

The Lever Arm Distance from T to $C = 30.00 - 3.90 = 26.10''$

$$M_n = (506)(26.10) = 13,206 \text{ in.-k} = 1101 \text{ ft-k}$$
$$M_u = (0.90)(1101) = 991 \text{ ft-k}$$

Checking ρ_{max} (Figure 4.8)

$$a = (0.85)(19.07) = 16.21''$$
$$A_c = (4)(30) + (12.20)(14) = 291 \text{ in.}^2$$
$$C = (0.85)(3)(291) = 742 \text{ k}$$

Maximum $T = \frac{3}{4}C = (\frac{3}{4})(742) = 556 \text{ k} >$ actual $T = (10.12)(50) = 506 \text{ k}$

OK ∎

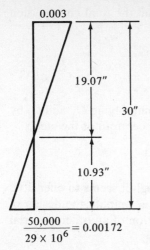

$$\frac{50{,}000}{29 \times 10^6} = 0.00172$$

Figure 4.8

4.3 ANOTHER METHOD FOR ANALYZING T BEAMS

In the preceding section a very important method of analyzing reinforced concrete beams was presented. It is a general method which is applicable to tensilely reinforced beams of any cross section, including T beams. T beams are so very common, however, that many designers prefer another method which is specifically designed for T beams.

First the value of a is determined as previously described in this chapter. Should it be less than the flange thickness h_f, we will have a rectangular beam and the rectangular beam formulas will apply. Should it be greater than the flange thickness h_f (as was the case for Example 4.2), the special method to be described here will be very useful.

The beam is divided into a set of rectangular parts consisting of the overhanging parts of the flange and the compression part of the web (see Figure 4.9).

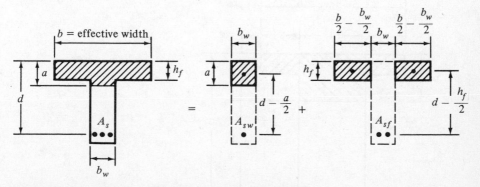

Figure 4.9 Separation of T beam into rectangular parts.

The total compression C_w in the web rectangle and the total compression in the overhanging flange C_f are computed:

$$C_w = 0.85f'_c ab_w$$

$$C_f = 0.85f'_c(b - b_w)(h_f)$$

but if $a < h_f$ replace h_f with a.

Then the nominal moment M_n is determined by multiplying C_w and C_f by their respective lever arms from their centroids to the centroid of the steel:

$$M_n = C_w\left(d - \frac{a}{2}\right) + C_f\left(d - \frac{h_f}{2}\right)$$

This procedure is illustrated in Example 4.3. Though it seems to offer little advantage in computing M_n, we will learn that it does simplify the design of T beams when $a > h_f$ because it permits a direct solution of an otherwise trial and error problem.

■ EXAMPLE 4.3

Repeat Example 4.2 using the value of a (9.60 in.) previously obtained and the alternate formulas just developed. Reference is made to Figure 4.10, the dimensions of which were taken from Figure 4.7.

SOLUTION (noting that $a > h_f$)

Computing C_w and C_f

$$C_w = (0.85)(3)(9.60)(14) = 342.7 \text{ k}$$

$$C_f = (0.85)(3)(30 - 14)(4) = 163.2 \text{ k}$$

Computing M_n and M_u

$$M_n = (342.7)\left(30 - \frac{9.60}{2}\right) + (163.2)\left(30 - \frac{4}{2}\right) = 13,205 \text{ in.-k} = 1100 \text{ ft-k}$$

$$M_u = \phi M_n = (0.90)(1100) = 990 \text{ ft-k} \qquad ■$$

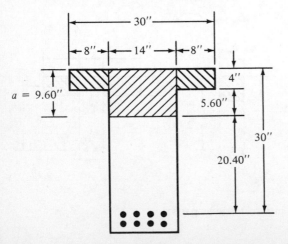

Figure 4.10

4.4 DESIGN OF T BEAMS

For the design of T beams, the flange has normally already been selected in the slab design, as it is the slab. The size of the web is normally not selected on the basis of moment requirements but probably is given an area based on shear requirements; that is, a sufficient area is used so as to provide a certain minimum shear capacity, as will be described in Chapter 7. It is also possible that the width of the web may be selected on the basis of the width estimated to be needed to put in the reinforcing bars. Sizes may also have been preselected as previously described in Section 3.13 to simplify formwork, for architectural requirements, or for deflection reasons. For the examples that follow (4.4 and 4.5), the values of d and b_w are given.

The flanges of most T beams are usually so large that the neutral axis probably falls within the flange and thus the rectangular beam formulas apply. Should the neutral axis fall within the web, a trial-and-error process is often used for the design. In this process a lever arm from the center of gravity of the compression block to the center of gravity of the steel is estimated to equal the larger of $0.9d$ or $d - (h_f/2)$ and from this value called z, a trial steel area is calculated ($A_s = M_n/f_y z$). Then by the process used in Example 4.2, the value of the estimated lever arm is checked. T beams are designed in Examples 4.4 and 4.5 by this process.

Example 4.6 presents a more direct approach for the case where $a > h_f$. This is the case where the beam is assumed to be divided into its rectangular parts.

■ EXAMPLE 4.4

Design a T beam for the floor system shown in Figure 4.11 for which b_w and d are given, $M_D = 50$ ft-k, $M_L = 100$ ft-k, $f'_c = 4000$ psi, $f_y = 50,000$ psi, and simple span = 20 ft.

SOLUTION

Effective Flange Width

(a) $\frac{1}{4} \times 20 = 5'0'' = \underline{60''}$
(b) $12 + (2)(8)(4) = \overline{76''}$
(c) $10' \ 0'' = 120''$

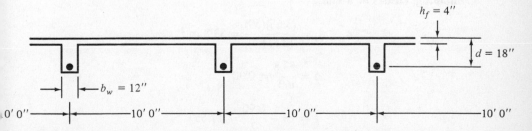

Figure 4.11

Moments

$$M_u = (1.4)(50) + (1.7)(100) = 240 \text{ ft-k}$$

$$M_n = \frac{240}{0.90} = 267 \text{ ft-k}$$

Assuming a Lever Arm z Equal to the Larger of $0.9d$ or $d - (h_f/2)$

$$z = (0.9)(18) = \underline{16.20''}$$

$$z = 18 - \frac{4}{2} = 16.00''$$

Trial Steel Area

$$A_s f_y z = M_n$$

$$A_s = \frac{(12)(267)}{(50)(16.20)} = 3.96 \text{ in.}^2$$

Computing Values of a and z

$$0.85 f_c' A_c = A_s f_y$$

$$(0.85)(4)(A_c) = (3.96)(50)$$

$$A_c = 58.2 \text{ in.}^2$$

$$a = \frac{58.2}{60} = 0.97''$$

Therefore, the neutral axis is in the flange.

$$z = 18 - \frac{0.97}{2} = 17.52''$$

Calculating A_s with This Revised z

$$A_s = \frac{(12)(267)}{(50)(17.52)} = 3.66 \text{ in.}^2$$

Computing Values of a and z

$$A_c = \frac{(3.65)(50)}{(0.85)(4)} = 53.8 \text{ in.}^2$$

$$a = \frac{53.8}{60} = 0.90''$$

$$z = 18 - \frac{0.90}{2} = 17.55''$$

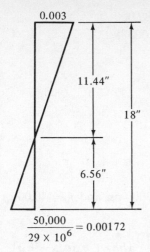

0.003

11.44"

18"

6.56"

$$\frac{50,000}{29 \times 10^6} = 0.00172$$

Figure 4.12

Calculating A_s with This Revised z

$$A_s = \frac{(12)(267)}{(50)(17.55)} = 3.65 \text{ in.}^2 \qquad \underline{\text{OK, close enough to previous value}}$$

Checking ρ_{max} (Figure 4.12)

$$a = (0.85)(11.44) = 9.72''$$
$$A_c = (4)(60) + (5.72)(12) = 308.6 \text{ in.}^2$$
$$C = (0.85)(4)(308.6) = 1049 \text{ k}$$
$$\text{Maximum } T = (\tfrac{3}{4})(1049) = 787 \text{ k} > (3.65)(50) = 182.5 \text{ k} \qquad \underline{\textbf{OK}}$$

Checking ρ_{min}   Section 10.5.1 of the ACI Code states that to compute ρ to check against $\rho_{min} = 200/f_y$, the width of the web is to be used when the web is in tension.

$$\rho = \frac{A_s}{b_w d} = \frac{3.65}{(12)(18)} = 0.0169$$

$$\rho_{min} = \frac{200}{50,000} = 0.004 < 0.0169 \qquad \underline{\textbf{OK}} \quad \blacksquare$$

It is possible to write an expression for the maximum amount of tensile steel permitted by the Code for a particular T beam. This can be accomplished by following exactly the procedure used for determining the maximum T value in Example 4.4. Reference is made to Figure 4.1 for the letters used.

$$C_{bal} = \frac{0.00300}{0.00300 + [f_y/(29 \times 10^6)]} d = \frac{87,000}{87,000 + f_y} d$$

$$a_{bal} = \beta_1 c_{bal}$$

$$C_{bal} = 0.85f_c'[bh_f + b_w(a_{bal} - h_f)]$$

$$C_{bal} = T_{bal}$$

$$A_{s\,max} = \frac{T_{bal}}{f_y}$$

$$A_{s\,max} = 0.75A_{s\,bal}$$

An expression for $A_{s\,max}$ for a T beam with $f_c' = 3000$ psi and $f_y = 40,000$ psi can be developed as follows:

$$A_{s\,max} = 0.75A_{s\,bal} = 0.75\frac{T_{bal}}{f_y} = 0.75\frac{C_{bal}}{f_y} = \frac{(0.75)(0.85f_c')}{f_y}[bh_f + b_w(a_{bal} - h_f)]$$

$$= \frac{(0.75)(0.85 \times 3000)}{40,000}\left[bh_f + b_w\left(0.85\frac{87,000}{87,000 + 40,000}d - h_f\right)\right]$$

$$A_{s\,max} = 0.0478[bh_f + b_w(0.582d - h_f)]$$

Following a similar procedure, maximum A_s values permitted by the Code for T beams can be computed for other concrete and steel grades, with the results as shown in Table 4.1. These expressiions are applicable regardless of whether the neutral axis falls in the flange or in the stem.

TABLE 4.1   MAXIMUM TENSILE STEEL PERMITTED
                    IN T BEAMS

Concrete and steel	Formula (units in square inches)
$f_c' = 3000$ psi $f_y = 40,000$ psi	$A_{s\,max} = 0.0478[bh_f + b_w(0.582d - h_f)]$
$f_c' = 4000$ psi $f_y = 40,000$ psi	$A_{s\,max} = 0.0638[bh_f + b_w(0.582d - h_f)]$
$f_c' = 3000$ psi $f_y = 50,000$ psi	$A_{s\,max} = 0.0382[bh_f + b_w(0.540d - h_f)]$
$f_c' = 4000$ psi $f_y = 50,000$ psi	$A_{s\,max} = 0.0510[bh_f + b_w(0.540d - h_f)]$
$f_c' = 3000$ psi $f_y = 60,000$ psi	$A_{s\,max} = 0.0319[bh_f + b_w(0.503d - h_f)]$
$f_c' = 4000$ psi $f_y = 60,000$ psi	$A_{s\,max} = 0.0425[bh_f + b_w(0.503d - h_f)]$

■ EXAMPLE 4.5

Design a T beam for the floor system shown in Figure 4.13, for which b_w and d are given, $M_D = 200$ ft-k, $M_L = 340$ ft-k, $f'_c = 3000$ psi, $f_y = 50{,}000$ psi, and simple span = 18 ft.

SOLUTION

Effective Flange Width

(a) $\frac{1}{4} \times 18' = 4'6'' = \underline{54''}$

(b) $15 + (2)(8)(3) = 63''$

(c) $6'0'' = 72''$

Moments

$$M_u = (1.4)(200) + (1.7)(340) = 858 \text{ ft-k}$$

$$M_n = \frac{858}{0.90} = 953 \text{ ft-k}$$

Assuming a Lever Arm z

$$z = (0.90)(24) = 21.6''$$

$$z = 24 - \tfrac{3}{2} = \underline{22.5''}$$

Trial Steel Area

$$A_s = \frac{(12)(953)}{(50)(22.5)} = 10.17 \text{ in.}^2$$

Checking Values of a and z

$$A_c = \frac{(50)(10.17)}{(0.85)(3)} = 199.4 \text{ in.}^2$$

The stress block extends down into the web, as shown in Figure 4.14.

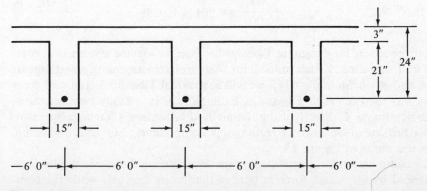

Figure 4.13

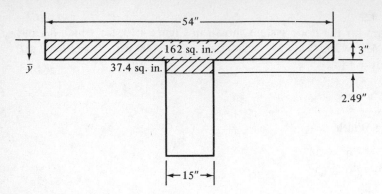

Figure 4.14

Computing the Distance $\bar{y}$ from the Top of the Flange to the Center of Gravity of A_c

$$\bar{y} = \frac{(162)(1.5) + (37.4)(4.25)}{199.4} = 2.02''$$

$$z = 24 - 2.02 = 21.98''$$

$$A_s = \frac{(12)(953)}{(50)(21.98)} = 10.41 \text{ in.}^2 \qquad \underline{\text{close enough; use}}$$

Calculating $A_{s\,max}$,

$$A_{s\,max} = 0.0382[bh_f + b_w(0.540d - h_f)]$$
$$= 0.0382[(54)(3) + (15)(0.540 \times 24 - 3)]$$
$$= 11.90 \text{ in.}^2 > 10.41 \text{ in.}^2 \qquad \underline{\text{OK}}$$

Checking ρ_{min},

$$\rho = \frac{10.41}{(15)(24)} = 0.0289$$

$$\rho_{min} = \frac{200}{50,000} = 0.004 < 0.0289 \qquad \underline{\text{OK}} \quad \blacksquare$$

Our procedure for designing T beams has been to assume a value of z, compute a trial steel area A_s, determine a for that steel area assuming a rectangular section, and so on. Should $a > h_f$, we will have a real T beam. A trial-and-error process was used for such a beam in Example 4.5. It is easily possible, however, to determine A_s directly using the method of Section 4.3 where the member was broken down into its rectangular components. For this discussion, reference is made to Figure 4.9.

The compression force provided by the overhanging flange rectangles must be balanced by the tensile force in part of the tensile steel A_{sf} while the compression force in the web is balanced by the tensile force in the remaining tensile steel A_{sw}.

For the overhanging flange we have

$$0.85f'_c(b - b_w)(h_f) = A_{sf}f_y$$

from which the required area of steel A_{sf} equals

$$A_{sf} = \frac{0.85f'_c(b - b_w)h_f}{f_y}$$

The design strength of these overhanging flanges is

$$M_{uf} = \phi A_{sf}f_y\left(d - \frac{h_f}{2}\right)$$

The remaining moment to be resisted by the web of the T beam and the steel required to balance that value are next determined.

$$M_{uw} = M_u - M_{uf}$$

The steel required to balance the moment in the rectangular web is obtained by the usual rectangular beam expression. The value $M_{uw}/\phi b_w d^2$ is computed, and ρ is determined from the appropriate Appendix table or the expression for ρ previously given in Section 3.14 of this book. Then

$$A_{sw} = \rho b_w d$$
$$A_s = A_{sf} + A_{sw}$$

■ **EXAMPLE 4.6**

Rework Example 4.5 using the rectangular component method just described.

SOLUTION

$$A_{sf} = \frac{(0.85)(3)(54 - 15)(3)}{50} = 5.97 \text{ in.}^2$$

$$M_{uf} = (0.9)(5.97)(50)(24 - \tfrac{3}{2}) = 6044 \text{ in.-k} = 503.72 \text{ ft-k}$$

$$M_{uw} = 858 - 503.72 = 354.28 \text{ ft-k}$$

Design a rectangular beam with $b = 15$ in. and $d = 24$ in. to resist 354.28 ft-k.

$$\frac{M_{uw}}{\phi b_w d^2} = \frac{(12)(354.28)(1000)}{(0.9)(15)(24)^2} = 546.7$$

$$\rho = 0.01246 \text{ from Table A.11}$$

$$A_{sw} = (0.01246)(15)(24) = 4.49 \text{ in.}^2$$

$$A_s = 5.97 + 4.49 = \underline{\underline{10.46 \text{ in.}^2}}$$ ■

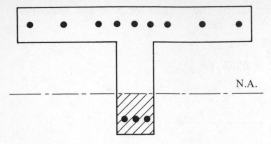

Figure 4.15 T beam with flange in tension and bottom (hatched) in compression (a rectangular beam).

4.5 DESIGN OF T BEAMS FOR NEGATIVE MOMENTS

When T beams are resisting negative moments, their flanges will be in tension and the bottom of their stems will be in compression, as shown in Figure 4.15. Obviously for such situations the rectangular beam design formulas will be used. Section 10.6.6 of the ACI Code requires that part of the flexural steel in the top of the beam in the negative-moment region be distributed over the effective width of the flange or over a width equal to one-tenth of the beam span, whichever is smaller. Should the effective width be greater than one-tenth of the span length, the Code requires that some additional longitudinal steel be placed in the outer portions of the flange. The intention of this part of the Code is to minimize the sizes of the flexural cracks that will occur in the top surface of the flange perpendicular to the stem of a T beam subject to negative moments.

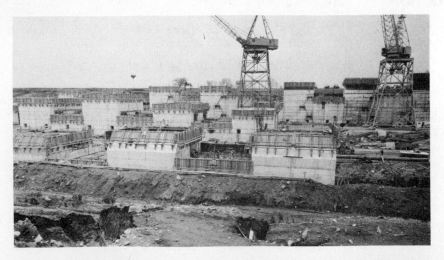

Construction of locks at Smithfield, Kentucky. (Courtesy Symons Corporation.)

New Comiskey Park, Chicago, Illinois. (Courtesy of Economy Forms Corporation.)

4.6 L SHAPED BEAMS

The author assumes for this discussion that L beams (or that is, edge T beams with a flange on one side only) are not free to bend laterally. Thus they will bend about their horizontal axes and will be handled as symmetrical sections exactly as T beams.

For L beams the effective width of the overhanging flange may not be larger than one twelfth the span length of the beam, 6 times the slab thickness nor one half the clear distance to the next web (ACI 8.10.3).

If an L beam is assumed to be free to deflect both vertically and horizontally it will be necessary to analyze it as an unsymmetrical section with bending about both the horizontal and vertical axes. An excellent reference on this topic is a book by MacGregor.[1]

4.7 COMPRESSION STEEL

The steel that is occasionally used on the compression sides of beams is called *compression steel*, and beams with both tensile and compressive steel are referred to as *doubly reinforced beams*. Compression steel is not normally required in sections designed by the strength method because the use of the full compressive strength of the concrete decidedly decreases the need for such reinforcement, as compared to designs made with the working-stress design method.

---

[1] MacGregor, J. G., *Reinforced Concrete Mechanics and Design* 1988 (Prentice–Hall, Englewood Cliffs, N.J.) pp. 159–162.

Occasionally, however, beams are limited to such small sizes by space or aesthetic requirements that compression steel is needed in addition to tensile steel. To increase the moment capacity of a beam beyond that of a tensilely reinforced beam with the maximum percentage of steel ($\frac{3}{4}\rho_b$), it is necessary to introduce another resisting couple in the beam. This is done by adding steel in both the compression and tensile sides of the beam. Not only does compressive steel increase the resisting moments of concrete sections, it also increases the amount of curvature that a member can take before flexural failure. This means that the ductility of such sections will be appreciably increased. Though expensive, compression steel makes beams tough and ductile, enabling them to withstand large moments, deformations, and stress reversals such as might occur during earthquakes. As a result, many building codes for earthquake zones require that certain minimum amounts of compression steel be included in flexural members.

Compression steel is very effective in reducing long-term deflections due to shrinkage and plastic flow. In this regard you should note the effect of compression steel on the long-term deflection expression in Section 9.5.2.5 of the Code (to be discussed in Chapter 5 of this text). Continuous compression bars are also helpful for positioning stirrups (by tying them to the compression bars) and keeping them in place during concrete placement and vibration.

Tests of doubly reinforced concrete beams have shown that even if the compression concrete crushes, the beam may very well not collapse if the compression steel is enclosed by stirrups. Once the compression concrete reaches its crushing strain, the concrete cover spalls or splits off the bars, much as in columns (see Chapter 8). If the compression bars are confined by closely spaced stirrups, the bars will not buckle until additional moment is applied. This additional moment cannot be considered in practice because beams are not practically useful after part of their concrete breaks off. (Would you like to use a building after some parts of the concrete beams have fallen on the floor?)

Section 7.11.1 of the ACI Code states that compression steel in beams must be enclosed by ties or stirrups, or by welded wire fabric of equivalent area. In Section 7.10.5.1 the Code states that the ties must be at least #3 in size for longitudinal bars #10 and smaller and at least #4 for larger longitudinal bars and bundled longitudinal bars. The ties may not be spaced farther apart than 16 bar diameters, 48 tie diameters, or the least lateral dimension of the beam cross section (Code 7.10.5.2).

In doubly reinforced beams an initial assumption is made that the compression steel yields as well as the tensile steel. (The tensile steel is always assumed to yield because of the ductile requirements of the ACI Code.) If the strain at the extreme fiber of the compression concrete is assumed to equal 0.00300 and the compression steel A'_s is located two-thirds of the distance from the neutral axis to the extreme concrete fiber, then the strain in the compression steel equals $\frac{2}{3} \times 0.003 = 0.002$. If this is greater than the strain in the steel at yield, as say $50,000/(29 \times 10^6) = 0.00172$ for 50,000-psi steel, the steel has yielded. It should be noted that actually the creep and shrinkage occurring in the compression concrete help the compression steel to yield.

Sometimes the neutral axis is quite close to the compression steel. As a matter of fact, in some beams with low steel percentages, the neutral axis may be right at the compression steel. For such cases the addition of compression steel is probably a waste of time and money.

When compression steel is used, the ultimate resisting moment of the beam is assumed to consist of two parts: the part due to the resistance of the compression concrete and the balancing tensile reinforcing and the part due to the ultimate moment capacity of the compression steel and the balancing amount of tensile steel. This situation is illustrated in Figure 4.16. In the expressions developed here, the effect of the concrete in compression, which is replaced by the compressive steel A'_s, is neglected. This omission will cause us to overestimate M_n by a very small and negligible amount (less than 1%). The first of the two resisting moments is illustrated in Figure 4.16(b):

$$M_{n1} = A_{s1} f_y \left(d - \frac{a}{2} \right)$$

The second resisting moment is that produced by the additional tensile and compressive steel (A_{s2} and A'_s), which is presented in Figure 4.16(c):

$$M_{n2} = A'_s f_y (d - d')$$

Up to this point it has been assumed that the compression steel has reached its yield stress. If such is the case, the values of A_{s2} and A'_s will be equal because the addition to T of $A_{s2} f_y$ must equal the addition to C of $A'_s f_y$ for equilibrium. If such is not the case, A'_s must be larger than A_{s2}, as will be described later in this section.

Combining the two values we obtain

$$M_n = A_{s1} f_y \left(d - \frac{a}{2} \right) + A'_s f_y (d - d')$$

$$M_u = \phi M_n$$

$$M_u = \phi \left[A_{s1} f_y \left(d - \frac{a}{2} \right) + A'_s f_y (d - d') \right]$$

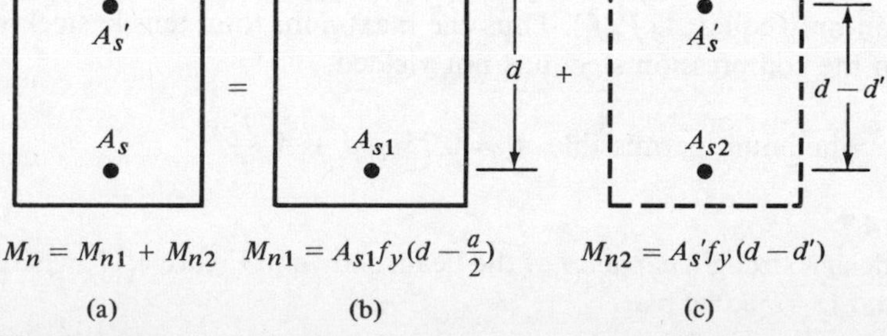

$$M_n = M_{n1} + M_{n2} \qquad M_{n1} = A_{s1} f_y (d - \frac{a}{2}) \qquad M_{n2} = A'_s f_y (d - d')$$

(a)   (b)   (c)

Figure 4.16

When the total percentage of tensile steel is equal to or less than $0.75\rho_b$ (where ρ_b is for a rectangular beam with tensile steel only), the compression steel will have little effect on the resisting moment of a doubly reinforced section because with such a tensile-yielding failure situation, the lever arm of the internal couple is not affected very much by the presence of compression steel. This is generally not the case when the tensile steel percentage is greater than $0.75\rho_b$. Nevertheless, for the problems of this chapter, the effect of the compression steel is considered regardless of the tensile steel percentage.

Examples 4.7 and 4.8 illustrate the calculations involved in determining design strengths of doubly reinforced sections. In each of these problems, the strain in the compression steel is checked to determine whether the steel has yielded. If not, as in the second of the two examples, a trial-and-error process is used to determine the stress. Based on the strains, an estimate is made of the stress and the value of A_{s2} is computed by the following expression, where f_s' is the estimated stress in the compression steel:

$$A_{s2}f_y = A_s'f_s'$$

A new value of a is calculated, the neutral axis location is determined, a new strain in the compression steel is computed, and then f_s' is calculated and compared with the estimate. This process can be repeated until there is a reasonably close correlation between the estimated value of f_s' and its calculated value. Once the trial-and-error process is understood, it is possible to set up a quadratic equation that will, upon solution, yield directly the correct position of the neutral axis, thus eliminating the trial-and-error work. Such an equation is presented at the end of Example 4.8.

At the end of each of these two examples, the maximum permissible area of tensile reinforcing that will ensure a tensile failure is calculated. Tests of doubly reinforced beams indicate they are quite ductile; as a result, *the Code (10.3.3) says that to ensure ductile behavior in beams with compression reinforcement only the part of the tensile steel that is balanced by compression in the concrete has to be limited by the 0.75 factor.* The maximum tensile steel area equals the value permitted if the section were singly reinforced $(0.75\rho_b bd)$ plus an area that will provide a force equal to the force produced by the compression steel. If the compression steel has yielded,

$$\text{Maximum permissible } A_s = 0.75\rho_b bd + A_s'$$

If the compression steel has not yielded, this expression needs to be revised. To ensure a tensile failure as required by the Code, $A_{s2}f_y$ can only equal $A_s'f_s'$ and thus A_{s2} can only equal $A_s'(f_s'/f_y)$. Thus the maximum total tensile steel is as follows when the compression steel has not yielded:

$$\text{Maximum permissible } A_s = 0.75\rho_b bd + A_s'\frac{f_s'}{f_y}$$

■ **EXAMPLE 4.7**

Determine the design strength capacity of the beam shown in Figure 4.17, where $f_c' = 3000$ psi and $f_y = 50,000$ psi.

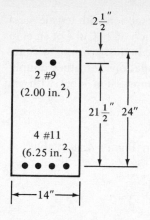

Figure 4.17

SOLUTION

Assuming that the Compression Steel Has Yielded and Thus $A_{s2} = A'_s$

$$A_{s2} = 2.00 \text{ in.}^2$$

$$A_{s1} = A_s - A_{s2} = 6.25 - 2.00 = 4.25 \text{ in.}^2$$

$$a = \frac{A_{s1}f_y}{0.85f'_c b} = \frac{(4.25)(50)}{(0.85)(3)(14)} = 5.95''$$

Locating Neutral Axis and Checking Strain in Compression Steel

$$c = \frac{5.95}{0.85} = 7.00''$$

$$\epsilon'_s = \frac{4.50}{7.00} \times 0.003 = 0.00193 > \frac{50,000}{29 \times 10^6} = 0.00172$$

Therefore, compression steel has yielded.

$$M_{n1} = A_{s1}f_y\left(d - \frac{a}{2}\right) = (4.25)(50)\left(24 - \frac{5.95}{2}\right) = 4468 \text{ in.-k} = 372.3 \text{ ft-k}$$

$$M_{n2} = A'_s f_y(d - d') = (2.00)(50)(24 - 2.5) = 2150 \text{ in.-k} = 179.2 \text{ ft-k}$$

$$M_n = M_{n1} + M_{n2} = 372.3 + 179.2 = 551.5 \text{ ft-k}$$

$$M_u = (0.90)(551.5) = 496.4 \text{ ft-k}$$

Checking the maximum amount of tensile steel permitted, noting that compression steel has yielded,

$$\text{Maximum } A_s = 0.75\rho_b bd + A'_s$$

$$= (0.75)(0.0275)(14)(24) + 2.00$$

$$= 8.93 \text{ in.}^2 > 6.25 \text{ in.}^2 \qquad \underline{\text{OK}} \quad \blacksquare$$

Note Should you compute the design strength for this section with tensile steel only (6.25 in.²), you will find it to be 460 ft-k. The addition of 2.00 in.² of steel in the top (a 32% increase in steel) will provide an increase in design strength equal to 36.4 ft-k (a 7.9% increase in moment). Similar calculations for other doubly reinforced beams will show that the addition of compression steel with no additional strength is quite uneconomical.

■ EXAMPLE 4.8

Compute the design strength of the section shown in Figure 4.18 with $f_c' = 4000$ psi and $f_y = 50,000$ psi.

SOLUTION

Computing Value of a (Assuming Compression Steel Yields)

$$A_{s2} = 1.20 \text{ in.}^2$$

$$A_{s1} = 5.06 - 1.20 = 3.86 \text{ in.}^2$$

$$a = \frac{(3.86)(50)}{(0.85)(4)(14)} = 4.05''$$

Locating Neutral Axis and Computing Strain in Compression Steel

$$c = \frac{4.05}{0.85} = 4.77''$$

$$\epsilon_s' = \left(\frac{2.27}{4.77}\right)(0.00300) = 0.00143 < \frac{50,000}{29 \times 10^6} = 0.00172$$

Therefore, compression steel has not yielded.

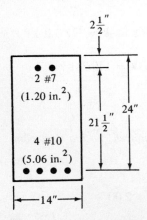

2 #7
(1.20 in.²)

4 #10
(5.06 in.²)

$2\frac{1}{2}''$

24"

$21\frac{1}{2}''$

—14"—

Figure 4.18

$$f'_s = \left(\frac{0.00143}{0.00172}\right)(50) = 41.57 \text{ ksi}$$

$$A'_s f'_s = A_{s2} f_y$$

$$(1.20)(41.57) = (A_{s2})(50)$$

$$A_{s2} = 1.00 \text{ in.}^2$$

Computing Value of a

$$A_{s1} = 5.06 - 1.00 = 4.06 \text{ in.}^2$$

$$a = \frac{(4.06)(50)}{(0.85)(4)(14)} = 4.26''$$

Locating Neutral Axis and Calculating Strain and Stress in Compression Steel

$$c = \frac{4.26}{0.85} = 5.01''$$

$$\epsilon'_s = \left(\frac{2.51}{5.01}\right)(0.00300) = 0.00150 < 0.00172$$

$$f'_s = \left(\frac{0.00150}{0.00172}\right)(50) = 43.60 \text{ ksi}$$

Change is not large; therefore, use 43.60 k/in.^{2}.

$$A'_s f'_s = A_{s2} f_y$$

$$(1.20)(43.60) = (A_{s2})(50)$$

$$A_{s2} = 1.05 \text{ in.}^2$$

$$A_{s1} = 5.06 - 1.05 = 4.01 \text{ in.}^2$$

$$a = \frac{(4.01)(50)}{(0.85)(4)(14)} = 4.21''$$

$$c = \frac{4.21}{0.85} = 4.95''$$

Computing Design Strength

$$M_u = \phi\left[A_{s1} f_y\left(d - \frac{a}{2}\right) + A'_s f'_s(d - d')\right]$$

$$M_u = 0.90\left[(4.01)(50)\left(24 - \frac{4.21}{2}\right) + (1.20)(43.6)(24 - 2.5)\right]$$

$$M_u = 4962 \text{ in.-k} = 413.6 \text{ ft-k}$$

Checking Maximum Amount of Tensile Steel Permitted, Noting That Compression Steel Has Not Yielded

$$\text{Maximum permissible } A_s = 0.75\rho_b bd + A_s'\frac{f_s'}{f_y}$$

$$= (0.75)(0.0367)(14)(24) + (1.20)\left(\frac{43.60}{50}\right)$$

$$= 10.30 \text{ in.}^2 > 5.06 \text{ in.}^2 \qquad \underline{\text{OK}} \quad \blacksquare$$

Alternate Method for Locating Neutral Axis

The expression to follow is written by equating the total compression in the compression concrete $(0.85f_c'A_c)$ and the compression steel $(A_s'f_s')$ to the total tension in the tensile steel (A_sf_y). The depth of the neutral axis c is the only unknown, and it can be determined by solving the following quadratic equation:

$$C + C' = T$$

$$(0.85)(4)(14)(0.85c) + (1.20)\left[0.003 \times 29,000\left(\frac{c-2.5}{c}\right)\right] = 5.06 \times 50$$

$$40.46c + 104.4 - \frac{261}{c} = 253$$

$$40.46c^2 + 104.4c - 261 = 253c$$

$$40.46c^2 - 148.6c = 261$$

$$c^2 - 3.67c = 6.45$$

$$c - 1.835 = \sqrt{6.45 + (-1.835)^2} = 3.13$$

$$c = 4.96'' \text{ as compared to } 4.95'' \text{ previously}$$
$$\text{obtained by trial and error}$$

4.8 DESIGN OF DOUBLY REINFORCED BEAMS

It should be remembered that sufficient tensile steel can be placed in most beams so that compression steel is not needed. But if it is needed, the design is usually quite straightforward. Examples 4.9 and 4.10 illustrate the design of doubly reinforced beams. The solutions follow the theory used for analyzing doubly reinforced sections.

■ **EXAMPLE 4.9**
Design a rectangular beam for $M_D = 250$ ft-k and $M_L = 400$ ft-k if $f_c' = 4000$ psi and $f_y = 60,000$ psi. The maximum permissible beam dimensions are shown in Figure 4.19.

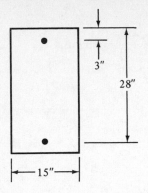

Figure 4.19

SOLUTION

$$M_u = (1.4)(250) + (1.7)(400) = 1030 \text{ ft-k}$$

$$M_n = \frac{1030}{0.90} = 1144 \text{ ft-k}$$

$\rho_{\max}$ if singly reinforced $= 0.0214$

$$A_{s1} = (0.0214)(15)(28) = 8.99 \text{ in.}^2$$

$$\frac{M_u}{\phi bd^2} = 1040.8 \text{ (from Table A.14 in Appendix)}$$

$$M_{u1} = (1040.8)(0.90)(15)(28)^2 = 11,015,827 \text{ in.-lb}$$
$$= 918 \text{ ft-k}$$

$$M_{n1} = \frac{918}{0.90} = 1020 \text{ ft-k}$$

$$M_{n2} = M_n - M_{n1} = 1144 - 1020 = 124 \text{ ft-k}$$

Checking to See Whether Compression Steel Has Yielded

$$a = \frac{(8.99)(60)}{(0.85)(4)(15)} = 10.58''$$

$$c = \frac{10.58}{0.85} = 12.45''$$

$$\epsilon'_s = \left(\frac{9.45}{12.45}\right)(0.00300) = 0.00228 > 0.00207$$

Therefore, compression steel has yielded.

$$\text{Theoretical } A_s' \text{ required} = \frac{M_{n2}}{(f_y)(d - d')} = \frac{(12)(124)}{(60)(28 - 3)} = 0.99 \text{ in.}^2$$

$$A_s' f_s' = A_{s2} f_y$$

$$A_{s2} = \frac{(0.99)(60)}{60} = 0.99 \text{ in.}^2$$

$$A_s = 8.99 + 0.99 = 9.98 \text{ in.}^2 \qquad \blacksquare$$

■ EXAMPLE 4.10

A beam is limited to the dimensions shown in Figure 4.20. If $M_D = 150$ ft-k, $M_L = 220$ ft-k, $f_c' = 4000$ psi, and $f_y = 60,000$ psi, select the reinforcing required.

SOLUTION

$$M_u = (1.4)(150) + (1.7)(220) = 584 \text{ ft-k}$$

$$M_n = \frac{584}{0.90} = 648.9 \text{ ft-k}$$

$$\rho_{max} \text{ if singly reinforced} = 0.0214$$

$$A_{s1} = (0.0214)(15)(20) = 6.42 \text{ in.}^2$$

Checking to See Whether Compression Steel Has Yielded

$$a = \frac{(6.42)(60)}{(0.85)(4)(15)} = 7.55''$$

$$c = \frac{7.55}{0.85} = 8.88''$$

$$\epsilon_s' = \left(\frac{4.88}{8.88}\right)(0.00300) = 0.00165 < \frac{60,000}{29 \times 10^6} = 0.00207$$

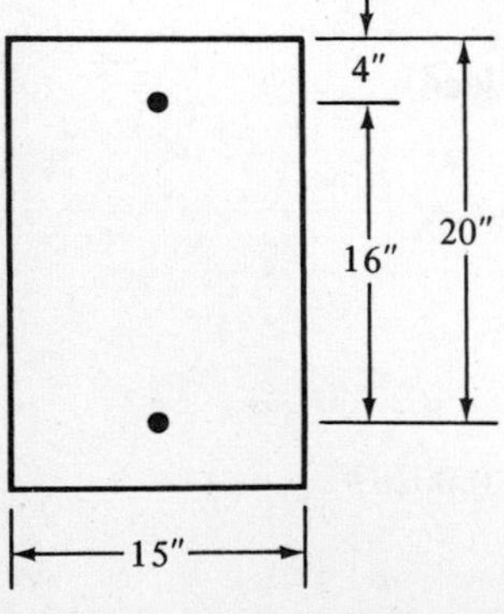

Figure 4.20

Therefore, compression steel has not yielded.

$$f'_s = \left(\frac{0.00165}{0.00207}\right)(60) = 47.83 \text{ ksi}$$

Resisting Moment of Singly Reinforced Section

$$\frac{M_u}{\phi bd^2} = 1040.8 \text{ (from Table A.14)}$$

$$M_{u1} = (1040.8)(0.90)(15)(20)^2 = 5,620,320 \text{ in.-lb} = 468.4 \text{ ft-k}$$

$$M_{n1} = \frac{468.4}{0.90} = 520.4 \text{ ft-k}$$

Determination of Steel Areas

$$M_{n2} = 648.9 - 520.4 = 128.5 \text{ ft-k}$$

$$\text{Theoretical } A'_s \text{ required} = \frac{M_{n2}}{(f'_s)(d - d'')} = \frac{(12)(128.5)}{(47.83)(20 - 4)} = 2.01 \text{ in.}^2$$

$$A'_s f'_s = A_{s2} f_y$$

$$A_{s2} = \frac{(2.01)(47.83)}{60} = 1.60 \text{ in.}^2$$

$$A_s = 6.42 + 1.60 = 8.02 \text{ in.}^2$$

4.9 COMPUTER EXAMPLES

■ EXAMPLE 4.11
Repeat Example 4.2 using the CONCRETE disk.

SOLUTION

```
ANALYSIS OF T-BEAMS
-------------------
Effective Flange Width  b   (in.)   = 30.0
Web Width               bw  (in.)   = 14.0
Depth to Bars           d   (in.)   = 30.0
Area of Steel           As (sq in)  = 10.1
Flange Thickness        hf  (in)    =  4.0
Concrete Strength       f'c (psi)   = 3000
Steel Yield Strength    fy  (psi)   = 50000


RESULTS
=======
Nominal Moment Capacity, Mn (ft-k) = 1098.9
Design Moment Capacity,  Mu (ft-k) =  989.1
```

■

■ EXAMPLE 4.12
Repeat Example 4.10 using the CONCRETE disk.

SOLUTION

```
DESIGN OF DOUBLY REINFORCED RECTANGULAR BEAMS
---------------------------------------------
Beam Width                b   (in.) = 15.0
Depth to Bottom Bars      d   (in.) = 20.00
Depth to Top Bars         d'  (in.) =  4.00
Ultimate Moment           Mu  (ft-k) = 584.0
Concrete Strength         f'c (psi) = 4000
Steel Yield Strength      fy  (psi) = 60000


RESULTS
=======
Use   8 # 9 bars in bottom
and   3 # 8 bars in top.
```

■

PROBLEMS

In Problems 4.1 to 4.11 determine the design strengths of the sections shown. Use f_y = 60,000 psi and f'_c = 3000 psi except for Problems 4.5 and 4.11, where different f'_c values are given. Check each section to see whether it is underreinforced.

Problem 4.1 (*Ans.* 284.4 ft-k)

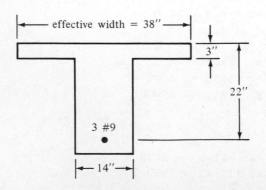

4.2 Repeat Problem 4.1 if 4 #11 bars are used.

4.3 Repeat Problem 4.1 if 8 #8 bars are used (*Ans.* 561.8 ft-k)

Problem 4.4

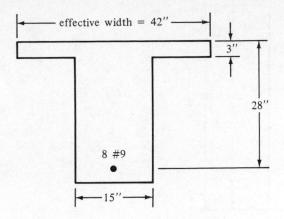

4.5 Repeat Problem 4.4 if f'_c = 5000 psi (*Ans.* 959.6 ft-k)

4.6 Repeat Problem 4.4 if 6 #10 bars are used.

Problem 4.7 (*Ans.* 406.3 ft-k)

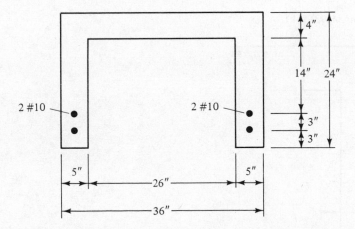

Problem 4.8

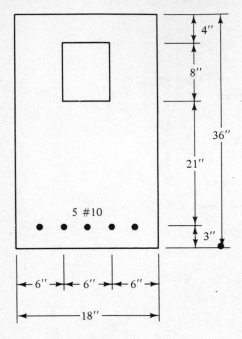

Problem 4.9 (*Ans.* 429.0 ft-k)

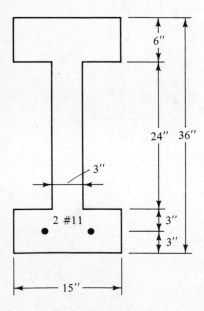

Problem 4.10

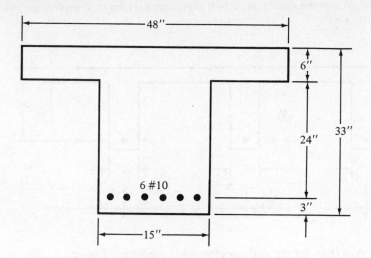

4.11 Compute ϕM_n, $f_y = 60,000$ psi, $f'_c = 4000$ psi. (*Ans.* 838.1 ft-k)

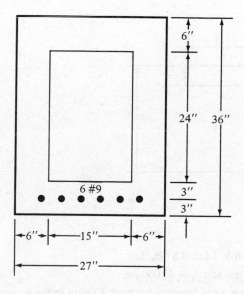

4.12 Calculate the design strength ϕM_n for one of the T beams shown if $f'_c = 3000$ psi, $f_y = 40,000$ psi, and the section has a 24-ft simple span. Is the steel percentage equal to or less than ρ_{max}?

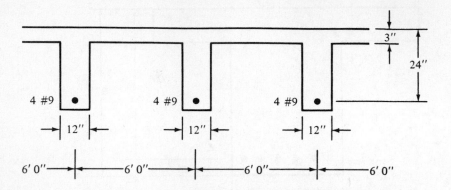

4.13 Prepare a flow chart for the analysis of tensilely reinforced T beams.

4.14 Determine the area of reinforcing steel required for the T beam shown if $f'_c = 3000$ psi, $f_y = 60,000$ psi, $M_u = 300$ ft-k and L = 30 ft. Clear distance between flanges = 3 ft.

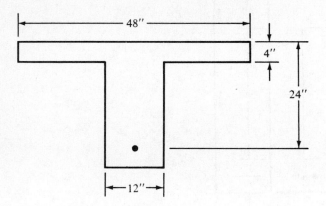

4.15 Repeat Problem 4.14 if $M_u = 500$ ft-k. (*Ans.* 4.87 in.²)

4.16 Repeat Problem 4.14 if $f'_c = 3000$ psi and $f_y = 40,000$ psi.

4.17 Determine the amount of reinforcing steel required for each T beam in the accompanying illustration if $f_y = 50,000$ psi, $f'_c = 3000$ psi, simple span = 20 ft, clear distance between stems = 3 ft, $M_D = 250$ ft-k, and $M_L = 325$ ft-k. (*Ans.* 8.54 in.²)

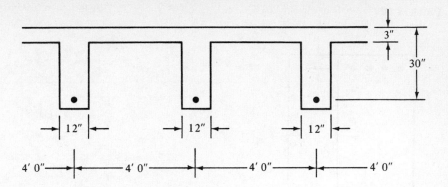

4.18 Select the tensile reinforcing needed for the T beams if the reinforced concrete is assumed to weigh 150 lb/ft³ and if a live floor load of 100 lb/ft² is to be supported. Assume 40 ft simple spans and $f_y = 60,000$ psi and $f'_c = 3000$ psi.

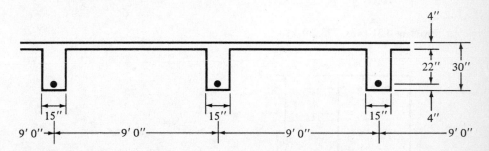

4.19 Prepare a flow chart for the design of tensilely reinforced T beams.

In Problems 4.20 to 4.26 compute the design strengths of the beams shown if $f_y = 60,000$ psi and $f'_c = 4000$ psi. Check the maximum permissible A_s in each case to ensure tensile failure.

Problem 4.20

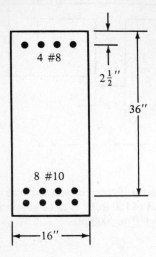

Problem 4.21 (*Ans.* 773.2 ft-k)

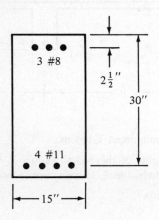

Problem 4.22

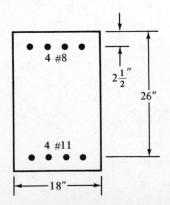

Problem 4.23 (*Ans.* 511.8 ft-k)

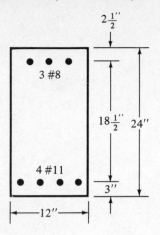

Problem 4.24

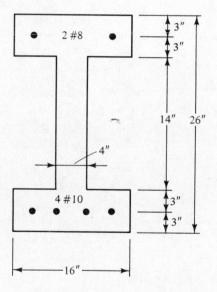

Problem 4.25 (*Ans.* 494.5 ft-k)

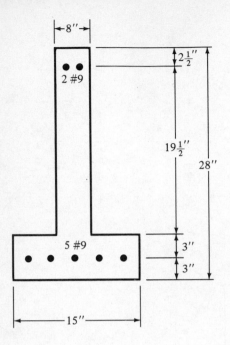

Problem 4.26

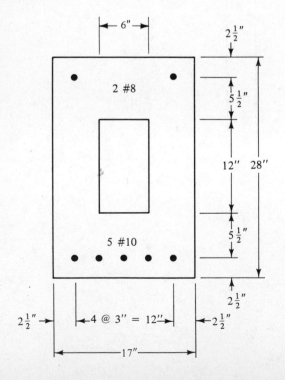

4.27 Compute the design strength of the beam shown in the accompanying illustration. How much can this permissible moment be increased if 4 #9 bars are added to the top $2\frac{1}{2}$ in. from the compression face if $f'_c = 3000$ psi and $f_y = 60,000$ psi? (*Ans.* 705.9 ft-k, 68.5 ft-k)

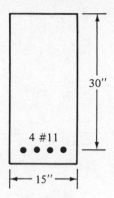

4.28 Prepare a flow chart for the analysis of doubly reinforced rectangular beams.

In Problems 4.29 to 4.32 determine the steel areas required for the sections shown in the accompanying illustrations. In each case the dimensions are limited to the values shown. If compression steel is required, assume it will be placed 3 in. from the compression face. $f'_c = 3000$ psi and $f_y = 60,000$ psi.

Problem 4.29 (*Ans.* $A_s = 9.19$ in.^{2}, $A'_s = 3.33$ in.^{2})

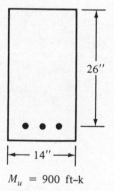

$M_u = 900$ ft–k

Problem 4.30

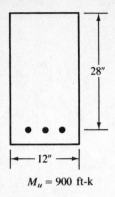

$M_u = 900$ ft-k

Problem 4.31 (*Ans. $A_s = 7.28$ in.^{2}, $A'_s = 3.03$ in.^{2}*)

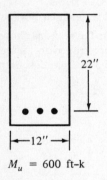

$M_u = 600$ ft-k

Problem 4.32

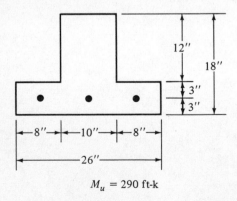

$M_u = 290$ ft-k

4.33 Prepare a flow chart for the design of doubly reinforced rectangular beams.

Solve Problems 4.34 to 4.39 using the enclosed computer disk.

4.34 Problem 4.1

4.35 Problem 4.3 (*Ans.* 561.9 ft-k)

4.36 Problem 4.10

4.37 Problem 4.17 (*Ans.* one ans. 6 #11 bars)

4.38 Problem 4.21

4.39 Problem 4.29 (*Ans.* one ans, 6 #11 bars in bottom, 4 #9 bars in top)

PROBLEMS WITH SI UNITS
In Problems 4.40 and 4.41 determine the design strengths of the beams shown in the accompanying illustrations if $f'_c = 27.6$ MPa and $f_y = 413.7$ MPa. Are the steel percentages in each case equal to or less than ρ_{max}? $E_s = 200000$ MPa.

Problem 4.40

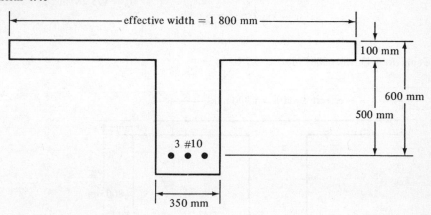

Problem 4.41 (*Ans.* 1 737 kN·m)

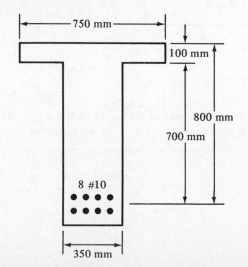

In Problems 4.42 and 4.43 determine the area of reinforcing steel required for the T beams shown if $f'_c = 20.7$ MPa and $f_y = 344.8$ MPa.

Problem 4.42

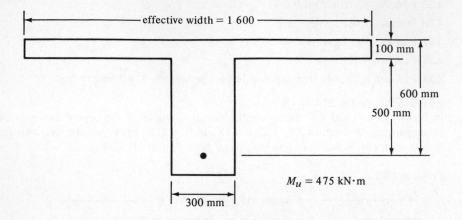

Problem 4.43 (*Ans.* 5081 mm² 8 #9 bars)

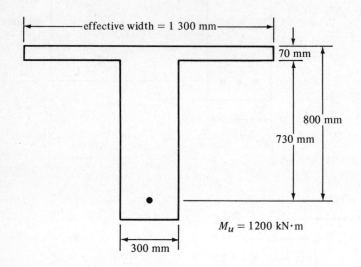

In Problems 4.44 to 4.46 compute the design strengths of the beams shown if $f_y = 344.8$ MPa and $f'_c = 20.7$ MPa. Check the maximum permissible A_s in each case to ensure ductile failure. $E_s = 200000$ MPa.

Problem 4.44

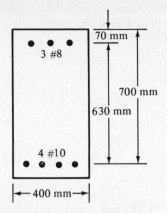

Problem 4.45 (*Ans.* 736.4 kN·m)

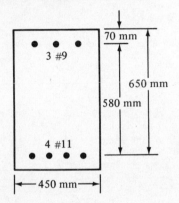

Problem 4.46

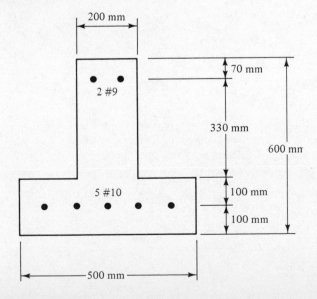

In Problems 4.47 and 4.48 determine the steel areas required for the sections shown in the accompanying illustrations. In each case the dimensions are limited to the values shown. If compression steel is required, assume it will be placed 70 mm from the compression face. $f'_c = 27.6$ MPa, $f_y = 413.7$ MPa, and $E_s = 200000$ MPa.

Problem 4.47 (*Ans.* $A_s = 6198$ mm^2, $A'_s = 1704$ mm^2)

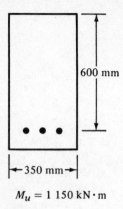

$M_u = 1\,150$ kN·m

Problem 4.48

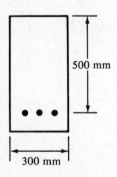

$M_u = 800$ kN·m

5
Serviceability

5.1 INTRODUCTION

Today the structural design profession is concerned with a *limit states* philosophy. The term *limit state* is used to describe a condition at which a structure or some part of a structure ceases to perform its intended function. There are actually two categories of limit states, strength and serviceability.

Strength limit states are based on the safety or load-carrying capacity of structures and include buckling, fracture, fatigue, overturning, and so on. Chapters 3 and 4 have been concerned with the bending limit state of various members.

Serviceability limit states refer to the performance of structures under normal service loads and are concerned with the uses and/or occupancy of structures including such items as deflections, cracking, and vibrations. You will note that these items may disrupt the use of structures but do not involve collapse.

Vertical vibration for bridge and building floors, as well as lateral and torsional vibration in tall buildings, can be quite annoying to users of these structures. Vibrations, however, are not usually a problem in the average size reinforced concrete building, but we should be on the lookout for the situations where they can be objectionable.

This chapter is concerned with serviceability limits for deflections and crack widths. The ACI Code concentrates on very specific requirements relating to the strength limit states of reinforced concrete members but allows the designer some freedom of judgment in the serviceability areas. This doesn't mean that the serviceability limit states are not significant, but by far the most important consideration (as in all structural specifications) is the life and property of the public. As a result, public safety is not left up to the judgment of the individual designer.

Georgia Dome, Atlanta, Georgia. (Courtesy of Economy Forms Coporation.)

5.2 IMPORTANCE OF DEFLECTIONS

The adoption of the strength design method in recent years, together with the use of higher-strength concretes and steels, has permitted the use of relatively slender members. As a result, deflections and deflection cracking have become more severe problems than they were a few decades ago.

The magnitudes of deflections for concrete members can be quite important. Excessive deflections of beams and slabs may cause sagging floors, excessive vibrations, and even interference with the proper operation of supported machinery. Such deflections may damage partitions and cause poor fitting of doors and windows. In addition, they may damage a structure's appearance or frighten the occupants of the building, even though the building may be perfectly safe. Any structure used by people should be quite rigid and relatively vibration-free so as to provide a sense of security.

Perhaps the most common type of deflection damage in reinforced concrete structures is the damage to light masonry partitions. They are particularly subject to injury due to concrete's long-term creep. When the floors above and below deflect, the relatively rigid masonry partitions do not bend easily and are often severely damaged. On the other hand, the more flexible gypsum board partitions are much more adaptable to such distortions.

5.3 CONTROL OF DEFLECTIONS

Deflections of reinforced concrete members are usually controlled by limiting the member thicknesses in some proportion to their span lengths or by providing maximum permissible computed deflections for various situations.

Minimum Thicknesses

Table 3.1 of Chapter 3, which is Table 9.5(a) of the ACI Code, provides a set of minimum thicknesses for beams and one-way slabs to be used, unless actual deflection calculations indicate that lesser thicknesses are permissible. These minimum thickness values, which were developed primarily on the basis of experience over many years, should be used only for beams and slabs that are not supporting or attached to partitions or other members likely to be damaged by deflections.

Maximum Deflections

If the designer chooses not to meet the minimum thicknesses given in Table 3.1, he or she must compute deflections. If this is done, the values determined may not exceed the values specified in Table 5.1, which is Table 9.5(b) of the ACI Code.

TABLE 5.1 MAXIMUM PERMISSIBLE COMPUTED DEFLECTIONS

Type of member	Deflection to be considered	Deflection limitation
Flat roofs not supporting or attached to nonstructural elements likely to be damaged by large deflections	Immediate deflection due to live load L	$\dfrac{\ell^a}{180}$
Floors not supporting or attached to nonstructural elements likely to be damaged by large deflections	Immediate deflection due to live load L	$\dfrac{\ell}{360}$
Roof or floor construction supporting or attached to non-structural elements likely to be damaged by large deflections	That part of the total deflection occurring after attachment of nonstructural elements (sum of the long-time deflection due to all sustained loads and the immediate deflection due to any additional live load)^c	$\dfrac{\ell^b}{480}$
Roof or floor construction supporting or attached to non-structural elements not likely to be damaged by large deflections		$\dfrac{\ell^d}{240}$

^{a} Limit not intended to safeguard against ponding. Ponding should be checked by suitable calculations of deflection, including added deflections due to ponded water and considering long-term effects of all sustained loads, camber, construction tolerances, and reliability of provisions for drainage.

^{b} Limit may be exceeded if adequate measures are taken to prevent damage to supported or attached elements.

^{c} Long-time deflection shall be determined in accordance with ACI Sections 9.5.2.5 or 9.5.4.2 but may be reduced by amount of deflection calculated to occur before attachment of nonstructural elements. This amount shall be determined on the basis of accepted engineering data relating to time-deflection characteristics of members similar to those being considered.

^{d} But not greater than tolerance provided for nonstructural elements. Limit may be exceeded if camber is provided so that total deflection minus camber does not exceed limit.

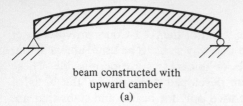

beam constructed with
upward camber
(a)

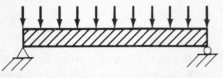

beam straight under dead load
plus some percentage of live load
(b)

Figure 5.1 Cambering.

Camber

The deflection of reinforced concrete members may also be controlled by cambering. The members are constructed of such a shape that they will assume their theoretical shape under some loading condition (usually dead load and perhaps some part of the live load). A simple beam would be constructed with a slight convex bend so that under certain gravity loads it would become straight as assumed in the calculations. (See Figure 5.1.) Some designers take into account both dead and live loads in figuring the amount of camber. Camber is usually only for the longer-span members.

5.4 CALCULATION OF DEFLECTIONS

Deflections for reinforced concrete members can be calculated with the usual deflection expressions, several of which are shown in Figure 5.2. A few comments should be made about the magnitudes of deflections in concrete members as determined by the expressions given in this figure. It can be seen that the ℄ deflection of a uniformly loaded simple beam [Figure 5.2(a)] is five times as large as the ℄ deflection of the same beam if its ends are fixed [Figure 5.2(b)]. Nearly all concrete beams and slabs are continuous, and their deflections fall somewhere in between the two extremes mentioned here.

Because of the very large deflection variations that occur with different end restraints, it is essential that those restraints be considered if realistic deflection calculations are to be made. For most practical purposes it is sufficiently accurate to calculate the ℄ deflection of a member as though it is simply supported and to subtract from that value the deflection caused by the average of the negative moments at the member ends.

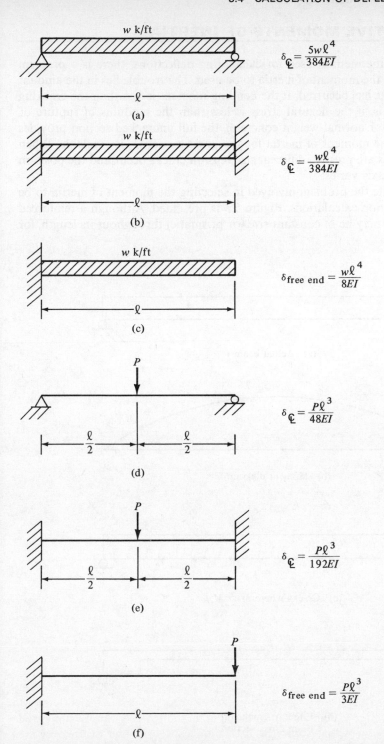

$$\delta_{\mathbb{C}} = \frac{5w\ell^4}{384EI}$$

(a)

$$\delta_{\mathbb{C}} = \frac{w\ell^4}{384EI}$$

(b)

$$\delta_{\text{free end}} = \frac{w\ell^4}{8EI}$$

(c)

$$\delta_{\mathbb{C}} = \frac{P\ell^3}{48EI}$$

(d)

$$\delta_{\mathbb{C}} = \frac{P\ell^3}{192EI}$$

(e)

$$\delta_{\text{free end}} = \frac{P\ell^3}{3EI}$$

(f)

Figure 5.2 Some deflection expressions.

5.5 EFFECTIVE MOMENTS OF INERTIA

Regardless of the method used for calculating deflections, there is a problem in determining the moment of inertia to be used. The trouble lies in the amount of cracking that has occurred. If the bending moment is less than the cracking moment (that is, if the flexural stress is less than the modulus of rupture of about $7.5\sqrt{f'_c}$ for normal-weight concrete), the full uncracked section provides rigidity, and the moment of inertia for the gross section I_g is available. When larger moments are present, different size tension cracks occur and the position of the neutral axis varies.

To illustrate the problem involved in selecting the moment of inertia to be used for deflection calculations, Figure 5.3 is presented. Although a reinforced concrete beam may be of constant size (or prismatic) throughout its length, for

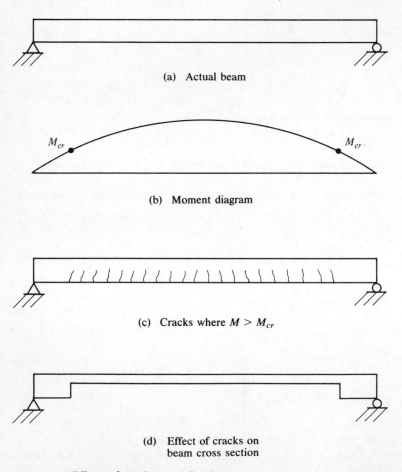

(a) Actual beam

(b) Moment diagram

(c) Cracks where $M > M_{cr}$

(d) Effect of cracks on
beam cross section

Figure 5.3 Effects of cracks on deflections.

deflection calculations it will behave as though it is composed of segments of different-size beams.[1]

For the portion of a beam where the moment is less than the cracking moment, M_{cr}, the beam can be assumed to be uncracked and the moment of inertia can be assumed to equal I_g. When the moment is greater than M_{cr}, the tensile cracks that develop in the beam will, in effect, cause the beam cross section to be reduced, and the moment of inertia may be assumed to equal the transformed value, I_{cr}. It is as though the beam consists of the segments shown in Figure 5.3(d).

The problem is even more involved than indicated by Figure 5.3. It is true that at cross sections where tension cracks are actually located, the moment of inertia is probably close to the transformed I_{cr}, but in between cracks it is perhaps closer to I_g. Furthermore, diagonal tension cracks may exist in areas of high shear, causing other variations. As a result, it is difficult to decide what value of I should be used.

A concrete section that is fully cracked on its tension side will have a rigidity of anywhere from one-third to three-fourths of its rigidity if it is uncracked. At different sections along the beam, the rigidity varies depending on the moment present. It is easy to see that an accurate method of calculating deflections must take these variations into account.

If it is desired to obtain the immediate deflection of an uncracked prismatic member, the moment of inertia may be assumed to equal I_g along the length of the member. Should the member be cracked at one or more sections along its length or if its depth varies along the span, a more exact value of I needs to be used.

Section 9.5.2.3 of the Code gives a moment of inertia that is to be used for deflection calculations. This moment of inertia is an average value and is to be used at any point in a simple beam where the deflection is desired. It is referred to as I_e, the effective moment of inertia, and is based on an estimation of the probable amount of cracking caused by the varying moment throughout the span:[2]

$$I_e = \left(\frac{M_{cr}}{M_a}\right)^3 (I_g) + \left[1 - \left(\frac{M_{cr}}{M_a}\right)^3\right] I_{cr} \quad \text{(ACI Equation 9-7)}$$

In this expression, I_g is the gross moment of inertia (without considering the steel) of the section, M_{cr} is the cracking moment $= f_r I_g / y_t$, with $f_r = 7.5\sqrt{f'_c}$ for normal-weight concrete (different for lightweight concrete, as per Section 9.5.2.3 of the Code), M_a is the maximum service-load moment occurring for the condition under consideration, and I_{cr} is the transformed moment of inertia of the cracked section.

---

[1] Leet, K., 1991, *Reinforced Concrete Design*, 2nd. ed (New York: McGraw-Hill), pp. 51–52.

[2] Branson, D. E., 1965, "Instantaneous and Time-Dependent Deflections on Simple and Continuous Reinforced Concrete Beams," HPR Report No. 7, Part I, Alabama Highway Department, Bureau of Public Roads, August 1963, pp. 1–78.

5.6 LONG-TERM DEFLECTIONS

With I_e and the appropriate deflection expressions, instantaneous or immediate deflections are obtained. Long-term or sustained loads, however, cause large increases in these deflections due to shrinkage and creep. The factors affecting deflection increases include humidity, temperature, curing conditions, compression steel content, ratio of stress to strength, and the age of the concrete at the time of loading.

If concrete is loaded at an early age, its long-term deflections will be greatly increased. Excessive deflections in reinforced concrete structures can very often be traced to the early application of loads. The creep strain after about five years (after which creep is negligible) may be as high as four or five times the initial strain when loads were first applied seven to ten days after the concrete was placed, while the ratio may only be two or three when the loads were first applied three or four months after concrete placement.

Because of the several factors mentioned in the last two paragraphs, the magnitudes of long-term deflections can only be estimated. The Code (9.5.2.5) states that to estimate the increase in deflection due to these causes, the part of the instantaneous deflection that is due to sustained loads may be multiplied by the empirically derived factor λ at the end of this paragraph and the result added to the instantaneous deflection.[3]

$$\lambda = \frac{\xi}{1 + 50\rho'} \qquad \text{(ACI Equation 9-10)}$$

In this expression, which is applicable to both normal and lightweight concrete, ξ is a time-dependent factor that may be determined from Table 5.2.

Should times differing from the values given in Table 5.2 be used, values of ξ may be selected from the curve of Figure 5.4.

TABLE 5.2 TIME FACTOR FOR
SUSTAINED LOADS
(ACI CODE 9.5.2.5)

Duration of sustained load	Time-dependent factor ξ
5 years or more	2.0
12 months	1.4
6 months	1.2
3 months	1.0

---

[3] Branson, D.E., 1971, "Compression Steel Effect on Long-Time Deflections," *Journal ACI*, 68 (8), pp. 555–559.

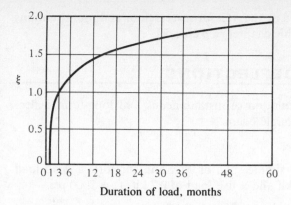

Figure 5.4 Multipliers for long-time deflections
(ACI Commentary Figure 9.5, 2.5).

The effect of compression steel on long-term deflections is taken into account in the λ expression with the term ρ'. It equals A'_s/bd and is to be computed at midspan for simple and continuous spans and at the supports for cantilevers.

The full dead load of a structure can be classified as a sustained load, but the type of occupancy will determine the percentage of live load that can be called sustained. For an apartment house or for an office building, perhaps only 20% to 25% of the service live load should be considered as being sustained, whereas perhaps 70% to 80% of the service live load of a warehouse might fall into this category.

A study by the ACI indicates that under controlled laboratory conditions, 90% of the specimens had deflections between 20% below and 30% above the values calculated in this section.[4] The reader should realize, however, that field conditions are not lab conditions, and deflections in actual structures will vary much more than those occurring in the lab specimens. Despite the use of plans and specifications and field inspection, it is difficult to control field work adequately. Construction crews may add a little water to the concrete to make it more workable. They may not obtain satisfactory mixing and compaction of the concrete, with the result that voids and honeycomb occur. Finally, the forms may be removed before the concrete has obtained its full design strength. If this is the case, the moduli of rupture and elasticity will be low and excessive cracks may occur in beams that would not have occurred if the concrete had been stronger. All of these factors can cause reinforced concrete structures to deflect appreciably more than is indicated by the usual computations.

As previously mentioned, the deflections calculated as described in this chapter should not exceed certain limits depending on the type of structure.

---

[4] ACI Committee 435. 1972, "Variability of Deflections of Simply Supported Reinforced Concrete Beams," *Journal ACI*, 69(1), p. 29.

Maximum deflections permitted by the ACI for several floor and roof situations were previously presented in this chapter in Table 5.1.

5.7 SIMPLE-BEAM DEFLECTIONS

Example 5.1 presents the calculation of instantaneous and long-term deflections for a uniformly loaded simple beam.

■ EXAMPLE 5.1

The beam of Figure 5.5 has a simple span of 20 ft and supports a dead load including its own weight of 1 klf and a live load of 0.7 klf. $f'_c = 3000$ psi.

(a) Calculate the instantaneous deflection for $D + L$.
(b) Calculate the deflection for the same loads after five years assuming that 30% of the live load is sustained.

SOLUTION

(a) Instantaneous deflection:

$$I_g = (\tfrac{1}{12})(12)(20)^3 = 8000 \text{ in.}^4$$

$$M_{cr} = \frac{f_r I_g}{y_t} = \frac{(7.5\sqrt{3000})(8000)}{10} = 328{,}633 \text{ in.-lb} = 27.4 \text{ ft-k}$$

$$M_a = \frac{(1.7)(20)^2}{8} = 85 \text{ ft-k}$$

By transformed-area calculations,

$$x = 6.78''$$

$$I_{cr} = 4067 \text{ in.}^4$$

$$I_e = \left(\frac{27.4}{85}\right)^3 (8000) + \left[1 - \left(\frac{27.4}{85}\right)^3\right]4067 = 4198 \text{ in.}^4$$

$$E_c = 57{,}000\sqrt{3000} = 3.122 \times 10^6 \text{ psi}$$

$$\delta = \frac{5wl^4}{384 E_c I_e} = \frac{(5)(\tfrac{1700}{12})(12 \times 20)^4}{(384)(3.12 \times 10^6)(4198)} = \underline{0.468''*}$$

(b) Long-term deflection: The initial deflection due to the sustained loads is equal to

$$\frac{1 + (0.3)(0.7)}{1.7}(0.468) = 0.333 \text{ in.*}$$

---

* The author really got carried away in this chapter when he calculated deflections to three digits. We cannot expect this kind of accuracy, and one digit is probably realistic.

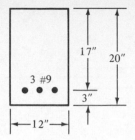

Figure 5.5

This value is multiplied by the following λ value (noting that $\xi = 2.0$ from Table 5.2) and added to the deflection obtained in part (a):

$$\lambda = \frac{\xi}{1 + 50\rho'} = \frac{2.0}{1 + (50)(0)} = 2.0$$

Total long-term deflection $= 0.468 + (2.0)(0.333) = \underline{1.134''}*$ ∎

5.8 CONTINUOUS-BEAM DEFLECTIONS

For the following discussion a continuous T beam subject to both positive and negative moments is considered. As shown in Figure 5.6, the effective moment of inertia used for calculating deflections varies a great deal throughout the member. For instance, at the center of the span at section 1–1 where the positive moment is largest, the web is cracked and the effective section consists of the hatched section plus the tensile reinforcing in the bottom of the web. At section 2–2 in the figure, where the largest negative moment occurs, the flange is cracked and the effective section consists of the hatched part of the web (including any compression steel in the bottom of the web) plus the tensile bars in the top. Finally, near the points of inflection, the moment will be so low that the beam will probably be uncracked and thus the whole cross section is effective, as shown for section 3–3 in the figure. (For this case I is usually calculated only for the web and the effect of the flanges is neglected as shown in Figure 5.10).

From the preceding discussion it is obvious that to theoretically calculate the deflection in a continuous beam, it is necessary to use a deflection procedure that takes into account the varying moment of inertia along the span. Such a procedure would be very lengthy, and it is doubtful that the results so obtained would be within $\pm 20\%$ of the actual values. For this reason the ACI Code (9.5.2.4) permits the use of a constant moment of inertia throughout the member equal to the average of the I_e values computed at the critical positive- and negative-moment sections. It should also be noted that the multipliers at the two sections should be averaged for long-term deflections.

Example 5.2 illustrates the calculation of deflections for a continuous member. Although much of the repetitious math is omitted from the solution given

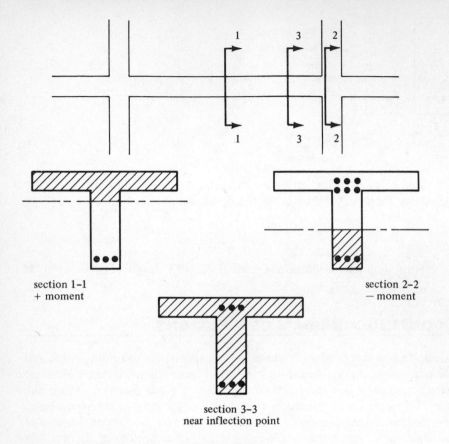

Figure 5.6 Deflections for a continuous T beam.

herein, you can see that the calculations are still very lengthy and you will understand why approximate deflection calculations are commonly used for continuous spans.

■ EXAMPLE 5.2

Determine the instantaneous and 5-year deflections at the midspan of the continuous T beam shown in Figure 5.7(a). The member supports a dead load including its own weight of 1.5 k/ft and a live load of 2.5 k/ft, of which 50% is assumed to be sustained. $f'_c = 3000$ psi and $n = 9$. The moment diagram for full dead and live loads is shown in Figure 5.7(b), and the beam cross section is shown in Figure 5.7(c).

SOLUTION

For Positive-Moment Region

1. Locating centroidal axis for uncracked section and calculating gross moment of inertia I_g and cracking moment M_{cr} for the positive-moment

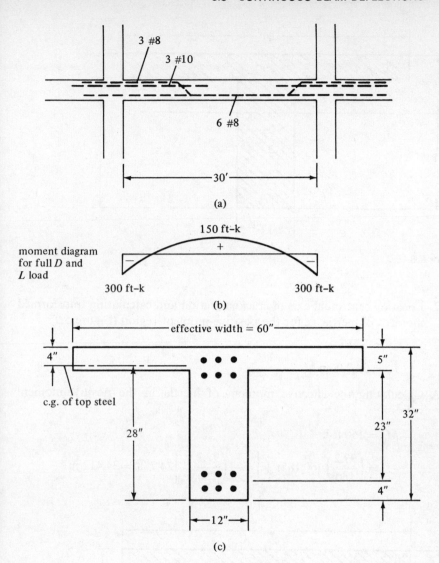

Figure 5.7

region (Figure 5.8):

$$\bar{y} = 10.81''$$

$$I_g = 60{,}185 \text{ in.}^4$$

$$M_{cr} = \frac{(7.5)(\sqrt{3000})(60{,}185)}{21.19} = 1{,}166{,}754 \text{ in.-lb} = 97.2 \text{ ft-k}$$

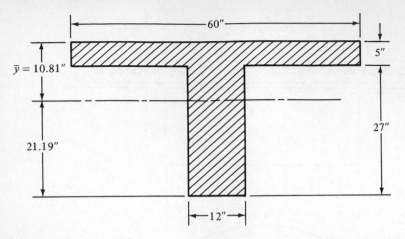

Figure 5.8

2. Locating centroidal axis of cracked section and calculating transformed moment of inertia I_{cr} for the positive-moment region (Figure 5.9):

$$x = 5.65''$$

$$I_{cr} = 24{,}778 \text{ in.}^4$$

3. Calculating the effective moment of inertia in the positive-moment region:

$$M_a = 150 \text{ ft-k}$$

$$I_e = \left(\frac{97.2}{150}\right)^3 (60{,}185) + \left[1 - \left(\frac{97.2}{150}\right)^3\right] 24{,}778 = 34{,}412 \text{ in.}^4$$

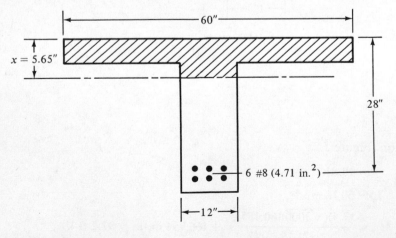

Figure 5.9

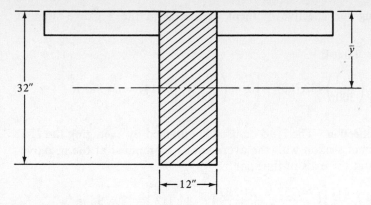

Figure 5.10

For Negative-Moment Region

1. Locating centroidal axis for uncracked section and calculating gross moment of inertia I_g and cracking moment M_{cr} for the negative-moment region, considering only the hatched rectangle shown in Figure 5.10:

$$\bar{y} = \tfrac{32}{2} = 16''$$

$$I_g = (\tfrac{1}{12})(12)(32)^3 = 32{,}768 \text{ in.}^4$$

$$M_{cr} = \frac{(7.5)(\sqrt{3000})(32{,}768)}{16} = 841{,}302 \text{ in.-lb} = 70.1 \text{ ft-k}$$

2. Locating centroidal axis of cracked section and calculating transformed moment of inertia I_{tr} for the negative-moment region (Figure 5.11):

$$x = 10.43''$$

$$I_{cr} = 24{,}147 \text{ in.}^4$$

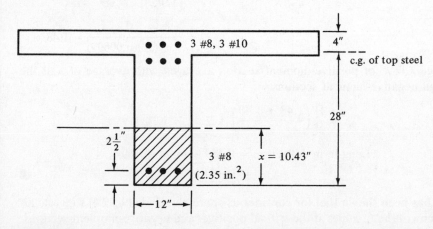

Figure 5.11

3. Calculating the effective moment of inertia in the negative-moment region:

$$M_a = 300 \text{ ft-k}$$

$$I_e = \left(\frac{70.1}{300}\right)^3 (32{,}768) + \left[1 - \left(\frac{70.1}{300}\right)^3\right] 24{,}147 = 24{,}257 \text{ in.}^4$$

Instantaneous Deflection The I_e to be used is obtained by averaging the I_e at the positive-moment section with the average of I_e computed at the negative-moment sections at the ends of the span:

$$\text{Average } I_e = \frac{1}{2}\left[\left(\frac{24{,}257 + 24{,}257}{2}\right) + 34{,}412\right] = 29{,}334 \text{ in.}^4$$

$$E_c = 57{,}000\sqrt{3000} = 3.122 \times 10^6 \text{ psi}$$

By the conjugate beam procedure, δ = moment at midspan of conjugate beam for the moment diagram of Figure 5.7(b) = $(8437.5 \text{ ft}^3\text{-k})/EI$:

$$\delta = \frac{(8437.5)(1728)(1000)}{(3.122 \times 10^6)(29{,}334)} = 0.159''$$

Long-Term Deflection

Initial deflection due to sustained loads $= \dfrac{1.5 + (0.5)(2.5)}{4.0}(0.159'') = 0.109''$

The long-term multipliers are

for positive-moment region: $\qquad \lambda = \dfrac{2.0}{1 + (50)(0)} = 2.0$

for negative-moment region: $\qquad e' = \dfrac{2.35}{(12)(28)} = 0.00699$

$$\lambda = \frac{2.0}{1 + (50)(0.00699)} = 1.48$$

Average $\lambda = \lambda$ for positive-moment section averaged with average of λ at the two end negative-moment sections

$$= \frac{1}{2}\left[\left(\frac{1.48 + 1.48}{2}\right) + 2.0\right] = 1.74$$

$$\text{Long-term deflection} = 0.159 + (1.74)(0.109)$$
$$= 0.349 \text{ in.} \qquad \blacksquare$$

It has been shown that for continuous spans the Code (9.5.2.4) suggests an averaging of the I_e values at the critical positive- and negative-moment sections. The ACI Commentary (R9.5.2.4) says that for approximate deflection calculations for continuous prismatic members it is satisfactory to use the midspan section properties because these properties which include the effect of cracking have the greatest effect on deflections. An alternate procedure is presented elsewhere.[5]

5.9 TYPES OF CRACKS

A few introductory comments concerning some of the several types of cracks which occur in reinforced concrete are presented in this section. The remainder of this chapter is concerned with the estimated widths of flexural cracks.

Flexural cracks are vertical cracks which extend from the tension sides of beams up to the region of their neutral axes. They are illustrated in Figure 5.12(a). Should beams have very deep webs (more than 3 or 4 ft), the cracks will be very closely spaced with some of them coming together above the reinforcing and some disappearing there. These cracks may be wider up in the middle of the beam than at the bottom.

Inclined cracks due to shear can develop in the webs of reinforced concrete beams either as independent cracks or as extensions of flexural cracks. Occasionally, inclined cracks will develop independently in a beam even though no flexure cracks are in that locality. These cracks, which are called *web–shear cracks* and which are illustrated in Figure 5.12(b), sometimes occur in the webs of prestressed sections, particularly those with large flanges and thin webs.

The usual type of inclined shear cracks are the *flexure–shear cracks* which are illustrated in Figure 5.12(c). They commonly develop in both prestressed and nonprestressed beams.

Torsion cracks, which are illustrated in Figure 5.12(d), are quite similar to shear cracks except they spiral around the beam. Should a plain concrete member be subjected to pure torsion, it will crack and fail along 45° spiral lines due to the diagonal tension corresponding to the torsional stresses. For a very effective demonstration of this type of failure you can take a piece of chalk in your hands and twist it until it breaks. Although torsion stresses are very similar to shear stresses, they will occur on all faces of a member. As a result, they add to the shear stresses on one side and subtract from them on the other.

Sometimes bond stresses between the concrete and the reinforcing lead to a splitting along the bars as shown in Figure 5.12(e).

Of course there are other types of cracks not illustrated here. Members which are loaded in axial tension will have transverse cracks through their entire cross sections. Cracks can also occur in concrete members due to shrinkage, temperature change, settlements, and so on. Considerable information concerning the development of cracks is available.[6]

---

[5] ACI Committee 435, 1978, "Proposed Revisions by Committee 435 to ACI Building Code and Commentary Provisions on Deflections," *Journal ACI*, 75(6), pp. 229–238.

[6] MacGregor, J. G., 1991, *Reinforced Concrete Mechanics and Design*, 2nd ed. (Englewood Cliffs, N.J.: Prentice-Hall), pp. 320–325.

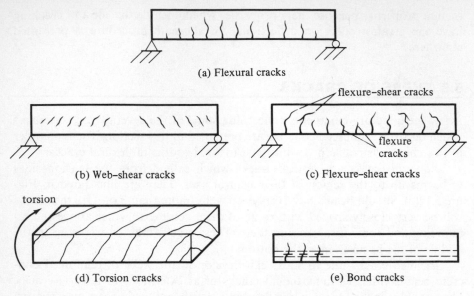

(a) Flexural cracks

(b) Web–shear cracks

(c) Flexure–shear cracks

(d) Torsion cracks

(e) Bond cracks

Figure 5.12 Some types of cracks in concrete members.

5.10 CONTROL OF FLEXURAL CRACKS

Cracks are going to occur in reinforced concrete structures because of concrete's low tensile strength. For members with low steel stresses at service loads, the cracks may be very small and in fact may not be visible except upon careful examination. Such cracks, called *microcracks*, are generally initiated by bending stresses.

When steel stresses are high at service load, particularly where high-strength steels are used, visible cracks will occur. These cracks should be limited to certain maximum sizes so that the appearance of the structure is not spoiled and so that corrosion of the reinforcing does not result. The use of high-strength bars and the strength method of design have made crack control a very important item indeed. Because the yield stresses of reinforcing bars in general use have increased from 40 ksi to 60 ksi and above, it has been rather natural for designers to specify approximately the same size bars as they are accustomed to using, but fewer of them. The result has been more severe cracking of members.

Although cracks cannot be eliminated, they can be limited to acceptable sizes by spreading out or distributing the reinforcement. In other words, smaller cracks will result if several small bars are used with moderate spacings rather than a few large ones with large spacings. Such a practice will usually result in satisfactory crack control even for Grades 60 and 75 bars.

The maximum crack widths that are acceptable vary from approximately 0.004 in. to 0.016 in., depending on the location of the member in question, the type of structure, the surface texture of the concrete, illumination, and other

Lake Point Tower, Chicago, Illinois. (Courtesy of Portland Cement Association.)

factors. Somewhat smaller values may be required for members exposed to very aggressive environments such as deicing chemicals and saltwater spray.

ACI Committee 224, in a report on cracking,[7] presented a set of approximately permissible maximum crack widths for reinforced concrete members subject to different exposure situations. These values are summarized in Table 5.3.

Definite data are not available as to the sizes of cracks above which bar corrosion becomes particularly serious. As a matter of fact, tests seem to indicate that concrete quality, cover thickness, amount of concrete vibration, and other items may be more important than crack sizes in their effect on corrosion.

Results of laboratory tests of reinforced concrete beams to determine crack sizes vary. The sizes are greatly affected by shrinkage and other time-dependent

7 ACI Committee 224, 1972, "Control of Cracking in Concrete Structures," *Journal ACI*, 69 no. 12, pp. 717–753.

TABLE 5.3   PERMISSIBLE CRACK WIDTHS

Members subjected to	Permissible crack widths (in.)
Dry air	0.016
Moist air, soil	0.012
Deicing chemicals	0.007
Seawater and seawater spray	0.006
Use in water-retaining structures	0.004

factors. The purpose of crack control calculations is not really to limit cracks to certain rigid maximum values but rather to use reasonable bar details, as determined by field and laboratory experience, that will in effect keep cracks within a reasonable range.

In 1968 the following equation was developed for the purpose of estimating the maximum widths of cracks that will occur in the tension faces of flexural members.[8] It is merely a simplification of the many variables affecting crack sizes.

$$w = 0.076\beta_h f_s \sqrt[3]{d_c A}$$

where w = the estimated cracking width in thousandths of inches
β_h = ratio of the distance to the neutral axis from the extreme tension concrete fiber to the distance from the neutral axis to the centroid of the tensile steel (values to be determined by the working-stress method)
f_s = steel stress, in kips per square inch at service loads (designer is permitted to use $0.6f_y$ for normal structures)
d_c = the cover of the outermost bar measured from the center of the bar
A = the effective tension area of concrete around the main reinforcing (having the same centroid as the reinforcing) divided by the number of bars)

For reinforced concrete beams with bars having yield stresses greater than 40,000 psi, the Code (10.6.4) in effect sets limiting values on crack sizes. This is done by requiring that member cross sections be so proportioned that the value of z computed by the expression at the end of this paragraph may not exceed 175 k/in. for members with interior exposure or 145 k/in. for members with exterior exposure. The Code uses the preceding expression, with β_h equal to 1.2 to establish the limiting value of z. The limitations of z to 175 k/in. and 145 k/in. correspond, respectively, to crack widths of 0.016 and 0.013 in.

$$z = f_s \sqrt[3]{d_c A} \qquad \text{(ACI Equation 10-4)}$$

---

[8] Gergely, P., and Lutz, L. A., 1968, "Maximum Crack Width in Reinforced Flexural Members." *Causes, Mechanisms and Control of Cracking in Concrete*, SP-20 (Detroit: American Concrete Institute), pp. 87–117.

This equation does not apply to beams with extreme exposure or to structures that are supposed to be watertight. Special consideration must be given to such situations. The value of z should be checked for both positive- and negative-moment bars. If the bars are not all the same size in a particular group, the number of bars should be considered to equal the total steel area divided by the area of the largest bar in the group, for purposes of determining the value of A to use in the equation.

It is felt that this the Gergely–Lutz equation works reasonably well for slabs if $\beta_h = 1.35$, as the ACI Commentary (10.6.4) recommends. In effect the ACI is saying that to keep the same maximum crack sizes in slabs as in beams, the permissible value of z should not exceed 1.2/1.35 times the 175-k/in. and 145-k/in. values previously mentioned.

Particular attention needs to be given to crack control for doubly reinforced beams, where it is common to use small numbers of large-diameter tensile bars. Application of the z equation to such beams will frequently yield excessive values, and hence a larger number of smaller bars will have to be specified.

Example 5.3 illustrates the application of the crack control expression to a beam.

■ EXAMPLE 5.3

The beam cross section shown in Figure 5.13 was selected using $\rho = 0.18f'_c/f_y$, $f_y = 60,000$ psi, and $f'_c = 4000$ psi. Will estimated crack sizes be within the requirements of the ACI Code if the member has exterior exposure? If not, revise the design.

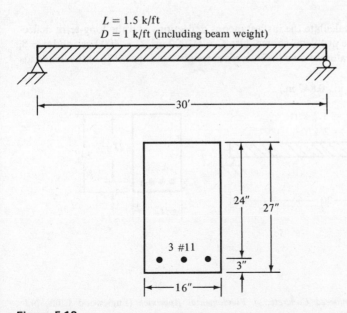

$L = 1.5$ k/ft
$D = 1$ k/ft (including beam weight)

30'

24" 27" 3"

3 #11

16"

Figure 5.13

SOLUTION

$$z = (0.6 \times 60)\sqrt[3]{(3)\left(\frac{6 \times 16}{3}\right)}$$

$$= 164.8 \text{ k/in.} > 145 \text{ k/in. permitted for exterior exposure} \qquad \underline{\text{no good}}$$

Replace the 3 #11 bars with 5 #9 bars and try again.

$$z = (0.6 \times 60)\sqrt[3]{(3)\left(\frac{6 \times 16}{5}\right)} = 139 \text{ k/in.} < 145 \text{ k/in.} \qquad \underline{\text{OK}}$$

Use 5 #9 ∎

The bond stress between the concrete and the reinforcing steel decidedly affects the size and spacing of cracks in the concrete. When bundled bars are used, there is appreciably less contact between the concrete and the steel as compared to the case where the bars are placed separately. To successfully apply the ACI expression for z, it is wise to consider this reduced contact surface.[9]

If reinforced concrete members are tested under carefully controlled laboratory conditions and cracks measured for certain leadings, considerable variations in crack sizes will occur. Consequently the calculations of crack widths described in this chapter should only be used to help the designer select good details for his or her reinforcing bars. The calculations are clearly not sufficiently accurate for comparison with field crack sizes.

PROBLEMS

In Problems 5.1 to 5.9 calculate the instantaneous deflections and the long-term deflections after 5 years for the problems shown. Use $f_y = 60,000$ psi, $f'_c = 4000$ psi, and $n = 8$, and assume that the D values shown include beam weights and that 30% of the live loads are sustained.

Problem 5.1 (*Ans.* 0.399 in., 0.847 in.)

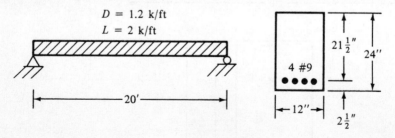

[9] Nawy, E. G., 1985, *Reinforced Concrete: A Fundamental Approach* (Englewood Cliffs, N.J.: Prentice–Hall), pp. 288–290.

Problem 5.2

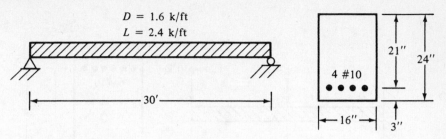

$D = 1.6$ k/ft
$L = 2.4$ k/ft

30'

4 #10

21"   24"

16"   3"

5.3 Repeat Problem 5.1 if 2 #9 bars are placed $2\frac{1}{2}$ in. from the top of the section. (*Ans.* 0.362 in., 0.656 in.)

5.4 Repeat Problem 5.2 if a 25-k concentrated live load is added at the ℄ of the span.

Problem 5.5 (*Ans.* 0.637 in., 1.375 in.)

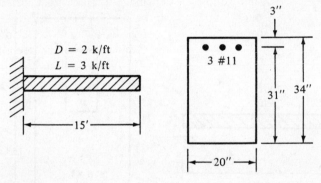

3"

$D = 2$ k/ft
$L = 3$ k/ft

3 #11

$31''$ 34"

15'

20"

Problem 5.6

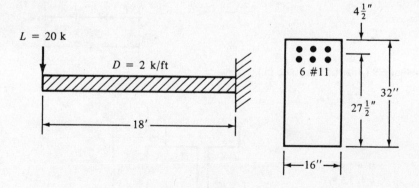

$4\frac{1}{2}''$

$L = 20$ k

$D = 2$ k/ft

6 #11

32"

$27\frac{1}{2}''$

18'

16"

Problem 5.7 (*Ans.* 1.244 in., 2.826 in.)

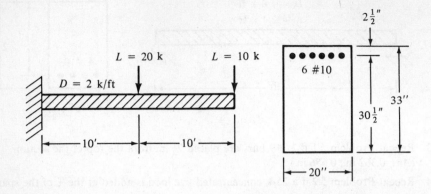

Problem 5.8

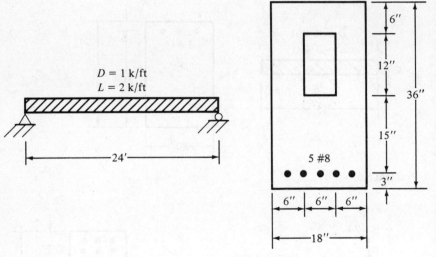

Problem 5.9 (*Ans.* 1.53 in., 3.06 in.)

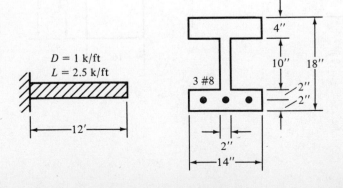

5.10 Select a rectangular beam section for the span and loads shown in the accompanying illustration. Use ρ_{max}, #10 bars, $f_c' = 4000$ psi, and $f_y = 60,000$ psi. Will the estimated crack sizes be within the requirements of the ACI Code if the member has interior exposure? If not, change the bar sizes.

$L = 2$ k/ft

$D = 1$ k/ft (including beam weight)

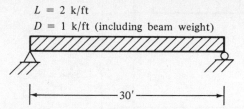

|← ——————— 30' ——————— →|

Problems 5.11 and 5.12. Check the ACI crack control provisions for the beams shown assuming exterior exposure and $f_y = 60,000$ psi. If the crack situation is not satisfactory, select smaller bars and recheck.

Problem 5.11 (*Ans.* $z = 157.6$ k/in., ∴ use 4 #9 bars)

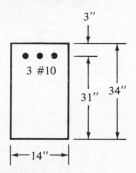

Problem 5.12

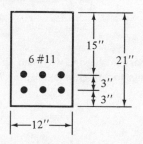

Problems 5.13, 5.14, and 5.15. Are the beams shown satisfactory as to the ACI Code crack control provisions for interior exposure. $f_y = 60,000$ psi.

Problem 5.13 (*Ans.* $z = 143.2$ k/in.   OK)

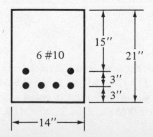

Problem 5.14

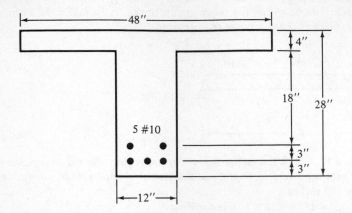

Problem 5.15 (*Ans. z* = 149.8 k/in. < 175 k/in.   OK)

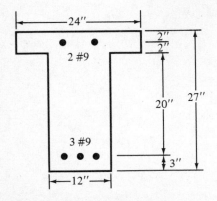

5.16 What is the maximum permissible spacing of #6 bars in the one-way slab shown which will satisfy the ACI Code crack requirements for exterior exposure? f_y = 60,000 psi.

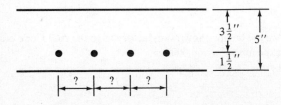

PROBLEMS WITH SI UNITS

In Problems 5.17 to 5.19 calculate the instantaneous deflections and the long-term deflections after 5 years for the problems shown. Use f'_c = 20.7 MPa, f_y = 344.8 MPa, f_r = 2.832 MPa, and n = 9. Assume that the *D* values shown include the beam weights and that 20% of the live loads are sustained. E_s = 200 000 MPa, E_c = 21,650 MPa.

Problem 5.17 (*Ans.* 11.36 mm, 23.69 mm)

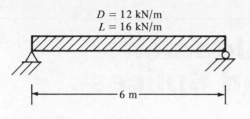

$D = 12$ kN/m
$L = 16$ kN/m

6 m

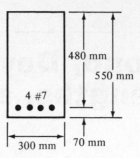

480 mm

550 mm

4 #7

300 mm   70 mm

Problem 5.18

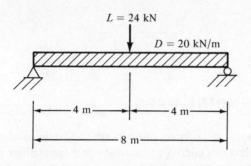

$L = 24$ kN

$D = 20$ kN/m

4 m   4 m

8 m

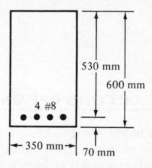

530 mm

600 mm

4 #8

350 mm   70 mm

Problem 5.19 (*Ans.* 12.02 mm, 22.97 mm)

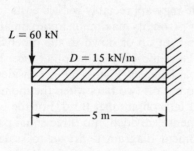

$L = 60$ kN

$D = 15$ kN/m

5 m

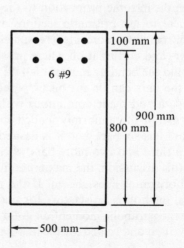

100 mm

6 #9

900 mm

800 mm

500 mm

6

Bond, Development Lengths, and Splices

6.1 CUTTING OFF OR BENDING BARS

The beams designed up to this point have been selected on the basis of maximum moments. These moments have occurred at or near span centerlines for positive moments and at the faces of supports for negative moments. At other points in the beams, the moments were less. Although it is possible to vary beam depths in some proportion to the bending moments, it is normally more economical to use prismatic sections and reduce or cut off some reinforcing when the bending moments are sufficiently small. Reinforcing steel is quite expensive, and cutting it off where possible may appreciably reduce costs.

Should the bending moment fall off 50% from its maximum, approximately 50% of the bars can be cut off or perhaps bent up or down to the other face of the beam and made continuous with the reinforcing in the other face. For this discussion the uniformly loaded simple beam of Figure 6.1 is considered This beam has six bars, and it is desired to cut off two bars when the moment falls off a third and two more bars when it falls off another third. For the purpose of this discussion, the maximum moment is divided into three equal parts by the horizontal lines shown. If the moment diagram is drawn to scale, a graphical method is satisfactory for finding the theoretical cutoff points.

For the parabolic moment diagram of Figure 6.1, the following expressions can be written and solved for the half-bar lengths x_1 and x_2 shown in the figure:

$$\frac{x_1^2}{(\ell/2)^2} = \frac{2}{6} \tag{6.1}$$

$$\frac{x_2^2}{(\ell/2)^2} = \frac{4}{6} \tag{6.2}$$

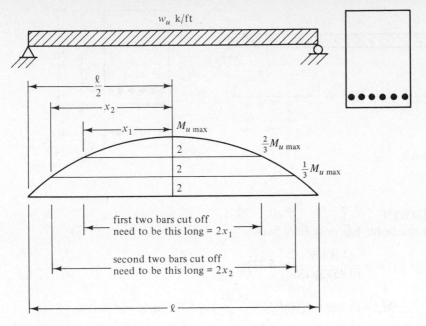

Figure 6.1

For different-shaped moment diagrams, other mathematical expressions would have to be written or a graphical method used.

Actually the permissible ultimate moment capacity,

$$M_u = \phi A_s f_y \left(d - \frac{a}{2} \right)$$

does not vary exactly in proportion to the area of the reinforcing bars as is illustrated in Example 6.1, because of variations in the depth of the compression block as the steel area is changed. The change is so slight, however, that for all practical purposes the moment capacity of a beam can be assumed to be directly proportional to the steel area. (It will be shown in this chapter that the moment capacities calculated as illustrated in this example problem will have to be reduced if sufficient lengths are not provided for the bars to develop their full stresses.)

■ EXAMPLE 6.1

For the uniformly loaded simple beam of Figure 6.2, determine the theoretical points on each end of the beam where two bars can be cut off and then determine the points where two more bars can be cut off. $f'_c = 3000$ psi, $f_y = 60,000$ psi.

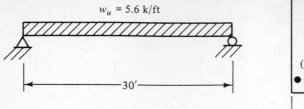

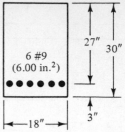

Figure 6.2

SOLUTION

When the beam has only four bars,

$$a = \frac{(4.00)(60)}{(0.85)(3)(18)} = 5.23''$$

$$M_u = (0.9)(4.00)(60)\left(27 - \frac{5.23}{2}\right) = 5267 \text{ in.-k} = 439 \text{ ft-k}$$

When the moment falls off to 439 ft-k, two bars can theoretically be cut off.
   When the beam has only two bars,

$$a = \frac{(2.00)(60)}{(0.85)(3)(18)} = 2.61''$$

$$M_u = (0.9)(2.00)(60)\left(27 - \frac{2.61}{2}\right) = 2775 \text{ in.-k} = 231 \text{ ft-k}$$

When the moment falls off to 231 ft-k, two more bars can theoretically be cut off.

   The moment at any section in the beam at a distance x from the left support is as follows, with reference being made to Figure 6.3:

$$M = 84 \times -(5.6x)\left(\frac{x}{2}\right)$$

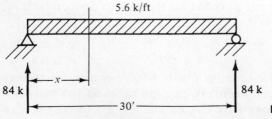

Figure 6.3

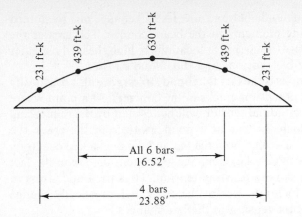

Figure 6.4

From this expression the location of the points in the beam where the moment is 439 ft-k and 231 ft-k can be determined. The results are shown in Figure 6.4.

Discussion   Should the approximate procedure have been followed (where bars are cut off purely on the basis of the ratio of the number of bars to the maximum moment, as was illustrated with Equations 6.1 and 6.2), the first two bars would have had lengths equal to 17.32 ft (as compared to the theoretically correct value of 16.52 ft) and the second two bar lengths equal to 24.50 ft (as compared to the theoretically correct value of 23.88 ft). It can then be seen that the approximate procedure yields fairly reasonable results.   ∎

In this section the theoretical points of cutoff have been discussed. As will be seen in subsequent sections of this chapter, the bars will have to be run additional distances because of variations of moment diagrams, anchorage requirements of the bars, and so on. The discussion of cutting off or bending bars is continued in Section 6.11.

6.2 BOND STRESSES

A basic assumption made for reinforced concrete design is that there must be absolutely no slippage of the bars in relation to the surrounding concrete. In other words, the steel and the concrete should stick together or *bond* together so they will act as a unit. If there is a slipping of the steel with respect to the concrete, there will be no transfer of stress from the concrete to the steel. As a result, the concrete beam will act as an unreinforced member and will be subject to sudden collapse as soon as the concrete cracks.

It is obvious that the magnitude of bond stresses will change in a reinforced concrete beam as the bending moments in the beam change. The greater the rate of bending moment change (occurring at locations of high shear) the greater will be the rate of change of bar tensions and thus bond stresses.

What may not be so obvious is the fact that bond stresses are also drastically affected by the development of tension cracks in the concrete. At a point where a crack occurs, all of the longitudinal tension will be resisted by the reinforcing bar. At a small distance along the bar at a point away from the crack the longitudinal tension will be resisted by both the bar and the uncracked concrete. In this small distance there can be a large change in bar tension due to the fact that the uncracked concrete is now resisting tension. Thus the bond stress in the surrounding concrete, which was zero at the crack, will drastically change within this small distance as the tension in the bar changes.

In the past it was common to compute the maximum bond stresses at points in the members and to compare them with certain allowable values obtained by tests. It is the practice today, however, to look at the problem from an ultimate standpoint, where the situation is a little different. Even if the bars are completely separated from the concrete over considerable parts of their length, the ultimate strength of the beam will not be affected if the bars are so anchored at their ends that they cannot pull loose.

The bonding of the reinforcing bars to the concrete is due to several factors, including the chemical adhesion between the two materials, the friction due to the natural roughness of the bars, and the bearing of the closely spaced rib-shaped deformations on the bar surfaces againt the concrete. The application of the force P to the bar shown in Figure 6.5 is considered in the discussion that follows.

When the force is first applied to the bar, the resistance to slipping is provided by the adhesion between the bar and the concrete. If plain bars were used, it would not take much tension in the bars to break this adhesion, particularly adjacent to a crack in the concrete. If this were to happen for a smooth surface bar, only friction would remain to keep the bar from slipping. There is also some Poisson's effect due to the tension in the bars. As they are tensioned they become a little smaller, enabling them to slip more easily. If we were to use straight, plain or smooth reinforcing bars in beams there would be very little bond strength and the beams would only be a little stronger than if there were no bars. The introduction of deformed bars was made so that in addition to the adhesion and friction there would also be a resistance due to the bearing of the concrete on the lugs or ribs (or deformations) of the bars as well as the so-called shear-friction strength of the concrete between the lugs.

Thus as we have described, smooth bars are not used in reinforced concrete. There are some exceptions, namely, anchor belts and stirrups (which are used for shear reinforcement). Although they are made from smooth bars, they do make use of mechanical anchorage such as (a) hooks on anchor bolts and stirrups and (b) washers and nuts on the tops of the anchor bolts.

As a result of these facts, reinforcing bars are made with rib-type deformations. The chemical adhesion and friction between the ribs are negligible and

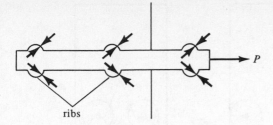

ribs

Figure 6.5 Bearing forces on bar and bearing of
bar ribs on concrete.

thus bond is primarily supplied by bearing on the ribs. Based on testing, the
crack patterns in the concrete show that the bearing stresses are inclined to the
axis of the bars from about 45° to 80° (the angle being appreciably affected by
the shape of the ribs).[1]

Internally the bars produce equal and opposite forces as shown in Figure
6.5. These internal forces are caused by the wedging action of the ribs bearing
against the concrete. They will cause tensile stresses in a cylindrical piece of
concrete around each bar. It's rather like a concrete pipe filled with water which
is pressing out against the pipe wall, causing it to be placed in tension. If the
tension becomes too high the pipe will split.

In a similar manner, if the bond stresses in a beam become too high the concrete will split around the bars and eventually the splits will extend to the side
and/or bottom of the beams. If either of these types of splits runs all the way
to the end of a bar, the bar will slip and the beam will fail. The closer the bars
are spaced together and the smaller the cover, the thinner will be the concrete
cylinder around each bar and the more likely that a bond splitting failure will
occur.

Figure 6.6 shows examples of bond failures which may occur for different
values of concrete cover and bar spacing. These are as shown by MacGregor.[2]

Splitting resistance along bars depends on quite a few factors, such as the
thickness of the concrete cover, the number of bars in the vicinity, the transverse
confining effect of stirrups, and so on. As a result of these variables, it is impossible to make comprehensive bond tests that are good for a wide range of structures. Nevertheless, the ACI by its equations has attempted to do just this, as
will be described in the sections to follow.

---

[1] Goto, Y., 1971, "Cracks Formed on Concrete Around Deformed Tensioned Bars," *ACI Journal,
Proceedings,* 68, p. 244.

[2] MacGregor, J. G., 1991, *Reinforced Concrete Mechanics and Design,* 2nd ed. (Englewood Cliffs,
N.J.: Prentice-Hall), p. 266.

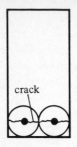 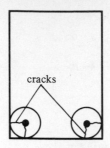 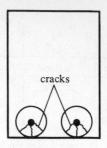

(a) Side cover
and one half
clear spacing
between bars
< bottom cover

(b) Cover on sides
and bottom equal
and < one half
clear spacing
between bars

(c) Bottom cover
< side cover and <
one half clear
spacing between
bars

Figure 6.6 Types of bond failures.

6.3 DEVELOPMENT LENGTHS

For this discussion, reference is made to the cantilever beam of Figure 6.7. It can be seen that both the maximum moment in the beam and the maximum stresses in the tensile bars occur at the face of the support. Theoretically a small distance back into the support the moment is zero, and thus it would seem that reinforcing bars would no longer be required. This is the situation pictured in Figure 6.7(a). Obviously if the bars were stopped at the face of the support, the beam would fail.

The bar stresses must be transferred to the concrete by bond between the steel and the concrete before the bars can be cut off. In this case the bars must be extended some distance back into the support and out into the beam to anchor them or develop their stresses. This distance, called the *development length* (ℓ_d), is shown in Figure 6.7(b). It can be defined as the minimum length

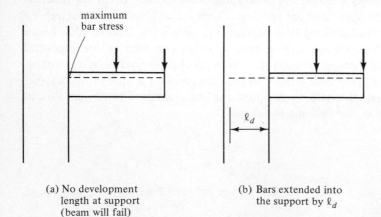

(a) No development
length at support
(beam will fail)

(b) Bars extended into
the support by ℓ_d

Figure 6.7

of embedment of bars that is necessary to permit them to be stressed to their yield point plus some extra distance to ensure member toughness. A similar discussion can be made for bars in other situations and in other types of beams.

As previously mentioned, the ACI for many years required designers to calculate bond stresses with a formula which was based on the change of moment in a beam. Then the computed values were compared to allowable bond stresses in the Code. Originally, bond strength was measured by means of pull-out tests. A bar would be cast in a concrete cylinder and a jack used to see how much force was required to pull it out. The fault with such a test is that the concrete is placed in compression, preventing the occurrence of cracks. In a flexural member, however, we have an entirely different situation due to the off-again-on-again nature of the bond stresses caused by the tension cracks in the concrete. In recent years, more realistic tests have been made with beams; the ACI Code development length expressions to be presented in this chapter are based primarily upon such tests at the National Bureau of Standards and the University of Texas.

To determine the development lengths required for reinforcing bars, the ACI first requires the calculation of basic development lengths needed to anchor rather widely spaced uncoated bars stressed to f_y, which are located in the bottoms of normal-weight concrete members and which provide relatively large cover. Secondly, the values obtained are multiplied by modification factors if the favorable conditions described are not met. For instance, the bars may be spaced close together, have small cover, or be epoxy-coated for corrosion protection, or the concrete may be made with a lightweight aggregate.

The Code (12.2) specifies certain expressions for determining minimum development lengths for bars in tension. (It will be noted that the required development lengths are the same for both the strength design and the alternate design methods.)

In tests it was found that the ultimate average bond force in beams (measured in pounds per inch) which would result in splitting equaled about $35\sqrt{f'_c}$ and this applied for small bars up through #11. The tests were generally made for beams reinforced with a single bar.[3-5]

The required development length for a bar is represented by ℓ_d. In the Code (12.2.2), however, expressions for basic development lengths represented by ℓ_{db} are given. These ℓ_{db} values are the development lengths required before adjustments or modifications are made to account for bar spacing, bundling, lightweight aggregates, coated bars, and so on.

[3] ACI Committee 408, 1966, "Bond Stress—The State of the Art," *ACI Journal, Proceedings*, 63, pp. 1161–1190. Discussion is on pp. 1569–1570.

[4] Ferguson, P. M. and Thompson, J. N., 1965, "Development Length for Large High Strength Reinforcing Bars in Bond," *ACI Journal, Proceedings*, 62, pp. 71–93. Discussion is on pp. 1153–1156.

[5] Orangon, C., Jirsa, J. O., and Gergely, P., 1977, "A Reevaluation of Test Data on Development Length and Splices," *ACI Journal, Proceedings*, 74, pp. 114–122. Discussion is on pp. 470–475.

Because reinforcing bars are usually spaced rather close together, the ACI, instead of assuming bar stresses to be developed equal to f_y, uses $1.25 f_y$ to provide a safety factor. Furthermore, instead of applying a ϕ factor of less than 1.0 as is done for flexure, shear, and so on, the Code increases the ℓ_d required by 15%. From this information the required basic development length ℓ_{db} for common bar sizes, including #11 and smaller bars and deformed wire, can be determined as follows:

$$(A_b)(1.25f_y)(1.15) = (\ell_{db})(35\sqrt{f'_c})$$

$$\ell_{db} = \frac{0.04A_b f_y}{\sqrt{f'_c}} \quad \text{for #11 and smaller bars and deformed wire}$$

Tests have shown that this expression overestimates the development length required for #14 and #18 bars. As a result, the Code assumes a 6% increase in bond strength from the commonly sized bars up to the #14 bar and assumes a 21% increase from the #14 to the #18 bars. The ℓ_{db} values for these large bars have also had their cross-sectional areas included with the following results:

$$\ell_{db} = \frac{0.085f_y}{\sqrt{f'_c}} \quad \text{for #14 bars}$$

$$\ell_{db} = \frac{0.125f_y}{\sqrt{f'_c}} \quad \text{for #18 bars}$$

These expressions provide the estimated minimum lengths needed to anchor uncoated bottom tension bars embedded in normal-weight concrete. Furthermore, they were developed under the very favorable conditions of large bar spacing and concrete cover thickness. Actual reinforced concrete members often do not meet these conditions and it is necessary to make some adjustments. The Code (12.2.3) requires that the basic ℓ_{db} values be multiplied successively by a series of modification factors. These factors, which are listed in Table 6.1, account for actual concrete cover, bar spacing, and the presence of any enclosing transverse reinforcing. The result of the multiplications is a modified basic length factor ℓ'_{db}:[6]

$$\ell'_{db} = (\ell_{db})(\text{applicable factors from Table 6.1})$$

You will discover that a rather large percentage of bars will fall into one of the four categories of condition 1 in the table. Thus a modification factor of 1.0 will usually be selected. Conditions 4 and 5 in the table (ACI 12.2.3.4 and 12.2.3.5) are situations which may permit the reduction of development lengths. The first of the two is a multiplication factor of 0.8 which *may* be used when very widely spaced bars are used (clear spacing at least $5d_b$ and clear cover at least $2.5d_b$). The second one provides a multiple of 0.75 which *may* be used

---

[6] Leet, K., 1991, *Reinforced Concrete Design*, 2nd ed. (New York: McGraw–Hill), pp. 243–246.

TABLE 6.1 Modification Factors for Development of Tension Reinforcing[a]

ACI section no.	Modification factor	Condition
12.2.3.1	1.0	1. For bars satisfying *any* one of the following conditions: (a) Bars in beams or columns with (1) minimum cover not less than specified in ACI 7.71, (2) transverse reinforcing which satisfies the tie requirements of ACI 7.10.5 or minimum stirrup requirements of ACI 11.5.4 and 11.5.5.3 along the development length, and (3) clear spacing between bars not less than $3d_b$. (b) Bars in beams or columns with (1) minimum cover not less than specified in ACI, 7.7.1 and (2) enclosed within transverse reinforcing of minimum area A_{tr} along the development length determined by the equation to follow in which s is the spacing of stirrups and N is the number of bars in a layer being spliced or developed at a critical section $$A_{tr} = \frac{d_b s N}{40} \quad \text{(ACI Equation 12-1)}$$ (c) Bars which are in the inner layer of reinforcing for a wall or slab and which have a clear spacing not less than $3d_b$. (d) Any bars with cover no less than $2d_b$ and with clear spacing no less than $3d_b$.
12.2.3.2	2.0	2. Bars with cover of d_b or less or with a clear spacing of $2d_b$ or less.
12.2.3.3	1.4	3. Bars not included in 1 or 2 above.
12.2.3.4	0.8	4. For #11 and smaller bars which have a clear spacing of no less than $5d_b$ and a cover from the face of the member to the edge bar measured in the plane of bars which is not less than $2.5d_b$, the factors in 1, 2, and 3 above may be multiplied by 0.8.
12.2.3.5	0.75	5. For reinforcement which is enclosed in spirals of not less than $\frac{1}{4}$-in. diameter and which have a pitch no greater than 4 in., within #4 or larger circular ties spaced at not more than 4 in. on center, or within #4 or larger ties or stirrups spaced not more than 4 in. on center and arranged so that alternate bars are supported in a corner of a tie or loop with an included angle not greater than 135°, the factors in 1, 2, and 3 may be multiplied by 0.75.
12.2.3.6	See condition to right	6. The value of the basic development length ℓ_{db} obtained by 1 through 5 above may not be less than $0.03 d_b f_y / \sqrt{f_c'}$

[a] These factors account for bar spacing, cover, and transverse steel.

173

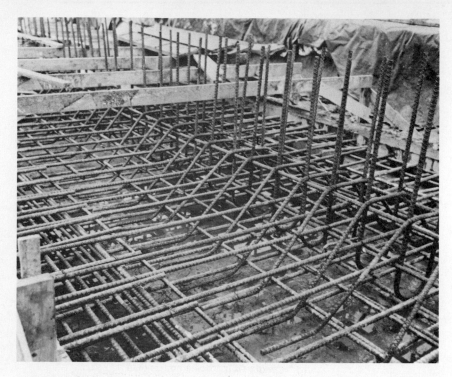

Lab Building, Portland Cement Association, Skokie, Illinois. (Courtesy of Portland Cement Association.)

when the reinforcement is enclosed within spirals or ties meeting the requirements of ACI Section 12.2.3.5. (*Ties* are individual pieces of wires or bars wrapped around the longitudinal bars of a member at intervals to hold them in position. *Spirals* are closely spaced wires or bars wrapped in a continuous spiral around the longitudinal bars of a member to hold them in position.)

Condition 6 in the table (ACI 12.2.3.6) provides an absolute minimum basic development length which is based on test results which indicate that even though we may have sufficient splitting resistance in the concrete the bar may pull out. This minimum value is $0.03d_b f_y/\sqrt{f'_c}$ and will control only for smaller bars and high-strength concretes.

For these expressions, f'_c is expressed in psi. In Section 12.1.2 of the Code a maximum permissible value for $\sqrt{f'_c}$ (its 100 psi) is given. This corresponds to an f'_c of 10,000 psi. A sufficient amount of research has not been performed to date to justify larger values and thus smaller development lengths.

There are other variables which affect development length such as top bars, epoxy coatings, lightweight aggregate concrete, and excess reinforcing. Modification factors for these factors are presented in Table 6.2. They are to be multiplied successively times ℓ'_{db}. These modifications are discussed in the paragraphs to follow.

$$\ell_d = (\ell'_{db})(\text{applicable modification factors from Table 6.2})$$

TABLE 6.2 OTHER MODIFICATION FACTORS*[a]*

ACI section no.	Modification factor	Condition
12.2.4.1	1.3	1. *Top reinforcement* (placed so there is at least 12 in. of fresh concrete cast below the development length or splice).
12.2.4.2	1.3 or $\dfrac{6.7\sqrt{F'_c}}{F_{ct}} \geq 1.0$	2. For *lightweight aggregate concrete* the modification factor is 1.3 or as determined by the expression to the left if f_{ct} (average splitting tensile strength of lightweight concrete) is available. The expression to the left is used.
12.2.4.3	1.5 or 1.2 as described to right	3. For *epoxy-coated bars* with cover less than $3d_b$ or for clear spacing less than $6d_b$ between bars, the modification factor is 1.5. For all other conditions use 1.2. For top reinforcing, the 1.3 value times the appropriate epoxy value need not exceed 1.7.
12.2.5	$\dfrac{\text{As required}}{\text{As furnished}}$	4. *For reinforcement used in excess of that required by analysis* (except where anchorage or development for f_y is specifically required or the reinforcing is designed for seismic forces as required by ACI 21.2.1.4), a modification factor equal to the ratio to the left may be used.

*[a] These account for top reinforcing, for lightweight aggregate concrete, for epoxy-coated bars, and for reinforcing in excess of that required by analysis.

Top bars are horizontal bars that have at least 12 in. of *fresh* concrete placed beneath them. These bars may have less bonding or anchorage to the concrete because of settlement of the concrete below the bars. Thus they will not be held as tightly in place, and there can be a considerable reduction in bond strength. For such cases we will use a 1.3 modifier.

Lightweight concrete has lower splitting strength than does regular concrete, and hence greater development lengths are considered necessary. Modifiers are specified in the Code and shown in the table. Some relaxation of the values given is permitted (Code 12.2.4.2) if the concrete mix is carefully controlled and if split-cylinder tests have been made which give good results. There is not a great deal of test data available for lightweight concrete.

In Chapter 1 we discussed the fact that epoxy-coated bars may often be used in situations where very corrosive environments exist. Tests have shown that the bond strength of such bars is substantially reduced, and as a result a modification factor of 1.5 is required for them. However, if the cover is $3d_b$ or larger and the clear spacing between bars is $6d_b$ or larger, there is less danger of a splitting failure of the concrete and the factor is reduced to 1.2 (ACI 12.2.4.3).

If more tensile steel is being used than is theoretically required, ℓ'_{db} may be multiplied by the ratio of the calculated steel area required divided by the actual

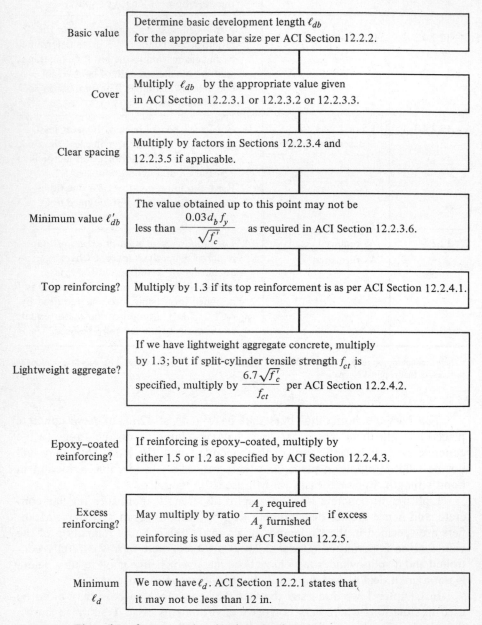

Figure 6.8 Flow chart for computing development length ℓ_d.

steel area furnished. This last modification does not apply if the full f_y development length is required as in tension lap splices (ACI 12.15.1). Using this reduction is not necessarily a good idea because we may later want to change the structure or add to it with the result that we may want to use the full bending capacity of the member. This could not be done if the reduction factor (A_s required/A_s furnished) was used in the original development length design.

The ℓ_d value obtained after the appropriate modification factors of Tables 6.1 and 6.2 are applied may not be taken as less than 12 in. Smaller values may very occasionally be computed for small bars and high-strength concrete.

When bundled bars are used, greater development lengths are needed because there is not a "core" of concrete between the bars to provide resistance to slipping. For such cases the Code (12.4) states that the development lengths are to be determined for individual bars and are then to be increased by 20% for three-bar bundles and by 33% for four-bar bundles.

Sometimes as the moment falls off in a member we may reduce the steel by using alternate long and short bars at different locations. The development lengths for these different bars occur at different locations. The ACI Commentary (R12.2.3.4) says that for calculations the clear spacing for the long bars is the clear distance between the long bars and that the clear distance for the short bars is the distance between the short bars.

A brief summary of these development length requirements is presented in flow-chart form in Figure 6.8. (This is a revised version of the flow chart given in Figure 12.2 of the ACI Commentary.)

6.4 EXAMPLE PROBLEMS

Examples 6.2 through 6.5 illustrate the calculations needed to determine development lengths for several reinforced concrete members.

■ EXAMPLE 6.2

The square reinforced concrete column footing shown in Figure 6.9 has #7 bottom tensile bars. These bars, which are spaced 6 in. on center, extend from the face of the wall (where the moment is maximum) 2 ft 6 in. to within 3 in. of the edge of the footing. If $f_y = 60,000$ psi and $f'_c = 3000$ psi, is their length sufficient for development purposes?

SOLUTION

Basic development length required $= \ell_{db} = \dfrac{(0.04)(0.60)(60,000)}{\sqrt{3000}} = 26.29$ in.

Table 6.1 Checking lateral spacing and edge distance of bars:

Clear lateral spacing $= 6.00 - \frac{7}{8} = 5.125$ in. $> 3d_b = 3 \times \frac{7}{8} = 2.625$ in.

Clear cover $= 3 - (\frac{1}{2})(\frac{7}{8}) = 2.56$ in. $> 2d_b = 2 \times \frac{7}{8} = 1.75$ in.

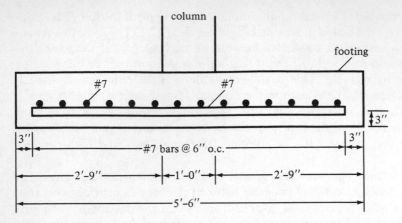

Figure 6.9 Column footing.

Both satisfy condition 1(d). ∴ Modification factor = 1.0 Also satisfies ACI 12.2.2.3.2.

$$\ell_{db} = (1.0)(0.8)(26.29) = 21.03 \text{ in.}$$

$$< \frac{0.03 d_b f_y}{\sqrt{f'_c}} = \frac{(0.03)(0.875)(60,000)}{\sqrt{3000}} = 28.76 \text{ in.}$$

$$\therefore \ell'_{db} = 28.76 \text{ in.}$$

Table 6.2 There are no other applicable modification factors. Thus

$$\ell_d = \ell'_{db} = 28.76 \text{ in.} \qquad \underline{\text{say 29 in.}}$$

which is less than the 2 ft 6 in. furnished. <u>OK</u> ∎

■ EXAMPLE 6.3

Determine the development length required for the #8 bars shown in the light-weight aggregate beam of Figure 6.10 if $f_y = 50,000$ psi and $f'_c = 3000$ psi. The steel required by analysis for the beam was 2.2 in.[2]

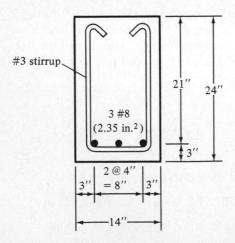

Figure 6.10

SOLUTION

Basic development length required $= \ell_{db} = \dfrac{(0.04)(0.785)(50,000)}{\sqrt{3000}} = 28.66$ in.

Table 6.1 Checking clear lateral spacing and clear edge distance of bar:

Clear lateral spacing $= 4.00 - 1.0 = 3.0$ in. $= 3d_b = 3.0$ in.

Clear cover $= 3.00 - (\tfrac{1}{2})(1.0) = 2.50$ in. $> 2d_b = 2.0$ in.

Both satisfy condition 1(d) in Table 6.1. $\therefore$ Modification factor $= 1.0$.

$$\ell'_{db} = (1.0)(28.66) = 28.66 \text{ in.}$$

$$> \dfrac{(0.03)(1.0)(50,000)}{\sqrt{3000}} = 27.39 \text{ in.}$$

Table 6.2

Modification factor for lightweight aggregate concrete $= 1.3$

Modification factor for excess reinforcement $= \dfrac{2.2}{2.35} = 0.94$

$\ell_d = (1.3)(0.94)(28.66) = 35.02$ in. $\qquad$ <u>Use 3 ft 0 in.</u> ■

■ EXAMPLE 6.4

Calculate the development length required for the #14 epoxy-coated top bars of the beam of Figure 6.11 if the required area of steel was 4.32 in.^{2} and if $f_y = 60,000$ psi and $f'_c = 4000$ psi.

SOLUTION

Basic development length required $= \ell_{db} = \dfrac{(0.085)(60,000)}{\sqrt{4000}} = 80.64$ in.

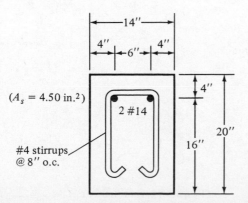

$(A_s = 4.50 \text{ in.}^2)$

2 #14

#4 stirrups @ 8" o.c.

14"

4" 6" 4"

4"

20"

16"

Figure 6.11

Table 6.1 Checking lateral spacing of bars:

Clear lateral spacing $= 6.00 - (1)(1.693) = 4.307$ in.

$< 3d_b = (3)(1.693) = 5.079$ in.

Transverse steel furnished by the two vertical prongs of the stirrups $= (2)(0.20) = 0.40$ in.^{2}

$$A_{tr} = \frac{(1.693)(8)(2)}{40} = 0.677 \text{ in.}^2 > 0.40 \text{ in.}^2$$

$\therefore$ None of the requirements of (a), (b), (c), or (d) of condition 1 are met.

Modification factor $= 1.4$ (and not 2.0 since cover $> d_b$ and clear spacing $> 2d_b$)

$$\ell'_{db} = (1.4)(80.64) = 112.90 \text{ in.} > \frac{(0.03)(1.693)(60,000)}{\sqrt{4000}} = 48.18 \text{ in.}$$

Table 6.2

Modification factor for top bars $= 1.3$

Modification factor for epoxy-coated bars $= 1.5$

Product of top bar and epoxy factors

$= (1.3)(1.5) = 1.95$ but need not be > 1.7 says ACI (12.2.4.3)

$$\ell_d = (1.7)(112.9)\left(\frac{4.32}{4.50} \text{ for excess reinf.}\right) = 184 \text{ in.} \quad \underline{\text{Use 15 ft 4 in.}} \quad \blacksquare$$

■ EXAMPLE 6.5

Compute the development length required for the bundled bars of Figure 6.12 if $f_y = 60,000$ psi and $f'_c = 3000$ psi.

SOLUTION

$$\ell_{db} = \frac{(0.04)(60,000)(0.785)}{\sqrt{3000}} = 34.40 \text{ in.}$$

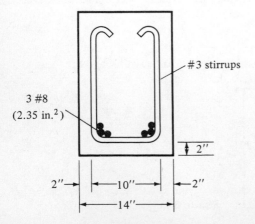

Figure 6.12 Bundled bars.

Table 6.1 Diameter d_b of a single bar of area 2.35 in.^{2}:

$$\frac{\pi d_b^2}{4} = 2.35$$

$$d_b = 1.73 \text{ in.}$$

Clear lateral spacing $= 10.00 - (2)(\frac{3}{8}) - (2)(1.73) = 5.79$ in.
$$> 3d_b = (3)(1.73) = 5.19 \text{ in.}$$

Distance from side to c.g. of bars $= 2 + \frac{3}{8} + (\frac{1}{2})(1.73)$
$$= 3.24 \text{ in.} < 2d_b = (2)(1.73) = 3.46 \text{ in.}$$

Since clear spacing $> 2d_b$ and cover $> d_b$, the modification factor $= 1.4$

$$\ell_{db} = (1.4)(34.40) = 48.16 \text{ in.} < \frac{(0.03)(1.73)(60,000)}{\sqrt{3000}} = 56.85 \text{ in.}$$

Table 6.2 No applicable modification factors.

Bundled bars Development length must be increased by 20% to account for a three-bar bundle (ACI 12.4.1).

$$\ell_d = (1.20)(56.85) = 68.22 \text{ in.} \qquad \underline{\text{Use 5 ft 8 in.}} \quad \blacksquare$$

6.5 HOOKS

When sufficient space is not available to anchor *tension* bars by running them straight for their required development lengths, as described in Sections 6.3 and 6.4 of this text, hooks may be used. (Hooks are considered to be useless for compression bars for development length purposes.)

Figure 6.13 shows details of the standard 90° and 180° hooks specified in Sections 7.1 and 7.2 of the ACI Code. Either the 90° hook with an extension of 12 bar diameters ($12d_b$) at the free end or the 180° hook with an extension of 4 bar diameters ($4d_b$) but not less than $2\frac{1}{2}$ in. may be used at the free end. The radii and diameters shown are measured on the inside of the bends.

The dimensions given for hooks were developed to protect members against splitting of the concrete or bar breakage no matter what concrete strengths, bar sizes, or bar stresses are used. Actually, hooks do not provide an appreciable increase in anchorage strength because the concrete in the plane of the hook is somewhat vulnerable to splitting. This means that adding more length (i.e., more than the specified $12d_b$ or $4d_b$ values) onto bars beyond the hooks doesn't really increase their anchorage strengths.

The basic development length required when a standard hook is being used is referred to as ℓ_{hb} by the ACI. For reinforcing bars with $f_y = 60,000$ psi, ℓ_{hb} is equal to $1200d_b/\sqrt{f_c'}$.

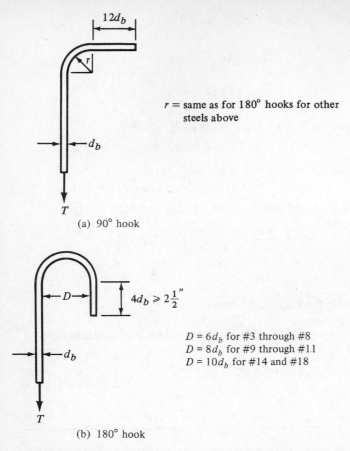

r = same as for 180° hooks for other steels above

(a) 90° hook

$D = 6d_b$ for #3 through #8
$D = 8d_b$ for #9 through #11
$D = 10d_b$ for #14 and #18

(b) 180° hook

Figure 6.13 Hooks.

The actual development length to be used is called ℓ_{dh} and equals ℓ_{hb} times any applicable modification factors but may be not less than $8d_b$ or 6 in. (ACI 12.5.1), whichever is greater. The development length ℓ_{dh} is measured from the critical section of the bar to the outside end of the hooks, as shown in Figure 6.14.

The modification factors that may have to be multiplied by ℓ_{hb} to determine ℓ_{dh} are listed in Section 12.5.3 of the Code and are summarized here. These values apply only for cases where standard hooks are used. The effect of hooks with larger radii is not covered by the Code. For the design of hooks, no distinction is made between top bars and other bars. (It is difficult to distinguish top from bottom anyway when hooks are involved.)

1. If the reinforcing has an f_y other than 60,000 psi, the value of ℓ_{hb} is to be multiplied by $f_y/60,000$ (ACI 12.5.3.1).

2. When 90° hooks and #11 or smaller bars are used and when $2\frac{1}{2}$ in. or more of side cover normal to the hook is present, together with at least 2 in. of cover for the bar extension, multiply by 0.7 (ACI 12.5.3.2).

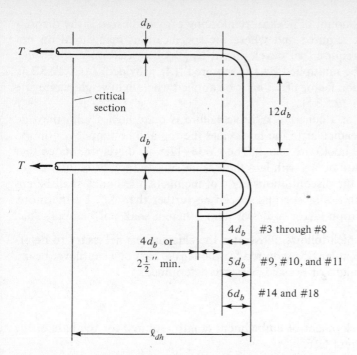

Figure 6.14 Hooked-bar details for development of standard hooks.

3. When hooks made of #11 or smaller bars are enclosed vertically and horizontally within ties or stirrup ties spaced no further apart than $3d_b$ (where d_b is the diameter of the hooked bar), multiply by 0.8. This situation is shown in Figure 6.15. (Detailed dimensions are given for stirrup and tie hooks in Section 7.1.3 of the ACI Code.)

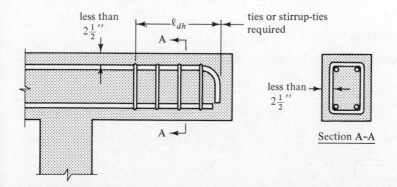

Figure 6.15 Concrete cover per Section 12.5.4 of ACI Code.

4. Where the amount of flexural reinforcing provided exceeds the theoretical amount required and where the specifications being used do not specifically require that development lengths be based on f_y, the value of ℓ_{hb} may be multiplied by $(A_s \text{ required})/(A_s \text{ provided})$ (ACI 12.5.3.4).
5. A modification factor of 1.3 *must* be applied when lightweight aggregate is used (ACI 12.5.3.5).
6. The danger of a concrete splitting failure is quite high if both the side cover (perpendicular to the hook) and the top and bottom cover (in the plane of the hook) are small. The Code (12.5.4), therefore, states that when standard hooks with less than $2\frac{1}{2}$ in. side and top or bottom cover are used at the discontinuous ends of members, the hooks shall be enclosed within ties or stirrups spaced no further than $3d_b$. Furthermore, the modification factor of 0.8 of item 3 herein shall not be applicable.

Example 6.6 which follows, illustrates the calculations necessary to determine the development lengths required for the tensile bars of a cantilever beam. The lengths for straight or hooked bars are determined.

■ EXAMPLE 6.6

Determine the development or embedment length required for the bars of the beam shown in Figure 6.16.

   (a) If the bars are straight.
   (b) If a 180° hook is used.
   (c) If a 90° hook is used.

The 6 #9 bars shown are considered to be top bars. $f_c' = 4000$ psi and $f_y = 40,000$ psi. Assume side, top, and bottom cover > 2.5 in. and clear spacing > $3d_b$.

SOLUTION

   (a) If bars are straight,

$$\ell_{db} = \frac{(0.04)(1.0)(40,000)}{\sqrt{4000}} = 25.30 \text{ in.}$$

Referring to Table 6.1

Modification factor = 1.0 since it meets condition 1(d) in the Table.

$$\ell'_{db} = (1.0)(25.30) = 25.30 \text{ in.} > \frac{(0.03)(1.128)(40,000)}{\sqrt{4000}} = 21.40 \text{ in.}$$

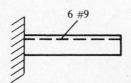

6 #9

Figure 6.16

Referring to Table 6.2

$$\text{Modification factor for top bars} = 1.3$$

$$\ell_d = (1.3)(25.30) = 32.89 \text{ in.}$$ Use 2 ft 9 in.

(b) If 180° hooks are used (Figure 6.17):

$$\ell_{hb} = \frac{(1200)(1.128)}{\sqrt{4000}} = 21.4''$$

$$\ell_{dh} = (21.4)\left(\frac{40{,}000}{60{,}000}\right) = 14.27'' > 8d_b = 9.02 \text{ in.}$$ Use $14\frac{1}{2}''$

(c) If 90° hooks are used (Figure 6.18):

$$\ell_{hb} = 21.4''$$

$$\ell_{dh} = (21.4)\left(\frac{40{,}000}{60{,}000}\right)(0.7) = 9.99'' > 8d_b = 9.02 \text{ in.}$$ Use 10'' ∎

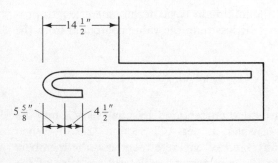

Figure 6.17

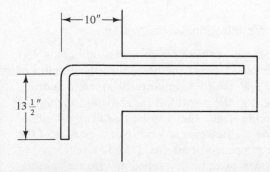

Figure 6.18

6.6 DEVELOPMENT LENGTHS FOR WELDED WIRE FABRIC

Section 12.7 of the ACI Code provides minimum development lengths required for deformed welded wire fabric, whereas Section 12.8 provides values for plain welded wire fabric. In these sections, values are given for a basic development length ℓ_{db} (with at least one cross wire within the development length located not less than 2 in. from the critical section), for modification factors for various items, and for minimum values.

6.7 DEVELOPMENT LENGTHS FOR COMPRESSION BARS

There is not a great deal of experimental information available about bond stresses and needed embedment lengths for compression steel. It is obvious, however, that embedment lengths will be smaller than those required for tension bars. For one reason, there are no tensile cracks present to encourage slipping. For another, there is some bearing of the end of the bars on concrete, which also helps develop the load.

The Code (12.3.2) states that the minimum basic development length provided for compression bars may not be less than the value computed from the following expression:

$$\ell_{db} = \frac{0.02 d_b f_y}{\sqrt{f'_c}} \geq 0.0003 d_b f_y$$

If more compression steel is used than required by analysis, ℓ_{db} may be multiplied by $(A_s$ required$)/(A_s$ furnished$)$ as per ACI Section 12.3.3.1. When bars are enclosed in spirals for any kind of concrete members, the members become decidedly stronger due to the confinement or lateral restraint of the concrete. The normal use of spirals is in spiral columns, which are discussed in Chapter 8. Should compression bars be enclosed by spirals of not less than $\frac{1}{4}$-in. diameter and with a pitch not greater than 4 in., the value of ℓ_{db} may be multiplied by 0.75 (ACI 12.3.3.2). In no case can the development length be less than 8 in. Thus

$$\ell_d = \ell_{db} \times \text{applicable modification factors} \geq 8.0 \text{ in.}$$

An introductory development length problem for compression bars is presented in Example 6.7. The forces in the bars at the bottom of the column of Figure 6.19 are to be transferred down into a reinforced concrete footing. Dowels such as these are usually bent at their bottoms (as shown in the figure) and set on the main footing reinforcing where they can be tied firmly in place. The bent or hooked parts of the dowels, however, do not count as part of the required development lengths for compression bars (ACT 12.5.5).

In a similar fashion the dowel forces must be developed up into the column. In Example 6.7 the required development lengths up into the column and down

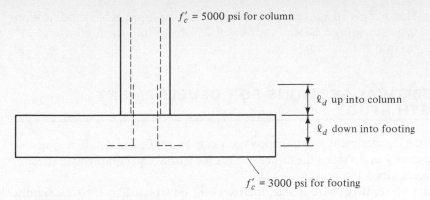

Figure 6.19

into the footing are different because the f'_c values for the footing and the column are different in this case. The topic of dowels and force transfer from walls and columns to footings is discussed in some detail in Chapter 11. (The development lengths determined in this example are for compression bars as would normally be the case at the base of columns. If uplift is possible, however, it will be necessary to consider tension development lengths, which could very well control.)

■ **EXAMPLE 6.7**

The forces in the column bars of Figure 6.19 are to be transferred into the footing with #9 dowels. Determine the development lengths needed for the dowels (a) down into the footing and (b) up into the column if $f_y = 60,000$ psi.

SOLUTION

(a) Down into footing,

$$\ell_{db} = \frac{0.02d_b f_y}{\sqrt{f'_c}} = \frac{(0.02)(1.128)(60,000)}{\sqrt{3000}} = 24.71'' \leftarrow$$

$$\ell_{db} = (0.0003)(1.128)(60,000) = 20.30''$$

$$\ell_{db} = 8''$$

Hence $\ell_d = 24.71''$ (say 25″) as there are no applicable modification factors.

(b) Up into column,

$$\ell_{db} = \frac{(0.02)(1.128)(60,000)}{\sqrt{5000}} = 19.14''$$

$$\ell_{db} = (0.003)(1.128)(60,000) = 20.30'' \leftarrow$$

$$\ell_{db} = 8''$$

Hence $\ell_d = 20.30''$, say $21''$, as there are no applicable modification factors. (*Answer:* Extend the dowels $25''$ down into the footing and $21''$ up into the column.)   ■

6.8 CRITICAL SECTIONS FOR DEVELOPMENT LENGTH STUDY

Before the development length expressions can be applied in detail, it is necessary to clearly understand the critical points for tensile and compressive stresses in the bars along the beam.

First of all, it is obvious that the bars will be stressed to their maximum values at those points where maximum moments occur. Thus those points must be no closer in either direction to the bar ends than the ℓ_d values computed.

There are, however, other critical points for development lengths. As an illustration, a critical situation occurs whenever there is a tension bar whose neighboring bars have just been cut off or bent over to the other face of the beam. Theoretically, if the moment is reduced by a third, one-third of the bars are cut off or bent and the remaining bars would be stressed to their yield points, and the full development lengths would be required for those bars.

This could bring up another matter in deciding the development length required for the remaining bars. The Code (12.10.3) requires that bars that are cut off or bent be extended a distance beyond their theoretical cutoff points by d or 12 bar diameters, whichever is greater. In addition, the point where the other bars are bent or cut off must also be at least a distance ℓ_d from their points of maximum stress. Thus these two items might very well cause the remaining bars to have a stress less than f_y, thus permitting their development lengths to be reduced somewhat. A conservative approach is normally used, however, in which the remaining bars are assumed to be stressed to f_y.

6.9 EFFECT OF COMBINED SHEAR AND MOMENT ON DEVELOPMENT LENGTHS

The ACI Code does not specifically consider the fact that shear affects the flexural tensile stress in the reinforcing. The Code (12.10.3) does require bars to be extended a distance beyond their theoretical cutoff points by a distance no less than d or 12 bar diameters, whichever is larger. The Commentary (R12.10.3) states that this extension is required to account for the fact that the locations of maximum moments may shift due to changes in loading, support settlement, and other factors. It can be shown that a diagonal tension crack in a beam without stirrups can shift the location of the computed tensile stress a distance approximately equal to d towards the point of zero moment. When stirrups are present, the effect is still there but somewhat less severe.

The combined effect of shear and bending acting simultaneously on a beam may produce premature failure due to over-stress in the flexural reinforcing. A great deal of work on this topic has been done by Professor Charles Erdei.[7-9] His work demonstrates that web reinforcing participates in resisting bending moment. He shows that the presence of inclined cracks increases the force in the tensile reinforcing at all points in the shear span except in the region of maximum moment. The result is just as though we have a shifted moment diagram and leads us to the thought that we should be measuring ℓ_d from the shifted moment diagram rather than from the basic one. He clearly explains the moment shift and the relationship between development length and the shift in the moment diagram.

The late Professor P. M. Ferguson[10] stated that whether or not we decide to use the shifted moment concept it is nevertheless desirable to stagger the cut-off points of bars (and it's better to bend them than to cut them).

6.10 EFFECT OF SHAPE OF MOMENT DIAGRAM ON DEVELOPMENT LENGTHS

A further consideration of development lengths will show the necessity of considering the slope of the moment diagram. To illustrate this point, the uniformly loaded beam of Figure 6.20 with its parabolic moment diagram is considered. *It is further assumed that the length of the reinforcing bars on each side of the beam centerline equals the computed development length ℓ_d.* The discussion to follow will prove that this distance is not sufficient to properly develop the bars for this moment diagram.[11]

At the centerline of the beam of Figure 6.20, the moment is assumed to equal M_u and the bars are assumed to be stressed to f_y. Thus the development length of the bars on either side of the beam centerline must be no less than ℓ_d. If one then moves along this parabolic moment diagram on either side to a point where the moment has fallen off to a value of $M_u/2$, it is correct to assume a required development length from this point equal to $\ell_d/2$.

The preceding discussion clearly shows that the bars will have to be extended further out from the centerline than ℓ_d. For the moment to fall off 50%, one must move more than halfway toward the end of the beam.

[7] Erdei, C. K., 1961, "Shearing Resistance of Beams by the Load-Factor Method," *Concrete and Constructional Engineering*, 56(9), pp. 318–319.

[8] Erdei, C. K., 1962, "Design of Reinforced Concrete for Combined Bending and Shear by Ultimate Lead Theory," *Journal of the Reinforced Concrete Association*, 1(1).

[9] Erdei, C. K., 1963, "Ultimate Resistance of Reinforced Concrete Beams Subjected to Shear and Bending," *European Concrete Committees Symposium on Shear*, Wiesbaden, West Germany, pp. 102–114.

[10] Ferguson, P. M., 1979, *Reinforced Concrete Fundamentals*, 4th ed. (New York: John Wiley & Sons) p. 187.

[11] Ferguson, P. M., 1979, *Reinforced Concrete Fundamentals* 4th ed. (New York: John Wiley & Sons), pp. 191–193.

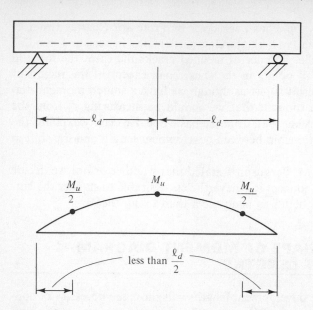

Figure 6.20

6.11 CUTTING OFF OR BENDING BARS (CONTINUED)

In this section a few concluding remarks are presented concerning the cutting off of bars, a topic that was introduced in Section 6.1. The last several sections have presented considerable information that affects the points where bars may be cut off. A summary of the previously mentioned requirements is presented here, together with some additional information. Initially, a few comments are presented concerning shear.

When some of the tensile bars are cut off at a point in a beam, a sudden increase in the tensile stress will occur in the remaining bars. For this increase to occur there must be a rather large increase in strain in the beam. Such a strain increase quite possibly may cause large tensile cracks to develop in the concrete. If large cracks occur, there will be a reduced beam cross section left to provide shear resistance—and thus a greater possibility of shear failure.

To minimize the possibility of a shear failure, Section 12.10.5 of the ACI Code states that *at least one* of the following conditions must be met if bars are cut off in a tension zone.

1. The shear at the cutoff point not exceeding two-thirds of the permissible shear in the beam (including the strength of any shear reinforcing provided.)

2. An area of shear reinforcing provided in excess of that required for shear and torsion for a distance equal to $\frac{3}{4}d$ from the cutoff point. The minimum area of this reinforcing and its maximum spacing are provided in Section 12.10.5.2 of the Code.

3. When #11 or smaller bars are used, the continuing bars should provide twice the area of steel required for flexure at the cutoff point and the shear should not exceed three-fourths of the permissible shear.

The structural designer is seldom involved with a fixed moment diagram—the loads move and the moment diagram changes. Therefore, the Code (12.10.3) says that reinforcing bars should be continued for a distance of 12 bar diameters or the effective depth of the member, whichever is greater (except at the supports of simple spans and the free ends of cantilevers), beyond their theoretical cutoff points.

Los Angeles County water project. (Courtesy of The Burke Company.)

There are various other rules for development lengths which apply specifically to positive moment reinforcing, negative moment reinforcement, and continuous beams. These are addressed in Chapter 13 of this text. Another item presented in that chapter which is usually of considerable interest to students is the rules of thumb which are frequently used in practice to establish cutoff and bend points.

Another rather brief development length example is presented in Example 6.8. A rectangular section and satisfactory reinforcing have been selected for the given span and loading condition. It is desired to determine where two of the four bars may be cut off considering both moment and development length.

■ **EXAMPLE 6.8**

The rectangular section with four #9 bars shown in Figure 6.21(b) has been selected for the span and loading shown in Figure 6.21(a). Determine the cutoff point for two of the bars considering both the actual moment diagram and the required development length.

The permissible moment capacities (ϕM_n) for this beam have been computed to equal 392.7 ft-k with four bars and 218.2 ft-k with two bars. In addition, ℓ_d for the bars has been determined to equal 43.28 in. or 3 ft 8 in.

SOLUTION

The solution for this problem is primarily shown in Figure 6.22. The two bars begin at the left end of the beam. As no development length is available the design strength of the member is zero. As we move from A to B on the diagram a distance equal to ℓ_d, the permissible moment capacity will increase on a straight line from 0 to 218.2 ft-k. From B to C it will remain equal to 218.2 ft-k.

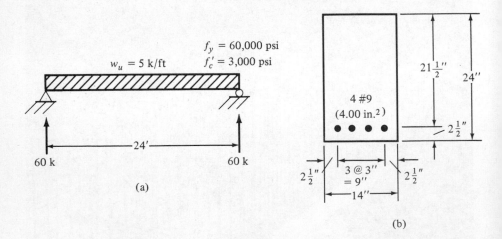

(a)

(b)

Figure 6.21

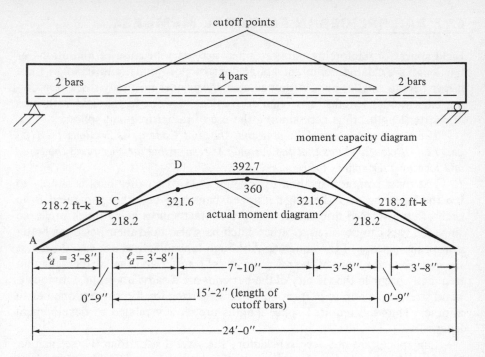

Figure 6.22

At C we reach the cutoff point of the bars, and from C to D (a distance equal to ℓ_d) the permissible moment capacity increases from 218.2 ft-k to 392.7 ft-k.

At no point may the design strength be less than the computed required moment. We can then see that point C can be located where the actual moment equals 218.2 ft-k. With reference to Figure 6.21(a) the moment at C is computed as

$$(60x) - (5)(x)\left(\frac{x}{2}\right) = 218.2$$

$$x = 4.47 \text{ ft} \qquad \underline{\text{say 4 ft 5 in. from both ends}}$$

By the time we reach point D (3 ft 8 in. to the right of C and 8 ft 1 in. from the left support), the required moment capacity is

$$M_u = (60)(8.083) - (8.083)(5)\left(\frac{8.083}{2}\right) = 321.6 \text{ ft-k}$$

Earlier in this section, reference was made to ACI Section 12.10.5, which considered shear at cutoff points. It is assumed that this beam will be properly designed for shear as described in the next chapter of this book and will meet the ACI requirements.

Obviously the permissible moment capacity of the beam at any point must be larger than the required moment capacity at that point. Figure 6.22 shows the beam with the bar lengths and the two moment diagrams. ∎

6.12 BAR SPLICES IN FLEXURAL MEMBERS

Field splices of reinforcing bars are often necessary because of limitations of bar lengths available, requirements at construction joints, and changes from larger bars to smaller bars. Although steel fabricators normally stock reinforcing bars in 60-ft lengths, it is often convenient to work in the field with bars of shorter lengths, thus necessitating the use of rather frequent splices.

The reader should carefully note that the ACI Code in its Sections 1.2.1(h) and 12.14.1 clearly states that the designer is responsible for specifying the types and locations for splices for reinforcement.

The most common method of splicing #11 or smaller bars is simply to lap the bars one over the other. Lapped bars may be either separated from each other or placed in contact, with the contact splices being much preferred since the bars can be wired together. Such bars also hold their positions better during the placing of the concrete. Although lapped splices are easy to make, the complicated nature of the resulting stress transfer and the local cracks that frequently occur in the vicinity of the bar ends are disadvantageous. Obviously, bond stresses play an important part in transferring the forces from one bar to another. Thus the required splice lengths are closely related to development lengths.

Lap splices are not very satisfactory for several situations. These include: (1) where they would cause congestion, (2) where the laps would be very long as they are for #9 to #11 Grade 60 bars, (3) where #14 or #18 bars are used because the Code (12.14.2) does not permit them to be lap spliced, and (4) where very long bar lengths would be left protruding from existing concrete structures for purposes of future expansion. For such situations, other types of splices such as those made by welding or by mechanical devices may be used. Welded splices, from the view of stress transfer, are the best splices, but they may be expensive and may cause metallurgical problems. The ACI Code (12.14.3.3) states that welded splices must be accomplished by butting the bars against each other and welding them together so that the connection will be able to develop at least 125% of the specified yield strength of the bars. Splices not meeting this strength requirement can be used at points where the bars are not stressed to their maximum tensile stresses. It should be realized that welded splices are usually the most expensive due to the high labor costs and due to the costs of proper inspection.

Mechanical connectors usually consist of some type of sleeve splice, which fits over the ends of the bars to be joined and into which a metallic grout filler is placed to interlock the grooves inside the sleeve with the bar deformations. From the standpoint of stress transfer, good mechanical connectors are next best to welded splices. They do have the disadvantage that some slippage may occur in the connections, and as a result there may be some concrete cracks in the area of the splices.

Before the specific provisions of the ACI Code are introduced, a few general comments are presented that should explain the background for these provisions. These remarks are taken from a paper by George F. Leyh of the CRSI.[12]

1. Splicing of reinforcement can never reproduce exactly the same effect as continuous reinforcing.
2. The goal of the splice provisions is to require a ductile situation where the reinforcing will yield before the splices fail. Splice failures occur suddenly without warning and with dangerous results.
3. Lap splices fail by splitting of the concrete along the bars. If some type of closed reinforcing is wrapped around the main reinforcing (such as ties and spirals described for columns in Chapter 8), the chances of splitting are reduced and smaller splice lengths are needed.
4. When stresses in reinforcement are reduced at splice locations, the chances of splice failure are correspondingly reduced. For this reason, the Code requirements are less restrictive where stresses are low.

Splices should be located away from points of maximum tensile stress. Furthermore, all of the bars should not be spliced at the same locations—that is, the splices should be staggered. Should two bars of different diameters be lap spliced, the lap length used should equal the value required for the smaller bar or the development length required for the larger bar, whichever is greater.

The length of lap splices for bundled bars must be equal to the required lap lengths for individual bars of the same size, but increased by 20% for three-bar bundles and 33% for four-bar bundles (ACI Code 12.4) because there is a smaller area of contact between the bars and the concrete, and thus less bond. Furthermore, individual splices within the bundles are not permitted to overlap each other.

6.13 TENSION SPLICES

The Code (12.15) divides tension lap splices into two classes, A and B. The class of splice used is dependent upon the level of stress in the reinforcing and upon the percentage of steel which is spliced at a particular location.

Class A splices are those where the reinforcing is lapped for a minimum distance of $1.0\ell_d$ (but not less than 12 in.) and where one-half or less of the reinforcing is spliced at any one location.

Class B splices are those where the reinforcing is lapped for a minimum distance of $1.3\ell_d$ (but not less than 12 in.) and where all the reinforcing is spliced at the same location.

The Code (12.15.2) states that lap splices for deformed bars and deformed wire in tension must be Class B unless (1) the area of reinforcing provided is equal to two or more times the area required by analysis over the entire length

---

[12] Proceedings of the PCA-ACI Teleconference on ACI 318-71 Building Code Requirements, 1972 (Skokie, Ill.: Portland Cement Association), p. 14-1.

of the splice and (2) one-half or less of the reinforcing is spliced within the required lap length.

In calculating the ℓ_d to be multiplied by 1.0 or 1.3, the reduction for excess reinforcing furnished should not be used because the class of splice (A or B) already reflects any excess reinforcing at the splice location (see ACI Commentary R12.15.1).

6.14 COMPRESSION SPLICES

Compression bars may be spliced by lapping, by end bearing, and by welding or mechanical devices. The Code (12.16.1) says that the minimum splice length of such bars should be the development length ℓ_d but may be not less than $0.0005f_y d_b$ for bars with f_y of 60,000 psi or less than $(0.0009f_y - 24)d_b$ for steels with higher f_y values or less than 12 in. Should the concrete strengths be less than 3000 psi, it is necessary to increase the computed laps by one-third. Reduced values are given in the Code for cases where the bars are enclosed by ties or spirals (12.17.2.4).

The required length of lap splices for compression bars of different sizes is the larger of the computed splice length of the smaller bars or the development length of the larger bars. It is permissible to lap splice # 14 and # 18 compression bars to # 11 and smaller bars (12.16.2).

The transfer of forces between bars that are always in compression can be accomplished by end bearing according to Section 12.17.4 of the Code. For such transfer to be permitted the bars must have their ends square cut (within $1\frac{1}{2}°$ of a right angle) and they must be suitably confined (by closed ties, closed stirrups, or spirals). Section 12.17.4 further states that when end-bearing splices are used in columns, in each face of the column additional reinforcement has to be added that is capable of providing a tensile strength at least equal to 25% of the yield strength of the vertical reinforcement provided in that face.

The Code (12.14.2.1) with one exception prohibits the use of lap splices for #14 or #18 bars. When column bars of those sizes are in compression it is permissible to connect them to footings by means of dowels of smaller sizes with lap splices as described in Section 15.8.2.3 of the Code.

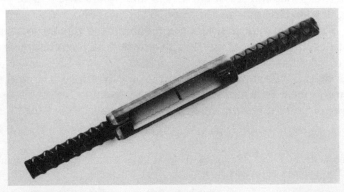

Extruded coupler splice. (Courtesy Dywidag Systems International, USA, Inc.)

PROBLEMS

6.1   For the cantilever beam shown, determine the point where two bars can be cut off from the standpoint of the calculated moment strength ϕM_n of the beam. $f_y = 60,000$ psi, $f'_c = 3000$ psi. (*Ans.* 9.088 ft from free end)

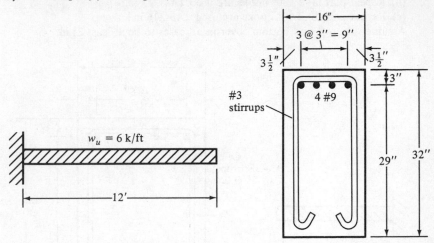

In Problems 6.2 to 6.8 determine the development lengths required for the following tension bar situations.

6.2   #9 bottom bars in regular-weight concrete. $f_y = 60,000$ psi, $f'_c = 4000$ psi. Clear spacing between bars $= 3\frac{7}{8}$ in. Clear cover $= 2\frac{3}{8}$ in.

6.3   #6 bottom bars in lightweight aggregate concrete. $f_y = 60,000$ psi. $f'_c = 3000$ psi. Clear spacing between bars $= 2\frac{1}{4}$ in. Clear cover $= 2\frac{1}{8}$ in. (*Ans.* 32 in.)

6.4   #7 top epoxy-coated bars in regular-weight concrete. $f_y = 60,000$ psi. $f'_c = 3000$ psi. Clear spacing between bars $= 2\frac{1}{8}$ in. Clear cover $= 2\frac{1}{16}$ in. #3 stirrups spaced @ 6 in. O.C.

6.5   #18 bottom bars in regular-weight concrete. $f_y = 60,000$ psi. $f'_c = 5000$ psi. Clear spacing between bars $= 4.0$ in. Clear cover $= 2.0$ in. #4 stirrups @ 8 in. O.C. (*Ans.* 212 in.)

6.6   #8 epoxy-coated top bars in lightweight aggregate. $f_y = 40,000$ psi. $f'_c = 3000$ psi. Clear spacing between bars $= 1\frac{1}{2}$ in. Clear cover of bars $= 1\frac{1}{2}$ in. #3 stirrups @ 6 in. O.C.

6.7   The #11 bundled bars shown in regular-weight concrete. $f_y = 60,000$ psi. $f'_c = 3000$ psi. (*Ans.* 96 in.)

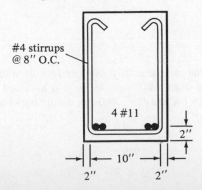

6.8 The bars of Problem 6.1.

6.9 **(a)** Determine the development length required for the #8 bars shown if normal weight concrete is used and the bars are straight. $f'_c = 3000$ psi, $f_y = 60,000$ psi. (*Ans.* 45 in.)

(b) Repeat part (a) if 180° hooks are used. (*Ans.* 22 in.)

(c) Repeat part (b) if 90° hooks are used. (*Ans.* 16 in.)

Assume side, top, and bottom cover in all cases to be at least $2\frac{1}{2}$ in.

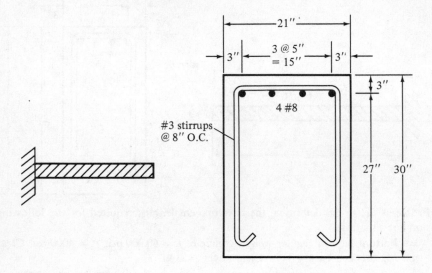

6.10 Are the #8 bars shown anchored sufficiently with their 90° hooks. $f'_c = 3000$ psi, $f_y = 60$ ksi. The stress in the bars is 50 ksi. Side and top cover is $2\frac{1}{2}$ in. on bar extensions.

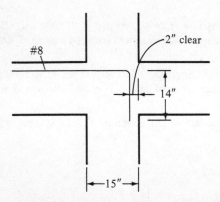

6.11 The required bar area for the wall footing shown is 0.70 in.² per foot of width; #8 bars 12 in. on center (0.78 in.²/ft) are used. Maximum moment is assumed to occur at the face of the wall. If $f_y = 60,000$ psi and $f'_c = 3000$ psi, do the bars have sufficient development lengths? (*Ans.* No)

Problem 6.11

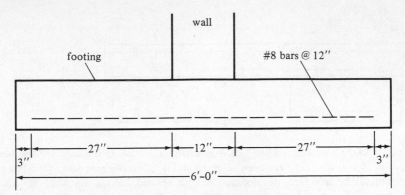

6.12 The beam shown is subjected to an M_u of 300 ft-k at the support. If cover is 1.5 in, $f_y = 60,000$ psi and $f'_c = 3000$ psi, do the following: (a) select #9 bars to be placed in 1 row, (b) determine the development lengths required if straight bars are used, and (c) determine the development lengths needed if 180° hooks are used.

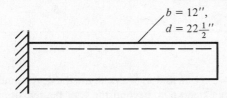

6.13 In the figure shown the lower column bars are #9 and the upper ones are #8. The bars are enclosed by ties spaced 12 in. on center. If $f_y = 60,000$ psi and $f'_c = 3000$ psi what is the minimum lap splice length needed. (*Ans*. 34 in.)

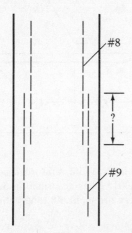

6.14 Calculations show that 2.76 in.² of top or negative steel is required for the beam shown. Three #9 bars have been selected. Are the 4' 6" development lengths shown satisfactory if $f'_c = 3000$ psi and $f_y = 60,000$ psi? Bars are spaced 3" o.c. with 2" side and top cover.

Problem 6.14

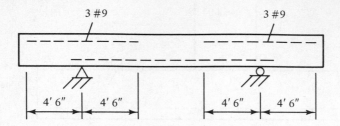

6.15 Calculations show that 3.65 in.² of top or negative steel is required for the beam shown. If four #9 bars have been selected and if f'_c = 3000 psi and f_y = 60,000 psi, determine the minimum development length needed for the standard 90° hooks shown. Assume bars have 3″ side and top clear cover. (*Ans.* 16 in.)

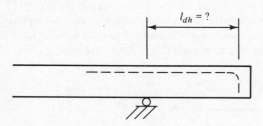

6.16 If f_y = 60,000 psi, f'_c = 3000 psi, w_D = 1.5 k/ft, and w_L = 5 k/ft, are the development lengths satisfactory? Assume the bars extend 6 in. beyond the ₵ of the reactions.

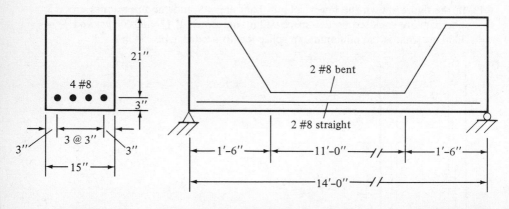

6.17 Determine the compression lap splices needed for a 14 in. × 14 in. reinforced concrete column with ties (whose effective area exceeds 0.0015 hs as described in Section 12.17.2.4 of the Code) for the cases to follow. There are eight #8 longitudinal bars equally spaced around the column.

(a) f'_c = 4000 psi and f_y = 60,000 psi (*Ans.* ℓ_d = 30 in.)

(b) f'_c = 2500 psi and f_y = 50,000 psi (*Ans.* ℓ_d = 28 in.)

7

Shear and Diagonal Tension

7.1 INTRODUCTION

As repeatedly mentioned in the preceding chapters of this book, the objective of today's reinforced concrete designer is to produce ductile members that provide warning of impending failure. To carry out this goal the Code provides design shear values that have larger safety factors against shear failures than

Massive reinforced concrete members. (Courtesy of Bethlehem Steel Corportion.)

do those provided for bending failures. Thus such members will fail in bending under loads that are appreciably smaller than those that would cause shear failures. As a result, those underreinforced members will fail ductilely. They may crack and sag a great deal if overloaded, but they will not fall apart as they might if shear failures were possible. When shear failures are possible before flexural failures, they can be quite dangerous because they can occur without warning.

7.2 SHEAR STRESSES IN CONCRETE BEAMS

Although no one has ever been able to accurately determine the resistance of concrete to pure shearing stress, the matter is not very important because pure shearing stress is probably never encountered in concrete structures. Furthermore, according to engineering mechanics, if pure shear is produced in a member, a principal tensile stress of equal magnitude will be produced on another plane. Because the tensile strength of concrete is less than its shearing strength, the concrete will fail in tension before its shearing strength is reached.

You have previously learned that in elastic homogeneous beams, where stresses are proportional to strains, two kinds of stresses occur (bending and shear) and they can be calculated with the following expressions:

$$f = \frac{Mc}{I}$$

$$v = \frac{VQ}{Ib}$$

An element of a beam not located at an extreme fiber or at the neutral axis is subject to both bending and shear stresses. These stresses combine into inclined compressive and tensile stresses, called *principal stresses*, which can be determined from the following expression:

$$f_p = \frac{f}{2} \pm \sqrt{\left(\frac{f}{2}\right)^2 + v^2}$$

The direction of the principal stresses can be determined with the formula to follow, in which α is the inclination of the stress to the beam's axis:

$$\tan 2\alpha = 2\frac{v}{f}$$

Obviously, at different positions along the beam the relative magnitudes of v and f change, and thus the directions of the principal stresses change. It can be seen from the preceding equation that at the neutral axis the principal stresses will be located at a 45° angle with the horizontal.

You understand by this time that tension stresses in concrete are a serious matter. Diagonal principal tensile stresses, called *diagonal tension*, occur at different places and angles in concrete beams and they must be carefully considered. If they reach certain values, additional reinforcing, called *web reinforcing*, must be supplied.

The discussion presented up to this point relating to diagonal tension applies rather well to plain concrete beams. If, however, reinforced concrete beams are being considered, the situation is quite different because the longitudinal bending tension stresses are resisted quite satisfactorily by the longitudinal reinforcing. These bars, however, do not provide significant resistance to the diagonal tension stresses.

7.3 SHEAR STRENGTH OF CONCRETE

A great deal of research has been done on the subject of shear and diagonal tension for nonhomogeneous reinforced concrete beams, and many theories have been developed. Despite all this work and all the resulting theories, no one has been able to provide a clear explanation of the failure mechanism involved. As a result, design procedures are based primarily on test data.

If V_u is divided by the effective beam area $b_w d$, the result is what is called an *average shearing stress*. This stress is not equal to the diagonal tension stress but merely serves as an indicator of its magnitude. Should this indicator exceed a certain value, shear or web reinforcing is considered necessary. In the ACI Code the basic shear equations are presented in terms of shear forces and not shear stresses. In other words, the average shear stresses described in this paragraph are multiplied by the effective beam areas to obtain total shear forces.

For this discussion V_n is considered to be the nominal or theoretical shear strength of a member. This strength is provided by the concrete and by the shear reinforcement

$$V_n = V_c + V_s$$

The permissible shear strength of a member, ϕV_n, is equal to ϕV_c plus ϕV_s, which must at least equal the factored shear force to be taken, V_u:

$$V_u = \phi V_c + \phi V_s$$

The shear strength provided by the concrete, V_c, is considered to equal an average shear stress strength (normally $2\sqrt{f_c'}$) times the effective cross-sectional area of the member, $b_w d$, where b_w is the width of a rectangular beam or of the web of a T beam or an I beam.

$$V_c = 2\sqrt{f_c'}\, b_w d \qquad \text{(ACI Equation 11-3)}$$

Beam tests have shown some interesting facts about the occurrence of cracks at different average shear stress values. For instance, where large moments occur even though appropriate longitudinal steel has been selected, extensive flexural cracks will occur. As a result the uncracked area of the beam cross section will be greatly reduced and the nominal shear strength V_c can be as low as $1.9\sqrt{f_c'}\, b_w d$.

On the other hand, in regions where the moment is small the cross section will be either uncracked or slightly so and a large portion of the cross section is available to resist shear. For such a case, tests show that a V_c of about $3.5\sqrt{f'_c}\,b_w d$ can be developed before shear failure occurs.[1]

As a result of this information the Code (11.3.1.1) suggests that conservatively V_c (the shear force that the concrete can resist without web reinforcing) can go as high as $2\sqrt{f'_c}\,b_w d$. As an alternative, the following shear force (from Section 11.3.2.1 of the Code) may be used, which takes into account the effects of the longitudinal reinforcing and the moment and shear magnitudes. This value must be calculated separately for each point being considered in the beam.

$$V_c = \left(1.9\sqrt{f'_c} + 2500\rho_w \frac{V_u d}{M_u}\right)b_w d \le 3.5\sqrt{f'_c}\,b_w d \qquad \text{(ACI Equation 11-6)}$$

In this expression $\rho_w = A_s/b_w d$ and M_u is the factored moment occurring simultaneously, with V_u being the factored shear at the section considered. The quantity $V_u d/M_u$ may not be taken to be greater than unity in computing V_c by means of the above expression. The purpose of this requirement is to limit V_c near points of inflection.

From this expression it can be seen that V_c increases as the amount of reinforcing (represented by ρ_w) is increased. As the amount of steel is increased, the length and width of cracks will be reduced. If the cracks are kept narrower, more concrete is left to resist shear and there will be closer contact between the concrete on opposite sides of the cracks. Hence there will be more resistance to shear by friction (called *aggregate interlock*) on the two sides.

Though this more complicated expression for V_c can easily be used for computer designs, it is quite tedious to apply when hand-held calculators are used. The reason is that the values of ρ_w, V_u, and M_u are constantly changing as we move along the span, requiring the computation of V_c at numerous positions. As a result the alternate value $2\sqrt{f'_c}\,b_w d$ is normally used. If the same member is to be constructed many times, the use of the more complex expression may be justified.

7.4 LIGHTWEIGHT CONCRETE

If the shear strength of a lightweight concrete section is being determined, the term $\sqrt{f'_c}$ is to be replaced with $f_{ct}/6.7 \le \sqrt{f'_c}$, where f_{ct} is the split-cylinder strength of the concrete. If f_{ct} is not available the values of $\sqrt{f'_c}$ are to be multiplied by 0.75 for "all-lightweight concrete" and by 0.85 for "sand-lightweight" concrete. When partial sand replacement is used, linear interpolation between the values is permitted. These provisions are made in ACI Section 11.2.

---

[1] ACI-ASCE Committee 326, 1962, "Shear and Diagonal Tension," part 2, *Journal ACI*, 59, p. 277.

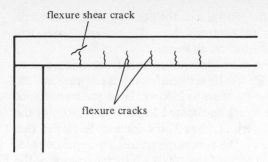

Figure 7.1

7.5 SHEAR CRACKING OF REINFORCED CONCRETE BEAMS

Inclined cracks can develop in the webs of reinforced concrete beams either as extensions of flexural cracks or occasionally as independent cracks. The first of these two types is the *flexure shear crack*, an example of which is shown in Figure 7.1. These are the ordinary types of shear cracks found in both pre-stressed and nonprestressed beams. For them to occur the moment must be larger than the cracking moment and the shear must be rather large. The cracks run at angles of about 45° with the beam axis and probably start at the top of a flexure crack. The approximately vertical flexure cracks shown are not dangerous unless there is a critical combination of shear stress and flexure stress occurring at the top of one of the flexure cracks.

Occasionally an inclined crack will develop independently in a beam, even though no flexure cracks are in that locality. Such cracks, which are called *web shear cracks*, will sometimes occur in the webs of prestressed sections, parti-cularly those with large flanges and thin webs. They also sometimes occur near the points of inflection of continuous beams or near simple supports. At such locations small moments and high shear often occur. These types of cracks will form near the middepth of sections and will move on a diagonal path to the tension surface. Web shear cracks are illustrated in Figure 7.2.

As a crack moves up to the neutral axis, the result will be a reduced amount of concrete left to resist shear—meaning that shear stresses will increase on the concrete above the crack.

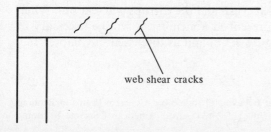

Figure 7.2

It will be remembered that at the neutral axis the bending stresses are zero and the shear stresses are at their maximum values. The shear stresses will therefore determine what happens to the crack there.

After a crack has developed, the member will fail unless the cracked concrete section can resist the applied forces. If web reinforcing is not present, the items that are available to transfer the shear are as follows: (1) the shear resistance of the uncracked section above the crack (estimated to be 20% to 40% of the total resistance), (2) the aggregate interlock, that is, the friction developed due to the interlocking of the aggregate on the concrete surfaces on opposite sides of the crack (estimated to be 33% to 50% of the total), (3) the resistance of the longitudinal reinforcing to a frictional force often called *dowel action* (estimated to be 15% to 25%), and (4) a tied-arch type of behavior that exists in rather deep beams produced by the longitudinal bars acting as the tie and by the uncracked concrete above and to the sides of the crack acting as the arch above.[2]

7.6 WEB REINFORCEMENT

When the factored shear V_u is high, it shows that serious cracks are going to occur unless some type of additional reinforcing is provided. This reinforcing is called *web reinforcing* and usually takes the form of stirrups that enclose the longitudinal reinforcing along the faces of the beam. The most common stirrups are ⊔ shaped, but they can be ⊔⊓⊔ shaped or perhaps have only a single vertical prong, as shown in Figure 7.3. Multiple stirrups such as the ones shown in Figure 7.3(f) are considered to inhibit splitting in the plane of the longitudinal bars. As a consequence, they are generally more desirable for wide beams than the ones shown in Figure 7.3(e). Sometimes it is rather convenient to use lap spliced stirrups such as the ones shown in Figure 7.3(h). These stirrups, which are described in ACI Section 12.13.5, are occasionally useful for deep members, particularly those with gradually varying depths. They are considered, however, to be unsatisfactory in seismic areas.

Bars called *hangers* (usually with about the same diameter as that of the stirrups) are placed on the compression sides of beams to support the stirrups, as illustrated in Figure 7.3. The stirrups are passed around the tensile steel and to meet anchorage requirements are run as far into the compression side of the beam as practical and are hooked around the hangers. Bending of the stirrups around the hangers reduces the bearing stresses under the hooks. If these bearing stresses are too high, the concrete will crush and the stirrups will tear out. When a significant amount of torsion is present in a member, it will be necessary to use closed stirrups as shown in Figure 7.3(b) and as discussed in Chapter 14.

---

[2] Taylor, H. P. J., 1974, "The Fundamental Behavior of Reinforced Concrete Beams in Bending and Shear". *Shear in Reinforced Concrete*. Vol. 1, SP-42 (Detroit: American Concrete Institute), pp. 43–47.

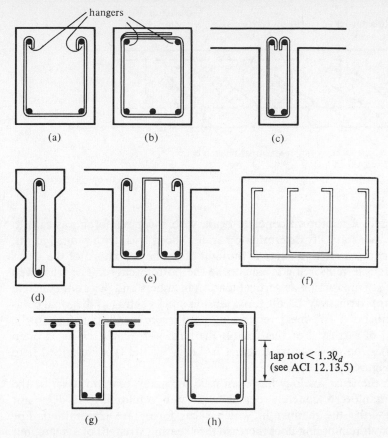

Figure 7.3 Types of stirrups.

The width of diagonal cracks is directly related to the strain in the stirrups. Consequently, the ACI 11.5.2 does not permit the design yield stress of the stirrups to exceed 60 ksi. This requirement limits the width of cracks that can develop. Such a result is important both from the standpoint of appearance and from the standpoint of aggregate interlock. When the width of cracks is limited, it enables more aggregate interlock to develop. A further advantage of a limited yield stress is that the anchorage requirements at the top of the stirrups are not quite as stringent as they would be for stirrups with greater yield strengths.

7.7 BEHAVIOR OF WEB REINFORCEMENT

The actual behavior of web reinforcement is not really understood, although several theories have been presented through the years. One theory, which has been widely used almost since the turn of the century, is the so-called truss

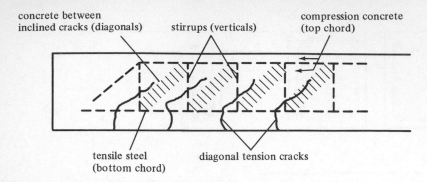

concrete between
inclined cracks (diagonals)   stirrups (verticals)   compression concrete
(top chord)

tensile steel   diagonal tension cracks
(bottom chord)

Figure 7.4 Truss analogy.

analogy, wherein a reinforced concrete beam with shear reinforcing is said to behave much as a statically determinate parallel chord truss with pinned joints. The flexural compression concrete is thought of as the top chord of the truss, whereas the tensile reinforcing is said to be the bottom chord. The truss web is made up of stirrups acting as vertical tension members and pieces of concrete between the approximately 45° diagonal tension cracks acting as diagonal compression members.[3,4] The shear reinforcing used is similar in its action to the web members of a truss. For this reason the term *web reinforcement* is often used when referring to shear reinforcing. A "truss" of the type described here is shown in Figure 7.4.

Although the truss analogy has been used for many years to describe the behavior of reinforced concrete beams with web reinforcing, it does not accurately describe the manner in which shear forces are transmitted. For example, the web reinforcing does increase the shearing strength of a beam, but it has little to do with shear transfer in a beam before inclined cracks form.

The Code requires web reinforcement for all major beams. In Section 11.5.5.1 a minimum area of web reinforcing is required for all concrete flexural members except slabs and footings, certain concrete floor joists, and in wide shallow beams whose total depths are not more than 10 in. or more than $2\frac{1}{2}$ times their flange thicknesses or one-half their web widths, whichever is greatest. Various tests have shown that shear failures do not occur before bending failures in shallow members. Shear forces are distributed across these wide sections. For joists the redistribution is via the slabs to adjacent joists.

Inclined or diagonal stirrups lined up approximately with the principal stress directions are more efficient in carrying the shears and preventing or delaying the formation of diagonal cracks. Such stirrups, however, are not usually considered to be very practical in the United States because of the great amount of labor costs required for positioning them. Actually, they can be

[3] Ritter, W., 1899, *Die Bauweise Hennebique* (Schweizerische Bauzeitung, XXXIII, no. 7).

[4] Mörsch, E., 1912, *Der Eisenbetonbau, seine Theorie and Anwendung* (Stuttgart: Verlag Konrad Wittwer).

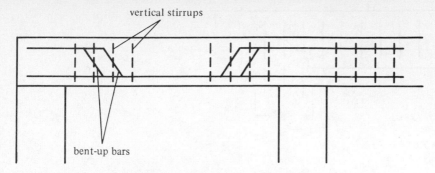

Figure 7.5 Bent-up bar web reinforcing.

rather practical for precast concrete beams where the bars and stirrups are preassembled into cages before being used and where the same beams are duplicated many times.

Bent-up bars (usually at 45° angles) are another satisfactory type of web reinforcing (see Figure 7.5). Although bent-up bars are very commonly used in flexural members in the United States, the average designer seldom considers the fact that they can resist diagonal tension. Two reasons for not counting their contribution to diagonal tension resistance are that there are only a few if any bent-up bars in a beam and that they may not be conveniently located for use as web reinforcement.

Welded wire fabric with wires perpendicular to the axis of a beam is another satisfactory type of web reinforcing.

7.8 SPACING OF VERTICAL STIRRUPS

The maximum shear V_u in a beam must not exceed the permissible shear capacity of the beam cross section ϕV_n, where ϕ is 0.85 and V_n is the nominal shear strength of the concrete and the shear reinforcing.

$$V_u \leq \phi V_n$$

The value of ϕV_n can be broken down into the permissible shear strength of the concrete ϕV_c plus the permissible shear strength of the shear reinforcing ϕV_s. The value of ϕV_c is provided in the Code for different situations and thus we are able to compute the required value of ϕV_s for each situation:

$$V_u \leq \phi V_c + \phi V_s$$

For this derivation an equals sign is used:

$$V_u = \phi V_c + \phi V_s$$

The purpose of stirrups is to minimize the size of diagonal tension cracks or to carry the diagonal tension stress from one side of the crack to the other. Very little tension is carried by the stirrups until after a crack begins to form.

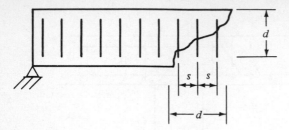

Figure 7.6

Before the inclined cracks begin to form, the strain in the stirrups is equal to the strain in the adjacent concrete. Because this concrete cracks at very low diagonal tensile stresses, the stresses in the stirrups at that time are very small, perhaps only 3 to 6 ksi. You can see that these stirrups do not prevent inclined cracks and that they really aren't a significant factor until the cracks begin to develop.

Tests made on reinforced concrete beams show that a beam cannot fail by the widening of the diagonal tension cracks until the stirrups going across the cracks have been stressed to their yield stresses. For the derivation to follow, it is assumed that a diagonal tension crack has developed and has run up into the compression zone but not all the way to the top as shown in Figure 7.6. It is further assumed that the stirrups crossing the crack have yielded.

The nominal shear strength of the stirrups V_s crossing the crack can be calculated from the following expression, where n is the number of stirrups crossing the crack and A_v is the cross-sectional area each stirrup has crossing the crack. (If a ⊔ stirrup is used, $A_v = 2$ times the cross-sectional area of the stirrup bar. If it is a ⊔⊔, $A_v = 4$ times the cross-sectional area of the stirrup bar.)

$$V_s = A_v f_y n$$

If it is conservatively assumed that the horizontal projection of the crack equals the effective depth d of the section (thus a 45° crack), the number of stirrups crossing the crack can be determined from the expression to follow, in which s is the center-to-center spacing of the stirrups:

$$n = \frac{d}{s}$$

Then

$$V_s = A_v f_y \frac{d}{s} \qquad \text{(ACI Equation 11-17)}$$

From this expression the required spacing of vertical stirrups is

$$s = \frac{A_v f_y d}{V_s}$$

and the value of V_s used here can be determined as follows:

$$V_u = \phi V_c + \phi V_s$$

$$V_s = \frac{V_u - \phi V_c}{\phi}$$

Going through a similar derivation, the following expression can be determined for the required area for inclined stirrups, in which α is the angle between the stirrups and the longitudinal axis of the member:

$$V_s = \frac{A_v f_y (\sin \alpha + \cos \alpha) d}{s} \qquad \text{(ACI Equation 11-18)}$$

And for a bent-up bar or a group of bent-up bars at the same distance from the support, we have

$$V_s = A_v f_y \sin \alpha \qquad \text{(ACI Equation 11-19)}$$

7.9 ACI CODE REQUIREMENTS

In this section a detailed list of the Code requirements controlling the design of web reinforcing is presented, even though some of these items have been previously mentioned in this chapter:

1. When the factored shear V_u exceeds one-half the permissible shear ϕV_c, the Code (11.5.5.1) requires the use of web reinforcing. The value of V_c is normally taken as $2\sqrt{f_c'} b_w d$, but the Code (11.3.2.1) permits the use of the following less conservative value:

$$V_c = \left(1.9\sqrt{f_c'} + 2500\rho_w \frac{V_u d}{M_u}\right) b_w d \leq 3.5\sqrt{f_c'} b_w d \qquad \text{(ACI Equation 11-6)}$$

As previously mentioned, M_u is the moment occurring simultaneously with V_u at the section in question. The value of $V_u d/M_u$ must not be taken greater than 1.0 in calculating V_c says the Code.

2. When shear reinforcing is required, the Code states that the amount provided must fall between certain clearly specified lower and upper limits. If the amount of reinforcing is too low, it may yield or even snap immediately after the formation of an inclined crack. As soon as a diagonal crack develops, the tension previously carried by the concrete is transferred to the web reinforcing. To prevent the stirrups (or other web reinforcing) from snapping at that time, their area is limited to the minimum value provided at the end of the next paragraph. The maximum value is presented in subsection 3 below.

In Section 11.5.5 the Code provides the least amount of web reinforcing permitted for all prestressed and nonprestressed concrete flexural members (other than the exceptions previously mentioned) for any sections of the beam

Placement of reinforcing bars for hemispherical dome of nuclear power plant reinforced concrete containment structure. (Courtesy of Bethlehem Steel Corporation.)

where V_u is greater than $\phi V_c/2$. The purpose of a minimum amount of web reinforcing is to provide an ultimate shear strength equivalent to a concrete shear strength of 50 psi. If such a value were available for a web width b_w and a length of beam equal to the stirrup spacing s, we could say

$$50 b_w s = f_y A_v$$

$$\min A_v = \frac{50 b_w s}{f_y} \qquad \text{(ACI Equation 11-14)}$$

This expression (ACI 11.5.5.3) provides the minimum area, which is to be used as long as the factored torsional moment T_u does not exceed $\phi(0.5\sqrt{f'_c}\,\Sigma x^2 y)$, the computation of which is described in Chapter 14.

Although you may not feel that the use of such minimum shear reinforcing is necessary, studies of earthquake damage in recent years have shown very large amounts of shear damage occurring in reinforced concrete structures, and it is felt that the use of this minimum value will greatly improve the resistance of such structures to seismic forces. Actually, a good many designers feel that the minimum area of web reinforcing should be used throughout beams and not just where V_u is greater than $\phi V_c/2$.

This requirement for a minimum amount of shear reinforcing may be waived if tests have been conducted that show that the required bending and shear strengths can be met without the shear reinforcing. (ACI 11.5.5.2)

3. As previously described, stirrups cannot resist appreciable shear unless they are crossed by an inclined crack. Thus to make sure that each 45° crack is intercepted by at least one stirrup, the maximum spacing of vertical stirrups permitted by the Code (11.5.4.1) is the lesser of $d/2$ or 24 in., provided that V_s is not greater than $4\sqrt{f_c'}\,b_w d$. Should V_s exceed this value, the maximum center-to-center spacing of stirrups is reduced to $d/4 \leq 12$ in. (Code 11.5.4.3). This closer spacing will lead to narrower inclined cracks.

Another advantage of limiting maximum spacing values for stirrups is that closely spaced stirrups will hold the longitudinal bars in the beam. They reduce the chance that the steel may tear or buckle through the concrete cover or possibly slip on the concrete.

Under no circumstances may V_s be allowed to exceed $8\sqrt{f_c'}\,b_w d$ (Code 11.5.6.8). The shear strength of a beam cannot be increased indefinitely by adding more and more shear reinforcing because the concrete will eventually disintegrate no matter how much shear reinforcing is added. The reader can understand the presence of an upper limit if he or she will think for a little while about the concrete above the crack. The greater the shear in the member that is transferred by the shear reinforcing to the concrete above, the greater will be the chance of a combination shear and compression failure of that concrete.

4. Section 11.1.2 of the Code states that the values of $\sqrt{f_c'}$ used for the design of web reinforcing may not exceed 100 psi except for certain cases listed in Section 11.1.2.1. In this section, permission is given to use a larger value if the minimum web reinforcing is equal to $f_c'/5000$ (but not more than 3) times the areas required in ACI Section 11.5.5.3 to 11.5.5.5. Members meeting these requirements for extra shear reinforcing have sufficient post-crack capacities to prevent diagonal tension failures.

5. Section 12.13 of the Code provides requirements about dimensions, embedment lengths, and so forth. For stirrups to develop their design strengths. they must be adequately anchored. Stirrups may be crossed by diagonal tension cracks at various points along their depths. Since these cracks may cross very close to the tension or compression edges of the members, the stirrups must be able to develop their yield strengths along the full extent of their lengths. It can then be seen why they should be bent around longitudinal bars of greater diameters than their own and extended beyond by adequate development lengths. Should there be compression reinforcing, the hooking of the stirrups around them helps to prevent them from buckling.

Stirrups should be carried as close as possible to the compression and tension faces of beams as the specified cover and longitudinal reinforcing will permit. Sections 7.1 and 12.13 of the Code specify the following stirrup details:

a. Stirrups with 90° bends and $6d_b$ extensions at their free ends may be used for #5 and smaller bars as shown in Figure 7.7(a). Tests have shown that 90° bends with $6d_b$ extensions should not be used for #6 or larger bars (unless f_y is 40,000 psi or less) because they tend to "pop out" under high loads.

b. If f_y is greater than 40,000 psi, #6, #7, and #8 bars with 90° bends may be used if the extensions are $12d_b$ [see Figure 7.7(b)].

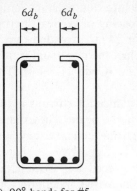

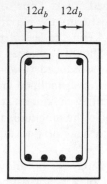

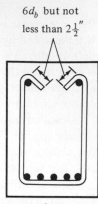

(a) 90° bends for #5 and smaller stirrups (also for #6, #7, and #8 stirrups with $f_y \leqslant 40,000$ psi

(b) 90° bends for #6, #7, and #8 stirrups with $f_y > 40,000$ psi

(c) 135° bends for #8 and smaller stirrups

Figure 7.7 Stirrup details.

c. Stirrups with 135° bends and $6d_b$ extensions may be used for #8 and smaller bars as shown in Figure 7.7(c).

6. When a beam reaction causes compression in the end of a member in the same direction as the external shear, the shearing strength of that part of the member is increased. Tests of such reinforced concrete members have shown that in general as long as a gradually varying shear is present (as with a uniformly loaded member), the first crack will occur at a distance d from the face of the support. It is therefore permissible, according to the Code (11.1.3.1), to decrease somewhat the calculated shearing force for a distance d from the face of the support. This is done by using a V_u in that range equal to the calculated V_u at a distance d from the face of the support. Should a concentrated load be applied in this region, no such shear reduction is permitted. Such loads will be transmitted directly to the support above the 45° cracks, with the result that we are not permitted a reduction in the end shear for design purposes.

Should the reaction tend to produce tension in this zone, no shear stress reduction is permitted because tests have shown that cracking may occur at the face of the support or even inside it. Figure 7.8 shows two cases where the end shear reduction is not permitted. In the situation shown in Figure 7.8(a) the critical section will be at the face of the support. In Figure 7.8(b) an I-shaped section is shown with the load applied to its tension flange. The loads have to be transferred across the inclined crack before they reach the support. Another crack problem like this one occurs in retaining wall footings and is discussed in Section 12.10 of this text.

7. Various tests of reinforced concrete beams of normal proportions with sufficient web reinforcing have shown that shearing forces have no significant

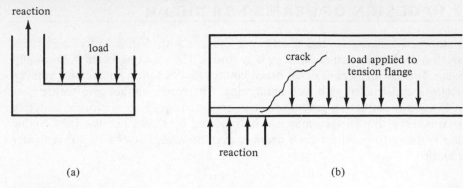

reaction

load

(a)

crack

load applied to
tension flange

reaction

(b)

Figure 7.8

effect on the flexural capacities of the beams. Experiments with deep beams, however, show that large shears will often keep those members from developing their full flexural capacities. As a result, the Code requirements given in the preceding paragraphs are not applicable to beams whose clear spans divided by their effective depths are less than five and where the loads are applied at the tops or compression faces. Members falling into this class include deep beams, short cantilevers, and corbels. Corbels are brackets that project from the sides of columns and are used to support beams and girders, as shown in Figure 7.9. They are quite commonly used in precast construction. Special web reinforcing provisions are made for such members in Section 11.8 of the Code and are considered in Section 7.12 of this chapter.

8. Section 8.11.8 of the ACI Code permits a shear of $1.1V_c$ for the ribs of joist construction as where we have closely spaced T beams with tapered webs. For the 10% increase in V_c the joist proportions must meet the provisions of ACI Section 8.11. In that section it is stated that the ribs must be no less than 4 in. wide, must have depths not more than 3.5 times the minimum width of the ribs, and may not have clear spacings between the ribs greater than 30 in.

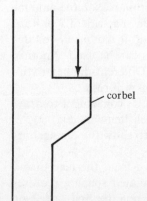

corbel

Figure 7.9

7.10 DESIGN OF BEAMS FOR SHEAR

Example 7.1 illustrates the selection of a beam with a sufficiently large cross section so that no web reinforcing is required. The resulting beam is unusually large. It is normally considered much better practice to use appreciably smaller sections constructed with web reinforcing. The reader should also realize that it is good construction practice to use some stirrups in all reinforced concrete beams (even though they may not be required by shear) because they enable the workers to build for each beam a cage of steel that can be conveniently handled.

■ EXAMPLE 7.1

Determine the minimum cross section required for a rectangular beam from a shear standpoint so that no web reinforcing is required by the ACI Code if $V_u = 38$ k and $f'_c = 4000$ psi. Use the conservative value of $V_c = 2\sqrt{f'_c}\,b_w d$.

SOLUTION

Shear strength provided by concrete is determined by the equation

$$\phi V_c = (0.85)(2\sqrt{4000}\,b_w d) = 107.52 b_w d$$

But the ACI Code 11.5.5.1 states that a minimum area of shear reinforcement is to be provided if V_u exceeds $\frac{1}{2}\phi V_c$:

$$38{,}000 = (\tfrac{1}{2})(107.52 b_w d)$$

$$b_w d = 706.8 \text{ in.}$$

<u>Use 22″ × 35″ beam $(d = 32.5″)$</u>   ■

The design of web reinforcing is illustrated by Examples 7.2 through 7.6. Maximum vertical stirrup spacings have been given previously, whereas no comment has been made about minimum spacings. Stirrups must be spaced far enough apart to permit the aggregate to pass through, and, in addition, they must be reasonably few in number so as to keep within reason the amount of labor involved in placing them. Accordingly, minimum spacings of 3 or 4 in. are normally used. Usually #3 stirrups are assumed; and if the calculated design spacings are less than $d/4$, larger-diameter stirrups can be used. Another alternative is to use ⊔⊓⊔ stirrups instead of ⊔ stirrups. Different-diameter stirrups should not be used in the same beam, or confusion will result.

As is illustrated in Examples 7.3, 7.5, and 7.6, it is quite convenient to draw the V_u diagram and carefully label it with values of such items as ϕV_c, $\phi V_c/2$, and V_u at a distance d from the face of the support and to show the dimensions involved.

Some designers place their first stirrup a distance d from the face of the support, while others place it one-half of the end-calculated spacing requirement from the face. The more common practice is to put the first one 2 or 3 in. from the support.

From a practical view, stirrups are usually spaced with center-to-center dimensions which are multiples of 3 or 4 in., to simplify the fieldwork. Though this procedure may require an additional stirrup or two, total costs should be less because of reduced labor costs. A common field procedure is to place chalk marks at 2-ft intervals on the forms and to place the stirrups by eye in between those marks. This practice is combined with a somewhat violent placing of the heavy concrete in the forms followed by vigorous vibration. These field practices should clearly show the student that it is foolish to specify odd theoretical stirrup spacings such as 4 @ $6\frac{7}{16}$ in. and 6 @ $5\frac{3}{8}$ in. because such careful positioning will not be achieved in the actual members. Thus the designer will normally specify stirrup spacings in multiples of whole inches and perhaps in multiples of 3 or 4 in.

With available computer programs it is easily possible to obtain theoretical arrangements of stirrups with which the least total amounts of shear reinforcing will be required. The use of such programs is certainly useful to the designer, but he or she needs to take the resulting values and revise them into simple economical patterns with simple spacing arrangements—as in multiples of 3 in., for example.

A summary of the steps required to design vertical stirrups is presented in Table 7.1. For each step the applicable section number of the Code is provided. The author has found this to be a very useful table for students to refer to while designing stirrups.

TABLE 7.1 SUMMARY OF STEPS INVOLVED IN VERTICAL
STIRRUP DESIGN

Is shear reinforcing necessary?

1. Draw V_u diagram.
2. Calculate V_u at a distance d from the support (with certain exceptions).   11.1.3.1 and
   Commentary (R11.1.3)
3. Calculate $\phi V_c = 2\phi\sqrt{f'_c}\,b_w d$ (or use the alternate method).   11.3.1.1
   11.3.2.1
4. Stirrups are needed if $V_u > \frac{1}{2}\phi V_c$ (with some exceptions for slabs, footings, shallow members, and joists).   11.5.5.1

Design of stirrups

1. Calculate theoretical stirrup spacing $s = A_v f_y d/V_s$ where $V_s = (V_u - \phi V_c)/\phi$.   11.5.6.2
2. Determine maximum spacing to provide minimum area of shear reinforcement $s = A_v f_y/50\,b_w$.   11.5.5.3
3. Compute maximum spacing: $d/2 \le 24$ in., if $V_s \le 4\sqrt{f'_c}\,b_w d$.   11.5.4.1
4. Compute maximum spacing: $d/4 \le 12$ in., if $V_s > 4\sqrt{f'_c}\,b_w d$.   11.5.4.3
5. V_s may not be $> 8\sqrt{f'_c}\,b_w d$.   11.5.6.8
6. Minimum practical spacing = approximately 3 or 4 in.

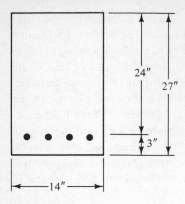

Figure 7.10

■ EXAMPLE 7.2

The beam shown in Figure 7.10 was selected using $f_y = 60,000$ psi and $f'_c = 3000$ psi. Determine the theoretical spacing of #3 $\sqcup$ stirrups for each of the following shears:

(a) $V_u = 14,000$ lb
(b) $V_u = 55,000$ lb
(c) $V_u = 70,000$ lb
(d) $V_u = 160,000$ lb

SOLUTION

(a) $V_u = 14,000$ lb

$$\phi V_c = (0.85)(2\sqrt{3000})(14)(24) = 31,286 \text{ lb}$$

$$\tfrac{1}{2}\phi V_c = 15,643 \text{ lb} > 14,000 \text{ lb}$$

$\therefore$ Stirrups not required

(b) $V_u = 55,000$ lb
Stirrups needed because $V_u > \tfrac{1}{2}\phi V_c$.
Theoretical spacing:

$$\phi V_c + \phi V_s = V_u$$

$$V_s = \frac{V_u - \phi V_c}{\phi} = \frac{55,000 - 31,286}{0.85} = 27,899 \text{ lb}$$

$$s = \frac{A_v f_y d}{V_s} = \frac{(2)(0.11)(60,000)(24)}{27,899} = 11.36'' \leftarrow$$

Maximum spacing to provide minimum A_v:

$$s = \frac{A_v f_y}{50 b_w} = \frac{(2)(0.11)(60,000)}{(50)(14)} = 18.86''$$

$$V_s = 27,899 < (4)(\sqrt{3000})(14)(24) = 73,614 \text{ lb}$$

$$\therefore \text{ Maximum } s = \frac{d}{2} = 12'' \qquad \underline{s = 11.36 \text{ in.}}$$

(c) $V_u = 70,000$ lb
Theoretical spacing:

$$V_s = \frac{V_u - \phi V_c}{\phi} = \frac{70,000 - 31,286}{0.85} = 45,546 \text{ lb}$$

$$s = \frac{A_v f_y d}{V_s} = \frac{(2)(0.11)(60,000)(24)}{45,546} = 6.96'' \leftarrow$$

Maximum spacing to provide minimum A_v:

$$s = \frac{A_v f_y}{50 b_w} = \frac{(2)(0.11)(60,000)}{(50)(14)} = 18.86''$$

$$V_s = 45,546 < (4)(\sqrt{3000})(14)(24) = 73,614 \text{ lb}$$

$$\therefore \text{ Maximum } s = \frac{d}{2} = 12'' \qquad \underline{s = 6.96 \text{ in.}}$$

(d) $V_u = 160,000$ lb

$$V_s = \frac{160,000 - 31,286}{0.85} = 151,428 \text{ lb}$$

$$151,428 \text{ lb} > (8)(\sqrt{3000})(14)(24) = 147,228 \text{ lb}$$

V_s may not be taken $> 8\sqrt{f_c'} b_w d$ $\qquad \therefore$ Need larger beam ■

■ **EXAMPLE 7.3**
Select #3 ⊔ stirrups for the beam shown in Figure 7.11, for which $D = 4$ k/ft and $L = 6$ k/ft. $f_c' = 4000$ psi and $f_y = 60,000$ psi.

SOLUTION

V_u at left end $= 7(1.4 \times 4 + 1.7 \times 6) = 110.6 \text{ k} = 110,600 \text{ lb}$

V_u at a distance d from face of support $= \left(\dfrac{84 - 22.5}{84}\right)(110,600) = 80,975 \text{ lb}$

$\phi V_c = \phi 2 \sqrt{f_c'} b_w d = (0.85)(2)(\sqrt{4000})(15)(22.5) = 36,287 \text{ lb}$

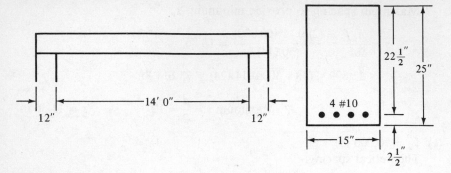

Figure 7.11

These values are shown in Figure 7.12.

$$V_u = \phi V_c + \phi V_s$$

$$\phi V_s = V_u - \phi V_c = 80{,}975 - 36{,}287 = 44{,}688 \text{ lb}$$

$$V_s = \frac{44{,}688}{0.85} = 52{,}574 \text{ lb}$$

Maximum spacing of stirrups $= d/2 = 11.25''$, since V_s is $< 4\sqrt{f'_c}b_w d = 85{,}382$ lb. Maximum theoretical spacing at left end:

$$s = \frac{A_v f_y d}{V_s} = \frac{(2)(0.11)(60{,}000)(22.5)}{52{,}574} = 5.65''$$

Maximum spacing to provide minimum A_v of stirrups:

$$s = \frac{A_v f_y}{50 b_w} = \frac{(2)(0.11)(60{,}000)}{(50)(15)} = 17.6''$$

For convenience, theoretical spacings are calculated at different points along the span and are listed in the following table:

Distance from face of support	V_u	$V_s = \dfrac{V_u - \phi V_c}{\phi}$	Theoretical $s = \dfrac{A_v f_y d}{V_s}$
0 to $d = 1.875'$	80,975 lb	52,574 lb	5.65"
2'	79,000	50,251	5.91"
3'	63,200	31,662	9.38"
4'	47,400	13,074	Maximum 11.25"

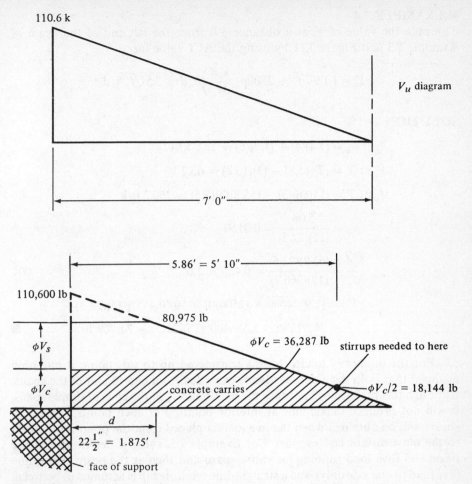

Figure 7.12

Spacings selected:

$$1 \ @ \ 3 \text{ in.} = \ \ 3 \text{ in.}$$
$$6 \ @ \ 6 \text{ in.} = 36 \text{ in.}$$
$$\underline{4 \ @ \ 9 \text{ in.} = 36 \text{ in.}}$$
$$6 \text{ ft } 3 \text{ in.} \qquad \underline{\text{symmetric about } \mathcal{C}} \quad \blacksquare$$

As previously mentioned, it is a good practice to space stirrups at multiples of 3 or 4 in. on center. As an illustration it is quite reasonable to select for Example 7.3 the following spacings: 1 @ 3 in., 6 @ 6 in., and 4 @ 9 in.

In Example 7.4, which follows, the value of V_c for the beam of Example 7.3 is computed by the alternate method of Section 11.3.2.1 of the Code.

■ **EXAMPLE 7.4**

Compute the value of V_c at a distance 3 ft from the left end of the beam of Example 7.3 and Figure 7.11 by using the ACI value for

$$V_c = \left(1.9\sqrt{f'_c} + 2500\rho_w \frac{V_u d}{M_u}\right)b_w d \leq 3.5\sqrt{f'_c}b_w d$$

SOLUTION

$$w_u = (1.4)(4) + (1.7)(6) = 15.8 \text{ k/ft}$$

$$V_u \text{ at } 3' = (7)(15.8) - (3)(15.8) = 63.2 \text{ k}$$

$$M_u \text{ at } 3' = (110.6)(3) - (15.8)(3)(1.5) = 260.7 \text{ ft-k}$$

$$\rho_w = \frac{5.06}{(15)(22.5)} = 0.0150$$

$$\frac{V_u d}{M_u} = \frac{(63.2)(22.5)}{(12)(260.7)} = 0.455 < 1.0 \qquad \underline{\text{OK}}$$

$$V_c = [1.9\sqrt{4000} + (2500)(0.0150)(0.455)](15)(22.5)$$

$$= \underline{46,315} \text{ lb} < 3.5\sqrt{4000}(15)(22.5) = 74,709 \text{ lb} \qquad ■$$

For the uniformly loaded beams considered up to this point, it has been assumed that both dead and live loads extended from end to end of the spans. Although this practice will produce the maximum V_u at the ends of simple spans, it will not produce maximums at interior points. For such points, maximum shears will be obtained when the live load is placed from the point in question to the most distant end support. For Example 7.5, shear is determined at the beam end (live load running for entire span) and then at the beam center line (live load to one side only), and a straight-line relationship is assumed in between. Though the ACI does not comment on the variable positioning of live load to produce maximum shears, such a practice is certainly prudent.

■ **EXAMPLE 7.5**

Select #3 ⊔ stirrups for the beam of Example 7.3, assuming the live load is placed to produce maximum shear at beam ℄, although this is not required by the Code.

SOLUTION

Maximum V_u at left end $= (7)(1.4 \times 4 + 1.7 \times 6) = 110.6 \text{ k} = 110,600 \text{ lb}$

For maximum V_u at ℄, the live load is placed as shown in Figure 7.13

$$V_u \text{ at } ℄ = 57.05 - (7)(1.4 \times 4) = 17.85 \text{ k} = 17,850 \text{ lb}$$

$$V_c = 2\sqrt{4000}(15)(22.5) = 42,691 \text{ lb}$$

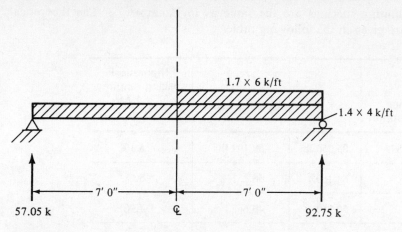

1.7 × 6 k/ft

1.4 × 4 k/ft

7' 0"

7' 0"

57.05 k

92.75 k

Figure 7.13

V_u at a distance d from face of support = 85,756 lb as determined by proportions from Figure 7.14:

$$V_u = \phi V_c + \phi V_s$$

$$\phi V_s = V_u - \phi V_c = 85,756 - (0.85)(42,691) = 49,470 \text{ lb at left end}$$

$$V_s = \frac{49,470}{0.85} = 58,200 \text{ lb}$$

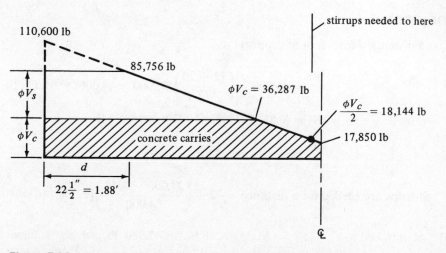

stirrups needed to here

110,600 lb

85,756 lb

ϕV_s

$\phi V_c = 36,287$ lb

$\dfrac{\phi V_C}{2} = 18,144$ lb

ϕV_c

concrete carries

17,850 lb

d

$22\frac{1}{2}'' = 1.88'$

Figure 7.14

The limiting spacings are the same as in Example 7.3. The theoretical spacings are given in the following table:

Distance face of support	V_u	$V_s = \dfrac{V_u - \phi V_c}{\phi}$	Theoretical spacing required $s = \dfrac{A_v f_y d}{V_s}$
0 to 1.875	85,756 lb	58,199 lb	5.10″
2′	84,100	56,251	5.28″
3′	70,850	40,662	7.30″
4′	57,600	25,074	Maximum 11.25″

Spacing selected:

$$
\begin{aligned}
1 \ @ \ 3 \text{ in.} &= 3 \text{ in.}\\
6 \ @ \ 6 \text{ in.} &= 36 \text{ in.}\\
5 \ @ \ 9 \text{ in.} &= 45 \text{ in.}\\
\hline
&= 7 \text{ ft } 0 \text{ in.} \qquad \text{symmetrical about } \mathcal{C} \ \blacksquare
\end{aligned}
$$

■ EXAMPLE 7.6

Select spacings for #3 ⊔ stirrups for a T beam with $b_w = 10$ in. and $d = 20$ in. for the V_u diagram shown in Figure 7.15, with $f_y = 60,000$ psi and $f_c' = 3000$ psi.

SOLUTION

V_u at a distance d from face of support

$$= 44,000 + \left(\frac{72-20}{72}\right)(68,000 - 44,000) = 61,333 \text{ lb}$$

$$\phi V_c = (0.85)(2\sqrt{3000})(10)(20) = 18,623 \text{ lb}$$

$$\frac{\phi V_c}{2} = \frac{18.623}{2} = 9311 \text{ lb}$$

Stirrups are needed for a distance $= 72 + \left(\dfrac{24,000 - 9311}{24,000}\right)(72) = 116''$

ϕV_s at left end $= V_u - \phi V_c = 61,333 - 18,623 = 42,710$ lb, which is larger than $\phi 4\sqrt{f_c'}b_w d = (0.85)(4\sqrt{3000}(10)(20) = 37,245$ lb but less than $\phi 8\sqrt{f_c'}b_w d$. Therefore the maximum spacing of stirrups in that range is $d/4 = 5''$. The point

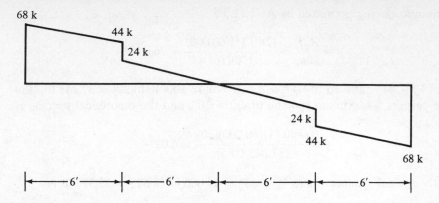

Figure 7.15

where $V_u - \phi V_c$ is equal to 37,245 is 36" from the left end, as shown in Figure 7.16.

Theoretical spacing of #3 stirrups where $V_u - \phi V_c$ is 42,710 lb is as follows:

$$V_s = \frac{42,710}{0.85} = 50,247 \text{ lb}$$

$$s = \frac{A_v f_y d}{V_s} = \frac{(2)(0.11)(60,000)(20)}{50,247} = 5.25"$$

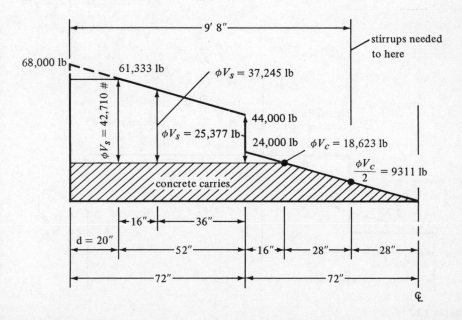

Figure 7.16

Maximum spacing permitted by ACI 11.5.5.3:

$$s = \frac{A_v f_y}{50 b_w} = \frac{(2)(0.11)(60,000)}{(50)(10)} = 26.4''$$

After $V_u - \phi V_c$ falls to $\phi 4\sqrt{f_c'} b_w d = (0.85)(4\sqrt{3000})(10)(20) = 37,245$ lb, the Code permits a maximum spacing of $d/2 = 10''$, and the theoretical spacing is

$$s = \frac{(2)(0.11)(60,000)(20)}{(37,245/0.85)} = 6.02''$$

The theoretical spacing where $V_u - \phi V_c = 44,000 - 18,623 = 25,377$ lb is

$$s = \frac{(2)(0.11)(60,000)(20)}{(25,377/0.85)} = 8.84''$$

A summary of the results of the preceding calculations is shown in Figure 7.17, where the solid dark line represents the maximum stirrup spacings permitted by the Code and the dash line represents the calculated theoretical spacings required for $V_u - \phi V_c$.

From this information the author selected the following spacings:

$$
\begin{array}{rcl}
1 \ @ \ 2 \ \text{in.} & = & 2 \ \text{in.} \\
17 \ @ \ 4 \ \text{in.} & = & 68 \ \text{in.} \\
6 \ @ \ 8 \ \text{in.} & = & 48 \ \text{in.} \\
\hline
& & 9 \ \text{ft} \ 10 \ \text{in.} \qquad \text{symmetrical about } \mathcal{C} \ \blacksquare
\end{array}
$$

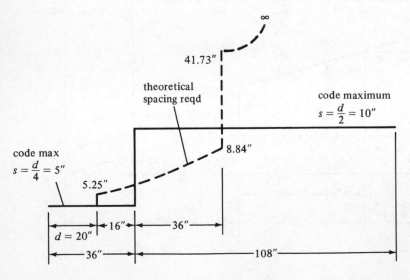

Figure 7.17

7.11 ECONOMICAL SPACING OF STIRRUPS

When stirrups are required in a reinforced concrete member, the Code specifies maximum permissible spacings varying from $d/4$ to $d/2$. On the other hand, it is usually thought that stirrup spacings less than $d/4$ are rather uneconomical. Many designers use a maximum of three different spacings in a beam. These are $d/4$, $d/3$, and $d/2$. It is easily possible to derive a value of ϕV_s for each size and style of stirrups for each of these spacings.[5]

Note that the number of stirrups is equal to d/n and that if we use spacings of $d/4$, $d/3$, and $d/2$ we can see that n equals 4, 3, or 2. Then the value ϕV_s can easily be calculated for any particular spacing, size, and style of stirrup. For instance, for #3 ⊔ stirrups spaced at $d/2$ with $\phi = 0.85$ and $f_y = 60$ ksi,

$$\phi V_s = \frac{\phi A_v f_y d}{s} = \frac{(0.85)(2 \times 0.11)(60)(d)}{d/2} = 22.4 \text{ k}$$

The values shown in Table 7.2 were computed in this way for 60-ksi stirrups.

For an example using this table, reference is made to the beam and V_u diagram of Example 7.3 which was shown in Figure 7.12 where 60-ksi #3 ⊔ stirrups were selected for a beam with a d of $22\frac{1}{2}$ in. For our closest spacing $d/4$, we can calculate $\phi V_c + 44.9 = 36.287 + 44.9 = 81.19$ k. Similar calculations are made for $d/3$ and $d/2$ spacings and we obtain, respectively, 69.99 k and 58.69 k. The shear diagram is repeated in Figure 7.18, and the preceding values are located on the diagram by proportions or by scaling.

From this information we can see that we can use $d/4$ for the first 2.60 ft, $d/3$ for the next 0.73 ft, and $d/2$ for the remaining 2.60 ft. Then the spacings are smoothed (preferably to multiples of 3 in.). Also for this particular beam we

TABLE 7.2 VALUES FOR 60-ksi STIRRUPS

s	ϕV_s for #3 ⊔ stirrups (kips)	ϕV_s for #4 ⊔ stirrups (kips)
$d/2$	22.4	40.8
$d/3$	33.7	61.2
$d/4$	44.9	81.6

---

[5] Neville B. B., ed., 1984, *Simplified Design Reinforced Concrete Buildings of Moderate Size and Height* (Skokie, Ill.: Portland Cement Association), pp. 3-12–3-16.

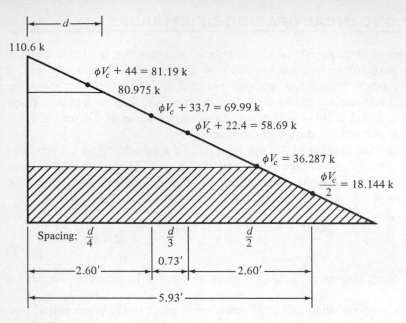

Figure 7.18

would probably use the $d/4$ spacing on through the 0.73-ft section and then use $d/2$ the rest of the required distance.

7.12 SHEAR FRICTION

If a crack occurs in a reinforced concrete member (whether caused by shear, flexure, shrinkage, etc.) and if the concrete pieces on opposite sides of the crack are prevented from moving apart, there will be a great deal of resistance to slipping along the crack due to the rough and irregular concrete surfaces. If reinforcement is provided across the crack to prevent relative displacement along the crack, shear will be resisted by friction between the faces, by resistance to shearing off of protruding portions of the concrete, and by dowel action of the reinforcing crossing the crack. The transfer of shear under these circumstances is called shear friction.

Shear friction failures are most likely to occur in short, deep members subject to high shears and small bending moments. These are the situations where the most nearly vertical cracks will occur. If moment and shear are both large, diagonal tension cracks will occur at rather large angles from the vertical. This is the situation that was discussed earlier in Sections 7.1 through 7.11.

A short cantilever member having a ratio of clear span to depth (a/d) of 1.0 or less is often called a *bracket* or *corbel*. One such member is shown in Figure 7.19(a). The shear-friction concept provides a convenient method for

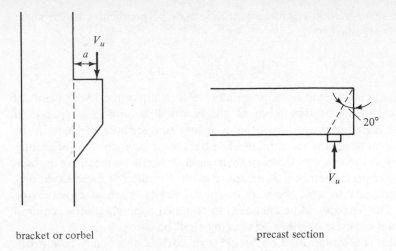

bracket or corbel   precast section

Figure 7.19 Possible shear friction failures.

designing for cases where diagonal tension design is not applicable. The most common locations where shear-friction design is used are for brackets, corbels, and for precast connections, but it may also be applied to the interfaces between concretes cast at different times, to the interfaces between concrete and steel sections, and so on.

When brackets or corbels or short, overhanging ends or precast connections support heavy concentrated loads, they are subject to possible shear-friction failures. The dashed lines in Figure 7.19 show the probable locations of these failures. It will be noted that for the end-bearing situations the cracks tend to occur at angles of about 20° from the direction of the force application.

Space is not taken in this chapter to provide an example of shear-friction design, but a few general remarks are presented concerning them. (In Section 11.13 of this text a numerical shear-friction example is presented in relation to the transfer of horizontal forces at the base of a column to a footing.) It is first assumed that a crack has occurred as shown by the dashed lines of Figure 7.19. As slip begins to occur along the cracked surface, the roughness of the concrete surfaces tends to cause the opposing faces of the concrete to separate.

As the concrete separates, it is resisted by the tensile reinforcement (A_{vf}) provided across the crack. It is assumed that this steel is stretched until it yields. (An opening of the crack of 0.01 in. would be more than sufficient to develop the yield strength of the bars.) The clamping force developed in the bars $A_{vf}f_y$ will provide a frictional resistance equal to $A_{vf}f_y\mu$, where μ is the coefficient of friction (values of which are provided for different situations in Section 11.7.4.3 of the Code).

Then the design shear strength of the member must at least equal the shear, to be taken as

$$V_u = \phi A_{vf} f_y \mu$$

And the shear-friction reinforcement across or perpendicular to the shear crack may be obtained by

$$A_{vf} = \frac{V_u}{\phi f_y \mu}$$

This reinforcing should be appropriately placed along the shear plane. If there is no calculated bending moment, the bars will be uniformly spaced. If there is a calculated moment, it will be necessary to distribute the bars in the flexural tension area of the shear plane. The bars must be anchored sufficiently on both sides of the crack to develop their yield strength by means of embedment, hooks, or other methods. Since space is often limited in these situations, it is often necessary to weld the bars to special devices such as crossbars or steel angles. The bars should be anchored in confined concrete (that is, column ties or external concrete or other reinforcing shall be used).

When beams are supported on brackets or corbels, there may be a problem with shrinkage and expansion of the beams, producing horizontal forces on the bracket or corbel. When such forces are present, the bearing plate under the concentrated load should be welded down to the tensile steel. Based on various tests, the ACI Code (11.9.3.4) says that the horizontal force used must be at least equal to $0.2V_u$.

The presence of direct tension across a cracked surface obviously reduces the shear-transfer strength. Thus direct compression will increase its strength. As a result, Section 11.7.7 of the Code permits the use of a permanent compressive load to increase the shear-friction clamping force. A typical corbel design and its reinforcing are shown in Figure 7.20.

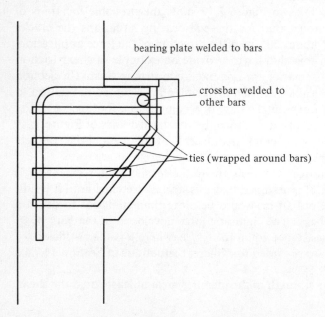

bearing plate welded to bars

crossbar welded to other bars

ties (wrapped around bars)

Figure 7.20

7.13 SHEAR STRENGTH OF MEMBERS SUBJECTED TO AXIAL FORCES

Reinforced concrete members subjected to shear forces can at the same time be loaded with axial compression or axial tension forces due to wind, earthquake, gravity loads applied to horizontal or inclined members, shrinkage in restrained members, and so on. These forces can affect the shear design of our members. Compressive loads tend to prevent cracks from developing. As a result, they provide members with larger compressive areas and thus greater shear strengths. On the other hand, tensile forces exaggerate cracks and reduce shear resistances because they will fairly well eliminate shear resistance due to aggregate interlock.

The design for shear in members with shear and axial loads is identical with the designs previously presented in this chapter except that V_c is revised up or down depending on whether the loads are compressive or tensile in nature. The revised V_c values are given in Section 11.3 of the ACI Code as follows:

Axial compression loads

$$V_c = 2\left(1 + \frac{N_u}{2000 A_g}\right)\sqrt{f_c'}\, b_w d \qquad \text{(ACI Equation 11-4)}$$

where N_u is the factored axial load and has a positive sign and where N_u/A_g is in psi.

Axial tension loads

$$V_c = 2\left(1 + \frac{N_u}{500 A_g}\right)\sqrt{f_c'}\, b_w d \qquad \text{(ACI Equation 11.9)}$$

where N_u is negative.

7.14 SHEAR DESIGN PROVISIONS FOR DEEP BEAMS

There are some special shear design provisions given in Section 11.8 of the Code for deep flexural members that are loaded on their top or compression faces. Such a member is shown in Figure 7.21(a). For shear considerations a

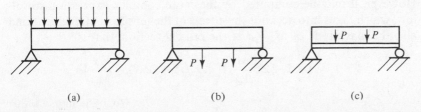

(a)   (b)   (c)

Figure 7.21

deep member is one for which ℓ_n/d is less than 5, where ℓ_n is defined as the clear span of a beam measured from face to face of supports. (In Section 10.7.1 of the Code, another definition of deep members is provided. There for flexural considerations, members with an overall depth-to-clear-span ratio greater than 4/5 for simple spans or 2/5 for continuous spans are said to be deep.)

If the loads are applied through the sides or the bottom (as where beams are framing into its sides or bottom) of the member as illustrated in Figure 7.21(b) and 7.21(c), the usual shear design provisions described earlier in this chapter are to be followed whether the member is deep or not.

The angles at which inclined cracks develop in deep flexural members (measured from the vertical) are usually much smaller than 45°—on some occasions being very nearly vertical. As a result, web reinforcing when needed has to be more closely spaced than for beams of regular depths. Furthermore, the web reinforcing needed is in the form of both horizontal and vertical reinforcing. These almost vertical cracks indicate that the principal tensile forces are primarily horizontal, and thus horizontal reinforcing is particularly effective in resisting them.

The Code (11.8.5) states that the shear to be used for design of deep members is to be calculated at a distance $0.15\ell_n$ from the face of the supports of uniformly loaded beams and at a distance $0.50a$ but not greater than d for beams supporting concentrated loads. The letter a, which represents not the depth of the compression stress block but rather the distance between the concentrated load and the face of the support, is called the *shear span*. The shear obtained as described herein is to be used for calculating the spacing of the shear reinforcing, and that spacing is to be used throughout the span.

The detailed provisions of the Code in its Section 11.8 relating to shear design for deep beams are summarized as follows:

1. The shear strength V_n of deep flexural members may not be greater than $8\sqrt{f'_c}b_w d$ when ℓ_n/d is less than 2; if it is between 2 and 5, V_n may not be greater than

$$V_n = \frac{2}{3}\left(10 + \frac{\ell_n}{d}\right)\sqrt{f'_c}b_w d \qquad \text{(ACI Equation 11-28)}$$

2. Unless a more detailed analysis is made, the shear strength of a deep beam may be taken as

$$V_c = 2\sqrt{f'_c}b_w d \qquad \text{(ACI Equation 11-29)}$$

However, it may be computed by the more complex expression to follow, which takes into account the effects of the tensile reinforcement and also the magnitude of $M_u/V_u d$ at the critical section; that is,

$$V_c = \left(3.5 - 2.5\frac{M_u}{V_u d}\right)\left(1.9\sqrt{f'_c} + 2500\rho_w \frac{V_u d}{M_u}\right)b_w d$$

$$\text{(ACI Equation 11-30)}$$

In the preceding expression for V_c, the first term within parentheses may not be greater than 2.5 and V_c may not be larger than $6\sqrt{f'_c}\,b_w d$.

3. When V_u is greater than ϕV_c, shear reinforcing is necessary and should be selected by the usual procedure, except that V_s is to be computed from the following expression:

$$V_s = \left[\frac{A_v}{s}\left(\frac{1+\dfrac{\ell_n}{d}}{12}\right) + \frac{A_{vh}}{s_2}\left(\frac{11-\dfrac{\ell_n}{d}}{12}\right)\right]f_y d \qquad \text{(ACI Equation 11-31)}$$

In this expression, A_v is the area of the shear reinforcing perpendicular to the flexural tensile reinforcing within a distance s, and A_{vh} is the area of the shear reinforcing parallel to the flexural reinforcing within a distance s_2. The term s_2 represents the spacing of shear or torsional reinforcing in a direction perpendicular to the longitudinal reinforcing—or the spacing of horizontal reinforcing in a wall.

4. The area of the shear reinforcing A_v may not be less than $0.0015b_w s$, and s may not be greater than $d/5$ or 18 in.
5. The area of the shear reinforcing A_{vh} may not be less than $0.0025b_w s_2$, and s_2 may not be greater than $d/3$ or 18 in.

7.15 INTRODUCTORY COMMENTS ON TORSION

Until recent years the safety factors required by design codes for proportioning members for bending and shear were very large and the resulting large members could almost always be depended upon to resist all but the very largest torsional moments. Today, however, with the smaller members selected using the strength design precedure this is no longer true and torsion needs to be considered much more frequently.

Torsion may be very significant for curved beams, spiral staircases, beams that have large loads applied laterally off center, and even in spandrel beams running between exterior building columns. These latter beams support the edges of floor slabs, floor beams, curtain walls, and facades, and they are loaded laterally on one side. Several situations where torsion can be a problem are shown in Figure 7.22.

When plain concrete members are subjected to pure torsion they will crack along 45° spiral lines when the resulting diagonal tension exceeds the design strength of the concrete. Although these diagonal tension stresses produced by twisting are very similar to those caused by shear, they will occur on all faces of a member. As a result, they add to the stresses caused by shear on one side of the beam and subtract from them on the other.

Reinforced concrete members subjected to large torsional forces may fail quite suddenly if they are not provided with torsional reinforcing. The addition of torsional reinforcing does not change the magnitude of the torsion that will

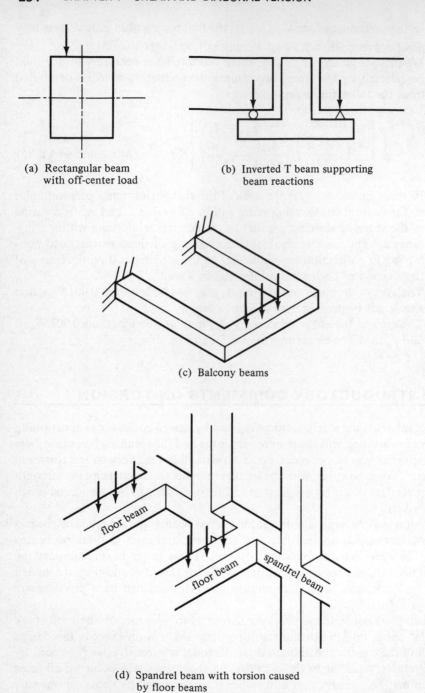

(a) Rectangular beam
with off-center load

(b) Inverted T beam supporting
beam reactions

(c) Balcony beams

(d) Spandrel beam with torsion caused
by floor beams

Figure 7.22 Some situations where torsion stresses may be significant.

cause diagonal cracks, but it does prevent the members from tearing apart. As a result, they will be able to resist substantial torsional moments without failure. Tests have shown that both longitudinal bars and closed stirrups or spirals are necessary to intercept the numerous diagonal tension cracks that occur on all surfaces of beams subject to appreciable torsional forces. There must be a longitudinal bar in each corner of the stirrups to resist the horizontal components of the diagonal tension caused by torsion.

Chapter 14 is devoted entirely to torsion in reinforced concrete members of various cross sections. If factored torsional moments are less than $\phi(0.5\sqrt{f'_c}\,\Sigma x^2 y)$, they will not appreciably reduce the shear or flexural strengths of the members. If higher, however, it will be necessary to consider the interaction of shear and flexural stresses. The $\Sigma x^2 y$ term is explained in Chapter 14.

7.16 COMPUTER EXAMPLE

In Example 7.7 the computer disk CONCRETE is used to determine the required stirrup spacing at one end of the beam of Example 7.3. Though the spacing is computed at only one point the user can immediately determine the spacing at any other point by revising the shear. The cursor is moved to the line entitled "Maximum Shear," a new value is inputted and the F5 key is pressed.

■ EXAMPLE 7.7
Determine the required spacing of #3 ⊔ stirrups at the left end of the beam of Example 7.3 using the enclosed computer disk.

SOLUTION

```
DESIGN OF SHEAR REINFORCING
---------------------------
Beam Width                    bw  (in)  = 15.0
Depth to Bars                 d   (in.) = 22.5
Maximum Shear                 Vu  (lb)  = 80,975
Stirrup Bar Number                        #3
How Many Vertical Prongs in Stirrup     = 2
Concrete Strength             f'c (psi) =  4000
Steel Yield Strength          fy  (psi) =  60,000
Normal Concrete

RESULTS
=======
Required theoretical spacing 5.6 in. stirrup spacing when Vu is
80,975
```
■

PROBLEMS

In Problems 7.1 to 7.3 compute ϕV_n for the sections shown if f_y of stirrups is 60 ksi and $f'_c = 3$ ksi.

Problem 7.1 (*Ans.* 64,762 lbs)

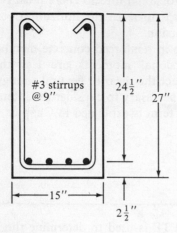

Problem 7.2

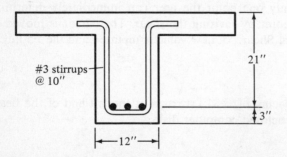

Problem 7.3 (*Ans.* 59,553 lbs)

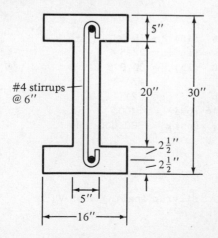

7.4   If $f'_c = 3$ ksi, $V_u = 40$ k, and $b = \frac{2}{3}d$, select rectangular beam section if no web reinforcing is used.

In problems 7.5 to 7.13 for the beams and loads given, select stirrup spacings if $f'_c = 3000$ psi and $f_y = 60,000$ psi. The dead loads shown include beam weights. Do not consider movement of live loads unless specifically requested. Assume $\#3 \sqcup$ stirrups.

Problem 7.5   (*One ans.* i $\#3 \sqcup$ @ 3 in., 10 @ 12 in.)

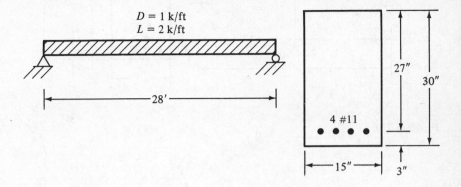

Problem 7.6

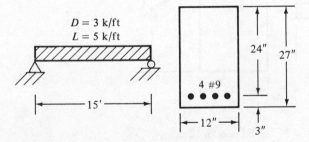

7.7   Repeat Problem 7.6 if live load positions are considered to cause maximum end shear and maximum ℄ shear. (*One ans.* 1 $\#3 \sqcup$ @ 3 in., 7 @ 6 in., 5 @ 9 in.)

Problem 7.8

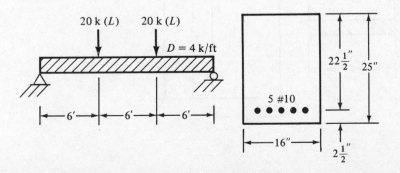

Problem 7.9 (*One ans.* 1 #3 ⊔ @ 3 in., 8 @ 12 in.)

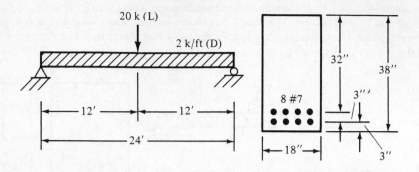

Problem 7.10

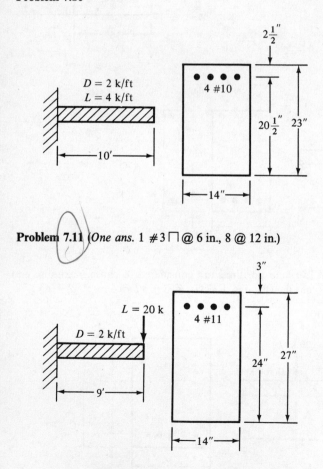

Problem 7.11 (*One ans.* 1 #3 ⊓ @ 6 in., 8 @ 12 in.)

Problem 7.12

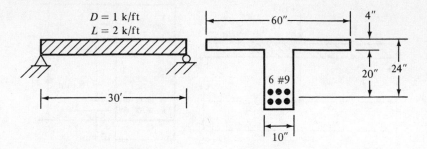

7.13 If the beam of Problem 7.8 has a factored axial compression load of 120 k in addition to other loads, calculate ϕV_c and redesign the stirrups. (*One ans.* 1 #3 ⊔ @ 3 in., 6 @ 6 in., 4 @ 9 in.)

In Problems 7.14 to 7.15 repeat the problems given using the enclosed computer disk.

7.14 If V_u equals 56,400 lb, determine the theoretical spacing of #3 ⊔ stirrups for the beam of Problem 7.5.

7.15 If V_u equals 79,600 lb, determine the spacing of #4 ⊓ stirrups for the beam of Problem 7.10. (*Ans.* 7.8 in.)

7.16 Prepare a flow chart for the design of stirrups for rectangular, T or I beams.

PROBLEMS WITH SI UNITS
In Problems 7.17 to 7.20, for the beams and loads given, select stirrup spacings if $f'_c = 20.7$ MPa and $f_y = 275.8$ MPa. The dead loads shown include beam weights. Do not consider movement of live loads. Shear stress carried by concrete, V_c, shall not exceed $0.166\sqrt{f'_c}\,b_w d$.

Problem 7.17 (*One ans.* 1 #3 ⊔ @ 140 mm, 10 @ 280 mm)

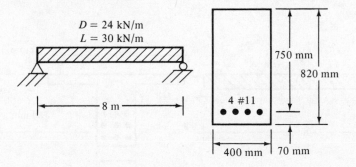

Problem 7.18

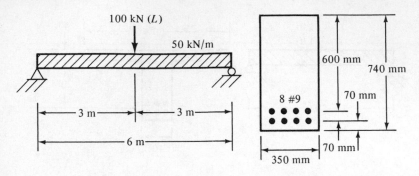

Problem 7.19 (*One ans.* 1 #3 ⊓ @ 120 mm, 11 @ 250 mm)

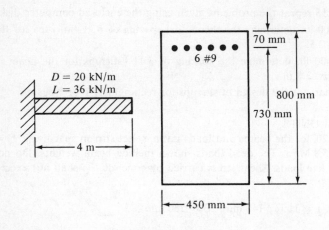

Problem 7.20

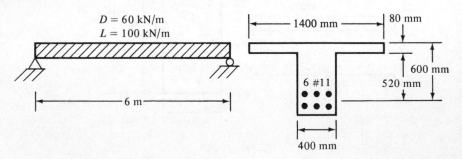

8

Introduction to Columns

8.1 GENERAL

An introductory discussion of reinforced concrete columns is presented in this chapter, with particular emphasis given to short, stocky columns subjected to small bending moments. Such columns are often said to be "axially loaded." In Chapter 9, short, stocky columns with large bending moments are discussed, while long or slender columns are considered in Chapter 10.

Concrete columns can be roughly divided into the following three categories:

Short compression blocks or pedestals. If the height of an upright compression member is less than three times its least lateral dimensions, it may be considered to be a pedestal. The ACI (2.1 and 10.15) states that a pedestal may be designed with unreinforced or plain concrete with a maximum permissible compressive stress equal to $0.85\phi f'_c$, where ϕ is 0.70. Should the compressive stress be greater than this value, the pedestal will have to be designed as a reinforced concrete column.

Short reinforced concrete columns. Should a reinforced concrete column fail due to initial material failure, it is classified as a short column. The load that it can support is controlled by the dimensions of the cross section and the strength of the materials of which it is constructed. We think of a short column as being a rather stocky member with little flexibility.

Long or slender reinforced concrete columns. As slenderness ratios are increased, bending deformations will increase as will the resulting

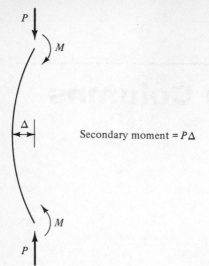

Figure 8.1 Secondary or $P\Delta$ moment.

secondary moments. If these moments appreciably weaken the member, it will be referred to as a *long* or *slender* column.

When a column is subjected to *primary moments* (those moments caused by applied loads, joint rotations, etc.), the axis of the member will deflect laterally, with the result that additional moments equal to the column load times the lateral deflection will be applied to the column. These moments are called *secondary moments* or *$P\Delta$ moments* and are illustrated in Figure 8.1.

A column that has large secondary moments is said to be a slender column, and it is necessary to size its cross section for the sum of both the primary and secondary moments. It is the intent of the ACI to permit columns to be designed as short columns if the secondary or $P\Delta$ effect does not reduce their strength by more than 5%. Effective slenderness ratios are described and evaluated in Chapter 10 and are used to classify columns as being short or slender. When the ratios are larger than certain values (depending on whether the columns are braced or unbraced laterally), they are classed as slender columns.

In 1970 the ACI Committee estimated that about 40% of all unbraced columns and about 90% of those braced against sidesway had their strengths reduced by 5% or less by $P\Delta$ effects and thus should be classified, as short columns.[1] These percentages are probably decreasing year by year, however, due to the increasing use of slenderer columns designed by the strength method using stronger materials and with a better understanding of column buckling behavior.

---

[1] American Concrete Institute, 1972, *Notes on ACI 318-71 Building Code with Design Applications*, Skokie, Illinois, page 10-2.

8.2 TYPES OF COLUMNS

A plain concrete column can support very little load, but its load-carrying capacity will be greatly increased if longitudinal bars are added. Further substantial strength increases may be made by providing lateral restraint for these longitudinal bars. Under compressive loads, columns tend not only to shorten lengthwise but also to expand laterally due to the Poisson effect. The capacity of such members can be greatly increased by providing lateral restraint in the form of closely spaced closed ties or helical spirals wrapped around the longitudinal reinforcing.

Reinforced concrete columns are referred to as *tied* or *spiral* columns depending on the method used for laterally bracing or holding the bars in place. If the column has a series of closed ties, as shown in Figure 8.2(a), it is referred to as a *tied column*. These ties are effective in increasing the column strength. They prevent the longitudinal bars from being displaced during construction

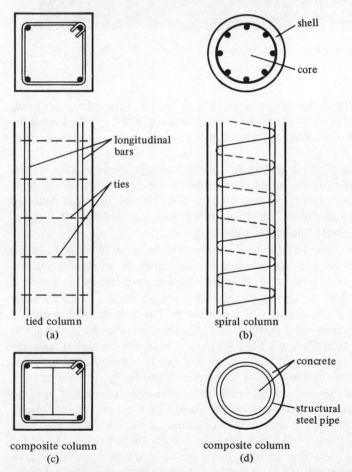

Figure 8.2 Types of columns.

Column forms. (Courtesy of Economy Forms Corporation.)

and they resist the tendency of the same bars to buckle outwards under load, which would cause the outer concrete cover to break or spall off. Tied columns are ordinarily square or rectangular, but they can be octagonal, round, L-shaped, and so forth.

The square and rectangular shapes are commonly used because of the simplicity of constructing the forms. Sometimes, however, when they are used in open spaces, circular shapes are very attractive. The forms for round columns are often made from cardboard or plastic tubes, which are peeled off and discarded once the concrete has sufficiently hardened.

If a continuous helical spiral made from bars or heavy wire is wrapped around the longitudinal bars, as shown in Figure 8.2(b), the column is referred to as a *spiral column*. Spirals are even more effective than ties in increasing a column's strength. The closely spaced spirals do a better job of holding the longitudinal bars in place, and they also confine the concrete inside and greatly increase its resistance to axial compression. As the concrete inside the spiral tends to spread out laterally under the compressive load, the spiral that restrains it is put into hoop tension, and the column will not fail until the spiral yields or breaks, permitting the bursting of the concrete inside. Spiral columns are normally round, but they also can be made into rectangular, octagonal, or other shapes. For such columns, circular arrangements of the bars are still used. Spirals, though adding to the resilience of columns, appreciably increase costs. As a result they are usually used only for large heavily loaded columns and for columns in seismic areas due to their considerable resistance to earthquake loadings. Spirals very effectively increase the ductility and toughness of columns, but they are several times as expensive as ties.

Composite columns, illustrated in Figures 8.2(c) and 8.2(d), are concrete columns that are reinforced longitudinally by structural steel shapes, which may or may not be surrounded by structural steel bars, or they may consist of structural steel tubing filled with concrete (commonly called *lally columns*).

8.3 AXIAL LOAD CAPACITY OF COLUMNS

In actual practice there are no perfect axially loaded columns, but a discussion of such members provides an excellent starting point for explaining the theory involved in designing real columns with their eccentric loads. Several basic ideas can be explained for purely axially loaded columns, and the strengths obtained provide upper theoretical limits that can be clearly verified with actual tests.

It has been known for several decades that the stresses in the concrete and the reinforcing bars of a column supporting a long-term load cannot be calculated with any degree of accuracy. You might think that such stresses could be determined by multiplying the strains by the appropriate moduli of elasticity. But this idea does not work too well practically because the modulus of elasticity of the concrete is changing during loading due to creep and shrinkage. So it can be seen that the parts of the load carried by the concrete and by the steel vary with the magnitude and duration of the loads. For instance, the larger the percentage of dead loads and the longer they are applied, the greater the creep in the concrete and the larger the percentage of load carried by the reinforcement.

Though stresses cannot be predicted in columns in the elastic range with any degree of accuracy, several decades of testing have shown that the ultimate strength of columns can be estimated very well. Furthermore, it has been shown that the proportions of live and dead loads, the length of loading, and other such items have little effect on the ultimate strength. It does not even matter whether the concrete or the steel approaches its ultimate strength first. If one of the two materials is stressed close to its ultimate strength, its large deformations will cause the stress to increase quicker in the other material.

For these reasons, only the ultimate strength of columns is considered here. At failure the theoretical ultimate strength or nominal strength of a short axially loaded column is quite accurately determined by the expression that follows, in which A_g is the gross concrete area and A_{st} is the total cross-sectional area of longitudinal reinforcement including bars and steel shapes:

$$P_n = 0.85f'_c(A_g - A_{st}) + f_y A_{st}$$

Concrete that is placed in forms on a construction job, vibrated, and cured under field conditions is not the same as concrete carefully placed in test cylinders, thoroughly vibrated or compacted, and cured under ideal conditions. The 0.85 factor in the preceding expression is used to account for this difference.

8.4 FAILURE OF TIED AND SPIRAL COLUMNS

Should a short, tied column be loaded until it fails, parts of the shell or covering concrete will spall off and, unless the ties are quite closely spaced, the longitudinal bars will buckle almost immediately as their lateral support (the covering concrete) is gone. Such failures may often be quite sudden, and apparently they have occurred rather frequently in structures subjected to earthquake loadings.

When spiral columns are loaded to failure, the situation is quite different. The covering concrete or shell will spall off but the core will continue to stand, and if the spiral is closely spaced, the core will be able to resist an appreciable amount of additional load beyond the load that causes spalling. The closely spaced loops of the spiral together with the longitudinal bars form a cage that very effectively confines the concrete. As a result, the spalling off of the shell of a spiral column provides a warning that failure is going to occur if the load is further increased.

American practice is to neglect any excess capacity after the shell spalls off since it is felt that once the spalling occurs the column will no longer be useful—at least from the viewpoint of the occupants of the building. For this reason

Round spiral columns. (Courtesy of Economy Forms Corporation.)

the spiral is designed so that it is just a little stronger than the shell that is assumed to spall off. The spalling gives a warning of impending failure, and then the column will take a little more load before it fails. Designing the spiral so that it is just a little stronger than the shell does not increase the column's ultimate strength much, but it does result in a more gradual or ductile failure.

The strength of the shell is given by the following expression, where A_c is the area of the core, which is considered to have a diameter that extends from out to out of the spiral:

$$\text{Shell strength} = 0.85f'_c(A_g - A_c)$$

It can be shown, by considering the estimated hoop tension that is produced in spirals due to the lateral pressure from the core and by tests, that spiral steel is at least twice as effective in increasing the ultimate column capacity as is longitudinal steel.[2,3] Therefore, the strength of the spiral can be computed approximately by the following expression, in which ρ_s is the percentage of spiral steel:

$$\text{Spiral strength} = 2\rho_s A_c f_y$$

Equating these expressions and solving for the required percentage of spiral steel; we obtain

$$0.85f'_c(A_g - A_c) = 2\rho_s A_c f_y$$

$$\rho_s = 0.425 \frac{(A_g - A_c)f'_c}{A_c f_y} = 0.425 \frac{[(A_g/A_c) - 1]f'_c}{f_y}$$

To make the spiral a little stronger than the spalled concrete, the Code (10.9.3) specifies the minimum spiral percentage to be as follows:

$$\rho_s = 0.45 \frac{[(A_g/A_c) - 1]f'_c}{f_y} \qquad \text{(ACI Equation 10-5)}$$

Once the required percentage of spiral steel is determined, the spiral may be selected with the expression to follow, in which ρ_s is written in terms of the volume of the steel in one loop:

$$\rho_s = \frac{\text{volume of spiral in one loop}}{\text{volume of concrete core for a pitch } s}$$

$$= \frac{V_{\text{spiral}}}{V_{\text{core}}}$$

$$= \frac{a_s \pi(D_c - d_b)}{(\pi D_c^2/4)s} = \frac{4a_s(D_c - d_b)}{sD_c^2}$$

[2] Park, A., and Paulay, T., 1975, *Reinforced Concrete Structures* (New York: John Wiley & Sons), pp. 25, 119–121.

[3] Considere, A., 1902, "Compressive Resistance of Concrete Steel and Hooped Concrete, Part I," *Engineering Record*, December 20, pp. 581–583; "Part II," December 27, pp. 605–606.

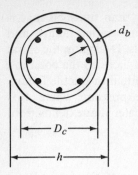

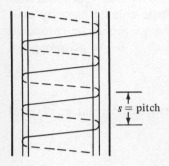

Figure 8.3

In this expression, D_c is the diameter of the core out to out of the spiral, a_s is the cross-sectional area of the spiral bar, and d_b is the diameter of the spiral bar. Here reference is made to Figure 8.3. The designer can assume a spiral diameter and solve for the pitch required. If the results do not seem reasonable, he or she can try another diameter. The pitch used must be within the limitations listed in the next section of this chapter. Actually, Table A.16 (see Appendix), which is based on this expression, permits the designer to select spirals directly.

8.5 CODE REQUIREMENTS FOR CAST-IN-PLACE COLUMNS

The ACI Code specifies quite a few limitations on the dimensions, reinforcing, lateral restraint, and other items pertaining to concrete columns. Some of the most important limitations are listed in the paragraphs to follow.

1. The percentage of longitudinal reinforcement may not be less than 1% of the gross cross-sectional area of a column (ACI Code 10.9.1). It is felt that if the amount of steel is less than 1%, there will be a distinct possibility of a sudden nonductile failure, as might occur in a plain column. The 1% minimum steel value will also lessen creep and shrinkage and provide some bending strength for the column. Actually the Code (10.8.4) does permit the use of less than 1% steel if the column has been made larger than is necessary to carry the loads because of architectural or other reasons. In other words, a column can be designed with 1% longitudinal steel to support the factored load and then more concrete can be added with no increase in reinforcing and no increase in calculated load-carrying capacity. In no circumstances, however, may the steel area be less than 0.005 times the area of the concrete actually provided.

2. The maximum percentage of steel may not be greater than 8% of the gross cross-sectional area of the column (ACI Code 10.9.1). This maximum value is given to prevent too much crowding of the bars. Practically it is rather difficult to fit more than 5% or 6% steel into the forms and still get the concrete down into the forms and around the bars. When the percentage of steel is high, the chances of having honeycomb in the concrete is decidedly increased. If this happens, there can be a substantial reduction in the column's load-carrying capacity. Usually the percentage of reinforcement should not exceed 4% when the bars are to be lap-spliced. It is to be remembered that if the percentage of steel is very high, the bars may be bundled.

3. The minimum numbers of longitudinal bars permissible for compression members (ACI 10.9.2) are as follows: 4 for bars within rectangular or circular ties, 3 for bars within triangular shaped ties, and 6 for bars enclosed within spirals.

4. The Code does not directly provide a minimum column cross-sectional area, but to provide the necessary cover outside of ties or spirals and to provide the necessary clearance between longitudinal bars from one face of the column to the other it is obvious that minimum widths or diameters of about 8 to 10 in. are necessary. To use as little rentable floor space as possible small columns are frequently desirable. In fact thin columns may often be enclosed or "hidden" in walls.

5. When tied columns are used, the ties shall not be less than #3, provided that the longitudinal bars are #10 or smaller. The minimum size is #4 for longitudinal bars larger than #10 and for bundled bars. The center-to-center spacing of ties shall not be more than 16 times the diameter of the longitudinal bars, 48 times the diameter of the ties, or the least lateral dimension of the column. The ties must be arranged so that every corner and alternate longitudinal bar have lateral support provided by the corner of a tie having an included angle not greater than 135°. No bars can be located a greater distance than 6 in. clear on either side from such a laterally supported bar. These requirements are given by the ACI Code in its Section 7.10.5. Figure 8.4 shows tie arrangements for several column cross sections. Some of the arrangements with interior ties such as the last one shown in the figure are rather expensive. Should longitudinal bars be arranged in a circle, round ties may be placed around them and the bars do not have to be individually tied or restrained otherwise (7.10.5.3). The ACI also states (7.10.3) that the requirements for lateral ties may be waived if tests and structural analysis show that the columns are sufficiently strong without them and that such construction is feasible.

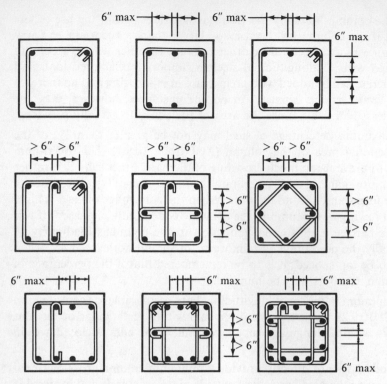

Figure 8.4 Typical tie arrangements.

There is little evidence available concerning the behavior of spliced bars and bundled bars. For this reason, Section R7.10.5 of the Commentary states that it is advisable to provide ties at each end of lap-spliced bars and provides recommendations concerning the placing of ties in the region of end-bearing splices and offset bent bars.

Ties should not be placed more than one-half a spacing above the top of a footing or slab and not more than one-half a spacing below the lowest reinforcing in a slab or drop panel (see Fig. 15.1). Where beams frame into a column from all four directions, the last tie may be below the lowest reinforcing in any of the beams.

6. The Code (7.10.4) states that the clear spacing of spirals may not be less than 1 in. or greater than 3 in. Should splices be necessary in spirals, they are to be provided by welding or by lapping the spiral bars or wires by the larger of 48 diameters of 12 in. Special spacer bars are used to hold the spirals in place and at the desired pitch until the concrete hardens. These spacers consist of vertical bars with small hooks. Spirals are supported by the spacers, not by the longitudinal bars. Section R7.10.4 of the ACI Commentary provides minimum numbers of spacers required for different-size columns.

8.6 SAFETY PROVISIONS FOR COLUMNS

The values of ϕ to be used for columns as specified in Section 9.3.2 of the Code are well below those used for flexure and shear (0.90 and 0.85, respectively). A value of 0.70 is specified for tied columns and 0.75 for spiral columns because of their greater toughness.

The failure of a column is generally a more severe matter than is the failure of a beam because a column generally supports a larger part of a structure than does a beam. In other words, if a column fails in a building, a larger part of the building will fall down than if a beam fails. This is particularly true for a lower-level column in a multistory building. As a result, lower ϕ values are desirable for columns.

There are other reasons for using lower ϕ values in columns. As an example, it is more difficult to do as good a job in placing the concrete for a column than it is for a beam. The reader can readily see the difficulty of getting concrete down into narrow column forms and between the longitudinal and lateral reinforcing. As a result, the quality of the resulting concrete columns is probably not as good as that of beams and slabs.

The failure strength of a beam is normally dependent on the yield stress of the tensile steel—a property that is quite accurately controlled in the steel mills. On the other hand, the failure strength of a column is closely related to the concrete's ultimate strength, a value that is quite variable. The length factors also drastically affect the strength of columns and thus make the use of lower ϕ factors necessary.

It seems impossible for a column to be perfectly axially loaded. Even if loads could be perfectly centered at one time, they would not stay in place. Furthermore, columns may be initially crooked or have other flaws with the result that lateral bending will occur. Wind and other lateral loads cause columns to bend, and the columns in rigid frame buildings are subjected to moments even when the frame is supporting gravity loads alone.

8.7 DESIGN FORMULAS

In the pages that follow, the letter e is used to represent the eccentricity of the load. The reader may not understand this term because he or she has analyzed a structure and has computed an axial load P_u and a bending moment M_u but no specific eccentricity e for a particular column. The term e represents the distance the axial load P_u would have to be off center of the column to produce M_u. Thus

$$P_u e = M_u$$

or

$$e = \frac{M_u}{P_u}$$

Royal Towers, Baltimore, Maryland. (Courtesy Simpson Timber Company)

Despite the facts mentioned in the preceding paragraph, there are many situations where there are no calculated moments for the columns of a structure. For many years the Code specified that such columns had to be designed for certain minimum moments even though no calculated moments were present. This was accomplished by requiring designers to assume certain minimum eccentricities for their column loads. These minimum values were 1 in. or 0.05h, whichever was larger, for spiral columns and 1 in. or 0.10h for tied columns (the term h represents the outside diameter of round columns or the total depth of square or rectangular columns). A moment equal to the axial load times the minimum eccentricity was used for design.

In today's Code, minimum eccentricities are not specified, but the same objective is accomplished by requiring that theoretical axial load capacities be multiplied by a factor called α, which is equal to 0.85 for spiral columns and 0.80 for tied columns. Thus as shown in Section 10.3.5 of the Code, the axial load capacity of columns may not be greater than the following values:

For spiral columns ($\phi = 0.75$)

$$P_u = \alpha\phi P_n = 0.85\phi[0.85f'_c(A_g - A_{st}) + f_y A_{st}] \qquad \text{(ACI Equation 10–1)}$$

For tied columns ($\phi = 0.70$)

$$P_u = \alpha\phi P_n = 0.80\phi[0.85f'_c(A_g - A_{st}) + f_y A_{st}] \qquad \text{(ACI Equation 10–2)}$$

It is to be clearly understood that the preceding expressions are to be used only when the moment is quite small or when there is no calculated moment.

The equations presented here are applicable only for situations where the moment is sufficiently small so that e is less than 0.10h for tied columns or less than $0.05h$ for spiral columns. Short columns can be completely designed with these expressions as long as the e values are under the limits described. Should the e values be greater than the limiting values and/or should the columns be classified as long ones, it will be necessary to use the procedures described in the next two chapters.

8.8 COMMENTS ON ECONOMICAL COLUMN DESIGN

Reinforcing bars are quite expensive, and thus the percentage of longitudinal reinforcing used in reinforced concrete columns is a major factor in their total costs. This means that under normal circumstances a small percentage of steel should be used (perhaps in the range of 1.5% to 3%). This can be accomplished by using larger column sizes and/or higher-strength concretes.

Higher-strength concretes can be more economically used in columns than in beams. Under ordinary loads, only 30% to 40% of a beam cross section is in compression, while the remaining 60% to 70% is in tension and thus assumed to be cracked. This means that if a high-strength concrete is used for a beam, 60% to 70% of it is wasted. For the usual column, however, the situation is quite different because a much larger percentage of its cross section is in compression. As a result, it is quite economical to use high-strength concretes for columns. Although some designers have used concretes with ultimate strengths as high as 19,000 psi (as at Two Union Square in Seattle) for column design with apparent economy, the use of 4000- to 6000-psi columns is the normal rule when higher strengths are specified for columns.

In general, tied columns are more economical than spiral columns particularly if square or rectangular cross sections are to be used. Of course, spiral columns, high-strength concretes, and high percentages of steel save floor space.

As few different column sizes as possible should be used throughout a building. In this regard it is completely uneconomical to vary a column size from floor to floor to satisfy the different loads it must support. This means that the designer may select a column size for the top floor of a multistory building (using as small a percentage of steel as possible) and then continue to use that same size vertically for as many stories as possible by increasing the steel percentage floor by floor as required. Furthermore, it is desirable to use the same column size as much as possible on each floor level. This consistency of sizes will provide appreciable savings in labor costs.

The usual practice for the columns of multistory reinforced concrete buildings is to use one-story-length vertical bars tied together in preassembled cages. This is the preferred procedure when the bars are #11 or smaller, where all the bars can be spliced at one location just above the floor line. For columns where staggered splice locations are required (as for larger-size bars), the number of splices can be reduced by using preassembled two-story cages of reinforcing.

In fairly short buildings the floor slabs are often rather thin, and thus deflections may be a problem. As a result, rather short spans and thus close column spacings may be used. As buildings become taller, the floor slabs will probably be thicker to help provide lateral stability. For such buildings, slab deflections will not be as much of a problem, and the columns may be spaced further apart.

Even though the columns in tall buildings may be spaced at fairly large intervals, they still will occupy expensive floor space. For this reason, many designers try to place many of their columns on the building perimeters so they will not use up the valuable interior space. In addition, the omission of interior columns provides more flexibility for the users for placement of partitions and also makes large open spaces available.

8.9 DESIGN OF AXIALLY LOADED COLUMNS

As a brief introduction to columns, the design of three axially loaded short columns is presented in this section. Moment and length effects are completely neglected. Examples 8.1 and 8.3 present the design of axially loaded square tied columns, while Example 8.2 illustrates the design of a similarly loaded round spiral column. Table A.17 of the Appendix provides several properties for circular columns which are particularly useful for designing round columns.

■ EXAMPLE 8.1

Design a square tied column to support an axial dead load D of 130 k and an axial live load L of 180 k. Assume 2% longitudinal steel is desired, $f'_c = 4000$ psi, and $f_y = 60,000$ psi.

SOLUTION

$$P_u = (1.4)(130) + (1.7)(180) = 488 \text{ k}$$

Selecting Column Dimensions

$$P_u = \phi 0.80[0.85f'_c(A_g - A_{st}) + f_y A_{st}]$$

$$488 = (0.70)(0.80)[(0.85)(4)(A_g - 0.02A_g) + (60)(0.02A_g)]$$

$$A_g = 192.3 \text{ in.}^2 \qquad \underline{\text{Use } 14 \times 14(A_g = 196 \text{ in.}^2)}$$

Selecting Longitudinal Bars Substituting into column equation with known A_g and solving for A_{st}, we obtain

$$488 = (0.70)(0.80)[(0.85)(4)(196 - A_{st}) + 60A_{st}]$$

$$A_{st} = 3.62 \text{ in.}^2 \qquad \underline{\text{Use } 6 \text{ #7 bars } (3.61 \text{ in.}^2)}$$

Design of Ties (Assuming #3 Bars)

Spacing:
(a) $48 \times \frac{3}{8} = 18''$
(b) $16 \times \frac{7}{8} = 14'' \leftarrow$
(c) Least dim. $= 14'' \leftarrow$ Use #3 ties at 14''

A sketch of the column cross section is shown in Figure 8.5. ■

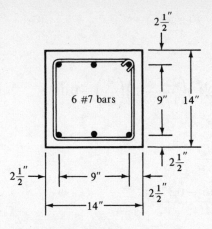

Figure 8.5

■ EXAMPLE 8.2

Design a round spiral column to support an axial dead load D of 180 k and an axial live load L of 300 k. Assume that 2% longitudinal steel is desired, $f'_c = 4000$ psi, and $f_y = 60,000$ psi.

SOLUTION

$$P_u = (1.4)(180) + (1.7)(300) = 762 \text{ k}$$

Selecting Column Dimensions and Bar Sizes

$$P_u = \phi 0.85[0.85f'_c(A_g - A_{st}) + f_y A_{st}]$$

$$762 = (0.75)(0.85)[(0.85)(4)(A_g - 0.02A_g) + (60)(0.02A_g)]$$

$A_g = 263.7 \text{ in.}^2$ <u>Use 18″ diameter column (255 in.²)</u>

$$762 = (0.75)(0.85)[(0.85)(4)(255 - A_{st}) + 60A_{st}]$$

$A_{st} = 5.80 \text{ in.}^2$ <u>Use 6 #9 bars (6.00 in.²)</u>

A sketch of the column cross section is shown in Figure 8.6.

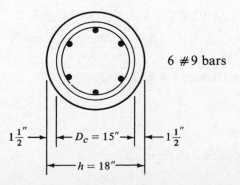

6 #9 bars

Figure 8.6

Design of Spiral

$$A_c = \frac{(\pi)(15)^2}{4} = 177 \text{ in.}^2$$

$$\text{Minimum } \rho_s = (0.45)\left(\frac{A_g}{A_c} - 1\right)\frac{f'_c}{f_y} = (0.45)\left(\frac{255}{177} - 1\right)\left(\frac{4}{60}\right) = 0.0132$$

Assume a #3 spiral.

$$\rho_s = \frac{4a_s(D_c - d_b)}{sD_c^2}$$

$$0.0132 = \frac{(4)(0.11)(15 - 0.375)}{(s)(15)^2}$$

$$s = 2.17'' \hspace{8em} \text{say } 2''$$

(Checked with Table A.16 in Appendix.) ■

■ **EXAMPLE 8.3**

Design a square tied column to support an axial dead load D of 600 kN and an axial live load L of 400 kN. Assume 2% longitudinal steel is desired, with $f'_c = 20.7$ MPa and $f_y = 413.7$ MPa.

SOLUTION

$$P_u = (1.4)(600) + (1.7)(400) = 1520 \text{ kN}$$

Selecting Column Dimensions

$$P_u = \phi 0.80[0.85f'_c(A_g - A_{st}) + f_y A_{st}]$$

$$(1520)(10)^3 = (0.70)(0.80)[(0.85)(20.7)(102\,400 - A_{st}) + (413.7)(A_{st})]$$

$$A_g = 106\,371 \text{ mm}^2 \hspace{3em} \underline{\text{Use } 320 \text{ mm} \times 320 \text{ mm} = 102\,400 \text{ mm}^2}$$

Selecting Longitudinal Bars

$$(1520)(10)^3 = (0.70)(0.80)[(0.85)(20.7)(102\,400 - A_{st}) + (413.7)(A_{st})]$$

$$A_{st} = 2304 \text{ mm}^2 \hspace{3em} \underline{\text{Use } 6 \text{ #7 bars } (2322 \text{ mm}^2)}$$

Design of Ties (Assuming #3 Bars with Diameter = 9.52 mm)

Spacing: **(a)** $(48)(9.52) = 457$ mm
 (b) $(16)(22.22) = 356$ mm
 (c) Least dimension = 320 mm ←

A sketch of the column cross section is shown in Figure 8.7. ■

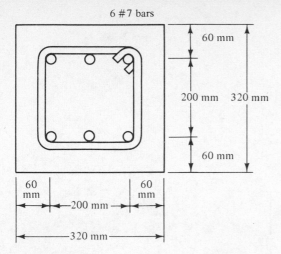

Figure 8.7

8.10 DESIGN OF REINFORCED CONCRETE COLUMNS USING THE ALTERNATE DESIGN METHOD

Should columns be designed using the alternate or working-stress method, the strength procedure is to be used. The combined flexural and axial load capacity of a particular column is to be taken as equal to 40% of that computed by the strength procedure (ACI Appendix A6.1). Slenderness is to be handled by the same method used in strength design, with the term P_u replaced by 2.5 times the design axial load and ϕ taken equal to 1.0 (ACI Appendix A6.2).

PROBLEMS

For Problems 8.1 to 8.4 compute the load-bearing capacity of the concentrically loaded short columns. $f_y = 60,000$ psi and $f'_c = 4000$ psi.

8.1 A 22-in. square column reinforced with 8 #11 bars. (*Ans.* 1317.7 k)

Problem 8.2

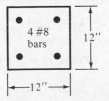

Problem 8.3 (*Ans.* 664.8 k)

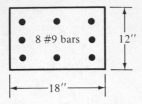

Problem 8.4

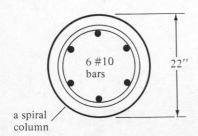

a spiral
column

In Problems 8.5 to 8.10 design columns for axial load only. Include the design of ties
or spirals and a sketch of the cross sections selected, including bar arrangements. All
columns are assumed to be short and are not exposed to the weather.

8.5 Square tied column; $P_D = 300$ k, $P_L = 400$ k, $f'_c = 4000$ psi, and $f_y = 60,000$ psi.
Assume $\rho_g = 2\%$. (*One ans.* 21″ × 21″ column with 6 #11 bars)

8.6 Repeat Problem 8.5, if ρ_g is to be 5%.

8.7 Round spiral column; $P_D = 180$ k, $P_L = 300$ k, $f'_c = 3500$ psi, and $f_y = 50,000$ psi.
Assume $\rho_g = 3\%$. (*One ans.* 18-in.-diameter column with 6 #11 bars)

8.8 Round spiral column; $P_D = 250$ k, $P_L = 350$ k, $f'_c = 5000$ psi, $f_y = 60,000$ psi, and
$\rho_g = 4\%$.

8.9 Smallest possible square tied column; $P_D = 260$ k, $P_L = 320$ k, $f'_c = 4000$ psi, and
$f_y = 60,000$ psi. (*One ans.* 15″ × 15″ column with 12 #10 bundled bars)

8.10 Design a rectangular tied column with the long side equal to two times the length of
the short side. $P_D = 250$ k, $P_L = 400$ k, $f'_c = 4000$ psi, $f_y = 60,000$ psi, and $\rho_g = 2\%$.

PROBLEMS WITH SI UNITS

In Problems 8.11 to 8.13 design columns for axial load only for the conditions described.
Include the design of ties or spirals and a sketch of the cross sections selected, including bar
arrangements. All columns are assumed to be short.

8.11 Square tied column; $P_D = 600$ kN, $P_L = 800$ kN, $f'_c = 20.7$ MPa, and $f_y = 344.8$
MPa. Assume $\rho_g = 0.02$. (*One ans.* 400 mm × 400 mm with 6 #9 bars)

8.12 Smallest possible square tied column; $P_D = 700$ kN, $P_L = 900$ kN, $f'_c = 27.6$ MPa,
and $f_y = 413.7$ MPa.

8.13 Round spiral column; $P_D = 500$ kN, $P_L = 650$ kN, $f'_c = 17.2$ MPa, $f_y = 275.8$ MPa,
and $\rho_g = 0.03$. (*One ans.* 400-mm diameter, 6 #9 bars and #3 spiral @ 50 mm)

9

Design of Short Columns Subject to Axial Load and Bending

9.1 AXIAL LOAD AND BENDING

All columns are subjected to some bending as well as axial forces and they need to be proportioned to resist both. The so-called "axial load" formulas presented in Chapter 8 do take into account some moments because they include the effect of small eccentricities with the 0.80 and 0.85 factors. These values are approximately equivalent to the assumption of actual eccentricities of $0.10h$ for tied columns and $0.05h$ for spiral columns.

Columns will bend under the action of moments, and those moments will tend to produce compression on one side of the columns and tension on the other. Depending on the relative magnitudes of the moments and axial loads, there are several ways in which the sections might fail. Figure 9.1 shows a column supporting a load P_n. In the various parts of the figure the load is placed at greater and greater eccentricities (thus producing larger and larger moments) until finally in part (f) the column is subject to such a large bending moment that the effect of the axial load is negligible. Each of the six cases shown is briefly discussed in the paragraphs to follow, where the letters (a) through (f) correspond to those same letters in the figure. Failure of the column is assumed to occur when the compressive strain at any point reaches 0.003.

(a) *Large axial load with negligible moment.* For this situation, failure will occur by the crushing of the concrete, with all reinforcing bars in the column having reached their yield stress in compression.

(b) *Large axial load and small moment such that the entire cross section is in compression.* When a column is subject to a small bending moment (that is, when the eccentricity is small), the entire column will be in

259

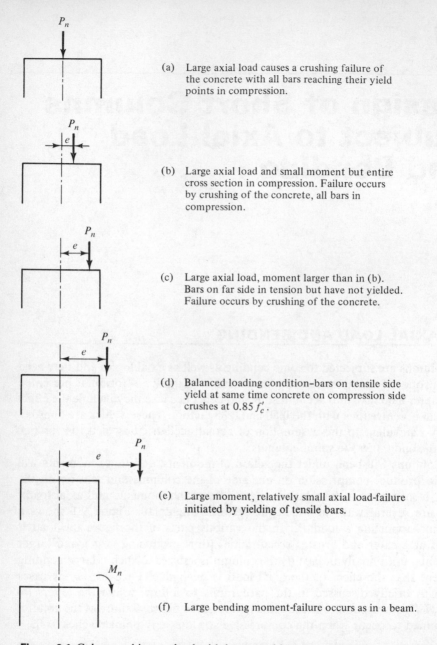

(a) Large axial load causes a crushing failure of the concrete with all bars reaching their yield points in compression.

(b) Large axial load and small moment but entire cross section in compression. Failure occurs by crushing of the concrete, all bars in compression.

(c) Large axial load, moment larger than in (b). Bars on far side in tension but have not yielded. Failure occurs by crushing of the concrete.

(d) Balanced loading condition–bars on tensile side yield at same time concrete on compression side crushes at $0.85 f'_c$.

(e) Large moment, relatively small axial load-failure initiated by yielding of tensile bars.

(f) Large bending moment-failure occurs as in a beam.

Figure 9.1 Column subject to load with larger and larger eccentricities.

compression but the compression will be higher on one side than on the other. The maximum compressive stress in the column will be $0.85f'_c$, and failure will occur by the crushing of the concretre with all the bars in compression.

Pennsylvania Southern Expressway, Philadelphia,
Pennsylvania. (Courtesy of Economy Forms Corporation.)

(c) *Eccentricity larger than in case (b) such that tension begins to develop on
one side of the column.* If the eccentricity is increased somewhat from the
preceding case, tension will begin to develop on one side of the column
and the steel on that side will be in tension but less than the yield stress.
On the far side the steel will be in compression. Failure will occur by
crushing of the concrete on the compression side.

(d) *A balanced loading condition.* As we continue to increase the eccentricity, a condition will be reached at which the reinforcing bars on the
tension side will reach their yield stress at the same time that the concrete on the opposite side reaches its maximum compression $0.85f'_c$.
This situation is called the *balanced loading condition.*

(e) *Large moment with small axial load.* If the eccentricity is further
increased, failure will be initiated by the yielding of the bars on the tensile
side of the column.

(f) *Large moment with no appreciable axial load.* For this condition, failure
will occur as it does in a beam.

9.2 THE PLASTIC CENTROID

The eccentricity of a column load is the distance from the load to the *plastic centroid* of the column. The plastic centroid represents the location of the resultant force produced by the steel and the concrete. It is the point in the column cross section through which the resultant column load must pass to produce uniform strain at failure. For locating the plastic centroid, all concrete is assumed to be stressed in compression to $0.85f'_c$ and all steel to f_y in compression. For symmetrical sections the plastic centroid coincides with the centroid of the column cross section, while for nonsymmetrical sections it can be located by taking moments.

Example 9.1 illustrates the calculations involved in locating the plastic centroid for an unsymmetrical cross section. The ultimate load P_n is determined by computing the total compressive forces in the concrete and the steel and adding them together. Then P_n is assumed to act downward at the plastic centroid at a distance $\bar{x}$ from one side of the column and moments are taken on that side of the column of the upward compression forces acting at their centroids and the downward P_n.

■ EXAMPLE 9.1

Determine the plastic centroid of the T-shaped column shown in Figure 9.2 if $f'_c = 4000$ psi and $f_y = 60,000$ psi.

SOLUTION

The plastic centroid falls on the x axis as shown in Figure 9.2 due to symmetry. The column is divided into two rectangles, the left one being $16'' \times 6''$ and the right one $8'' \times 8''$. C_1 is assumed to be the total compression in the left concrete

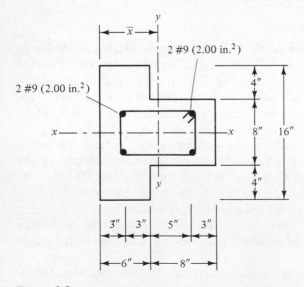

Figure 9.2

rectangle, C_2 the total compression in the right rectangle, and C_s the total compression in the reinforcing bars.

$$C_1 = (16)(6)(0.85)(4) = 326.4 \text{ k}$$

$$C_2 = (8)(8)(0.85)(4) = 217.6 \text{ k}$$

In computing C_s, the concrete where the bars are located is subtracted; that is,

$$C_s = (4.00)(60 - 0.85 \times 4) = 226.4 \text{ k}$$

$$\text{Total compression} = P_n = 326.4 + 217.6 + 226.4 = 770.4 \text{ k}$$

Taking Moments About Left Edge of Column

$$-(326.4)(3) - (217.6)(10) - (226.4)(7) + (770.4)(\bar{x}) = 0$$

$$\bar{x} = 6.15''$$

9.3 DEVELOPMENT OF INTERACTION DIAGRAMS

Should an axial compressive load be applied to a short concrete member, it will be subjected to a uniform strain or shortening as as shown in Figure 9.3(a). If a moment with zero axial load is applied to the same member, the result will be bending about the member's neutral axis such that strain is proportional to the distance from the neutral axis. This linear strain variation is shown in Figure 9.3(b). Should axial load and moment be applied at the same time, the resulting strain diagram will be a combination of two linear diagrams and will itself be linear as illustrated in Figure 9.3(c). As a result of this linearity, we can assume certain numerical values of strain in one part of a column and determine strains at other locations by straight line interpolation.

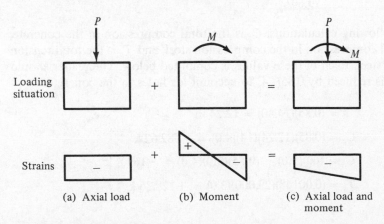

Figure 9.3 Column strains.

As the axial load applied to a column is changed, the moment which the column can resist will change. In this section the author shows how an interaction curve of nominal axial load and moment values can be developed for a particular column.

Assuming the concrete on the compression edge of the column will fail at a strain of 0.003, a strain can be assumed on the far edge of the column and the values of P_n and M_n can be computed by statics. Then holding the compression strain at 0.003 on the far edge, we can assume a series of different strains on the other edge and calculate P_n and M_n for each.[1] Eventually a sufficient number of values will be obtained to plot an interaction curve such as the one shown in Figure 9.8. Example 9.2 illustrates the calculation of P_n and M_n for a column for one set of assumed strains.

■ **EXAMPLE 9.2**

It is assumed that the column of Figure 9.4 has a strain on its compression edge equal to -0.00300 and has a tensile strain of $+0.00200$ on its other edge. Determine the values of P_n and M_n which cause this strain distribution if $f_y = 60$ ksi and $f'_c = 4$ ksi.

SOLUTION

Determine the values of c and of the steel strains ϵ'_s, and ϵ_s by proportions with reference to the strain diagram shown in Figure 9.5:

$$c = \left(\frac{0.00300}{0.00300 + 0.00200}\right)(24) = 14.40 \text{ in.}$$

$$\epsilon'_s = \left(\frac{11.90}{14.40}\right)(0.00300) = 0.00248 > 0.00207 \qquad \therefore \text{ yields}$$

$$\epsilon_s = \left(\frac{7.10}{9.60}\right)(0.00200) = 0.00148$$

In the following calculations, C_c is the total compression in the concrete, C'_s is the total compression in the compression steel, and T_s is the total tension in the tensile steel. Each of these values is computed below. The reader should note that C'_s is reduced by $0.85f'_cA'_s$ to account for holes in the concrete.

$$a = (0.85)(14.40) = 12.24 \text{ in.}$$

$$C_c = (0.85)(12.24)(14)(4.0) = -582.62 \text{ k}$$

$$C'_s = (60)(3.0) - (0.85)(3.0)(4.0) = -169.8 \text{ k}$$

$$T_s = (0.00148)(29,000)(3.0) = +128.76 \text{ k}$$

---

[1] Leet, K. 1991, *Reinforced Concrete Design*, 2nd ed. (New York: McGraw-Hill), pp. 316–317.

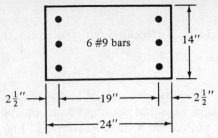

Figure 9.4

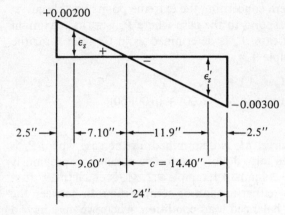

Figure 9.5

By statics, P_n and M_n are determined with reference made to Figure 9.6, where the values of C_c, C_s', and T are shown.

$\Sigma V = 0$

$$-P_n + 169.8 + 582.62 - 128.76 = 0$$

$$P_n = 623.7 \text{ k}$$

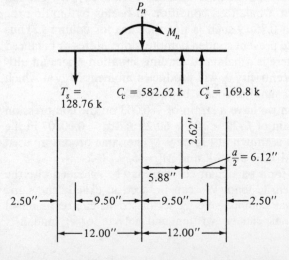

Figure 9.6

$\Sigma M = 0$ about tensile steel

$$(623.66)(9.50) + M_n - (582.62)(15.38) - (169.8)(19.00) = 0$$

$$M_n = 6262.13 \text{ in.-k} = 521.8 \text{ ft-k} \qquad \blacksquare$$

In this manner a series of P_n and M_n values are determined to correspond with a strain of -0.003 on the compression edge and varying strains on the far column edge. The resulting values are plotted on a curve as shown in Figure 9.8.

A few remarks are made here concerning the extreme points on this curve. One end of the curve will correspond to the case where P_n is at its maximum value and M_n is zero. For this case, P_n is determined as in Chapter 8 for the axially loaded column of Example 9.2:

$$\begin{aligned} P_n &= 0.85f'_c(A_g - A_s) + A_s f_y \\ &= (0.85)(4.0)(14 \times 24 - 6.00) + (6.00)(60) \\ &= 1482 \text{ k} \end{aligned}$$

On the other end of the curve, M_n is determined for the case where P_n is zero. This is the procedure used for a doubly reinforced member as previously described in Chapter 4. For the column of Example 9.2, M_n is equal to 297 ft-k.

A column normally fails by either tension or compression. In between the two extremes lies the so-called balanced load condition, where we may have a simultaneous tension and compression failure. In Chapter 3 the term "balanced section" was used in referring to a section whose compression concrete strain reached 0.003 at the same time as the tensile steel reached its yield strain at f_y/E_s. In a beam this situation theoretically occurred when the steel percentage equaled ρ_b.

For columns the definition of balanced loading is the same as it was for beams—that is, a column that has a strain of 0.003 on its compression side at the same time that its tensile steel on the other side has a strain of f_y/E_s. Though it is easily possible to prevent a balanced condition in beams by limiting the maximum steel percentage to $0.75\rho_b$, such is not the case for columns. Thus for columns it is not possible to prevent sudden compression failures or balanced failures. For every column there is a balanced loading situation where an ultimate load P_{bn} placed at an eccentricity e_b will produce a moment M_{bn}, at which time the balanced strains will simultaneously be reached.

At the balanced condition we have a strain of -0.003 on the compression edge of the column and a strain of $f_y/29 \times 10^3 = 60/29 \times 10^3 = 0.00207$ in the tensile steel. This information is shown in Figure 9.7. The same procedure used in Example 9.2 is used to find $P_n = 504.4 \text{ k}$ and $M_n = 559.7$ ft-k.

The curve for P_n and M_n for a particular column may be extended into the range where P_n becomes a tensile load. We can proceed in exactly the same fashion as we did when P_n was compressive. A set of strains can be assumed and the usual statics equations can be written and solved for P_n and M_n.

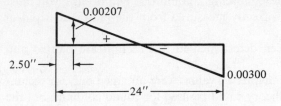

Figure 9.7

Several different sets of strains were assumed for the column of Figure 9.4 and then the values of P_n and M_n were determined. The results were plotted at the bottom of Figure 9.8 and were connected with the dashed line which is labeled tensile loads.

Because axial tension and bending are not very common for reinforced concrete columns, the tensile load part of the curves is not shown in subsequent figures in this chapter. You will note that the largest tensile value of P_n will occur when the moment is zero. For that situation, all of the column steel has yielded and all of the concrete has cracked. Thus P_n will equal the total steel area A_s times the yield stress. For the column of Figure 9.4,

$$P_n = A_s f_y = (6.0)(60) = 360 \text{ k}$$

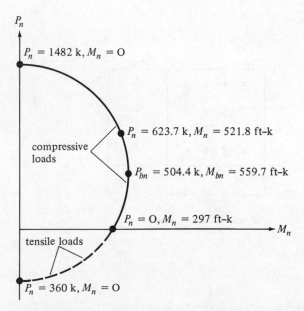

Figure 9.8 Interaction curve for the column of Figure 9.4.

On some occasions, members subject to axial tension and bending have unsymmetrical arrangements of reinforcing. Should this be the case, you must remember that eccentricity is correctly measured from the plastic centroid of the section.

The values which have been determined in this section are plotted and shown in Figure 9.8.

It will be noted that in this chapter P_n values were obtained only for rectangular tied columns. The same theory could be used for round columns, but the mathematics would be somewhat complicated because of the circular layout of the bars, and the calculations of distances would be rather tedious. Several approximate methods have been developed that greatly simplify the mathematics. Perhaps the best known of these is the one proposed by Whitney, in which equivalent rectangular columns are used to replace the circular ones.[2] This method gives results that correspond quite closely with test results.

In Whitney's method the area of the equivalent column is made equal to the area of the actual circular column and its depth in the direction of bending is 0.80 times the outside diameter of the real column. One-half the steel is assumed to be placed on one side of the equivalent column and one-half on the other. The distance between these two areas of steel is assumed to equal two-thirds of the diameter (D_s) of a circle passing through the center of the bars in the real column. These values are illustrated in Figure 9.9. Once the equivalent column is established, the calculations for P_n and M_n are made as for rectangular columns.

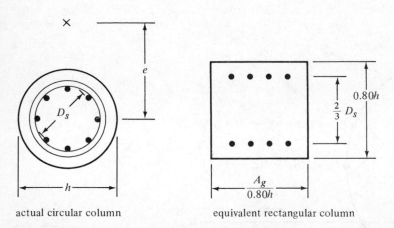

actual circular column   equivalent rectangular column

Figure 9.9

[2] Whitney, Charles S., 1942, "Plastic Theory of Reinforced Concrete Design," *Transactions ASCE*, 107, pp. 251–326.

9.4 USE OF INTERACTION DIAGRAMS

We have seen that by statics the values of P_n and M_n for a given column with a certain set of strains can easily be determined. To prepare an interaction curve with a hand calculator for just one column, however, is quite tedious. Imagine the work involved in a design situation where various sizes, concrete strengths, and steel percentages need to be considered. Consequently, designers resort almost completely to computer programs, computer-generated interaction diagrams, or tables for their column calculations. The remainder of this chapter is primarily concerned with computer-generated interaction diagrams such as the one of Figure 9.10. As we have seen, such a diagram is drawn for a column as the load changes from one of a pure axial nature through varying combinations of axial loads and moments and on to a pure bending situation.

Interaction diagrams are obviously useful for studying the strengths of columns with varying proportions of loads and moments. Any combination of loading that falls inside the curve is satisfactory, whereas any combination falling outside the curve represents failure.

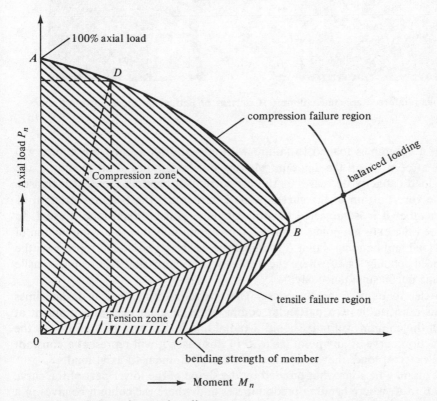

Figure 9.10 Column interaction diagram.

Massive reinforced concrete columns. (Courtesy of Bethlehem Steel Corporation.)

If a column is loaded to failure with an axial load only, the failure will occur at point A on the diagram. Moving out from point A on the curve, the axial load capacity decreases as the proportion of bending moment increases. At the very bottom of the curve, point C represents the bending strength of the member if it is subjected to moment only with no axial load present. In between the extreme points A and C, the column fails due to a combination of axial load and bending. Point B is called the *balanced point* and represents the balanced loading case, where theoretically a compression failure and tensile yielding occur simultaneously.

Refer to point D on the curve. The horizontal and vertical dashed lines to this point indicate a particular combination of axial load and moment at which the column will fail. Should a radial line be drawn from point 0 to the interaction curve at any point (as to D in this case), it will represent a constant eccentricity of load, that is, a constant ratio of moment to axial load.

You may be somewhat puzzled by the shape of the lower part of the curve from B to C, where bending predominates and where the column behaves in a manner very much like an underreinforced beam. From A to B on the curve the moment capacity of a section increases as the axial load decreases, but just

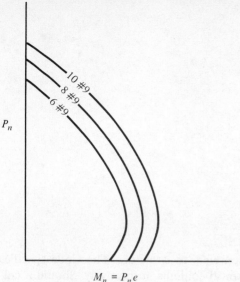

$M_n = P_n e$

Figure 9.11

the opposite occurs from B to C. A little thought on this point, however, shows the result is quite logical after all. The part of the curve from C to B represents the range of tensile failures. Any axial compressive load in that range tends to reduce the stresses in the tensile bars, with the result that a larger moment can be resisted.

In Figure 9.11 an interaction curve is drawn for the 14″ by 24″ column with 6 #9 bars considered in Section 9.3. If 8 #9 bars had been used in the same dimension column, another curve could be generated as shown in the figure; if 10 #9 bars were used, still another curve would result. The shape of the new diagrams would be the same as for the 6 #9 curve, but the values of P_n and M_n would be larger.

9.5 CODE MODIFICATIONS OF COLUMN INTERACTION DIAGRAMS

Before column interaction diagrams can be used for practical analysis and design, they must have three modifications made to them as specified in the ACI Code. These modifications are described in the following paragraphs.

(a) The diagrams described up to this point have been prepared for ultimate column loads or P_n values. The Code, however, (9.3.2) specifies strength reduction or ϕ factors (0.70 for tied columns and 0.75 for spiral columns) that must be multiplied by P_n values to obtain design values. Thus column design interaction curves must reflect the effect of ϕ factors, as shown in Figure 9.12.

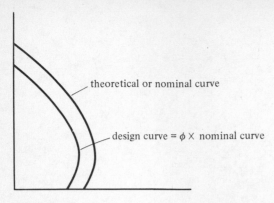

Figure 9.12

(b) The second modification also refers to ϕ factors. The Code specifies values of 0.70 and 0.75 for tied and spiral columns, respectively. Should a column have quite a large moment and a very small axial load so that it falls on the lower part of the curve between points B and C (see Figure 9.10), the use of these small ϕ values may be a little unreasonable. For instance, for a member in pure bending (point C on the same curve) the required ϕ is 0.90, but if the same member has a very small axial load added, ϕ would immediately fall to 0.70 or 0.75. Therefore, the Code (9.3.2.2) states that for members with f_y not exceeding 60,000 psi with symmetrical reinforcing and with $(h - d' - d_s)/h$ not less than 0.7 (see Figure 9.13), the value of ϕ may be increased linearly from 0.70 or 0.75 to 0.90 as ϕP_n decreases from $0.10 f'_c A_g$ to zero. For other members ϕ may be increased linearly from 0.70 or 0.75 to 0.9 as ϕP_n decreases from $0.10 f'_c A_g$ or ϕP_b (whichever is smaller) to zero. The effect of this modification is to produce a break in the lower part of a design interaction curve as illustrated in Figure 9.14. Should we have a tensile axial load, ϕ will always be 0.9.

(c) As described in Chapter 8, maximum permissible column loads were specified for columns no matter how small their e values. As a result, the upper

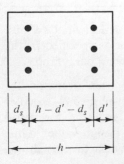

Figure 9.13

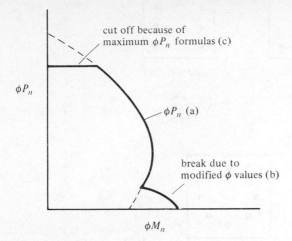

Figure 9.14 Shape of column design interaction curve.

part of each design interaction curve is shown as a horizontal line representing the appropriate value of

$$\phi P_{n\,\text{max}} \text{ for tied columns} = 0.80\phi[0.85f'_c(A_g - A_{st}) + f_y A_{st}]$$

$$\phi P_{n\,\text{max}} \text{ for spiral columns} = 0.85\phi[0.85f'_c(A_g - A_{st}) + f_y A_{st}]$$

It is to be remembered that these formulas were developed to be approximately equivalent to loads applied with eccentricities of $0.10h$ for tied columns and $0.05h$ for spiral columns.

Each of the three modifications described here is indicated on the design curve of Figure 9.14, and the letter in parentheses corresponds to the letter used in describing the modification in the preceding paragraphs.

9.6 ANALYSIS OF ECCENTRICALLY LOADED COLUMNS USING INTERACTION DIAGRAMS

If interaction diagrams were prepared as described in the preceding sections, it would be necessary to have a diagram for each different column cross section, for each different set of concrete and steel grades, and for each different bar arrangement. The result would be an astronomical number of diagrams. The number can be tremendously reduced, however, if the diagrams are plotted with ordinates of $\phi P_n/A_g$ (instead of P_n) and with abscissas of $\phi P_n e/A_g h$ (instead of M_n). Thus each interaction diagram can be used for cross sections with widely varying dimensions. The ACI has prepared interaction curves in this manner

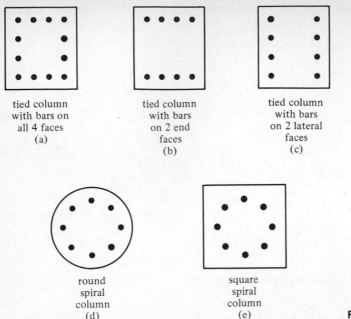

tied column
with bars on
all 4 faces
(a)

tied column
with bars
on 2 end
faces
(b)

tied column
with bars
on 2 lateral
faces
(c)

round
spiral
column
(d)

square
spiral
column
(e)

Figure 9.15

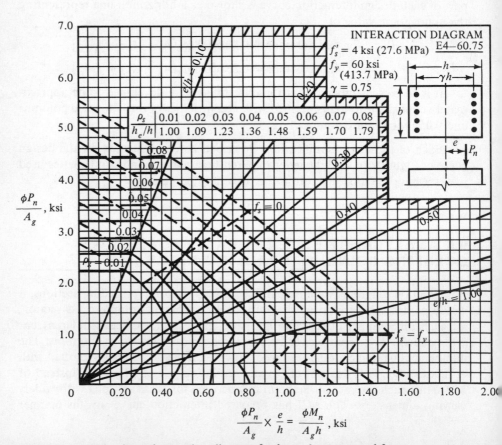

Figure 9.16 ACI column interaction diagram for bars along two end faces.

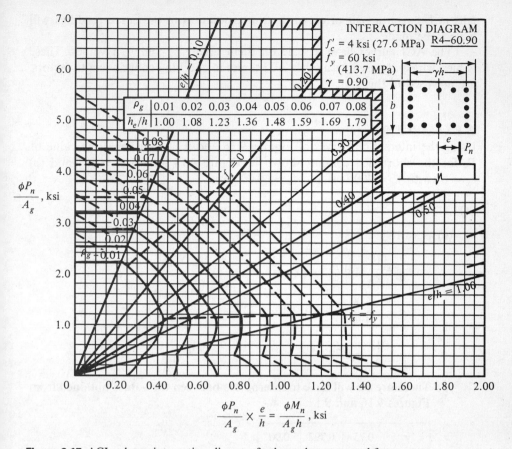

Figure 9.17 ACI column interaction diagram for bars along two end faces.

for the different cross section and bar arrangement situations shown in Figure 9.15 and for different grades of steel and concrete.[3]

Two of the ACI diagrams are given in Figures 9.16 and 9.17, while the Appendix (Graphs 2–13) presents several other ones for the situations given in parts (a), (b), and (d) of Figure 9.15.

Example 9.3 shows the use of interaction curves for determining the value of P_n for the column of Problem 9.2. To analyze a particular column the value of e/h can be calculated as well as the value of ρ. As shown in Figures 9.16 and 9.17, it is necessary to compute the value of γ, which is equal to the distance from the center of the bars on one side of the column to the center of the bars on the other side of the column divided by h, the depth of the column (both values being taken in the direction of bending). Usually the value of γ obtained

---

[3] *Design Handbook*, 1978, vol. 2, Columns, ACI Publication SP-17a (78), Detroit.

falls in between a pair of curves and interpolation of the curve readings will have to be made.

The ACI curves have a scale on them giving ρ_g and h_e/h values. These values, which are used with other ACI tables in checking column length factors, are not used in this text.

■ EXAMPLE 9.3

Using the interaction curves of Figures 9.16 and 9.17, determine the value of P_n for the short tied column shown in Figure 9.18 for (a) $e_x = 18''$ and (b) $e_x = 8''$ with $f'_c = 4000$ psi and $f_y = 60,000$ psi.

SOLUTION

(a) $e_x = 18''$:

$$\frac{e}{h} = \frac{18}{24} = 0.75$$

$$\rho = \frac{6.00}{(14)(24)} = 0.0179$$

$$\gamma = \frac{19}{24} = 0.792$$

Therefore we will have to interpolate between the values obtained from Figures 9.16 and 9.17.

γ	0.75	0.792	0.90
$\dfrac{\phi P_n}{A_g}\dfrac{e}{h}$	0.500	0.525	0.590

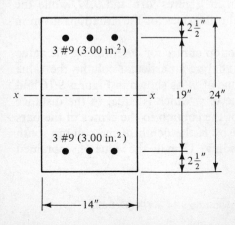

3 #9 (3.00 in.²)

3 #9 (3.00 in.²)

x ——————x $2\frac{1}{2}''$ 19"   24" $2\frac{1}{2}''$

14"

Figure 9.18

$$\frac{\phi P_n}{A_g}\frac{e}{h} = 0.50$$

$$P_n = \frac{(14)(24)(24)(0.525)}{(0.70)(18)} = \underline{336 \text{ k}}$$

(b) $e_x = 8''$:

$$\frac{e}{h} = \frac{8}{24} = 0.333$$

$$\rho = \frac{6.00}{(14)(24)} = 0.0179$$

$$\gamma = \frac{19}{24} = 0.792$$

γ	0.75	0.792	0.90
$\dfrac{\phi P_n}{A_g}\dfrac{e}{h}$	0.490	0.500	0.525

$$P_n = \frac{(14)(24)(24)(0.500)}{(0.70)(8)} = \underline{720.0 \text{ k}} \qquad \blacksquare$$

Example 9.4 shows the application of the column interaction curves to a round spiral column.

■ EXAMPLE 9.4
Using the column interaction curves in the Appendix, determine the value of P_n for the short spiral column shown in Figure 9.19. $f'_c = 4000$ psi and $f_y = 60{,}000$ psi.

SOLUTION

$$\frac{e}{h} = \frac{6}{20} = 0.30$$

$$\rho = \frac{4.81}{314} = 0.0153$$

$$\gamma = \frac{15}{20} = 0.75$$

$$\frac{\phi P_n}{A_g}\frac{e}{h} = 0.39$$

$$P_n = \frac{(314)(20)(0.39)}{(0.75)(6)} = \underline{544.3 \text{ k}} \qquad \blacksquare$$

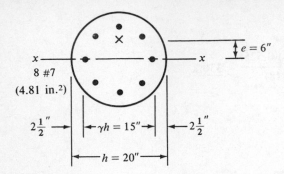

Figure 9.19

Another comment should be made concerning the use of these curves, although it does not affect the solution of the preceding examples. For tensile failures below the balanced points on the curves, a reduction in the axial load would reduce the moment capacity. As a result, the designer may theoretically have to check two loading situations for columns in this range. He or she would have to check the column for the largest P_n and M_n values and also would have to check for the smallest axial load value that could occur at the same time as the maximum M_n.

9.7 DESIGN OF ECCENTRICALLY LOADED COLUMNS USING INTERACTION DIAGRAMS

Examples 9.5 and 9.6 illustrate the application of the column interaction diagrams to the design of short columns subject to axial load and to bending about one axis. The column size is estimated as described in the next two paragraphs, the values of e/h and $\phi P_n e/A_g h$ are computed, and the value of ρ is determined from the appropriate diagram.

Although there are several methods available for selecting column sizes, the trial-and-error method illustrated in Example 9.6 is about as good as any. With this procedure the designer estimates what he or she thinks is a reasonable column size and then determines the steel percentage required for that column size from the interaction diagram. If it is felt that the ρ determined is unreasonably large or small, another column size can be selected and the new required ρ selected from the diagrams, and so on.

A slightly different approach is used in Example 9.6, where the average compression stress at ultimate load across the column cross section is assumed to equal some value—say, 0.5 to $0.6f'_c$. This value is divided into P_u to determine the column area required. Then cross-section dimensions are selected and the value of ρ determined from the interaction curves. Again if the percentage obtained seems unreasonable, the column size can be revised and a new steel percentage obtained.

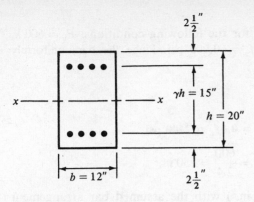

Figure 9.20

■ EXAMPLE 9.5

The 12 × 20 tied column of Figure 9.20 is to be used to support the following: $P_D = 100$ k, $P_L = 110$ k, $M_D = 60$ ft-k, and $M_L = 80$ ft-k. If $f'_c = 4000$ psi and $f_y = 60,000$ psi, select the reinforcing bars to be used in two end faces only by referring to the interaction curves.

SOLUTION

$$P_u = (1.4)(100) + (1.7)(110) = 327 \text{ k}$$

$$P_n = \frac{327}{0.70} = 467.1 \text{ k}$$

$$M_u = (1.4)(60) + (1.7)(80) = 220 \text{ ft-k}$$

$$M_n = \frac{220}{0.70} = 314.3 \text{ ft-k}$$

$$e = \frac{(12)(314.3)}{467.1} = 8.07$$

$$\frac{e}{h} = \frac{8.07}{20} = 0.404$$

$$\frac{\phi P_n}{A_g} \frac{e}{h} = \frac{(0.70)(467.1)}{240} \left(\frac{8.07}{20}\right) = 0.550$$

$$\gamma = \frac{15}{20} = 0.75$$

$$\rho = 0.020 \qquad \text{from Figure 9.16 or Appendix Graph 8}$$

$$A_s = (0.020)(20 \times 12) = 4.80 \text{ in.}^2$$

(Use 8 #7 bars = 4.81 in.²) ■

■ EXAMPLE 9.6

Design a short square tied column for the following conditions: $P_u = 600$ k, $M_u = 140$ ft-k, $f'_c = 4000$ psi, and $f_y = 60,000$ psi. Place the bars uniformly around the faces of the column.

SOLUTION

Assume Average Compression Stress = $0.6f'_c = 2400$ psi

$$A_g \text{ required} = \frac{600}{2.4} = 250 \text{ in.}^2$$

Try 16×16 column ($A_g = 256$ in.^{2}) with the assumed bar arrangement shown in Figure 9.21.

Select ρ from Column Interaction Diagrams

$$\gamma = \frac{11}{16} = 0.6875$$

$$e = \frac{(12)(140)}{600} = 2.80 \text{ in.}$$

$$\frac{e}{h} = \frac{2.80}{16} = 0.175$$

$$P_n = \frac{P_u}{\phi} = \frac{600}{0.7} = 857.14 \text{ k}$$

$$\frac{\phi P_n}{A_g} \frac{e}{h} = \frac{(0.7)(857.14)}{256}(0.175) = 0.410$$

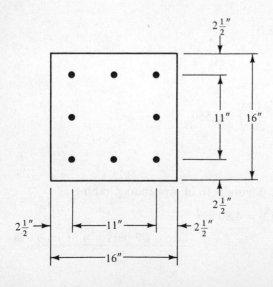

Figure 9.21

By interpolation between charts,

$$\rho = 0.0312$$

$$A_s = (0.0312)(256) = 7.99 \text{ in.}^2$$

(Use 8 #9 bars = 8.00 in.²) ■

9.8 SHEAR IN COLUMNS

Section 11.3.1.2 of the Code provides the following value for the shear strength of members subject to axial compression:

$$V_c = 2\left(1 + \frac{N_u}{2000A_g}\right)\sqrt{f_c'} \, b_w d \qquad \text{(ACI Equation 11-4)}$$

In this expression, N_u equals the factored axial load on the member acting simultaneously with V_u. The value N_u/A_g is to be expressed in psi. Should the actual shear V_c be greater than $\phi V_c/2$, it will be necessary to select tie spacings using the stirrup spacing procedures described in Chapter 7. The results will be closer tie spacings than required by the usual column rules. Although column shear forces are usually small for interior columns, they may be quite large in exterior columns, particularly those bent in double curvature.

9.9 BIAXIAL BENDING

Many columns are subjected to biaxial bending, that is, bending about both axes. Corner columns in buildings where beams and girders frame into the columns from both directions are the most common cases, but there are others, such as where columns are cast monolithically as part of frames in both directions or where columns are supporting heavy spandrel beams. Bridge piers are almost always subject to biaxial bending.

Circular columns have polar symmetry and thus the same ultimate capacity in all directions. The design process is the same, therefore, regardless of the directions of the moments. If there is bending about both the x and y axes, the biaxial moment can be computed by combining the two moments or their eccentricities as follows:

$$M_u = \sqrt{(M_{ux})^2 + (M_{uy})^2}$$

or

$$e = \sqrt{(e_x)^2 + (e_y)^2}$$

For shapes other than circular ones, it is necessary to consider the three-dimensional interaction effects. Whenever possible, it is desirable to make columns subject to biaxial bending circular in shape. Should it be necessary to use square or rectangular columns for such cases, the reinforcing should be placed uniformly around the perimeters.

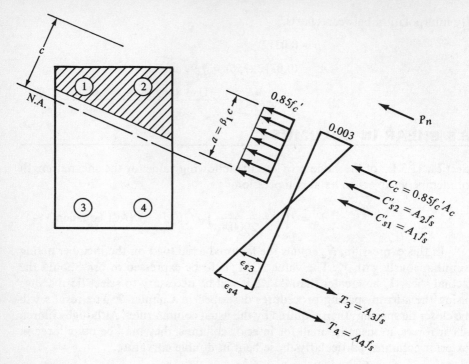

Figure 9.22

You might quite logically think that you could determine P_n for a biaxially loaded column by using static equations, as was done in Example 9.2. Such a procedure will lead to the correct answer, but the mathematics involved is so complicated due to the shape of the compression side of the column that the method is not a practical one. Nevertheless, a few comments are made about this type of solution and reference is made to Figure 9.22.

An assumed location is selected for the neutral axis, and the appropriate strain triangles are drawn as shown in the figure. The usual equations are written with $C_c = 0.85f'_c$ times the shaded area A_c and with each bar having a force equal to its cross-sectional area times its stress. The solution of the equation yields the load that would establish that neutral axis—but the designer usually starts with certain loads and eccentricities and does not know the neutral axis location and, furthermore, the neutral axis is probably not even perpendicular to the resultant $e = \sqrt{(e_x)^2 + (e_y)^2}$.

For column shapes other than circular ones, it is desirable to consider three-dimensional interaction curves such as the one shown in Figure 9.23. In this figure the curve labeled M_{nxo} represents the interaction curve if bending occurs about the x axis only, while the one labeled M_{nyo} is the one if bending occurs about the y axis only.

In this figure, for a constant P_n, the hatched plane shown represents the contour of M_n for bending about any axis.

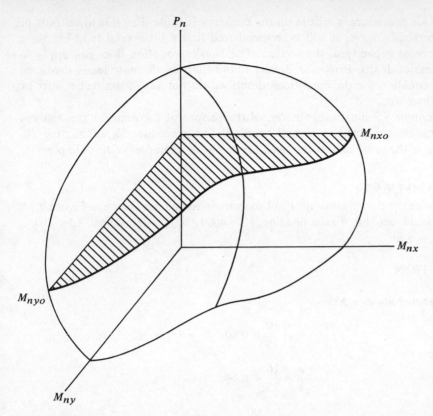

Figure 9.23

The Code does not provide empirical equations for the design of columns subject to biaxial bending. However, for such cases there is available, a so-called reciprocal interaction equation developed by Professor Boris Bresler of the University of California at Berkeley that provides satisfactory results.[4] In this equation, which follows, P_{no} is the pure axial load capacity of the section normally taken as $0.85f'_cA_g + A_sf_y$, P_{nx} is the axial load capacity when the load is placed at an eccentricity e_x with $e_y = 0$, and P_{ny} is the axial load capacity at e_y with $e_x = 0$.

$$\frac{P_n}{P_{nx}} + \frac{P_n}{P_{ny}} - \frac{P_n}{P_{no}} = 1.0$$

The Bresler equation works rather well as long as P_n is at least as large as $0.10P_{no}$. Should P_n be less than $0.10P_{no}$, it is satisfactory to neglect the axial force completely and design the section as a member subject to biaxial bending

---

[4] Bresler, B., 1960, "Design Criteria for Reinforced Concrete Columns Under Axial Load and Biaxial Bending," *Journal ACI*, 57, p. 481.

only. This procedure is a little on the conservative side. For this lower part of the interaction curve, it will be remembered that a little axial load increases the moment capacity of the section. The Bresler equation does not apply to axial tension loads. Professor Bresler found that the ultimate loads predicted by his equation for the conditions described do not vary from test results by more than 10%.

Example 9.7 illustrates the use of the reciprocal theorem for the analysis of a column subjected to biaxial bending. The procedure for calculating P_{nx} and P_{ny} is the same as the one used for the prior examples of this chapter.

■ EXAMPLE 9.7

Determine the design capacity P_u of the short tied column shown in Figure 9.24, which is subjected to biaxial bending. $f'_c = 4000$ psi, $f_y = 60,000$ psi, $e_x = 16$ in., and $e_y = 8$ in.

SOLUTION

For Bending about x Axis

$$\gamma = \frac{20}{25} = 0.80$$

$$\frac{e}{h} = \frac{16}{25} = 0.64$$

$$\rho = \frac{8.00}{(15)(25)} = 0.0213$$

By interpolation from interaction diagrams with bars on all four faces.

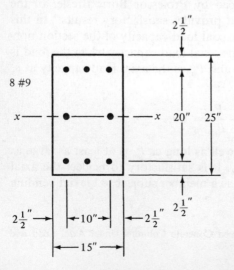

8 #9

Figure 9.24

$$\frac{\phi P_{nx}}{A_g} \times \frac{e}{h} = 0.523$$

$$P_{nx} = \frac{(15)(25)(25)(0.523)}{(0.7)(16)} = 437.8 \text{ k}$$

For Bending about y Axis

$$\gamma = \frac{10}{15} = 0.667$$

$$\frac{e}{h} = \frac{8}{15} = 0.533$$

$$\rho = \frac{8.00}{(15)(25)} = 0.0213$$

By interpolation,

$$\frac{\phi P_{ny}}{A_g} \times \frac{e}{h} = 0.487$$

$$P_{ny} = \frac{(15)(25)(15)(0.487)}{(0.7)(8)} = 489.2 \text{ k}$$

Determining Axial Load Capacity of Section

$$P_{no} = (0.85)(4.0)(15 \times 25) + (8.00)(60) = 1755 \text{ k}$$

Using the Bresler Expression to Determine P_n

$$\frac{P_n}{P_{nx}} + \frac{P_n}{P_{ny}} - \frac{P_n}{P_{no}} = 1.0$$

$$\frac{P_n}{437.8} + \frac{P_n}{489.2} - \frac{P_n}{1755} = 1.0$$

$$4.01 P_n + 3.59 P_n - P_n = 1755$$

$$\underline{P_n = 265.9 \text{ k}} \qquad ■$$

If the moments in the weak direction (y axis here) are rather small as compared to bending in the strong direction (x axis), it is rather common to neglect the smaller moment. This practice is probably reasonable as long as e_y is less than about 20% of e_x, since the Bresler expression will show little reduction for P_n. For the example just solved, an e_y equal to 50% of e_x caused the axial load capacity to be reduced by 40.0%.

Example 9.8 illustrates the design of a column subject to biaxial bending. The Bresler expression, which is of little use in the proportioning of such members, is used to check the capacities of the sections selected by some other procedure. Exact theoretical designs of columns subject to biaxial bending are very complicated and, as a result, are seldom made in design offices. They are proportioned either by approximate methods or with computer programs.

During the past few decades several approximate methods have been introduced for the design of columns with biaxial moments. For instance, there are available quite a few design charts with which satisfactory designs may be made. The problems are reduced to very simple calculations in which coefficients are taken from the charts and used to magnify the moments about a single axis. Designs are then made with the regular uniaxial design charts.[5-7]

Not only are charts used, but also some designers use rules of thumb for making initial designs. One very simple method (it's not too good) involves the following steps: (1) the selection of the reinforcement required in the x direction considering P_n and M_{nx}, (2) the selection of the reinforcement required in the y direction considering P_n and M_{ny}, and (3) the determination of the total column steel area required by adding the areas obtained in steps (1) and (2). This method will occasionally result in large design errors on the unsafe side because the strength of the concrete is counted twice, once for the x direction and once for the y direction.[8]

Another approximate procedure that works fairly well for design office calculations is used for the last example of this chapter (Example 9.8). If this simple method is applied to square columns, the values of both M_{nx} and M_{ny} are assumed to act about both the x axis and the y axis (i.e., $M_x = M_y = M_{nx} + M_{ny}$). The steel is selected about one of the axes and is spread around the column, and the Bresler expression is used to check the ultimate load capacity of the eccentrically loaded column.

Should a rectangular section be used where the y axis is the weaker direction, it would seem logical to calculate $M_y = M_{nx} + M_{ny}$ and to use that moment to select the steel required about the y axis and spread the computed steel area over the whole column cross section. Although such a procedure will produce safe designs, the resulting columns may be rather uneconomical because they will often be much too strong about the strong axis. A fairly satisfactory approximation is to calculate $M_y = M_{nx} + M_{ny}$ and multiply it by b/h, and with that moment, design the column about the weaker axis.[9]

Example 9.8 illustrates the design of a short square column subject to biaxial-bending. The approximate method described in the last two paragraphs

[5] Parme, A. L., Nieves, J. M., and Gouwens, A., 1966, "Capacity of Reinforced Rectangular Columns Subject to Biaxial Bending," *Journal ACI*, 63 (11), pp. 911–923.

[6] Weber, D. C., 1966, "Ultimate Strength Design Charts for Columns with Biaxial Bending," *Journal ACI*, 63 (11), pp. 1205–1230.

[7] Row, D. G., and Paulay, T., 1973, "Biaxial Flexure and Axial Load Interaction in Short Reinforced Concrete Columns," *Bulletin of New Zealand Society for Earthquake Engineering*, 6 (2), pp. 110–121.

[8] Park, R., and Paulay, T., 1975, *Reinforced Concrete Structures* (New York: John Wiley & Sons), pp. 158–159.

[9] Fintel, M., ed., 1974, *Handbook of Concrete Engineering* (New York: Van Nostrand), pp. 27–28.

is used, and the Bresler expression is used for checking the results. If this had been a long column, it would have been necessary to magnify the design moments for slenderness effects regardless of the design method used.

■ EXAMPLE 9.8

Select the reinforcing needed for the short square tied column shown in Figure 9.25 for the following: $P_D = 100$ k, $P_L = 140$ k, $M_{DX} = 50$ ft-k, $M_{LX} = 70$ ft-k, $M_{DY} = 40$ ft-k, $M_L = 60$ ft-k, $f_c' = 4000$ psi, and $f_Y = 60,000$ psi.

SOLUTION

Computing Design Values

$$P_u = (1.4)(100) + (1.7)(140) = 378 \text{ k}$$

$$\frac{P_u}{f_c' A_g} = \frac{378}{(4)(484)} = 0.195 > 0.10 \qquad \text{(See ACI 9.3.2.2)}$$

Therefore, $\phi = 0.70$.

$$P_n = \frac{378}{0.70} = 540 \text{ k}$$

$$M_{ux} = (1.4)(50) + (1.7)(70) = 189 \text{ ft-k}$$

$$M_{nx} = \frac{189}{0.70} = 270 \text{ ft-k}$$

$$M_{uy} = (1.4)(40) + (1.7)(60) = 158 \text{ ft-k}$$

$$M_{ny} = \frac{158}{0.70} = 225.7 \text{ ft-k}$$

As a result of biaxial bending, the design moment about the x or y axis is assumed to equal $M_{nx} + M_{ny} = 270 + 225.7 = 495.7$ ft-k.

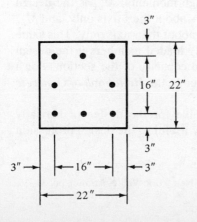

Figure 9.25

Determining Steel Required

$$e_x = e_y = \frac{(12)(495.7)}{540} = 11.02 \text{ in.}$$

$$\gamma = \frac{16}{22} = 0.727$$

$$\frac{e}{h} = \frac{11.02}{22} = 0.501$$

$$\frac{\phi P_n}{A_g} \times \frac{e}{h} = \frac{(0.70)(540)}{(22)(22)}(0.501) = 0.391$$

By interpolation from interaction diagrams with bars on all four faces,

$$\rho = 0.0123$$

$$A_s = (0.0123)(22)(22) = 5.95 \text{ in.}^2$$

<u>Use 8 #8 (6.28 in.²)</u>   ■

A review of the column with the Bresler expression gives a $P_n = 706$ k $>$ 540 k, which is satisfactory. Should the reader go through the Bresler equation here, her or she must remember to calculate the correct e_x and e_y values for use with the interaction diagrams. For instance,

$$e_x = \frac{(12)(270)}{540} = 6 \text{ in.}$$

When a beam is subjected to biaxial bending, the following approximate interaction equation may be used for design purposes:

$$\frac{M_x}{M_{ux}} + \frac{M_y}{M_{uy}} \le 1.0$$

In this expression M_x and M_y are the design moments, M_{ux} is the design moment capacity of the section if bending occurs about the x axis only, and M_{uy} is the design moment capacity if bending occurs about the y axis only. This same expression may be satisfactorily used for axially loaded members if the design axial load is about 15% or less of the axial load capacity of the section. For a detailed discussion of this subject you are referred to the *Handbook of Concrete Engineering*.[10]

There are numerous other methods available for the design of biaxially loaded columns. One method that is particularly useful to the design profession

---

[10] Fintel, M., ed., 1974 *Handbook of Concrete Engineering* (New York: Van Nostrand), pp. 34–36.

is the PCA Load Contour Method, which is recommended in the ACI *Design Handbook*.[11]

9.10 COMPUTER EXAMPLES

The computer module for the analysis or design of rectangular short tied columns with bars in two faces is used in the example to follow. Once the data is input the screen will (after an approximately 20 second delay) show a range of P_u and M_u values. Furthermore if you input a certain P_u value the design moment strength M_u will be supplied.

■ EXAMPLE 9.9

If P_u for the column of Example 9.3 is 241.9 k determine M_u using the computer disk enclosed.

```
ANALYSIS OR DESIGN OF SHORT SQUARE OR RECTANGULAR COLUMNS
-----------------------------------------------------------
Gross Column Width                          b (in.) = 14.0
Gross Column Depth                          h (in.) = 24.0
Distance from Column Edge to C.G. of Bars   d' (in.) =  2.5
Number of Bars and Bar #                            =  6 # 9
Concrete Strength                        f'c (psi) =  4000
Steel Yield Strength                      fy (psi) =  60000


RESULTS
=======
          Pu(K)    Mu(FT-K)
          829.9      0.0
          829.9     19.9
          829.9     41.5
          829.9     63.1
          817.6    181.0
          693.2    258.8
          593.1    307.0
          509.2    340.2
          436.6    365.3
          372.3    385.9
          332.7    389.3
```

(continued)

---

[11] *Design Handbook*, 1978, vol. 2, Columns, ACI Publications SP-17a(78), Detroit.

RESULTS (continued)
=======

Pu(K)	Mu(FT-K)
306.6	384.5
284.2	379.0
264.8	373.3
247.8	367.5
232.8	361.8
219.0	355.9
203.9	348.1
190.0	340.4
177.2	333.0
165.2	325.9
153.9	319.0
143.4	312.3
133.4	305.8
124.0	299.5
115.0	293.5
106.4	287.6
98.2	281.8
91.4	279.4
85.6	280.1
79.8	280.5
73.8	280.6
67.7	280.4
61.6	280.1
55.4	279.4
49.1	278.6
42.7	277.6
36.3	276.5
29.8	275.2
23.3	273.7
16.7	272.1
10.1	270.4
3.4	268.5
0.0	267.6

Design Pu (K) =   241.9
Design moment strength, Mu, will be (ft-k) =   365.3

9.11 APPLICATION OF COLUMN INTERACTION DIAGRAMS WHEN SI UNITS ARE USED

The ACI column diagrams shown in this text and used in the preceding sections of this chapter may be applied when SI units are being used just as they are for conventional units, but it is necessary to make use of a conversion factor.

If it is desired to determine the value of P_n for a given column where we enter the diagrams with e/h and ρ and pick off $\phi P_n/A_g$ or $\phi P_n e/A_g h$, we must divide the results by 0.006 895 or multiply them by 145.033. If we are using SI units and we wish to use the diagrams to determine the required value of ρ for a particular column, we enter the diagrams with $\phi P_n e/A_g h$ or $\phi P_n/A_g$ divided by 0.006 895—that is, multiplied by 145.033.

PROBLEMS

In Problems 9.1 and 9.2 locate the plastic centroids if $f'_c = 4000$ psi and $f_y = 60,000$ psi.
Problem 9.1 (*Ans.* 10.13 in. from left edge)

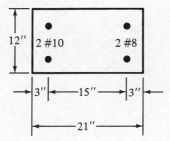

Problem 9.2

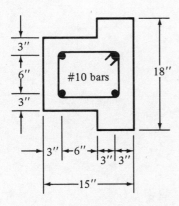

9.3 Using statics equations determine the values of P_n and M_n for the column assuming it is strained to -0.00300 on the right-hand edge and to $+0.00200$ on its left-hand edge. $f'_c = 4000$ psi, $f_y = 60,000$ psi. (*Ans.* $P_n = 475.6$ k, $M_n = 304.7$ ft-k)

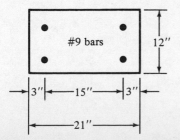

9.4   Repeat Problem 9.3 if the strain on the left edge is $+0.001$.

9.5   Repeat Problem 9.3 if the strain on the left edge is 0.000. (*Ans.* $P_n = 873.1$ k, $M_n = 155$ ft-k).

9.6   Repeat Problem 9.3 if the strain on the left edge is -0.001.

9.7   Repeat Problem 9.3 if the tensile steel on the left side yields—that is, a balanced loading situation. (*Ans.* $P_n = 369.2$ k, $M_n = 333.9$ ft-k)

For Problems 9.8 to 9.13 use the interaction diagrams in the Appendix to determine P_n values for the short columns shown if $f'_c = 4000$ psi and $f_y = 60,000$ psi.

Problem 9.8

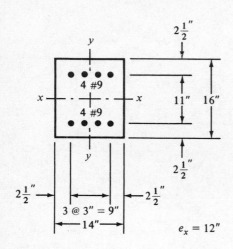

9.9   Repeat Problem 9.8 if $e_x = 8$ in. (*Ans.* 447 k)

Problem 9.10

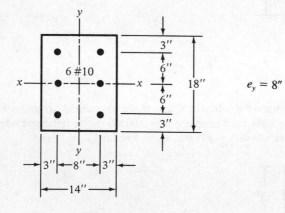

Problem 9.11 (*Ans.* 177 k)

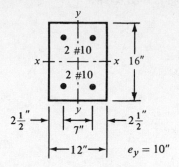

Problem 9.12

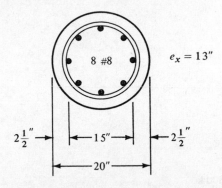

Problem 9.13 (*Ans.* 356 k)

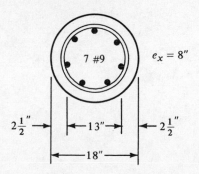

In Problems 9.14 to 9.16 use the interaction curves in the Appendix to select reinforcing for the short columns shown, with $f'_c = 4000$ psi and $f_y = 60{,}000$ psi. $f'_c = 4000$ psi for Problem 9.16.

Problem 9.14

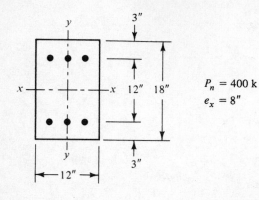

$P_n = 400$ k
$e_x = 8''$

Problem 9.15 (*One ans.* 6 #8)

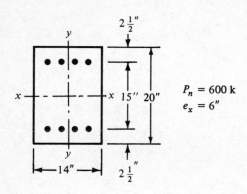

$P_n = 600$ k
$e_x = 6''$

Problem 9.16

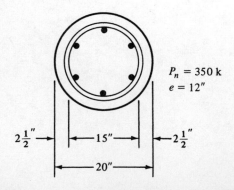

$P_n = 350$ k
$e = 12''$

In Problems 9.17 to 9.20 determine P_n values for the short columns shown if $f_y =$ 60,000 psi and $f'_c =$ 4000 psi.

Problem 9.17 (*Ans. $P_n = 201$ k*)

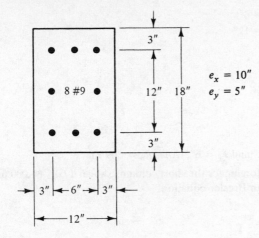

$e_x = 10''$
$e_y = 5''$

Problem 9.18

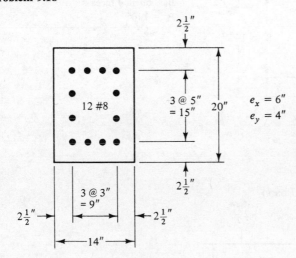

$e_x = 6''$
$e_y = 4''$

Problem 9.19 (*Ans. $P_n = 257$ k*)

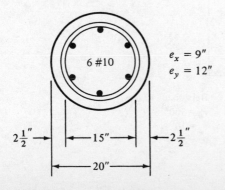

$e_x = 9''$
$e_y = 12''$

Problem 9.20

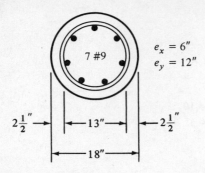

$e_x = 6''$
$e_y = 12''$

7 #9

$2\frac{1}{2}''$ ← 13'' → ← $2\frac{1}{2}''$

← 18'' →

9.21 Repeat Problem 9.20 if $e_x = 12''$ and $e_y = 6''$. (*Ans. $P_n = 209$ k*)

In Problems 9.22 and 9.23 select reinforcing for the short columns shown if $f_y = 60,000$ psi and $f'_c = 4000$ psi. Check results with Bresler equation.

Problem 9.22

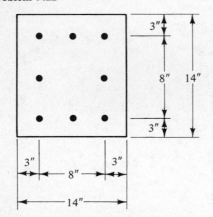

Bars on all 4 faces
$P_n = 140^k$
$e_x = 8''$
$e_y = 4''$

3''
8'' 14''
3''

3'' 8'' 3''
14''

Problem 9.23 (*One ans.* 8 #7 bars)

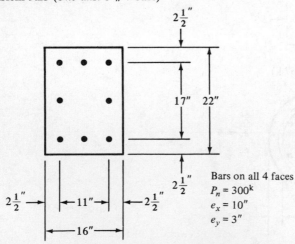

$2\frac{1}{2}''$

17'' 22''

$2\frac{1}{2}''$

$2\frac{1}{2}''$ → ← 11'' → ← $2\frac{1}{2}''$
← 16'' →

Bars on all 4 faces
$P_n = 300^k$
$e_x = 10''$
$e_y = 3''$

For Problems 9.24 to 9.26 use the enclosed computer disks.

9.24 If the column of Problem 9.8 is supporting a load $P_u = 233.3$ k how large can M_u be?

9.25 If the column of Problem 9.14 is supporting a load $P_u = 420$ k how large can M_u be if 6 #8 bars are used? (*Ans.* 139.1 ft-k)

9.26 If the column of Problem 9.14 is to support an axial load $P_u = 600$ k how many #11 bars need to be used to resist a design moment $M_u = 200$ ft-k?

9.27 Prepare a flow chart for the preparation of an interaction curve for axial compression loads and bending for a short rectangular tied column.

PROBLEMS WITH SI UNITS

In Problems 9.28 to 9.30 use the column interaction equations in the Appendix to determine P_n values for the short columns shown if $f'_c = 27.6$ MPa and $f_y = 413.7$ MPa. Remember to apply the conversion factor given in Section 9.11 when using the interaction curves.

Problem 9.28

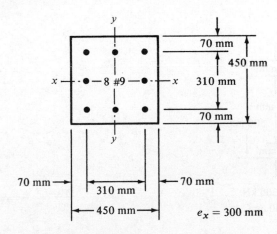

Problem 9.29 (*Ans.* 1860 kN)

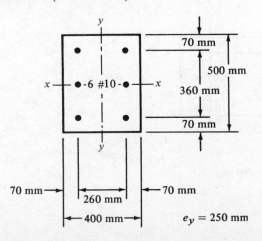

Problem 9.30

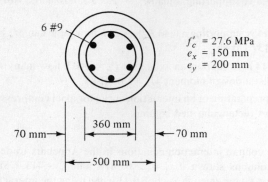

$f'_c = 27.6$ MPa
$e_x = 150$ mm
$e_y = 200$ mm

6 #9

70 mm→ |← 360 mm →| ←70 mm

|← 500 mm →|

In Problems 9.31 to 9.33 select reinforcing for the short columns shown if $f'_c = 27.6$ MPa and $f_y = 413.7$ MPa. Remember to apply the conversion factor given in Section 10.10 when using the interaction curves.

Problem 9.31 (*One ans.* 6 #7)

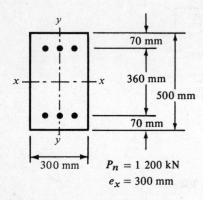

$P_n = 1\ 200$ kN
$e_x = 300$ mm

Problem 9.32

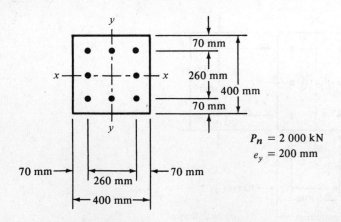

$P_n = 2\ 000$ kN
$e_y = 200$ mm

Problem 9.33 (*One ans.* 8 #8)

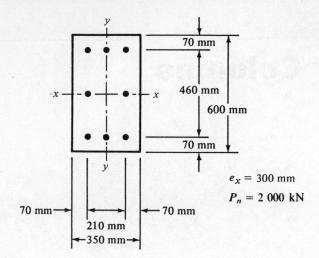

$e_x = 300$ mm

$P_n = 2\ 000$ kN

10

Slender Columns

10.1 INTRODUCTION

When a column bends or deflects laterally an amount Δ, its axial load will cause an increased column moment equal to $P\Delta$. This moment will be superimposed onto any moments already in the column. Should this $P\Delta$ moment be of such magnitude as to significantly reduce the axial load capacity of the column, the column will be referred to as a *slender column.*

In the ACI Code it is assumed that a significant reduction in axial loading capacity is something over 5%. If the reduction is 5% or less we will have a *short column,* but if it is more than 5% we will have a *slender column.* The Code provides slenderness ratios above which they feel axial load capacities will be reduced by more than 5%. These values are presented in Section 10.5 of this chapter.

10.2 SLENDERNESS EFFECTS

The slenderness of columns is based on their geometry and on their lateral bracing. As their slenderness increases, their bending stresses increase and thus buckling may occur. Reinforced concrete columns generally have small slenderness ratios. As a result they can usually be designed as short columns without strength reductions due to slenderness.

Section 10.10.1 of the Code states that the design of a compression member should desirably be based on a theoretical analysis of the structure that takes into account the effects of axial loads, moments, deflections, duration of loads, varying member sizes, end conditions, and so on. If such a theoretical procedure

Clemson University Library, Clemson, South Carolina. (Courtesy of Clemson University Communications Center.)

is not used, the Code (10.11.5) provides an approximate method for determining slenderness effects. This method, which is based on the factors just mentioned for an "exact" analysis, results in a moment magnifier δ, which is to be multiplied by the larger moment at the end of the column, and that value is used in design. If bending occurs about both axes, δ is to be computed separately for each direction and the values obtained multiplied by the respective moment values. Actually there are two possible types of moment magnifiers—one for loads that cause no appreciable sidesway and one for loads that do cause appreciable sidesway.

Several items involved in the calculation of δ values are discussed in the next several paragraphs. These include unsupported column lengths, effective length factors, radii of gyration, and the ACI Code requirements.

Unsupported Lengths

The length used for calculating the slenderness ratio of a column, ℓ_u, is its unsupported length. This length is considered to be equal to the clear distance between slabs, beams, or other members that provide lateral support to the column. If haunches or capitals are present, the clear distance is measured from the bottoms of the capitals or haunches.

Effective Length Factors

To calculate the slenderness ratio of a particular column, it is necessary to estimate its effective length. This is the distance between points of zero moment in the column. For this initial discussion it is assumed that no sidesway or joint translation is possible. Sidesway or joint translation means that one or both ends of a column can move laterally with respect to each other.

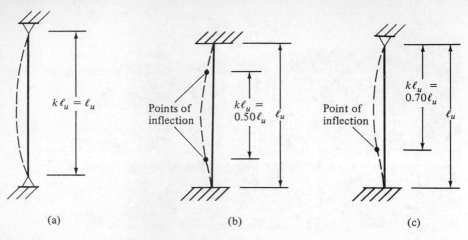

Figure 10.1 Effective lengths for columns in braced frames (sidesway prevented).

If there were such a thing as a perfectly pinned end column, its effective length would be its unsupported length, as shown in Figure 10.1(a). The *effective length factor k* is the number that must be multiplied by the column's unsupported length to obtain its effective length. For a perfectly pinned end column, $k = 1.0$.

Columns with different end conditions have entirely different effective lengths. For instance, if there were such a thing as a perfectly fixed end column, its points of inflection (or points of zero moment) would occur at its one-fourth points, and its effective length would be $\ell_u/2$, as shown in Figure 10.1(b). As a result, its k value would equal 0.5.

Obviously, the smaller the effective length of a particular column, the smaller its danger of buckling and the greater its load-carrying capacity. In Figure 10.1(c) is shown a column with one end fixed and one end pinned. The k factor for this column is theoretically 0.70.

The concept of effective lengths is simply a mathematical method of taking a column whatever its end and bracing conditions and replacing it with an equivalent pinned-end braced column. A complex buckling analysis could be made for a frame to determine the critical stress in a particular column. The k factor is determined by finding the pinned-end column with an equivalent length that provides the same critical stress. The k factor procedure is a method of making simple solutions for complicated frame buckling problems.

Reinforced concrete columns serve as parts of frames, and these frames are sometimes *braced* and sometimes *unbraced*. A braced frame is one for which sidesway or joint translation is prevented by means of bracing, shear walls, or lateral support from adjoining structures. An unbraced frame does not have any of these types of bracing supplied and must depend on the stiffness of its own members to prevent lateral buckling. For braced frames k values can never be greater than 1.0, but for unbraced frames the k values will always be greater than 1.0 because of sidesway.

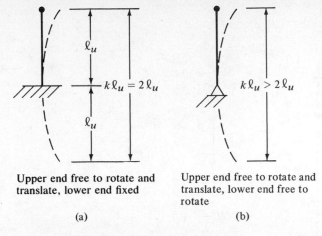

Upper end free to rotate and translate, lower end fixed

(a)

Upper end free to rotate and translate, lower end free to rotate

(b)

Figure 10.2 Columns for unbraced frames.

An example of an unbraced column is shown in Figure 10.2(a). The base of this particular column is assumed to be fixed, whereas its upper end is assumed to be completely free to both rotate and translate. The elastic curve of such a column will take the shape of the elastic curve of a pinned end column of twice its length. Its effective length will therefore equal $2\ell_u$, as shown in the figure. In Figure 10.2(b) another unbraced column case is illustrated.

The Code (10.11.2) states that the effective length factor is to be taken as 1.0 for compression members in frames braced against sidesway unless a theoretical analysis shows that a lesser value can be used. Should the member be in a frame not braced against sidesway, the Code states that the value of k is to be larger than 1.0 and is to be determined with proper consideration given to the effects of cracking and reinforcing on the column stiffness. ACI-ASCE Committee 441 suggests that it is not realistic to assume that k will be less than 1.2 for such columns, and therefore it seems logical to make preliminary designs with k equal to or larger than that value.

10.3 ALIGNMENT CHARTS

The primary procedure used for estimating effective lengths involves the use of the alignment charts shown in Figure 10.3.[1,2] The chart of part (a) of the figure is applicable to braced frames, whereas the one of part (b) is applicable to

[1] Johnston, B. G., editor, 1976, *Guide to Stability Design for Metal Structures*, 3rd ed. (New York: John Wiley & Sons), p. 420.

[2] Julian, O. G., and Lawrence, L. S., 1959, "Notes on J and L Nomograms for Determination of Effective Lengths," unpublished. These are also called the Jackson and Moreland Alignment Charts, after the firm with whom Julian and Lawrence are associated.

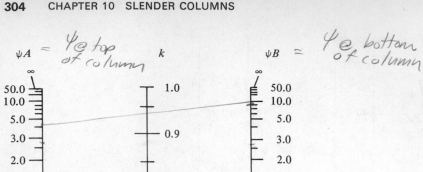

(a) Braced frames

Figure 10.3 Effective length factors. ψ = ratio of $\Sigma(EI/\ell_c)$ of compression members to $\Sigma(EI/\ell)$ of flexural members in a plane at one end of a compression member. k = effective length factor.

unbraced frames. The subject of braced and unbraced frames is addressed in the next section of this chapter.

To use the alignment charts for a particular column, ψ factors are computed at each end of the column. The ψ factor at one end of the column equals the sum of the stiffnesses $[\Sigma(EI/\ell)]$ of the columns meeting at that joint, including the column in question, divided by the sum of all the stiffnesses of the beams

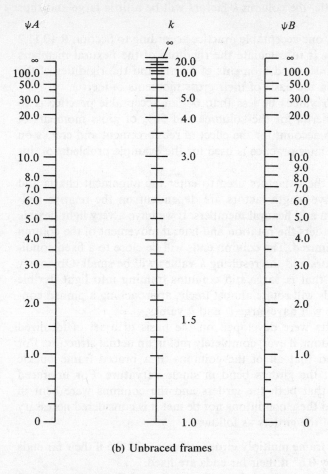

ψA   k   ψB

(b) Unbraced frames

Figure 10.3 continued

meeting at the joint. Should one end of the column be pinned, ψ is theoretically equal to ∞, and if fixed, $\psi = 0$. Since a perfectly fixed end is practically impossible to have, ψ is usually taken as 1.0 instead of 0 for assumed fixed ends. When column ends are supported by but not rigidly connected to a footing, ψ is theoretically infinity but usually is taken as about 10 for practical design.

One of the two ψ values is called ψ_A and the other is called ψ_B. After these values are computed, the effective length factor k is obtained by placing a straightedge between ψ_A and ψ_B. The point where the straightedge crosses the middle nomograph is k.

To calculate the ψ values it is necessary to use realistic moments of inertia. Usually the girders will be appreciably cracked on their tensile sides, whereas the columns will probably have only a few cracks. If the I values for the girders are underestimated a little, the column k factors will be a little large and thus on the safe side.

Based on these facts, one acceptable practice according to Section R.10.11.2 of the ACI Commentary is to calculate the rigidities of the flexural members on the basis of their transformed moments of inertia and the rigidities of the compression members on the basis of their gross moments of inertia.

Should column $k\ell_u/r$ values be less than 60, an acceptable practice is to use gross moments of inertia for the columns and 50% of gross moments of inertia for the girders (to account for the effect of reinforcement and cracks on relative stiffnesses). This later practice is used for the example problems of this chapter.

It can be seen that the ψ factors used to enter the alignment charts and thus the resulting effective length factors are dependent on the relative stiffnesses of the compression and flexural members. If we have a very light flexible column and large stiff girders the rotation and lateral movement of the column ends will be greatly minimized. The column ends will be close to a fixed condition, and thus the ψ values and the resulting k values will be small. Obviously if the reverse happens—that is, large stiff columns framing into light flexible girders—the column ends will rotate almost freely, approaching a pinned condition. Consequently we will have large ψ and k values.

The alignment charts were developed on the basis of a set of idealized conditions which are seldom, if ever, completely met in an actual structure. For instance, it was assumed that all of the columns in a braced frame buckle simultaneously and that the girders bend in single curvature. For unbraced frames it was assumed that both the girders and the columns were bent in double curvature. Should these conditions not be met it is considered necessary to modify the stiffness of the girders as follows:

1. For an unbraced frame multiply girder stiffnesses by 0.5 if their far ends are hinged and by 0.67 if their far ends are fixed.
2. For braced frames multiply girder stiffnesses by 1.5 if their far ends are hinged and by 2 if their far ends are fixed.

Instead of using the alignment charts for determining k values, the ACI Commentary (R10.11.2) provides an alternate method which involves the use of relatively simple equations. The equations (not shown here) were taken from the British Standard Code of Practice.[3]

10.4 IS A COLUMN BRACED OR UNBRACED?

The design of the columns in an unbraced frame is appreciably more complex than is the design of the columns in a braced frame. The first problem we face is deciding whether a frame is braced or unbraced. It is rare to find a column

---

[3] *Code of Practice for the Structural Use of Concrete* (CP110: Part 1), British Standards Institution, London, 1972, 154 pp.

which is completely braced or completely unbraced, and we will have to make a decision on the matter. The ACI Commentary (R10.11.2) provides information to help the designer in this regard. It states that a column may be considered braced if the bracing elements (shear walls, shear trusses, or other types of lateral bracing) provide a total stiffness resisting lateral movement of that story equal to at least six times the sum of the stiffnesses of all the columns within that story. If lateral stiffness of this magnitude is provided, lateral deflections will be negligible in their effect on column strength.

As an example, it is assumed that a frame is to be braced on a particular story with a structural wall (as perhaps the wall around the stairwell). For the columns to be considered braced by the wall, its computed stiffness EI in the direction of bending must be at least equal to six times the sum of the individual EI values of the columns on that level computed in the same direction.

$$EI \text{ of wall} \geq 6[\Sigma EI \text{ of columns}]$$

The ACI Commentary (R10.11.2) presents another method which may be used for deciding whether a frame is braced or unbraced. If the value of the so-called stability index Q which follows is ≤ 0.04 the Commentary states that the frame may be considered to be a braced one.

$$Q = \frac{\Sigma P_u \Delta_u}{H_u h_s}$$

where ΣP_u = total factored vertical load for all of the columns on the story in question

Δ_u = the elastically determined first-order lateral deflection due to H_u at the top of the story in question with respect to the bottom of that story

H_u = the total factored lateral load for the story in question (we've been calling it V_u)

h_s = the height of the story

Despite these suggestions from the ACI, the individual designer is going to have to make decisions as to what is adequate bracing and what is not, depending on the presence of structural walls and other bracing items. For the average-size reinforced concrete building, load eccentricities and slenderness values will be small and frames will be considered to be braced. Certainly, however, it is wise in questionable cases to err on the side of the unbraced.

10.5 IS IT A SLENDER COLUMN?

The effective slenderness ratio is normally used to determine whether a column is short or slender. In calculating $k\ell_u/r$, the radius of gyration r is equal to 0.25 times the diameter of a round column and 0.289 times the dimension of a rectangular column in the direction being considered. The ACI Code (10.11.3) permits the approximate value 0.30 to be used in place of 0.289, and this is done herein. For other sections the values of r can be computed from the properties of the gross sections.

For this discussion the moments at the ends of a column are referred to as M_{1b}, M_{2b}, and M_{2s} by the Code. M_{1b} is the value of the smaller factored end moment due to the loads that result in no appreciable sidesway. M_{2b} is the value of the larger factored end moment due to the loads that result in no appreciable sidesway. It may not be less than $P_u(0.6 + 0.03h)$ according to Section 10.11.5.4 of the Code. M_{2s} is the value of the larger factored end moment due to loads which result in appreciable sidesway but not less than $P_u(0.6 + 0.03h)$.

The ACI Commentary (R10.11.5.1) states that unless the lateral deflection in a column due to gravity loads is larger than $\ell_u/1500$, sway due to such loading may be neglected. For most reinforced concrete frames the dead and live loads will not cause lateral deflections larger than this value and thus they will be assumed to cause only M_b moments. On the other hand, wind and earthquake loads will cause appreciable sidesway or M_s moments.

Should a compression member be subjected to transverse loading between its supports, its largest moment may very well occur away from its ends. If this is the case, the largest moment will be M_{2b}.

The ACI Code 10.11.4.1 states that effects of slenderness may be neglected for columns braced against sidesway if

$$\frac{k\ell_u}{r} < 34 - 12\frac{M_{1b}}{M_{2b}}$$

In Section 10.11.4.2 the Code states that effects of slenderness may be neglected for unbraced columns if

$$\frac{k\ell_u}{r} < 22$$

For calculating the limiting slenderness ratios given for deciding whether columns are slender or short, the smaller moment M_{1b} is considered to have a positive sign if the member is bent in single curvature (C-shaped) and a negative sign if the member is bent in double curvature (S-shaped). M_{2b} is always considered to be positive. You can see that if the moments cause single curvature, the column will be in a worse situation as to the danger of buckling than if they cause double curvature (where the moments tend to oppose each other).

Should $k\ell_u/r$ for a particular column be larger than the applicable ratio, we will have a slender column. For such a column the effect of slenderness must be considered. This may be done by using approximate methods or by using a theoretical second-order analysis that takes into account the effect of deflections. If $k\ell_u/r > 100$ a theoretical second-order analysis must be made (Code 10.11.4.3).

A second-order analysis is one that takes into account the effect of deflections and also makes use of a reduced tangent modulus. The equations necessary for designing a column in this range are extremely complicated, and practically it is necessary to use column design charts or computer programs. Fortunately, most reinforced concrete columns have slenderness ratios less than 100.

Avoiding Slender Columns

The design of slender columns is appreciably more complicated than is the design of short columns. As a result, it may be wise to give some consideration to the use of certain minimum dimensions such that none of the columns will be slender. In this way they can be almost completely avoided in the average-size building.

If k is assumed equal to 1.0, slenderness can usually be neglected in braced frame columns if ℓ_u/h is kept to 10 or less on the first floor and 14 or less for the floors above the first one. To determine these values it was assumed that little moment resistance was provided at the footing–column connection and the first-floor columns were assumed to be bent in single curvature. Should the footing–column connection be designed to have appreciable moment resistance, the maximum ℓ_u/h value given above as 10 should be raised to about 14 or equal to the value used for the upper floors.

Should we have an unbraced frame and assume $k = 1.2$, it is probably necessary to keep ℓ_u/h to 6 or less. So for a 10-ft clear floor height it is necessary to use a minimum h of about 10 ft/6 = 1.67 ft = 20 in. in the direction of bending to avoid slender columns.[4]

Example 10.1 illustrates the selection of the k factor and the determination of the slenderness ratio for a column in an unbraced frame. For calculating I/L values the author used gross moments of inertia for the columns, half gross moments of inertia for the girders, and the full lengths of members center to center of supports.

■ EXAMPLE 10.1

(a) Using the alignment charts of Figure 10.3, calculate the effective length factor for column AB of the unbraced frame of Figure 10.4. Assume that the far ends of the girders are pinned and consider only bending in the plane of the frame.

(b) Compute the slenderness ratio of column AB. Is it a short or a slender column?

---

[4] Neville, G. B., editor, 1984, *Simplified Design Reinforced Concrete Buildings of Moderate Size and Height* (Skokie, Ill.: Portland Cement Association), pp. 5-10–5-12.

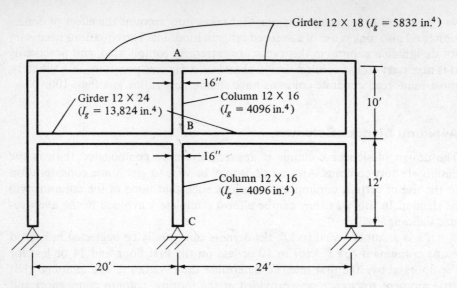

Figure 10.4

SOLUTION

(a) Effective length factor for column AB

$$\psi_A = \frac{\dfrac{4096}{12 \times 10}}{0.5\left[\dfrac{5832 \times \frac{1}{2}}{12 \times 20} + \dfrac{5832 \times \frac{1}{2}}{12 \times 24}\right]} = 3.06$$

$$\psi_B = \frac{\dfrac{4096}{12 \times 10} + \dfrac{4096}{12 \times 12}}{0.5\left[\dfrac{13,824 \times \frac{1}{2}}{12 \times 20} + \dfrac{13,824 \times \frac{1}{2}}{12 \times 24}\right]} = 2.37$$

From Figure 10.4,

$$\underline{\underline{k = 1.74}}$$

(b) Is it a slender column?

$$\ell_u = 10 \text{ ft} - \frac{9 + 12}{12} = 8.25 \text{ ft}$$

$$\frac{k\ell_u}{r} = \frac{(1.74)(12 \times 8.25)}{0.3 \times 16} = 35.89 > \text{Maximum } \frac{k\ell_u}{r} \text{ for a short column}$$

in an unbraced frame $= 22$

It's a slender column ∎

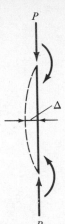

Figure 10.5 Moment magnification.

10.6 MAGNIFICATION OF COLUMN MOMENTS

When a column is subjected to moment along its unbraced length, it will be displaced laterally in the plane of bending. The result will be an increased or secondary moment equal to the axial load times the lateral displacement or eccentricity. In Figure 10.5 the load P causes the column moment to be increased by an amount $P\Delta$. This moment will cause Δ to increase a little more, with the result that the $P\Delta$ moment will increase which in turn will cause a further increase in Δ and so on until equilibrium is reached.

We could take the column moments, compute the lateral deflection, increase the moment by $P\Delta$, recalculate the lateral deflection and the increased moment, and so on. Although about two cycles would be sufficient, this would still be a tedious and impractical procedure.

It can be shown[5] that the increased moment can be estimated by multiplying the primary moment by $1/(1 - P/P_c)$, where P is the axial load and P_c is the Euler buckling load $\pi^2 EI/L^2$.

In Example 10.2 this expression is used to estimate the magnified moment in a laterally loaded column. It will be noted that in this problem the primary moment of 75 ft-k is estimated to increase by 7.4 ft-k. If we computed the deflection due to the lateral load we would get 0.445 in. For this value $P\Delta = (150)(0.445) = 66.75$ in.-k = 5.6 ft-k. This moment causes more deflection, which causes more moment and so on.

■ **EXAMPLE 10.2**

(a) Compute the primary moment in the column shown in Figure 10.6 due to the lateral 20-k load.

---

[5] Timoshenko, S. P., and Gere, J. M., 1961, *Theory of Elastic Stability*, 2nd ed. (McGraw-Hill, New York, pp. 319–356).

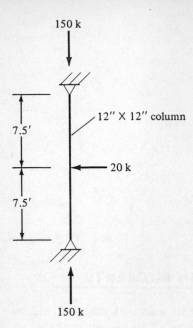

150 k

12" × 12" column

7.5'

← 20 k

7.5'

150 k

Figure 10.6

(b) Determine the estimated total moment including the secondary moment due to lateral deflection using the approximate magnification factor just presented. $E = 3.16 \times 10^3$ ksi. Assume $k = 1.0$ and $\ell_u = 15$ ft.

SOLUTION

(a) Primary moment due to lateral load:

$$M_u = \frac{(20)(15)}{4} = 75 \text{ ft-k}$$

(b) Total moment including secondary moment:

$$P_c = \text{Euler buckling load} = \frac{\pi^2 EI}{k\ell_u^2}$$

$$= \frac{(\pi^2)(3160)(1728)}{(1.0)(12 \times 15)^2} = 1663.4 \text{ k}$$

$$\text{Magnified moment} = 75 \, \frac{1}{1 - \dfrac{P}{P_c}}$$

$$= 75 \, \frac{1}{1 - \dfrac{150}{1663.4}} = 82.4 \text{ ft-k} \qquad \blacksquare$$

10.7 ACI MOMENT MAGNIFIER METHOD

As we have seen in the last section, it is possible to approximately calculate the increased moment due to lateral deflection by using the $1/(1 - P/P_c)$ expression. In the ACI Code (10.11.5.1) the factored design moment for slender columns is increased by using the following expression in which M_c is the magnified or increased moment:

$$M_c = \delta_b M_{2b} + \delta_s M_{2s} \qquad \text{(ACI Equation 10-6)}$$

where δ_b is a moment magnification factor that takes into account the effect of member curvature in a frame braced against sidesway and δ_s is a moment magnification factor that takes into account the lateral drift of a column caused by lateral and gravity loading if the frame is not braced against sidesway.

The determination of the moment magnifier involves the following calculations:

1. $E_c = 57,000\sqrt{f_c'}$.

2. I_g = gross inertia of the column cross section about the centroidal axis being considered.

3. $E_s = 29 \times 10^6$ psi.

4. I_{se} = moment of inertia of reinforcing about the centroidal axis of the section = Σ of each bar area times distance to centroidal axis squared.

5. β_d = ratio of maximum factored dead load to the maximum factored total load. This value is always assumed to have a positive value.

6. Next it is necessary to compute EI. The two expressions given for EI in the Code were developed so as to account for creep, cracks, and so on. If the column and bar sizes have already been selected, or estimated, EI can be computed with the following expression which is particularly satisfactory for columns with high steel percentages.

$$EI = \frac{\dfrac{E_c I_g}{5} + E_s I_{se}}{1 + \beta_d} \qquad \text{(ACI Equation 10-10)}$$

The alternate expression for EI which follows is probably the better expression to use when steel percentages are low. Notice also that this expression will be the one used if the reinforcing has not been previously selected.

$$EI = \frac{\dfrac{E_c I_g}{2.5}}{1 + \beta_d} \qquad \text{(ACI Equation 10-11)}$$

7. The Euler buckling load is computed:

$$P_c = \frac{\pi^2 EI}{(k\ell_u)^2}$$

8. For some moment situations in columns the amplification or moment magnifier expression provides moments that are too large. One such situation occurs when the moment at one end of the member is zero. For this situation the lateral deflection is actually about half of the deflection in effect provided by the amplification factor. Should we have approximately equal end moments that are causing reverse curvature bending, the deflection at middepth and the moment there are close to zero. As a result of these and other situations the Code provides modification factors to be used in the moment expression which will result in more realistic moment magnification.

For braced frames without transverse loads C_m can vary from 0.4 to 1.0 and is to be determined with the expression at the end of this paragraph. For all other cases it is to be taken as 1.0. (Remember the sign convention: M_{1b} is positive for single curvature and is negative for reverse curvature, and M_{2b} is always positive.)

$$C_m = 0.6 + 0.4 \frac{M_{1b}}{M_{2b}} \geq 0.4 \qquad \text{(ACI Equation 10-12)}$$

9. Finally the moment magnifiers can be computed as follows:

$$\delta_b = \frac{C_m}{1 - \dfrac{P_u}{\phi P_c}} \geq 1.0 \qquad \text{(ACI Equation 10-7)}$$

$$\delta_s = \frac{1}{1 - \dfrac{\Sigma P_u}{\phi \Sigma P_c}} \geq 1.0 \qquad \text{(ACI Equation 10-8)}$$

In the expressions for δ_s the term ΣP_u and ΣP_c are the summations for all of the columns on the story in question. The values of P_c are calculated using the $k\ell_u$ determined for columns in a frame subject to sidesway. All of the columns on a particular story are involved in the sway instability for that story. For frames that are braced against sidesway, it is necessary to calculate only δ_b while δ_s is to be taken as 1.

The remainder of this chapter is devoted to the presentation of numerical examples of the analysis and design of slender columns, making use of the preceding expressions.

10.8 DESIGN OF SLENDER COLUMNS IN BRACED FRAMES

Example 10.3 illustrates the design of a slender column in a frame braced against sidesway. As a result, the value of M_s is zero.

■ EXAMPLE 10.3

The tied column of Figure 10.7 has been approximately sized to the dimensions 12 in. × 15 in. It is to be used in a frame braced against sidesway. The column is bent in single curvature about its y axis and has an ℓ_u of 16 ft. If $k = 0.83$,

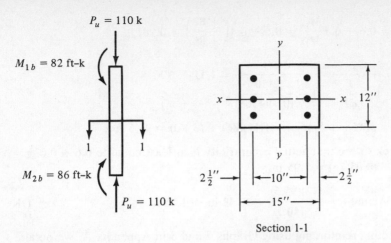

Section 1-1

Figure 10.7

$f_y = 60,000$ psi, and $f'_c = 4000$ psi, determine the reinforcing required. Consider only bending in the plane of the frame. Note also that the dead axial load P_D is 30 k.

SOLUTION

1. Is it a slender column?

$$\text{Max}\, \frac{k\ell_u}{r}\, \text{for short columns} = 34 - 12\frac{M_{1b}}{M_{2b}} = 34 - 12\left(\frac{+\,82}{+\,86}\right)$$
$$= 22.56$$

$$\text{Actual}\, \frac{k\ell_u}{r} = \frac{(0.83)(12 \times 16)}{0.3 \times 15} = 35.41 > 22.56$$

∴ It's a slender column

2. $E_c = 57,000\sqrt{f'_c} = 57,000\sqrt{4000} = 3,605,000$ psi $= 3.605 \times 10^3$ ksi

3. $I_g = (\frac{1}{12})(12)(15)^3 = 3375$ in.⁴

4. $\beta_d = \dfrac{\text{factored axial dead load}}{\text{factored axial total load}} = \dfrac{(1.4)(30)}{110} = 0.382$

5. Because reinforcing has not been selected, we must use the second EI expression:

$$EI = \frac{\dfrac{E_c I_g}{2.5}}{1 + \beta_d} = \frac{\dfrac{(3605)(3375)}{2.5}}{1 + 0.382} = 3.52 \times 10^6 \text{ k-in.}^2$$

6. $P_c = \dfrac{\pi^2 EI}{(k\ell_u)^2} = \dfrac{(\pi^2)(3.52 \times 10^6)}{(0.83 \times 12 \times 16)^2} = 1368$ k

7. $C_m = 0.6 + 0.4 \dfrac{M_{1b}}{M_{2b}} = 0.6 + 0.4\left(\dfrac{+82}{+86}\right) = 0.981$

8. $\delta_b = \dfrac{C_m}{1 - \dfrac{P_u}{\phi P_c}} = \dfrac{0.981}{1 - \dfrac{110}{(0.7)(1368)}} = 1.11$

9. $M_c = \delta_b M_{2b} + \delta_s M_{2s} = (1.11)(86) + (\delta_s)(0) = 95.5$ ft-k

10. Check to see that actual eccentricity is at least equal to $0.6 + 0.03h = 0.6 + (0.03(15) = 1.05$ in.:

$$\text{Actual } e = \frac{(12)(95.5)}{110} = 10.42 \text{ in.} > 1.05 \text{ in.} \qquad \underline{\underline{\text{OK}}}$$

11. Selecting reinforcing using Graphs 7 and 8 in Appendix A, we obtain

$$\frac{e}{h} = \frac{10.42}{15} = 0.695$$

$$\gamma = \frac{10}{15} = 0.667$$

$$\frac{\phi P_n}{A_g} \times \frac{e}{h} = \frac{110}{(12)(15)} \times 0.695 = 0.425$$

By interpolation,
$$\rho_g = 0.0158$$

$$A_s = (0.0158)(12 \times 15) = 2.84 \text{ in.}^2 \qquad \underline{\underline{\text{Use 4 \#8 bars}}} \quad \blacksquare$$

Discussion   Tests have shown that even though the lateral deflections in unbraced frames are rather small, their buckling loads are far less than they would be if the frames had been braced. As a result, the buckling strengths of the columns of an unbraced frame can be decidedly increased (perhaps as much as two or three times) by providing bracing.

In this example the author assumed that the frame was braced and yet we have said that frames are often in that gray area somewhere between being fully braced and fully unbraced. *Assuming a frame is fully braced clearly may be quite unconservative.*

10.9 DESIGN OF SLENDER COLUMNS IN UNBRACED FRAMES

The frame of Figure 10.8 is assumed to be unbraced in the plane of the frame. It supports a gravity load w_u and a short-term lateral load P_u. As a result, it is subject to both M_b and M_s moments. It will therefore be necessary to compute both δ_b and δ_s values if the column proves to be a slender one.

The M_s values are obviously caused by the lateral load in this case. The reader should realize, however, that if the gravity loads and/or the frame are unsymmetrical, additional M_s or sidesway moments will occur.

If we have an unbraced frame subjected to short-term lateral wind or earthquake loads, the columns will not have appreciable creep (which would increase lateral deflections and thus the $P\Delta$ moments). As a result, the Commentary (R10.11.5.2) suggests that β_d for this case be set equal to 0, which will in effect increase column stiffness. On the other hand, if we have a long-term lateral load or long-term lateral shear (as might occur in an unsymmetrical frame), β_d will be defined as

$$\beta_d = \frac{\text{maximum factored sustained lateral force}}{\text{maximum factored total lateral force}}$$

To illustrate the computation of the magnified moments needed for the design of a slender column in an unbraced frame, the author has chosen the very simple frame of Figure 10.8. He hopes thereby that the student will not become lost in the forest of numbers that would be necessary to handle a large frame.

The beam and columns of the frame have been tentatively sized as shown in the figure. In Example 10.4 the frame is analyzed for the following loading conditions:

$$U = 1.4D + 1.7L$$

$$U = 0.75(1.4D + 1.7L + 1.7W)$$

$$U = 0.9D + 0.75(1.7W)$$

In the example the magnification factors δ_b and δ_s for each loading case are computed as well as the magnified moments needed for design, Notice in the solution that different k values are used in determining δ_b and δ_s. The k for the δ_b calculation is determined from the alignment chart of Figure 10.3(a) for

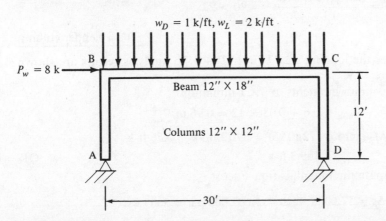

$w_D = 1$ k/ft, $w_L = 2$ k/ft

$P_w = 8$ k

B   C

Beam 12″ × 18″

12′

Columns 12″ × 12″

A   D

30′

Figure 10.8

braced frames, whereas the k for the δ_s calculation is determined from the alignment chart of part (b) of the figure for unbraced frames.

■ EXAMPLE 10.4

Determine the moments and axial forces which need to be used for the design of column CD of the unbraced frame of Figure 10.8. Consider only bending in the plane of the frame. The assumed sizes shown in the figure were used for the analyses given in the problem. $f_y = 60,000$ psi and $f'_c = 4000$ psi.

SOLUTION

1. Determine the effective length factor k using $\frac{1}{2}I_g$ for girder:

$$I_c \text{ for column} = (\tfrac{1}{12})(12)(12)^3 = 1728 \text{ in.}^4$$

$$\tfrac{1}{2}I_g \text{ for girder} = (\tfrac{1}{2})(\tfrac{1}{12})(12)(18)^3 = 2916 \text{ in.}^4$$

$$\psi_A = \frac{\dfrac{1728}{12}}{\dfrac{2916}{30}} = 1.48$$

$$\psi_B = \infty \text{ for pinned end} \qquad \text{(Practically, we might have used 10 instead.)}$$

$$k = 2.40$$

This is rather high k value. Usually if it's > 2.00 the designer will go back and try a larger column.)

2. Is it a slender column?

$$\ell_u = 12 - \tfrac{9}{12} = 11.25 \text{ ft}$$

$$\text{Max} \frac{k\ell_u}{r} \text{ to be short} = 22 \text{ for unbraced frames}$$

$$\frac{k\ell_u}{r} = \frac{(2.40)(12 \times 11.25)}{0.3 \times 12} = 90 > 22$$

$$\therefore \text{ It's a slender column}$$

3. Consider the loading case $U = 1.4D + 1.7L$ analysis results are shown in Figure 10.9.

   a. Are column moments $\geq$ ACI minimum?

   $$e_{min} = 0.6 + 0.03 \times 12 = 0.96 \text{ in.}$$

   $$M_{2b} \text{ min} = (72)(0.96) = 69.12 \text{ in.-k} = 5.76 \text{ ft-k}$$

   $$< 189.3 \text{ ft-k} \qquad\qquad \text{OK}$$

   b. Compute the magnification Factor δ_b:

   $$E_c = 57,000\sqrt{4000} = 3,605,000 \text{ psi} = 3605 \text{ ksi}$$

   $$\beta_d = \frac{(1.4)(1)}{(1.4)(1) + (1.7)(2)} = 0.292$$

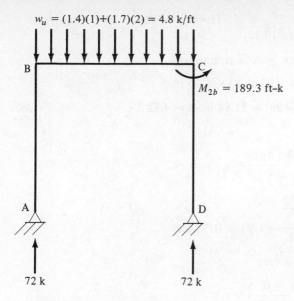

$w_u = (1.4)(1) + (1.7)(2) = 4.8$ k/ft

$M_{2b} = 189.3$ ft-k

72 k   72 k

Figure 10.9 Loading $1.4D + 1.7L$.

$$EI = \frac{\dfrac{(3605)(1728)}{2.5}}{1 + 0.292} = 1.93 \times 10^6 \text{ k-in.}^2$$

Assuming conservatively that $k = 1.0$ for calculating P_c, we obtain

$$P_c = \frac{(\pi^2)(1.93 \times 10^6)}{(1.0 \times 12 \times 11.25)^2} = 1045 \text{ k}$$

$$C_m = 0.6 + (0.4)\left(\frac{-0}{+189.3}\right) = 0.6$$

$$\delta_b = \frac{0.6}{1 - \dfrac{72}{0.7 \times 1045}} = 0.67 < 1.0 \qquad \qquad \underline{\underline{\text{Use } 1.0}}$$

c. Compute the magnification factor δ_s:
Using $k = 2.4$ for determining P_c, we obtain

$$P_c = \frac{(\pi^2)(1.93 \times 10^6)}{(2.40 \times 12 \times 11.25)^2} = 181.5 \text{ k}$$

$$\delta_s = \frac{1}{1 - \dfrac{\Sigma P_u}{\phi \Sigma P_c}} = \frac{1}{1 - \dfrac{(2)(72)}{0.7 \times 2 \times 181.5}} = 2.31$$

$$M_c = \phi_b M_{2b} + \phi_s M_{2s} = (1.0)(189.3) + (2.31)(0) = \underline{\underline{189.3 \text{ ft-k}}}$$

4. Consider the loading case $U = 0.75$ $(1.4D + 1.7L + 1.7W)$ (analysis results are shown in Figure 10.10.

   a. Are column moments $\geq$ ACI minimum?

$$e_{min} = 0.6 + 0.03 \times 12 = 0.96 \text{ in.}$$

$$M_{2b} \text{ min} = (54)(0.96) = 51.84 \text{ in.-k} = 4.32 \text{ ft-k} \qquad \underline{\underline{\text{OK}}}$$

   b. Computing δ_b:

$$\beta_d = \frac{(0.75)(1.4)(1)}{0.75[1.4 \times 1 + 1.7 \times 2]} = 0.292$$

$$EI = \frac{\dfrac{(3605)(1728)}{2.5}}{1 + 0.292} = 1.93 \times 10^6 \text{ k-in.}^2$$

$$P_c = 1045 \text{ k as before}$$

$$C_m = 0.6 + 0.4\left(\frac{-0}{+142}\right) = 0.6$$

$$\delta_b = \frac{0.6}{1 - \dfrac{54 + 4.1}{0.7 \times 1045}} = 0.65 < 1.0 \qquad \underline{\underline{\text{Use 1.0}}}$$

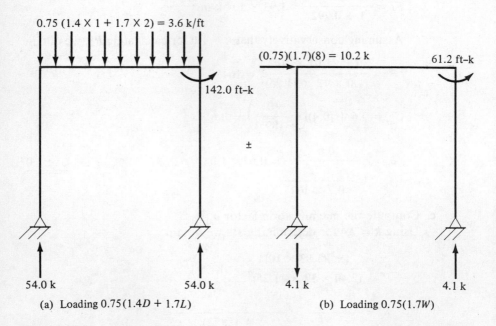

0.75 (1.4 × 1 + 1.7 × 2) = 3.6 k/ft

142.0 ft-k

$(0.75)(1.7)(8) = 10.2$ k

61.2 ft-k

±

54.0 k   54.0 k   4.1 k   4.1 k

(a) Loading $0.75(1.4D + 1.7L)$ (b) Loading $0.75(1.7W)$

Figure 10.10

c. Computing δ_s:

$\beta_d = 0$ since there is no dead load involved in this case

$$EI = \frac{\dfrac{(3605)(1728)}{2.5}}{1.0 + 0} = 2.49 \times 10^6 \text{ k-in.}^2$$

$$P_c = \frac{(\pi^2)(2.49 \times 10^6)}{(2.40 \times 12 \times 11.25)^2} = 234.1 \text{ k}$$

$$\delta_s = \frac{1}{1 - \dfrac{54 + 54 + 4.1 - 4.1}{0.7 \times 2 \times 234.1}} = 1.49$$

d. Calculate the magnified moment:

$$M_c = \delta_b M_{2b} + \delta_s M_{2s} = (1.0)(142) + (1.49)(61.2)$$
$$= \underline{\underline{233.2 \text{ ft-k}}}$$

5. Consider the loading case $0.9D + 0.75(1.7W)$ (analysis results are shown in Figure 10.11)

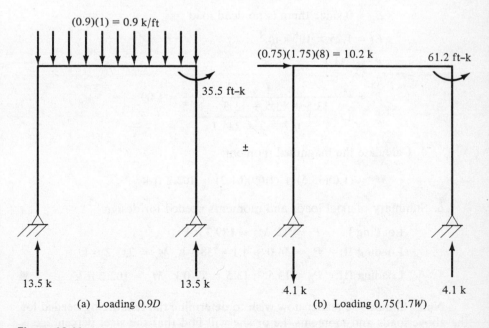

(a) Loading $0.9D$ (b) Loading $0.75(1.7W)$

Figure 10.11

a. Are column moments $\geq$ ACI minimum?

$$e_{min} = 0.6 + (0.03)(12) = 0.96 \text{ in.}$$

$$M_{2b}, \min = (13.5)(0.96) = 12.96 \text{ in.-k} = 1.08 \text{ ft-k} \qquad \underline{\underline{OK}}$$

b. Computing δ_b:

$\beta_d = 1.0$ since only D is acting

$$EI = \frac{\dfrac{(3605)(1728)}{2.5}}{1 + 1} = 1.25 \times 10^6 \text{ k-in.}^2$$

$$P_c = \frac{(\pi^2)(1.25 \times 10^6)}{(1.00 \times 12 \times 11.25)^2} = 676.9 \text{ k}$$

$$C_m = 0.6 + 0.4\left(\frac{-0}{35.5}\right) = 0.6$$

$$\delta_b = \frac{0.6}{1 - \dfrac{13.5 + 13.5}{0.7 \times 676.9}} = 0.64 < 1.0 \qquad \underline{\underline{\text{Use } 1.0}}$$

c. Computing δ_s:

$\beta_d = 0$ since there is no dead load

$EI = 1.25 \times 10^6 \text{ k-in.}^2$

$P_c = 234.1$

$$\delta_s = \frac{1}{1 - \dfrac{13.5 + 13.5 + 13.5 - 13.5}{0.7 \times 2 \times 234.1}} = 1.09$$

d. Calculate the magnified moment:

$$M_c = (1.0)(35.5) + (1.09)(61.2) = 102.2 \text{ ft-k}$$

6. Summary of axial loads and moments needed for design:

Loading I: $P_u = 72$, $M_c = 189.3$ ft-k

Loading II: $P_u = 54.0 + 4.1 = 58.1$ k, $M_c = 233.2$ ft-k

Loading III: $P_u = 13.5 + 13.5 = 27.0$ k, $M_c = 102.2$ ft-k   ■

Note: Should the reader now wish to determine the reinforcing needed for the above loads and moments, he or she will find that the steel percentage is much too high. As a result, a larger column will have to be used.

PROBLEMS

10.1 Using the alignment charts of Figure 10.3, determine the effective length factors for columns CD and DE. Assume that the beams are 12 in. × 18 in. and that the columns are 12 in. × 12 in. Use gross moments of inertia of columns and half gross moments of inertia of beams. Assume ψ_A and ψ_C = 10. (*Ans.* 0.89 and 0.76)

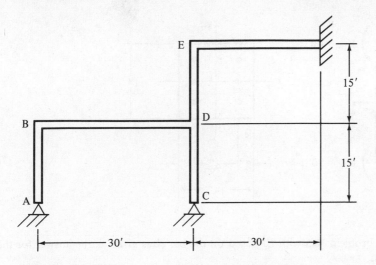

10.2 Repeat Problem 10.1 if the column bases are fixed.

10.3 Using the alignment charts of Figure 10.3, determine the effective length factors for columns AB and BC of the braced frame shown. Assume that all beams are 12 in. × 18 in. and that all columns are 12 in. × 12 in. Use gross moments of inertia of columns and half gross moments of inertia of beams. Assume far ends of beams are pinned and use ψ_A = 10. (*Ans.* 0.91 and 0.81)

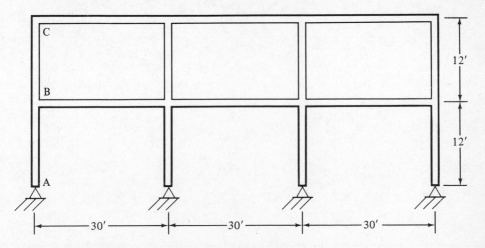

10.4 Repeat Problem 10.3 if the frame is unbraced.

10.5 The tied column shown is to be used in a braced frame. It is bent about its y axis with the factored moments shown, and ℓ_u is 15 ft. If $k = 1.0$, $f_y = 60,000$ psi, and $f_c' = 4000$ psi, select the reinforcing required. Assume $P_D = 40$ k. (*One ans.* 6 #8 bars)

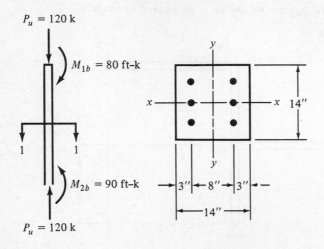

10.6 Repeat Problem 10.5 with *EI* based on the bar sizes given as the answer for that problem.

10.7 Repeat Problem 10.5 if the column is bent in reverse curvature and its length is 18 ft. Use 2nd *EI* expression. (*One ans.* 4 #8 bars)

10.8 Repeat Problem 10.5 if the frame is unbraced and if $M_{2s} = 10$ ft-k. Also assume $\sum P_c = 7200$ k and $\sum P_u = 1560$ k. Assume $k = 1.2$.

11

Footings

11.1 INTRODUCTION

Footings are those structural members used to support columns and walls and transmit their load to the underlying soils. Reinforced concrete is a material admirably suited for footings and is used as such for both reinforced concrete and structural steel buildings, bridges, towers, and other structures.

The permissible pressure on a soil beneath a footing is normally a few tons per square foot. The compressive stresses in the walls and columns of an ordinary structure may run as high as a few hundred tons per square foot. It is therefore necessary to spread these loads over sufficient soil areas to permit the soil to support the loads safely.

Not only is it desired to transfer the superstructure loads to the soil beneath in a manner that will prevent excessive or uneven settlements and rotations, but it is also necessary to provide sufficient resistance to sliding and overturning.

To accomplish these objectives, it is necessary to transmit the supported loads to a soil of sufficient strength and then to spread them out over an area such that the unit pressure is within a reasonable range. If it is not possible to dig a short distance and find a satisfactory soil, it will be necessary to use piles or caissons to do the job. These latter subjects are not considered to be within the scope of this text.

The closer a foundation is to the ground surface, the more economical it will be to construct. There are two reasons, however, that may keep the designer

Construction of foundations for Louisiana Super Dome and Convention Center (Courtesy United States Steel).

from using very shallow foundations. First, it is necessary to locate the bottom of a footing below the ground freezing level so that the soil does not freeze and expand in volume, causing vertical movement or heaving of the footing. This depth varies from about 3 to 6 feet in the northern states, with lesser values in southern states. Second, it is necessary to excavate a sufficient distance so that a satisfactory bearing material is reached, and this distance may on occasion be quite a few feet even where piles or caissons are not to be used.

11.2 TYPES OF FOOTINGS

Among the several types of reinforced concrete footings in common use are the wall, isolated, combined, raft, and pile cap types. These are briefly introduced in this section; the remainder of the chapter is used to provide more detailed information about the simpler types in this group.

1. A *wall footing* [Figure 11.1(a)] is simply an enlargement of the bottom of a wall that will sufficiently distribute the load to the foundation soil. Wall footings are normally used around the perimeter of a building and perhaps for some of the interior walls.

2. An *isolated or single-column footing* [Figure 11.1(b)] is used to support the load of a single column. These are the most commonly used footings, particularly where the loads are relatively light and the columns are not closely spaced.

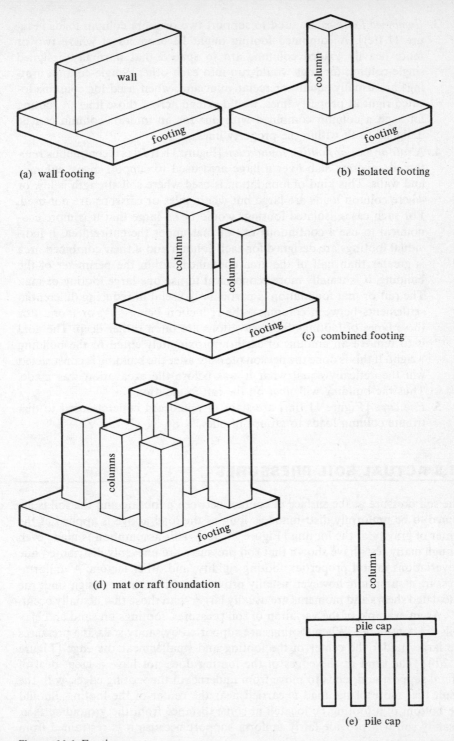

(a) wall footing

(b) isolated footing

(c) combined footing

(d) mat or raft foundation

(e) pile cap

Figure 11.1 Footings.

3. *Combined footings* are used to support two or more column loads [Figure 11.1(c)]. A combined footing might be economical where two or more heavily loaded columns are so spaced that normally designed single-column footings would run into each other. Single-column footings are usually square or rectangular and, when used for columns located right at property lines, would extend across those lines. A footing for such a column combined with one for an interior column can be designed to fit within the property lines.

4. A *raft or mat or floating foundation* [Figure 11.1(d)] is a continuous reinforced concrete slab over a large area used to support many columns and walls. This kind of foundation is used where soil strength is low or where column loads are large but where piles or caissons are not used. For such cases, isolated footings would be so large that it is more economical to use a continuous raft or mat under the entire area. If individual footings are designed for each column and if their combined area is greater than half of the area contained within the perimeter of the building, it is usually more economical to use one large footing or mat. The raft or mat foundation is particularly useful in reducing differential settlements between columns—the reduction being 50% or more. For these types of footings the excavations are often rather deep. The goal is to remove an amount of earth approximately equal to the building weight. If this is done the net soil pressure after the building is constructed will theoretically equal what it was before the excavation was made. Thus the building will *float* on the raft foundation.

5. *Pile caps* [Figure 11.1(e)] are slabs of reinforced concrete used to distribute column loads to groups of piles.

11.3 ACTUAL SOIL PRESSURES

The soil pressure at the surface of contact between a footing and the soil is assumed to be uniformly distributed as long as the load above is applied at the center of gravity of the footing [Figure 11.2(a)]. This assumption is made even though many tests have shown that soil pressures are unevenly distributed due to variations in soil properties, footing rigidity, and other factors. A uniform-pressure assumption, however, usually provides a conservative design since the calculated shears and moments are usually larger than those that actually occur.

As an example of the variation of soil pressures, footings on sand and clay soils are considered. When footings are supported by sandy soils, the pressures are larger under the center of the footing and smaller near the edge [Figure 11.2(b)]. The sand at the edges of the footing does not have a great deal of lateral support and tends to move from underneath the footing edges, with the result that more of the load is carried near the center of the footing. Should the bottom of a footing be located at some distance from the ground surface, a sandy soil will provide fairly uniform support because it is restrained from lateral movement.

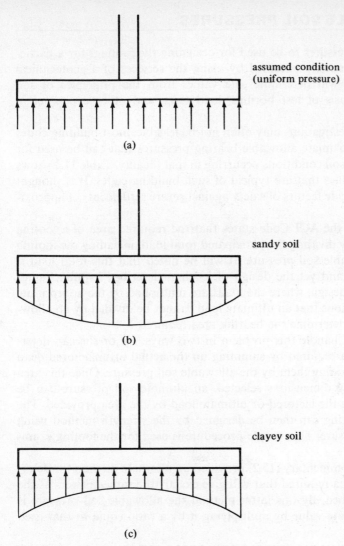

assumed condition
(uniform pressure)

(a)

sandy soil

(b)

clayey soil

(c)

Figure 11.2 Soil conditions

Just the opposite situation is true for footings supported by clayey soils The clay under the edges of the footing sticks to or has a cohesion with the surrounding clay soil. As a result, more of the load is carried at the edge of the footing than near the middle. [See Figure 11.2(c)].

The designer should clearly understand that the assumption of uniform soil pressure underneath footings is made for reasons of simplifying calculations and may very well have to be revised for some soil conditions.

Should the load be eccentrically applied to a footing with respect to the center of gravity of the footing, the soil pressure is assumed to vary uniformly in proportion to the moment, as illustrated in Section 11.12.

11.4 ALLOWABLE SOIL PRESSURES

The allowable soil pressures to be used for designing the footings for a particular structure are desirably obtained by using the services of a geotechnical engineer. He or she will determine safe values from the principles of soil mechanics on the basis of test borings, load tests, and other experimental investigations.

Because such investigations may often not be feasible, most building codes provide certain approximate allowable bearing pressures that can be used for the types of soils and soil conditions occurring in that locality. Table 11.1 shows a set of allowable values that are typical of such building codes. It is thought that these values provide factors of safety against severe settlements of approximately 3.

Section 15.2.2 of the ACI Code states that the required area of a footing is to be determined by dividing the anticipated total load, including the footing weight, by the allowable soil pressure. It will be noted that this total load is the unfactored load, and yet the design of footings described in this chapter is based on strength design, where the loads are multiplied by the appropriate load factors. It is obvious that an ultimate load cannot be divided by an allowable soil pressure to determine the bearing area required.

The designer can handle this problem in two ways. He or she can determine the bearing area required by summing up the actual or unfactored dead and live loads and dividing them by the allowable soil pressure. Once this area is determined and the dimensions selected, an ultimate soil pressure can be computed by dividing the factored or ultimate load by the area provided. The remainder of the footing can then be designed by the strength method using this ultimate soil pressure. This simple procedure is used for the footing examples here.

The 1971 ACI Commentary (15.2) provided an alternate method for determining the footing area required that will give exactly the same answers as the procedure just described. By this latter method the allowable soil pressure is increased to an ultimate value by multiplying it by a ratio equal to that used

TABLE 11.1 MAXIMUM ALLOWABLE SOIL PRESSURE

Class of material	Maximum allowable soil pressure (kip/ft²)
Rock	20% of ultimate crushing strength
Compact coarse sand, compact fine sand, hard clay, or sand clay	8
Medium stiff clay or sandy clay	6
Compact inorganic sand and silt mixtures	4
Loose sand	3
Soft sand clay or clay	2
Loose inorganic sand–silt mixtures	1
Loose organic sand–silt mixtures, muck, or bay mud	0

for increasing the service loads. For instance, the ratio for D and L loads would be

$$\text{Ratio} = \frac{1.4D + 1.7L}{D + L}$$

Or for $D + L + W$,

$$\text{Ratio} = \frac{0.75(1.4D + 1.7L + 1.7W)}{D + L + W}$$

The resulting ultimate soil pressure can be divided into the ultimate column load to determine the area required.

11.5 DESIGN OF WALL FOOTINGS

The theory used for designing beams is applicable to the design of footings with only a few modifications. The upward soil pressure under the wall footing of Figure 11.3 tends to bend the footing into the deformed shape shown. The footings will be designed as shallow beams for the moments and shears involved.

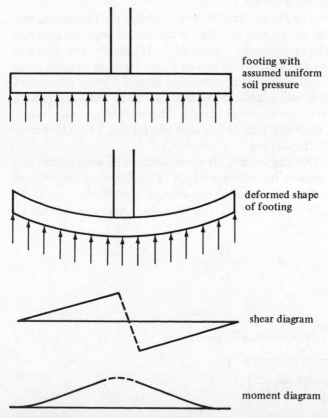

footing with
assumed uniform
soil pressure

deformed shape
of footing

shear diagram

moment diagram

Figure 11.3

In beams where loads are usually only a few hundred pounds per foot and spans are fairly large, sizes are almost always proportioned for moment. In footings, loads from the supporting soils may run several thousand pounds per foot and spans are relatively short. As a result, shears will almost always control depths.

It appears that the maximum moment in this footing occurs under the middle of the wall, but tests have shown that this is not correct because of the rigidity of such walls. If the walls are of reinforced concrete with their considerable rigidity, it is considered satisfactory to compute the moments at the faces of the walls (ACI Code 15.4.2). Should a footing be supporting a masonry wall with its greater flexibility, the Code states that the moment should be taken at a section halfway from the face of the wall to its center. (For a column with a steel base plate the critical section for moment is to be located halfway from the face of the column to the edge of the plate.)

To compute the bending moments and shears in a footing, it is necessary to compute only the net upward pressure q_u caused by the factored wall loads above. In other words, the weight of the footing and soil on top of the footing can be neglected. These items cause an upward pressure equal to their downward weights, and they cancel each other for purposes of computing shears and moments. In a similar manner, it is obvious that there are no moments or shears existing in a book lying flat on a table.

Should a wall footing be loaded until it fails in shear, the failure will not occur on a vertical plane at the wall face but rather at an angle of approximately 45° with the wall, as shown in Figure 11.4. Apparently the diagonal tension, which one would expect to cause cracks in between the two diagonal lines, is opposed by the squeezing or compression caused by the downward wall load and the upward soil pressure. Outside this zone the compression effect is negligible in its effect on diagonal tension. Therefore, shear may be calculated at a distance d from the face of the wall (ACI Code 11.1.3.1) due to the loads located outside the section.

The use of stirrups in footings is usually considered to be impractical and uneconomical. For this reason the effective depth of wall footings is selected such that V_u is limited to the permissible shear ϕV_c that the concrete can carry without web reinforcing, that is, $\phi 2\sqrt{f'_c}bd$. The following expression is used to select the depths of wall footings:

$$d = \frac{V_u}{(\phi)(2\sqrt{f'_c})(b)}$$

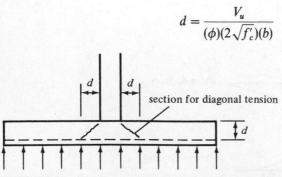

Figure 11.4

The design of wall footings is conveniently handled by using 12-in. widths of the wall, as shown in Figure 11.5. Such a practice is followed for the design of a wall footing in Example 11.1. It will be noted that Section 15.7 of the Code states that the depth of a footing above the bottom reinforcing bars may be no less than 6 in. for footings on soils and 12 in. for those on piles. Thus total minimum practical depths are at least 10 in. for the regular spread footings and 16 in. for pile caps.

The design of a wall footing is illustrated in Example 11.1. Although the example problems and homework problems of this chapter use various f'_c values, 3000-psi concrete is the most common and generally the most economical choice. Occasionally, when it is very important to minimize footing depths and weights, stronger concretes may be used. For most cases, however, the extra cost of higher-strength concrete will appreciably exceed the money saved with the smaller concrete volume.

The determination of a footing depth is a trial-and-error problem. The designer assumes an effective depth d, computes the d required for shear, tries another d, computes the d required for shear, and so on, until the assumed value and the calculated value are within about 1 in. of each other.

You probably get upset with the author when he assumes a footing size. You say "where in the world did you get that value?" The author thinks of what seems like a reasonable footing size to him and starts there. He computes the d required for shear and probably finds he's missed the assumed value quite a bit. He then tries another value roughly two-thirds to three-fourths of the way from the trial value to the computed value (for wall footings) and computes d. (For column footings he probably goes about half of the way from the trial value to the computed value.) Two trials are usually sufficient. The author has the advantage over you in that often in preparing the sample problems for this textbook he made some scratch-paper trials before he arrived at the assumed values used in the textbook.

■ EXAMPLE 11.1

Design a wall footing to support a 12-in.-wide reinforced concrete wall with a dead load $D = 20$ k/ft and a live load $L = 15$ k/ft. The bottom of the footing

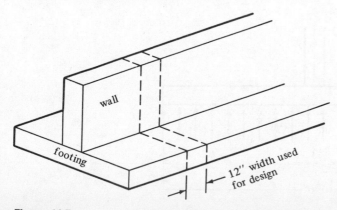

Figure 11.5

is 4 ft below the final grade, the soil weighs 100 lb/ft³, the allowable soil pressure q_a is 4 ksf, f_y = 60 ksi, and f'_c = 4 ksi.

SOLUTION

Assume a 12-in.-thick footing (d = 8.5 in). The cover is determined by referring to the Code (7.7.1), which says that for concrete cast against and permanently exposed to the earth, a minimum of 3 in. clear distance outside any reinforcing is required.

The footing weight is $\left(\frac{12}{12}\right)(150)$ = 150 psf and the soil fill on top of the footing is $\left(\frac{36}{12}\right)(100)$ = 300 psf. So 450 psf of the allowable soil pressure q_a is used to support the footing itself and the soil fill on top. The remaining soil pressure is available to support the wall loads. It is called q_e, the effective soil pressure

$$q_e = 4000 - \left(\tfrac{12}{12}\right)(150) - \left(\tfrac{36}{12}\right)(100) = 3550 \text{ psf}$$

$$\text{Width of footing required} = \frac{35}{3.55} = 9.86' \qquad \underline{\text{say 10'0''}}$$

Bearing Pressure for Strength Design

$$q_u = \frac{(1.4)(20) + (1.7)(15)}{10.00} = 5.35 \text{ ksf}$$

Depth Required for Shear at a Distance d from Face of Wall (See Figure 11.6)

$$V_u = \left(\frac{10.00}{2} - \frac{6}{12} - \frac{8.5}{12}\right)(5.35) = 20.29 \text{ k}$$

$$d = \frac{20,290}{(0.85)(2\sqrt{4000})(12)} = 15.73'' + 3.5'' = 19.23'' > 12'' \qquad \underline{\text{Try again}}$$

Assume 18-in. Footing (d = 14.5 in.)

$$q_e = 4000 - \left(\tfrac{18}{12}\right)(150) - \left(\tfrac{30}{12}\right)(100) = 3525 \text{ psf}$$

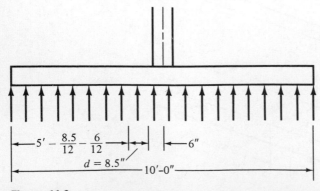

Figure 11.6

$$\text{Width required} = \frac{35}{3.525} = 9.93' \qquad \underline{\text{say } 10'0''}$$

Bearing Pressure for Strength Design

$$q_u = \frac{(1.4)(20) + (1.7)(15)}{10.00} = 5.35 \text{ ksf}$$

Depth Required for Shear

$$V_u = \left(\frac{10.00}{2} - \frac{6}{12} - \frac{14.5}{12}\right)(5.35) = 17.61 \text{ k}$$

$$d = \frac{17{,}610}{(0.85)(2\sqrt{4000})(12)} = 13.65'' + 3.5'' = 17.15'' \qquad \underline{\text{use } 18'' \text{ total depth}}$$

Steel Area (Using $d = 14.5$ in.) Taking moments at face of wall,

$$\text{Cantilever length} = \frac{10.00}{2} - \frac{6}{12} = 4.50'$$

$$M_u = (4.50)(5.35)(2.25) = 54.17 \text{ ft-k}$$

$$\frac{M_u}{\phi b d^2} = \frac{(12)(54{,}170)}{(0.9)(12)(14.5)^2} = 286$$

From Table A.14 (see Appendix), $\rho = 0.00499$. Thus

$$A_s = (0.00499)(12)(14.5) = 0.868 \text{ in.}^2 \qquad \underline{\text{Use } \#8 \text{ at } 10''}$$

Development Length

$\ell_{db} = 29.79''$ from Appendix Table A.15

$\ell'_{db} = (29.79)(0.8 \text{ modification factor as given in ACI Section } 12.2.3.4)$

$\qquad = 23.83''$ but may not be less than $\dfrac{(0.03)(1.00)(60{,}000)}{\sqrt{4000}} = 28.46''$

$\ell'_{db} = 28.46''$

$\ell_d = \ell'_{db}$ (because there are no applicable modification factors
  in Table 6.2 of Chapter 6 of this text) $= 28.46'' \qquad \text{say } 29''$

$$\left(\begin{array}{c}\text{Available development length} \\ \text{assuming bars are cut off} \\ \text{3'' from edge of footing}\end{array}\right) = \frac{10'0''}{2} - 6'' - 3''$$

$$= \left(\begin{array}{c}4'3'' \text{ from face of} \\ \text{wall at section of} \\ \text{maximum moment}\end{array}\right) > 29'' \qquad \underline{\underline{\text{OK}}}$$

(The bars should be extended to a point not less than 3 in. or more than 6 in. from the edge of the footing.)

Longitudinal Temperature and Shrinkage Steel

$$A_s = (0.0018)(12)(18) = 0.389 \text{ in.}^2 \qquad \underline{\text{Use \#5 at } 9''} \quad \blacksquare$$

11.6 DESIGN OF SQUARE ISOLATED FOOTINGS

Single-column footings usually provide the most economical column foundations. Such footings are generally square in plan, but they can just as well be rectangular or even circular or octagonal. Rectangular footings are used where such shapes are dictated by the available space or where the cross sections of the columns are very pronounced rectangles.

Most footings consist of slabs of constant thickness, such as the one shown in Figure 11.7(a), but if calculated thicknesses are greater than 3 or 4 ft, it may be economical to use stepped footings, as illustrated in Figure 11.7(b). The shears and moments in a footing are obviously larger near the column, with the result that greater depths are required in that area as compared to the outer parts of the footing. For very large footings, such as for bridge piers, stepped footings can give appreciable savings in concrete quantities.

Occasionally, sloped footings [Figure 11.7(c)] are used instead of the stepped ones, but labor costs can be a problem. Whether stepped or sloped, it is considered necessary to place the concrete for the entire footing in a single pour to ensure the construction of a monolithic structure, thus avoiding horizontal shearing weakness. If this procedure is not followed, it is desirable to use keys or shear friction reinforcing between the parts to ensure monolithic action. In addition, when sloped or stepped footings are used it is necessary to check stresses at more than one section in the footing. For example, steel area

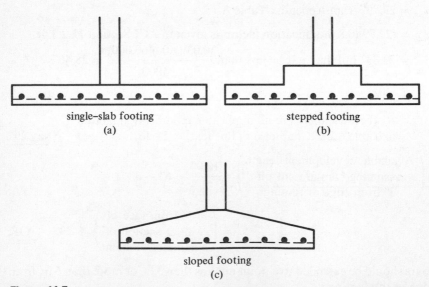

single–slab footing
(a)

stepped footing
(b)

sloped footing
(c)

Figure 11.7

and development length requirements should be checked at steps as well as at the faces of walls or columns.

Before a column footing can be designed, it is necessary to make a few comments regarding shears and moments. This is done in the paragraphs to follow, while a related subject, load transfer from columns to footings, is discussed in Section 11.8.

Shears

There are two shear conditions that must be considered in column footings, regardless of their shapes. The first of these is one-way or beam shear, which is the same as that considered in wall footings in the preceding section. For this discussion, reference is made to the footing of Figure 11.8. The total shear

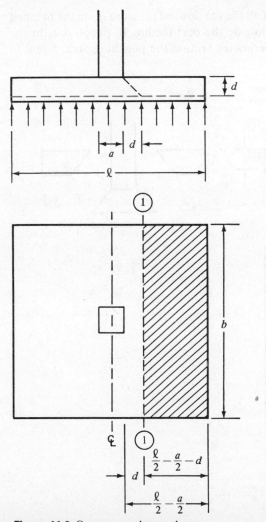

Figure 11.8 One-way or beam shear.

(V_{u1}) to be taken along section 1–1 equals the net soil pressure times q_u the hatched area outside the section. In the expression to follow, b is the whole width of the footing. The maximum value of V_{u1} if stirrups are not used equals ϕV_c, equals $\phi 2\sqrt{f'_c}\, bd$, and the maximum depth required is as follows:

$$d = \frac{V_{u1}}{\phi 2\sqrt{f'_c}\, b}$$

The second shear condition is two-way or punching shear, with reference being made to Figure 11.9. The compression load from the column tends to spread out into the footing, opposing diagonal tension in that area, with the result that a square column tends to punch out a piece of the slab, which has the shape of a truncated pyramid. The ACI Code (11.12.1.2) states that the critical section for two-way shear is located at a distance $d/2$ from the face of the column.

The shear force V_{u2} consists of all the net upward pressure q_u on the hatched area shown, that is, on the area outside the part tending to punch out. In the expressions to follow, b_o is the perimeter around the punching area, equal to

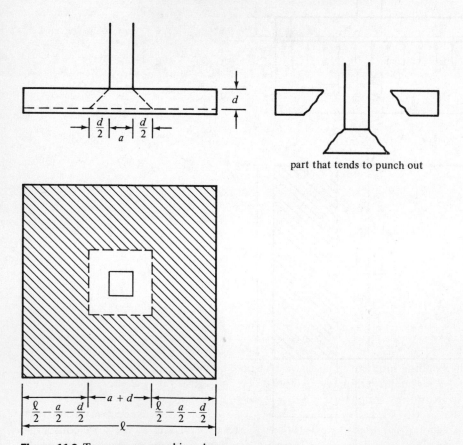

part that tends to punch out

Figure 11.9 Two-way or punching shear.

$4(a + d)$ in the figure. The nominal two-way shear strength of the concrete V_c is specified to be the smallest value obtained by substituting into the applicable equations which follow.

The first expression is the usual punching shear strength:

$$V_c = 4\sqrt{f_c'}\,b_o d \qquad \text{(ACI Equation 11-38)}$$

Tests have shown that when footing slabs are subjected to bending in two directions and when the long side of the loaded area is more than two times the length of the short side, the shear strength $V_c = 4\sqrt{f_c'}\,b_o d$ may be much too high. In the expression to follow, β_c is the ratio of the long side of the column to the short side of the column. For non-rectangular loaded areas, V_c can be obtained with the same expression if β_c is considered to equal the minimum length of the effective loaded area divided by the maximum width measured perpendicular to the length:

$$V_c = \left(2 + \frac{4}{\beta_c}\right)\sqrt{f_c'}\,b_o d \qquad \text{(ACI Equation 11-36)}$$

For columns where the ratio of b_o to d is quite large, another value of V_c may control. In the expression to follow, α_s is 40 for interior columns, 30 for edge columns and 20 for corner columns. These numbers correspond to 4-sided, 3-sided, and 2-sided perimeters, respectively.

$$V_c = \left(\frac{\alpha_s d}{b_o} + 2\right)\sqrt{f_c'}\,b_o d \qquad \text{(ACI Equation 11-37)}$$

The value of d required for two-way shear is the largest obtained from the following expressions:

$$d = \frac{V_{u2}}{\phi 4 \sqrt{f_c'}\,b_o}$$

$$d = \frac{V_{u2}}{\phi\left(2 + \dfrac{4}{\beta_c}\right)\sqrt{f_c'}\,b_o} \qquad \text{(not applicable unless } \beta_c \text{ is } > 2)$$

$$d = \frac{V_{u2}}{\phi\left(\dfrac{\alpha_s d}{b_o} + 2\right)\sqrt{f_c'}\,b_o}$$

Moments

The bending moment in a square reinforced concrete footing is the same about both axes due to symmetry. It should be noted, however, that the effective depth of the footing cannot be the same in the two directions because the bars in one direction rest on top of the bars in the other direction. The effective depth used for calculations might be the average for the two directions or, more conservatively, the value for the bars on top. This lesser value is used for the examples in this text. Although the result is some excess of steel in one direction, it is felt that the steel in either direction must be sufficient to resist the moment in that direction. It should be clearly understood that having an excess of steel in one direction will not make up for a shortage in the other direction.

The critical section for bending is taken at the face of a reinforced concrete column or halfway between the middle and edge of a masonry wall or at a distance halfway from the edge of the base plate and the face of the column if structural steel columns are used (Code 15.4.2).

The determination of footing depths by the procedure described here will often require several cycles of a trial and error procedure. There are, however, many tables and handbooks available with which footing depths can be accurately estimated. One of these is the previously mentioned *CRSI Handbook*. In addition, there are many rules of thumb used by designers for making initial thickness estimates, such as 20% of the footing width or the column diameter plus 3 in. and so on.

The flexural steel percentage calculated for footings is very often less than $200/f_y$. The ACI Code (10.5.3), however, states that in slabs of uniform thickness the minimum area and maximum spacing of reinforcing bars in the direction of bending shall be as required for shrinkage and temperature reinforcement. Most designers feel that the combination of high shears and low ρ values that often occur in footings is not good practice and thus use $200/f_y$ as a minimum.

Example 11.2 illustrates the design of an isolated column footing.

■ EXAMPLE 11.2

Design a square column footing for a 16-in. square tied interior column that supports a dead load $D = 200$ k and a live load $L = 160$ k. The column is reinforced with 8 #8 bars, the base of the footing is 5 ft below grade, the soil weight is 100 lb/ft³, $f_y = 60,000$ psi, $f'_c = 3000$ psi, and $q_a = 5000$ psf.

SOLUTION

Assume 24-in. Footing ($d = 19.5$ in.)

$$q_e = 5000 - (\tfrac{24}{12})(150) - (\tfrac{36}{12})(100) = 4400 \text{ psf}$$

$$A \text{ required} = \frac{360}{4.400} = 81.82 \text{ ft}^2$$

Use 9 ft 0 in. × 9 ft 0 in. footing $= 81.0 \text{ ft}^2$

Bearing Pressure for Strength Design

$$q_u = \frac{(1.4)(200) + (1.7)(160)}{81.0} = 6.81 \text{ ksf}$$

Depth Required for Two-Way or Punching Shear (Figure 11.10)

$$b_o = (4)(35.5) = 142 \text{ in.}$$

$$V_{u2} = (81.0 - 2.96^2)(6.81) = 491.94 \text{ k}$$

$$d = \frac{491,940}{(0.85)(4\sqrt{3000})(142)} = 18.60 \text{ in.} < 19.5 \text{ in.}$$

$$d = \frac{491,940}{0.85\left(\dfrac{40 \times 19.5}{142} + 2\right)(\sqrt{3000})(142)} = 9.33 \text{ in.} < 19.5 \text{ in.}$$

Depth Required for One-Way Shear (Figure 11.11)

$$V_{u1} = (9.00)(2.208)(6.81) = 135.33 \text{ k}$$

$$d = \frac{135,330}{(0.85)(2\sqrt{3000})(108)} = 13.46 \text{ in.} < 19.5 \text{ in.} \qquad \underline{\underline{\text{OK}}}$$

<u>Use 24-in. total depth</u>

Steel Area

$$M_u = (3.83)(9.00)(6.81)\left(\frac{3.83}{2}\right) = 449.5 \text{ ft-k}$$

$$\frac{M_u}{\phi b d^2} = \frac{(12)(449,500)}{(0.9)(108)(19.5)^2} = 145.94 \text{ psi}$$

$$\text{Use } \rho_{min} = \frac{200}{60,000} = 0.00333$$

$$A_s = (0.00333)(108)(19.5) = 7.01 \text{ in.}^2$$

<u>Use 9 #8 bars in both directions</u>

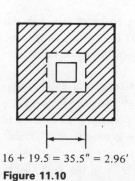

$16 + 19.5 = 35.5'' = 2.96'$

Figure 11.10

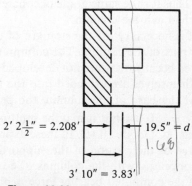

$2'\,2\frac{1}{2}'' = 2.208'$ ⟷ ⟷ $19.5'' = d$

1.48

$3'\,10'' = 3.83'$

Figure 11.11

Development Length

$\ell_{db} = 34.40$ in.   (from Appendix Table A.15)

$\ell'_{db} = (34.40)(0.8$ modification factor as given in ACI Section 12.2.3.4)

$$= 27.52 \text{ in. but may not be less than } \frac{(0.03)(1.00)(60,000)}{\sqrt{3000}} = 32.86 \text{ in.}$$

$\ell_d = \ell'_{db}$ because there are no other applicable modification factors in Table 6.2 of Chapter 6

Actual ℓ_d furnished $= 46 - 3 = 43$ in. > 32.86 in.   OK ∎

11.7 FOOTINGS SUPPORTING ROUND OR REGULAR POLYGON-SHAPED COLUMNS

Sometimes footings are designed to support round columns or regular polygon-shaped columns. If such is the case, Section 15.3 of the Code states that the column may be replaced with a square member having the same area as the round or polygonal one. Then the equivalent square is used for locating the critical sections for moment, shear, and development length.

11.8 LOAD TRANSFER FROM COLUMNS TO FOOTINGS

All forces acting at the base of a column must be satisfactorily transferred into the footing. Compressive forces can be transmitted directly by bearing, whereas uplift or tensile forces must be transferred by developed reinforcing.

A column transfers its load directly to the supporting footing over an area equal to the cross-sectional area of the column. The footing surrounding this contact area, however, supplies appreciable lateral support to the directly loaded part, with the result that the loaded concrete in the footing can support more load. Thus for the same grade of concrete, the footing can carry a larger bearing load than can the column.

It is necessary for the strength of the concrete alone to be checked in the lower part of the column. The column reinforcing bars at that point cannot be counted because they are not developed unless dowels are provided or unless the bars themselves are extended into the footing.

At the base of the column, the permitted bearing strength is $\phi(0.85 f'_c A_1)$ (where ϕ is 0.70), but it may be multiplied by $\sqrt{A_2/A_1} \leq 2$ for bearing on the footing (ACI Code 10.15). In these expressions, A_1 is the column area and A_2 is the area of the portion of the supporting footing that is geometrically similar and concentric with the columns. (See Figure 11.12.)

If the computed bearing force is higher than the smaller of the two allow-able values, it will be necessary to carry the excess with dowels or with column

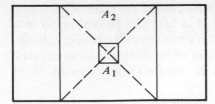

Figure 11.12

bars extended into the footing. Should the computed bearing force be less than the allowable value, no dowels or extended reinforcing are theoretically needed, but the Code (15.8.2.1) states that there must be a minimum area of dowels furnished equal to no less than 0.005 times the gross cross-sectional area of the column or pedestal.

The development length of the bars must be sufficient to transfer the compression to the supporting member, as per the ACI Code (12.3). In no case may the area of the designed reinforcement or dowels be less than the area specified for the case where the allowable bearing force was not exceeded. As a practical matter in placing dowels, it should be noted that regardless of how small a distance they theoretically need to be extended down into the footing, they are usually bent at their ends and set on the main footing reinforcing, as shown in Figure 11.13. There the dowels can be tied firmly in place and not be in danger of being pushed through the footing during construction, as might easily happen otherwise. The bent part of the bar does not count as part of the compression development length (ACI Code 12.5.5).

An alternative to the procedure described in the preceding paragraph is to place the footing concrete without dowels and then to push straight dowels down into the concrete while it is still in a plastic state. This practice is permitted by the Code in its Section 16.4.2 and is especially useful for plain concrete footings (to be discussed in Section 11.14 of this chapter). It is essential that the dowels be maintained in their correct position as long as the concrete is plastic. Before the engineer approves the use of straight dowels as described here, he or she must be satisfied that the dowels will be properly placed and the concrete satisfactorily compacted around them.

The Code normally does not permit the use of lapped splices for #14 and #18 compression bars because tests have shown that welded splices or other

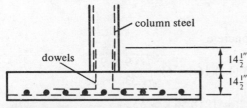

Figure 11.13

types of connections are necessary. Nevertheless, based on years of successful use the Code (15.8.2.3) states that #14 and #18 bars may be lap-spliced with dowels (no larger than #11) to provide for force transfer at the base of columns or walls or pedestals. The development lengths of these dowels however, shall be equal to those required for #14 or #18 bars or for the splice length of the dowels, whichever is greater.

If the computed development length of dowels is greater than the distance available from the top of the footing down to the top of the tensile steel, three possible solutions are available. One or more of the following alternatives may be selected:

1. A larger number of smaller dowels may be used. The smaller diameters will result in smaller development lengths.
2. A deeper footing may be used.
3. A cap or pedestal may be constructed on top of the footing to provide the extra development length needed.

Should bending moments or uplift forces have to be transferred to a footing such that the dowels would be in tension, the development lengths must satisfy the requirements for tension bars. For tension development length into the footing, a hook at the bottom of the dowel may be considered as being effective.

If there is moment or uplift, it will be necessary for the designer to conform to the splice requirements of Section 12.17 of the code in determining the distance the dowels must be extended up into the wall or column.

Examples 11.3 and 11.4 provide brief examples of column-to-footing load transfer calculations for vertical forces only. Consideration is given to lateral forces and moments in Sections 11.12 and 11.13 of this chapter.

■ EXAMPLE 11.3

Design for load transfer from the 16-in. × 16-in. column to the 9-ft-0-in. × 9-ft-0-in. footing of Example 11.2. $D = 200$ k, $L = 160$ k, $f'_c = 4000$ psi for both footing and column, and $f_y = 60,000$ psi.

SOLUTION

Bearing force at base of column $= (1.4)(200) + (1.7)(160) = 552$ k

Allowable bearing force in concrete at base of column

$$= \phi(0.85 f'_c A_1) = (0.70)(0.85)(4.0)(16 \times 16)$$
$$= 609.28 \text{ k} > 552 \text{ k}$$

Allowable bearing force in footing concrete

$$= \phi(0.85 f'_c A_1)\sqrt{\frac{A_2}{A_1}} = (0.70)(0.85)(4.0)(16 \times 16)\sqrt{\frac{9 \times 9}{1.33 \times 1.33}}$$

$$= (0.70)(0.85)(4.0)(16 \times 16)(\text{Use } 2) = 1218.56 \text{ k} > 552 \text{ k} \qquad \underline{\text{OK}}$$

Minimum A_s for dowels $= (0.005)(16 \times 16) = 1.28$ in.^{2} $\qquad$ Use 4 #6 bars

Development Lengths of Dowels

$$\ell_d = \frac{0.02 f_y d_b}{\sqrt{f_c'}} = \frac{(0.02)(60,000)(0.75)}{\sqrt{4000}} = 14.23'' \leftarrow$$

$$\ell_d = 0.0003 f_y d_b = (0.0003)(60,000)(0.75) = 13.50''$$

$$\ell_d = 8.00''$$

Use 4 #6 dowels extending $14\frac{1}{2}$ in. up into column and down into footing and set on top of reinforcing mat, as shown in Figure 11.13.   ∎

∎ EXAMPLE 11.4

Design for load transfer from a 14-in. × 14-in. column to a 13-ft-0-in. × 13-ft-0-in. footing with a P_u of 800 k, $f_c' = 3000$ psi in the footing and 5000 psi in the column, and $f_y = 60,000$ psi. The column has 8 #8 bars.

SOLUTION

Bearing force at base of column $= P_u = 800$ k

$\quad = \phi$ allowable bearing force in concrete $+ \phi$ strength of dowels

Allowable bearing force in concrete at base of column

$\quad = (0.70)(0.85)(5.0)(14 \times 14) = 583.1$ k < 800 k   <u>no good</u>

Allowable bearing force on footing concrete

$\quad = (0.70)(0.85)(3.0)(14 \times 14)(\text{Use } 2) = 699.72$ k < 800 k   <u>no good</u>

Therefore, the dowels must be designed for excess load.

$$\text{Excess load} = 800 - 583.1 = 216.9 \text{ k}$$

$$A_s \text{ of dowels} = \frac{216.9}{60} = \underline{3.62 \text{ in.}^2} \text{ or } (0.005)(14)(14) = 0.98 \text{ in.}^2$$

Use 6 #7 bars

Development Length of Dowels into Column

$$\ell_d = \frac{(0.02)(60,000)(0.875)}{\sqrt{5000}} = 14.85''$$

$$\ell_d = (0.0003)(60,000)(0.875) = 15.75'' \leftarrow$$

$$\ell_d = 8''$$

Development Length of Dowels into Footing (Different from Column Values Because f_c' Values are Different)

$$\ell_d = \frac{(0.02)(60,000)(0.875)}{\sqrt{3000}} = 19.17'' \leftarrow$$

$$\ell_d = (0.0003)(60,000)(0.875) = 15.75''$$

Use 6 #7 dowels extending 16 in. up into the column and 20 in. down into the footing.   ∎

11.9 RECTANGULAR ISOLATED FOOTINGS

As previously mentioned, isolated footings may be rectangular in plan if the column has a very pronounced rectangular shape or if the space available for the footing forces the designer into using a rectangular shape. Should a square footing be feasible, it is normally to be desired over a rectangular one because it will require less material and will be simpler to construct.

The design procedure is almost identical with the one used for square footings. After the required area is calculated and the lateral dimensions selected, the depths required for one-way and two-way shear are determined by the usual methods. One-way shear will very often control the depths for rectangular footings, whereas two-way shear normally controls the depths of square footings.

The next step is to select the reinforcing in the long direction. These bars are spaced uniformly across the footing, but such is not the case for the short-span reinforcing. With reference to Figure 11.14, it can be seen that the support provided by the footing to the column will be concentrated near the middle of the footing, and thus the moment in the short direction will be concentrated somewhat in the same area near the column.

As a result of this concentration effect, it seems only logical to concentrate a large proportion of the short-span reinforcing in this area. The Code (15.4.4.2) states that a certain minimum percentage of the total short-span reinforcing should be placed in a band width equal to the length of the shorter direction of the footing. The amount of reinforcing in this band is to be determined with the following expression, in which β is the ratio of the length of the long side to the width of the short side of the footing:

$$\frac{\text{Reinforcing in band width}}{\text{Total reinforcing in short direction}} = \frac{2}{\beta + 1} \qquad \text{(ACI Equation 15-1)}$$

The remaining reinforcing in the short direction should be uniformly spaced over the ends of the footing but should at least meet the shrinkage and temperature requirements of the ACI Code (7.12).

Example 11.5 presents the partial design of a rectangular footing in which the depths for one- and two-way shears are determined and the reinforcement selected.

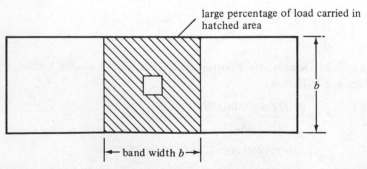

Figure 11.14

■ EXAMPLE 11.5

Design a rectangular footing for an 18-in. square edge column with a dead load of 185 k and a live load of 150 k. Make the length of the long side equal to twice the width of the short side. $f_y = 60,000$ psi, $f'_c = 4000$ psi, and $q_a = 4000$ psf. Assume the base of the footing is 5 ft 0 in. below grade.

SOLUTION

Assume 24-in. Footing ($d = 19.5$ in.)

$$q_e = 4000 - (\tfrac{24}{12})(150) - (\tfrac{36}{12})(100) = 3400 \text{ psf}$$

$$A \text{ required} = \frac{335}{3.4} = 98.5 \text{ ft}^2 \qquad \underline{\text{Use } 7'0'' \times 14'0'' = 98.0 \text{ ft}^2}$$

$$q_u = \frac{(1.4)(185) + (1.7)(150)}{98.0} = 5.24 \text{ ksf}$$

Checking Depth for One-Way Shear (Figure 11.15)

$$b = 84''$$

$$V_{u_1} = (7.0)(4.62)(5.24) = 169.5 \text{ k}$$

$$d = \frac{169,500}{(0.85)(2\sqrt{4000})(84)} = 18.77 + 4.5 = 23.27'' \qquad \underline{\text{Use } 24''}$$

Checking Depth for Two-Way Shear (Figure 11.16)

$$b_o = (4)(37.5) = 150''$$

$$V_{u_2} = [98.0 - (3.12)^2](5.24) = 462.5 \text{ k}$$

$$d = \frac{462,500}{(0.85)(4\sqrt{4000})(150)} = 14.34''$$

$$d = \frac{462,500}{0.85\left(\dfrac{30 \times 19.5}{150} + 2\right)(\sqrt{4000})(150)} = 9.72''$$

Design of Longitudinal Steel

$$M_u = (6.25)(7.0)(5.24)\left(\frac{6.25}{2}\right) = 716.4 \text{ ft-k}$$

$$\frac{M_u}{\phi b d^2} = \frac{(12)(716,400)}{(0.90)(84)(19.5)^2} = 299 \text{ psi}$$

$$p = 0.00523$$

$$A_s = (0.00523)(84)(19.5) = 8.57 \text{ in.}^2 \qquad \underline{\text{Use } 11 \text{ #8 bars}}$$

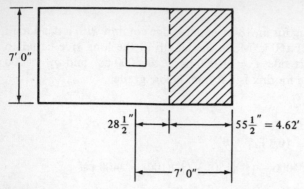

Figure 11.15

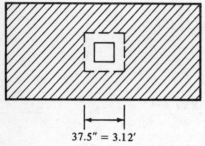

Figure 11.16

Design of Steel in Short Direction (Figure 11.17)

$$M_u = (2.75)(14.0)(5.24)\left(\frac{2.75}{2}\right) = 277.4 \text{ ft-k}$$

$$\frac{M_u}{\phi bd^2} = \frac{(12)(277,400)}{(0.90)(168)(19.5)^2} = 57.9 \text{ psi}$$

Use

$$p = \frac{200}{f_y} = \frac{200}{60,000} = 0.00333$$

$$A_s = (0.00333)(168)(19.5) = 10.91 \text{ in.}^2 \qquad \underline{\text{Use 18 } \#7 \text{ bars}}$$

$$\frac{\text{Reinforcing in band width}}{\text{Total reinforcing in short direction}} = \frac{2}{2+1} = \frac{2}{3}$$

$$\underline{\text{Use } \tfrac{2}{3} \times 18 = 12 \text{ bars in band width}} \quad \blacksquare$$

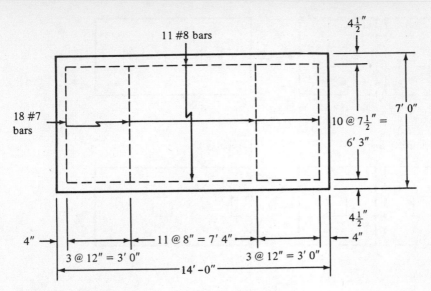

Figure 11.17 Two-way footing bar spacing diagram.

11.10 COMBINED FOOTINGS

Combined footings support more than one column. One situation where they may be used occurs when the columns are so close together that isolated individual footings would run into each other [Figure 11.18(a)]. Another frequent use of combined footings occurs where one column is very close to a property line, causing the usual isolated footing to extend across the line. For this situation the footing for the exterior column may be combined with the one for an interior column, as shown in Figure 11.18(b).

On some occasions where a column is close to a property line and where it is desired to combine its footing with that of an interior column, the interior column will be so far away as to make the idea impractical economically. For such a case, counterweights or deadmen may be provided to take care of the eccentric loading.

Because it is desirable to make bearing pressures uniform throughout the footing, the centroid of the footing should be made to coincide with the centroid of the column loads to attempt to prevent uneven settlements. This can be accomplished with combined footings that are rectangular in plan. Should the interior column load be greater than that of the exterior column, the footing may be so proportioned that its centroid will be in the correct position by extending the inward projection of the footing, as shown in the rectangular footing of Figure 11.18(b).

Other combined footing shapes that will enable the designer to make the centroids coincide are the trapezoid and strap or T footings shown in Figure

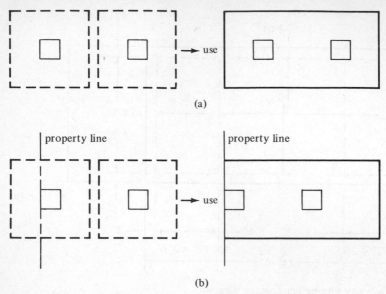

(a)

(b)

Figure 11.18

11.19(a) and 11.19(b). Footings with these shapes are usually economical when there are large differences between the magnitudes of the column loads or where the spaces available do not lend themselves to long rectangular footings. When trapezoidal footings are used, the longitudinal bars are usually arranged in a fan shape with alternate bars cut off some distance from the narrow end.

You probably realize that a problem arises in establishing the centroids of loads and footings when deciding whether to use service or factored loads. The required centroid of the footing will be slightly different for the two cases. The author determines the footing areas and centroids with the service loads (ACI Code 15.2.2), but the factored loads could be used with reasonable results, too. The important item is to be consistent throughout the entire problem.

The design of combined footings has not been standardized as have the procedures used for the previous problems worked in this chapter. For this reason, practicing reinforced concrete designers use slightly varying approaches. One of these methods is described in the paragraphs to follow.

First, the required area of the footing is determined for the service loads and the footing dimensions are selected such that the centroids coincide. Then the various loads are multiplied by the appropriate load factors and the shear and moment diagrams are drawn along the long side of the footing for these loads. After the shear and moment diagrams are prepared, the depth for one- and two-way shear is determined and the reinforcing in the long direction is selected.

In the short direction it is assumed that each column load is spread over a width in the long direction equal to the column width plus $d/2$ on each side if that much footing is available. Then the steel is designed and a minimum

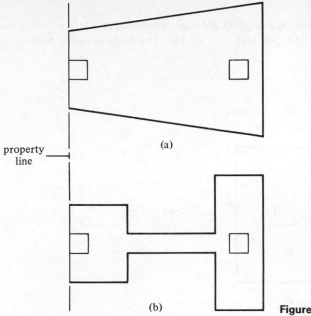

(a)

property
line

(b)

Figure 11.19

amount of steel for temperature and shrinkage is provided in the remaining part of the footing.

The ACI Code does not specify an exact width for these transverse strips, and designers may make their own assumptions as to reasonable values. The width selected will probably have very little influence on the transverse bending capacity of the footing, but it can affect appreciably its punching or two-way shear resistance. If the flexural reinforcing is placed within the area considered for two-way shear, this lightly stressed reinforcing will reduce the width of the diagonal shear cracks and will also increase the aggregate interlock along the shear surfaces.

Space is not taken here to design completely a combined footing, but Example 11.6 is presented to show those parts of the design that are different from the previous examples of this chapter. A comment should be made about the moment diagram. If the length of the footing is not selected so that its centroid is located exactly at the centroid of the column loads, the shear and moment diagrams will not close well at all since the numbers are very sensitive. Nevertheless, it is considered good practice to round off the footing lateral dimensions to the nearest 3 in. Another factor which keeps the moment diagram from closing is the fact that the average load factors at the columns will be different if the column loads are different. We could improve the situation a little by taking the total column factored loads and dividing the result by the total working loads to get an average load factor. This value (which works out to be 1.53 in Example 11.6) could then be multiplied by the total working load at each column and used for drawing the shear and moment diagrams.

■ EXAMPLE 11.6

Design a rectangular combined footing for the two columns shown in Figure 11.20. q_a = 5 ksf, f'_c = 3000 psi, and f_y = 60 ksi. The bottom of the footing is to be 6 ft below grade.

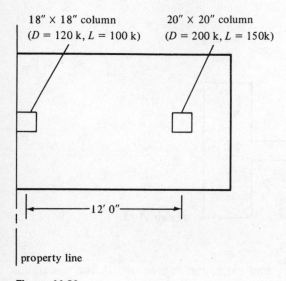

18" × 18" column
(D = 120 k, L = 100 k)

20" × 20" column
(D = 200 k, L = 150k)

—12' 0"—

property line

Figure 11.20

SOLUTION

Assume 30-in. Footing (d = 25.5 in.)

$$q_e = 5000 - (\tfrac{30}{12})(150) - (\tfrac{42}{12})(100) = 4275 \text{ psf}$$

$$A \text{ required} = \frac{570}{4.275} = 133.3 \text{ ft}^2$$

Locate Center of Gravity of Column Service Loads

$$X \text{ from c.g. of left column} = \frac{(350)(12)}{570} = 7.37'$$

$$\text{Distance from property line to c.g.} = 0.75 + 7.37$$
$$= 8.12'(2 \times 8.12 = 16.24', \text{ say } 16'3'')$$

Use 16'3" × 8'3" footing (A = 134 ft²).

$$q_u = \frac{(1.4)(320) + (1.7)(250)}{134} = 6.515 \text{ ksf}$$

Drawing Shear and Moment Diagrams (Figure 11.21)

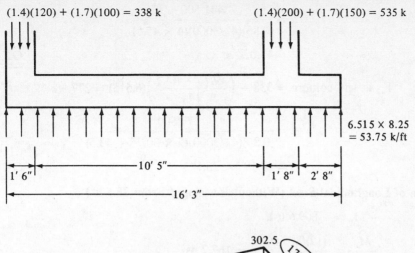

$$(1.4)(120) + (1.7)(100) = 338 \text{ k}$$

$$(1.4)(200) + (1.7)(150) = 535 \text{ k}$$

6.515 × 8.25
= 53.75 k/ft

10' 5"

1' 6"

1' 8"   2' 8"

16' 3"

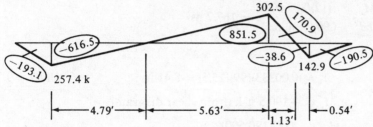

302.5

170.9

851.5

−616.5

−38.6

142.9   −190.5

−193.1

257.4 k

4.79'

5.63'

1.13'

0.54'

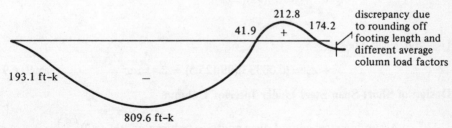

212.8

41.9   174.2

+

discrepancy due
to rounding off
footing length and
different average
column load factors

193.1 ft–k

−

809.6 ft–k

Figure 11.21

Depth Required for One-Way Shear

V_{u1} a distance d from interior face of right column

$$= 302.5 - \left(\frac{25.5}{12}\right)(53.75) = 188.3 \text{ k}$$

$$d = \frac{(188,300)}{(0.85)(2\sqrt{3000})(99)} = 20.42'' < 25.5'' \qquad \underline{\underline{\text{OK}}}$$

Depth Required for Two-Way Shear (ACI Equations 11-36 and 11-37 not shown as they do not control)

$$V_{u2} \text{ at right column} = 535 - \left(\frac{45.5}{12}\right)^2 (6.515) = 441.3 \text{ k}$$

$$d = \frac{441,500}{(0.85)(4\sqrt{3000})(4 \times 45.5)}$$

$$= 13.02'' < 25.5'' \qquad \underline{\underline{\text{OK}}}$$

$$V_{u2} \text{ at left column} = 338 - \left(\frac{30.75 \times 43.5}{144}\right)(6.515) = 277.5 \text{ k}$$

$$d = \frac{277,500}{(0.85)(4\sqrt{3000})(2 \times 30.75 + 43.5)}$$

$$= 14.19'' < 25.5'' \qquad \underline{\underline{\text{OK}}}$$

Design of Longitudinal Steel (Without Revising d From 25.5 in.)

$$-M_u = -809.6 \text{ ft-k}$$

$$\frac{M_u}{\phi bd^2} = \frac{(12)(809,600)}{(.90)(99)(25.5)^2} = 167.7 \text{ psi}$$

Use $200/f_y = 0.00333$.

$$-A_s = (0.00333)(99)(25.5) = 8.41 \text{ in.}^2 \qquad \text{say 9 \#9}$$

$$+M_u = +190.5 \text{ ft-k (from shear diagram)}$$

$$\frac{M_u}{\phi bd^2} = \frac{(12)(190,500)}{(0.90)(99)(25.5)^2} = 39.46 \text{ psi}$$

$$p = <\rho_{\min}$$

Use $200/f_y = 0.00333$.

$$+A_s = (0.00333)(99)(25.5) = 8.41 \text{ in.}^2 \qquad \text{say 9 \#9}$$

Design of Short-Span Steel Under Interior Column

Assuming steel spread over width = column width + $(2)\left(\dfrac{d}{2}\right)$

$$= 20 + (2)\left(\frac{25.5}{2}\right) = 45.5''$$

Referring to Figure 11.22 and calculating M_u:

$$M_u = (3.29)(64.85)\left(\frac{3.29}{2}\right) = 350.97 \text{ ft-k}$$

$$\frac{M_u}{\phi bd^2} = \frac{(12)(350,970)}{(0.90)(45.5)(25.5)^2} = 158.2 \text{ psi}$$

$$p = 0.00272 \text{ (by formula in Section 3.14 of this text)} < \frac{200}{f_y}$$

Use 0.00333

$$A_s = (0.00333)(45.5)(25.5) = 3.86 \text{ in.}^2 \qquad \blacksquare$$

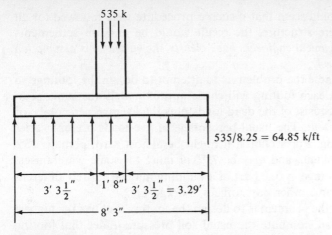

Figure 11.22

A similar procedure is used under the exterior column where the steel is spread over a width equal to 18 in. plus $d/2$ and not 18 in. plus $2(d/2)$, because sufficient room is not available on the property-line side of the column.

The author has treated this footing in the long direction just as he would treat a long beam. If such a procedure is followed, the reader may quite logically think that shear in the long direction should be handled just as it would be in a long beam. In other words, if the shear V_u exceeds $\phi V_c/2$, stirrups should be used. The Code does not address this particular point. Many designers will use stirrups in such situations where $V_u > \phi V_c/2$, particularly where the footings are deep and narrow. The author feels that footings should be designed very conservatively because the failure of a footing can result in severe damage to the structure above. Thus he feels stirrups should be designed for a combined footing as they would for a regular beam.

11.11 FOOTING DESIGN FOR EQUAL SETTLEMENTS

If three men are walking along a road carrying a log on their shoulders (a statically indeterminate situation) and one of them decides to lower his shoulder by 1 in., the result will be a drastic effect on the load supported by the other men. In the same way, if the footings of a building should settle by different amounts, the shears and moments throughout the structure will be greatly changed. In addition, there will be detrimental effects on the fitting of doors, windows, and partitions. Should all the footings settle by the same amount, however, these adverse effects will not occur. Thus equal settlement is the goal of the designer.

The footings considered in preceding sections have had their areas selected by taking the total dead plus live loads and dividing the sum by the allowable soil pressure. It would seem that if such a procedure were followed for all the footings of an entire structure, the result would be uniform settlements throughout—*but geotechnical engineers have clearly shown that this assumption may be very much in error.*

A better way to handle the problem is to attempt to design the footings so that the *usual loads* on each footing will cause approximately the same pressures. The usual loads consist of the dead loads plus the average percentage of live loads normally present. The usual percentage of live loads present varies from building to building. For a church it might be almost zero, perhaps 25% to 30% for an office building, and may be 75% or more for some warehouses. Furthermore, the percentage in one part of a building may be entirely different from that present for some other part (offices, storage, etc.).

One way to handle the problem is to design the footing that has the highest ratio of live to dead load, compute the usual soil pressure under that footing using dead load plus the estimated average percentage of live load, and then determine the areas required for the other footings so their usual soil pressures are all the same. It should be noted that the dead load plus 100% of the live load should not cause a pressure greater than the allowable soil pressure under any of the footings.

A student of soil mechanics will realize that the method of usual pressures, though not a bad design procedure, will not ensure equal settlements. This approach at best will only lessen the amounts of differential settlements. He or she will remember first that large footings tend to settle more than small footings, even though their soil pressures are the same, because the large footings exert compression on a larger and deeper mass of soil. There are other items that can cause differential settlements. Different types of soils may be present at different parts of the building; part of the area may be in fill and part in cut; there may be mutual influence of one footing on another; and so forth.

Example 11.7 illustrates the usual load procedure for a group of five isolated footings.

■ EXAMPLE 11.7

Determine the footing areas required for the loads given in Table 11.2 so that the usual soil pressures are equal. Assume that the usual live load percentage is 30% for all the footings, $q_e = 4$ ksf.

SOLUTION

The largest percentage of live load to dead load occurs for footing D.

$$\text{Area required for footing } D = \frac{100 + 150}{4} = 62.5 \text{ ft}^2$$

$$\text{Usual soil pressure under footing } D = \frac{100 + (0.30)(150)}{62.5} = 2.32 \text{ ksf}$$

TABLE 11.2   FOOTINGS

Footing	Dead load (k)	Live load (k)
A	150	200
B	120	100
C	140	150
D	100	150
E	160	200

Computing the areas required for the other footings and computing their soil pressures under total service loads, the results are as shown in Table 11.3. ∎

TABLE 11.3   AREAS AND SOIL PRESSURES

Footing	Usual load = D + 0.30L (k)	Area required = usual load ÷ 2.32 (ft²)	Total soil pressure (ksf)
A	210	90.5	3.87
B	150	64.7	3.40
C	185	79.7	3.64
D	145	62.5	4.00
E	220	94.8	3.80

11.12 FOOTINGS SUBJECTED TO LATERAL MOMENTS

Walls or columns often transfer moments as well as vertical loads to their footings. These moments may be due to wind, earthquake, lateral earth pressures, and so on. Such a situation is represented by the vertical load P and the bending moment M shown in Figure 11.23.

Because of this moment the resultant force will not coincide with the centroid of the footing. Of course, if the moment is constant in magnitude and direction, it will be possible to place the center of the footing under the resultant load and avoid the eccentricity, but lateral forces such as wind and earthquake can come from any direction and symmetrical footings will be needed.

The effect of the moment is to produce a uniformly varying soil pressure, which can be determined at any point with the expression

$$q = -\frac{P}{A} \pm \frac{Mc}{I}$$

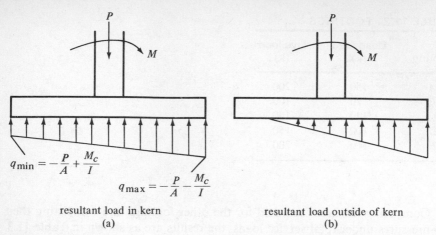

$$q_{min} = -\frac{P}{A} + \frac{Mc}{I}$$

$$q_{max} = -\frac{P}{A} - \frac{Mc}{I}$$

resultant load in kern
(a)

resultant load outside of kern
(b)

Figure 11.23

In this discussion the term *kern* is used. If the resultant force strikes the footing base within the kern, the value of $-P/A$ is larger than $+Mc/I$ at every point and the entire footing base is in compression, as shown in Figure 11.23(a). If the resultant strikes the footing base outside the kern, the value of $+Mc/I$ will at some points be larger than $-P/A$ and there will be uplift or tension. The soil cannot resist tension, and the pressure variation will be as shown in Figure 11.23(b). The location of the kern can be determined by replacing Mc/I with Pec/I, equating it to P/A, and solving for e.

Should the eccentricity be larger than this value, the method described for calculating soil pressures $[(-P/A) \pm Mc/I)]$ is not correct. To compute the pressure for such a situation it is necessary to realize that the centroid of the upward pressure must for equilibrium coincide with the centroid of the vertical component of the downward load. In Figure 11.24 it is assumed that the distance to this point from the right edge of the footing is a. Then the soil pressure will be spread over the distance $3a$ as shown. For a rectangular footing with dimensions $\ell \times b$, the total upward soil pressure is equated to the downward load and the resulting expression solved for q_{max} as follows:

$$(\tfrac{1}{2})(3ab)(q_{max}) = P$$

$$q_{max} = \frac{2P}{3a}$$

Example 11.8 shows that the required area of a footing subjected to a vertical load and a lateral moment can be determined by trial and error. The procedure is to assume a size, calculate the maximum soil pressure, compare it with the allowable pressure, assume another size, and so on.

Once the area has been established, the remaining design will be handled as it was for other footings. Although the shears and moments are not uniform, the theory of design is unchanged. The factored loads are computed, the bearing pressures are determined, and the shears and moments are calculated. For

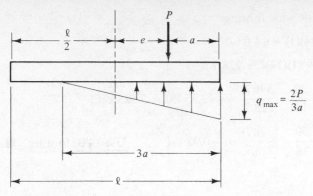

Figure 11.24

strength design the footing must be proportioned for the worst of these situations (from ACI Code Section 9.2):

$$U = 1.4D + 1.7L$$

$$U = 0.75(1.4D + 1.7L + 1.7W)$$

$$U = 0.9D + 1.3W$$

■ EXAMPLE 11.8
Determine the width needed for a wall footing to support loads: $D = 18$ k/ft and $L = 12$ k/ft. In addition, a lateral load of 6 k/ft is assumed to be applied 5 ft above the top of the footing. Assume the footing is 18 in. thick, its base is 4 ft below the final grade, and $q_a = 4$ ksf.

SOLUTION

First Trial

Neglecting moment $q_e = 4000 - (\frac{18}{12})(150) - (\frac{30}{12})(100) = 3525$ psf

$$\text{Width required} = \frac{18 + 12}{3.525} = 8.51' \qquad\qquad \underline{\text{try } 9'0''}$$

$$M = 6 \times \text{distance to footing base} = 6 \times 6.5 = 39 \text{ ft-k}$$

$$A = (9)(1) = 9 \text{ ft}^2$$

$$I = (\tfrac{1}{12})(1)(9)^3 = 60.75 \text{ ft}^4$$

$$q_{max} = -\frac{P}{A} - \frac{Mc}{I} = -\frac{30}{9} - \frac{(39)(4.5)}{60.75}$$

$$= -6.22 \text{ ksf} > 3.525 \text{ ksf} \qquad\qquad \underline{\text{no good}}$$

$$q_{min} = -\frac{P}{A} + \frac{Mc}{I} = -\frac{30}{9} + \frac{(39)(4.5)}{60.75} = -0.44 \text{ ksf}$$

Second trial Assume 14-ft wide footing:

$$A = (14)(1) = 14 \text{ ft}^2$$

$$I = (\tfrac{1}{12})(1)(14)^3 = 228.7 \text{ ft}^4$$

$$q_{max} = -\frac{30}{14} - \frac{(39)(7)}{228.7} = -3.33 \text{ ksf} < 3.525 \text{ ksf}$$

$$q_{min} = -\frac{30}{14} + \frac{(39)(7)}{228.7} = -0.95 \text{ ksf} \qquad \text{Use } 14'0'' \text{ footing} \quad\blacksquare$$

11.13 TRANSFER OF HORIZONTAL FORCES

When it is necessary to transfer horizontal forces from walls or columns to footings, the shear-friction method previously discussed in Section 7.12 of this text should be used. Sometimes shear keys (see Figure 12.1) are used between walls or columns and footings. This practice is of rather questionable value, however, because appreciable slipping has to occur to develop a shear key. A shear key may be thought of as providing an additional mechanical safety factor, but none of the lateral design force should be assigned to it.

The following example illustrates the consideration of lateral force transfer by the shear-friction concept.

■ EXAMPLE 11.9

A 14-in. × 14-in. column is supported by a 13-ft-0-in. × 13-ft-0-in. footing ($f'_c = 3000$ psi and $f_y = 60,000$ psi for both). For vertical compression force transfer, 6 #6 dowels (2.65 in.²) were selected extending 17 in. up into the column and 17 in. down into the footing. Design for a horizontal factored force V_u of 65 k acting at the base of the column. Assume that the footing concrete has not been intentionally roughened before the column concrete is placed. Thus $\mu = 0.6\lambda = (0.6)(1.0)$ for normal-weight concrete = 0.6.

SOLUTION

Determine Minimum Shear-Friction Reinforcement Required by ACI Equation 11-26

$$V_n = \frac{V_u}{\phi} = \frac{65}{0.85} = 76.47 \text{ k}$$

$$V_n = A_v f_y \mu$$

$$A_v = \frac{76.47}{(60)(0.6)} = 2.12 \text{ in.}^2 < 2.65 \text{ in.}^2 \qquad \underline{\text{OK}}$$

The 6 #6 dowels (2.65 in.²) present may also be used as shear-friction reinforcing. If their area had not been sufficient, it could have been increased and/or the value of μ could be increased significantly by intentionally roughening the concrete as described in Section 11.7.4.3 of the Code.

Check Tensile Development Lengths of These Dowels

$$\ell_d = 19.37 \text{ in. from Table A.15 in Appendix}$$

It may be possible to reduce this development length if more reinforcing is provided than required as perhaps $(2.12/2.65) \times 19.37 = 15.50$, but not less than $17''$ as required by compression.

Compute Maximum Shear Transfer Strength Permitted by the Code (11.7.5)

$$V_u \leq \phi 0.2 f'_c A_c, \text{ but not } > \phi(800 A_c)$$
$$\leq (0.85)(0.2)(3)(14 \times 14) = 99.96 \text{ k} > 65 \text{ k} \qquad \underline{\text{OK}}$$

but not $> (0.85)(800)(14 \times 14) = 133,280 \text{ lb} = 133.28 \text{ k}$ ■

11.14 PLAIN CONCRETE FOOTINGS

Occasionally, plain concrete footings are used to support light loads if the supporting soil is of good quality. Very often the widths and thicknesses of such footings are determined by rules of thumb such as "the depth of a plain footing must be equal to no less than the projection beyond the edges of the wall." In this section, however, a plain concrete footing is designed in accordance with the requirements of the ACI. The design of both plain concrete footings and pedestals is not covered in the ACI Code but is included in ACI Standard 318.1-83.

The ACI Standard states that when plain concrete footings are supported by soil, they cannot have an edge thickness less than 8 in. and they cannot be used on piles. The critical sections for shear and moment for plain concrete footings are the same as for reinforced concrete footings. Allowable bending and shear stresses are given for both strength design and allowable stress design. For strength design the maximum allowable flexural tension is $5.0\phi\sqrt{f'_c}$, and the average shear stress (V_u/bd) may not exceed $2\phi\sqrt{f'_c}$ for one-way shear, nor $(2 + 4/\beta_c)\phi\sqrt{f'_c}$ but not $>4\phi\sqrt{f'_c}$ for two-way shear, where $\phi = 0.65$ for flexure, shear, compression, and bearing.

A plain concrete footing will obviously require considerably more concrete than will a reinforced one. On the other hand the cost of purchasing reinforcing and placing it will be eliminated. Furthermore the use of plain concrete footings will enable us to save construction time in that we don't have to order the reinforcing and place it before the concrete can be poured. As a result plain concrete footings may be economical on more occasions than you would realize.

The author feels very strongly that even though plain footings are designed in accordance with the ACI requirements, they should at the very least be reinforced in the longitudinal direction with reinforcing to keep temperature and shrinkage cracks within reason and to enable the footing to bridge over soft spots in the underlying soil. Nevertheless, Example 11.10 presents the design of a plain concrete footing in accordance with the ACI Standard.

For concrete that is placed directly against the soil it is customary to neglect the bottom 2 and perhaps 3 in. of concrete in the strength calculations. This concrete is neglected to account for uneven excavation for the footing and for some loss of mixing water to the soil and other contamination.

■ EXAMPLE 11.10

Design a plain concrete footing for a 12-in. reinforced concrete wall that supports a dead load of 12 k/ft, including the wall weight, and a 6-k/ft live load. The base of the footing is to be 5 ft below the final grade, $f'_c = 3000$ psi, and $q_a = 4000$ psf.

SOLUTION

Assume 24-in. Footing

$$q_e = 4000 - (\tfrac{24}{12})(145) - (\tfrac{36}{12})(100) = 3410 \text{ psf}$$

$$\text{Width required} = \frac{18}{3.41} = 5.28 \qquad \text{say } 5'6''$$

Bearing Pressure for Strength Design

$$q_u = \frac{(1.4)(12) + (1.7)(6)}{5.5} = 4.91 \text{ ksf}$$

Checking Bending Stress Neglecting Bottom 3 in. of Footing

$$M_u = (2.25)(4.91)(1.125) = 12.43 \text{ ft-k}$$

$$I = (\tfrac{1}{12})(12)(21)^3 = 9261 \text{ in.}^4$$

$$f_t = \frac{(12)(12,430)(10.5)}{9261} = 169 \text{ psi} < (5)(0.65)\sqrt{3000} = 178 \text{ psi} \qquad \underline{\text{OK}}$$

Checking Shear Stress at a Distance d of 21 in. from Face of Wall

$$V_u = \left(\frac{5.5}{2} - \frac{6}{12} - \frac{21}{12}\right)(4.91) = 2.46 \text{ k}$$

$$v_u = \frac{2460}{(12)(21)} = 10 \text{ psi} < (2)(0.65)(\sqrt{3000}) = 71 \text{ psi} \qquad \underline{\text{OK}} \quad ■$$

11.15 COMPUTER EXAMPLES

In the example computer problems which follow, you are given the opportunity to apply development length modification factors. After doing this, you should keep pressing the ENTER key until the development length is complete. The computer will use successive modification factors each equal to 1.0 for each of the slots you don't use.

■ EXAMPLE 11.11
Repeat Example 11.1 using the enclosed computer disk.

```
DESIGN OF WALL FOOTINGS
-----------------------
Wall Thickness                                       t (in) =   12.0
Dead load per foot of wall                    Wd (lbs/ft) =   20000
Live load per foot of wall                    Wl (lbs/ft) =   15000
Distance from top of backfill to bottom of footing  z (in) =   48.0
Concrete Weight                                     (pcf) =  150.0
Soil Weight                                         (pcf) =  100.0
Allowable Soil Pressure                             (psf) =   4000
Concrete Strength                            f'c (psi) =   4000
Steel Yield Strength                          fy (psi) =  60000
Reinforced Concrete Wall


RESULTS
=======
Use: (1)  10.0 ft. wide footing
     (2)  14.5 in. effective depth
     (3)  # 8 bars @  10.0 in main reinforcing
     (4) 13 # 5 bars shrinkage and temperature reinforcing
```

■ EXAMPLE 11.12
Repeat Example 11.5 using the enclosed computer disk.

```
DESIGN OF SQUARE/RECTANGULAR FOOTINGS FOR SQUARE/RECTANGUAR COLUMNS
------------------------------------------------------------------------
Dead load                                      D (kips) =   185.0
Live load                                      L (kips) =   150.0
Column Side Dimension in Long Direction of Footing  t1 (in) =   18.0
Column Side Dimension in Short Direction of Footing t2 (in) =   18.0
Distance from Top of fill to Bottom of Footing   z (in) =   60.0
Allowable Soil Pressure                        qa (psf) =   4000
Concrete Weight                                    (pcf) =  150.0
Soil Weight                                        (pcf) =  100.0
Concrete Strength                             f'c (psi) =   4000
Steel Yield Strength                          fy (psi) =  60000
REQUIRED AREA (ft^2) =  97.0  Width B(ft) =  7.0  Length L(ft) =  14.0
Interior Column


RESULTS
=======
Use: (1)   7.0 ft. × 14.0 ft. footing
     (2)  19.5 in. effective depth
     (3) 11 # 8 bars in long direction
     (4) 18 # 7 bars in short direction
          with 12 of them in band width of 7 feet under column.
```

PROBLEMS

For all problems assume reinforced concrete weighs 150 lb/ft^3, plain concrete 145 lb/ft^3, and soil 100 lb/ft^3.

For Problems 11.1 to 11.5, design wall footings for the values given. The walls are to consist of reinforced concrete.

Problem	Wall thickness (in.)	D (k/ft)	L (k/ft)	f'_c (ksi)	f_y (ksi)	q_a (ksf)	Distance from bottom of footing to final grade (ft)
11.1	12	16	20	3	60	4	4
11.2	12	22	28	3	60	4	5
11.3	14	20	32	4	60	5	6
11.4	15	26	30	4	60	6	4
11.5	15	22	32	4	40	4	5

(Answer to Problem 11.1: 20 in. footing, #8 @ 10 in. main steel)

(Answer to Problem 11.3: 24 in. footing, #8 @ 8 in. main steel)

(Answer to Problem 11.5: 27 in. footing, #10 @ 7 in. main steel)

11.6 Repeat Problem 11.1 if a masonry wall is used.

In Problems 11.7 to 11.12, design square single-column footings for the values given.

Problem	Column size (in.)	D (k)	L (k)	f'_c (ksi)	f_y (ksi)	q_a (ksf)	Distance from bottom of footing to final grade (ft)	Column location
11.7	12 × 12	90	140	3	60	5	4	interior
11.8	12 × 12	100	160	3	60	4	5	interior
11.9	15 × 15	160	120	4	60	4	6	edge
11.10	15 × 15	180	220	3	40	4	4	corner
11.11	16 × 16	150	120	3	60	6	5	edge
11.12	Round, 18 dia.	175	190	3	60	5	6	interior

(Answer to Problem 11.7: 21 in. footing, with 8 #7 each direction)

(Answer to Problem 11.9: 20 in. footing, with 10 #7 each direction)

(Answer to Problem 11.11: 21 in. footing, with 8 #7 each direction)

11.13 Design for load transfer from an 18″ × 18″ column with 6 #8 bars ($D = 240$ k, $L = 300$ k) to an 8-ft-0-in. × 8-ft-0-in. footing. $f'_c = 3$ ksi for footing and 5 ksi for column. $f_y = 60$ ksi. (*Ans.* 6 #5, 14 in into footing, 12 in. into column)

11.14 Repeat Problem 11.7 if a rectangular footing with the length equal to twice the width is used.

11.15 Design a footing with a length equal to 1.5 times its width for the following: 12 in. × 12 in. edge column, $D = 120$ k, $L = 150$ k, $f'_c = 3000$ psi, $f_y = 60{,}000$ psi, $q_a = 4$ ksf and a distance from top of backfill to bottom of footing = 6 ft. (*Ans.* 7 ft 6 in. × 11 ft 3 in. with 8 #8 bars in long direction)

11.16 Design a footing limited to a maximum width of 7 ft 0 in. for the following: 15 in. × 15 in. interior column, $D = 200$ k, $L = 140$ k, $f'_c = 4000$ psi, $f_y = 60{,}000$ psi, $q_a = 4$ ksf and a distance from top of backfill to bottom of footing = 5 ft.

11.17 Design a rectangular combined footing for the two columns shown in the accompanying illustration. The bottom of the footing is to be 5 ft below the final grade, $f'_c = 3.5$ ksi, $f_y = 50$ ksi, and $q_a = 5$ ksf (*Ans.* 26 in. footing, 10 ft 3 in. × 12 ft 0 in., 11 #9 bars long direction)

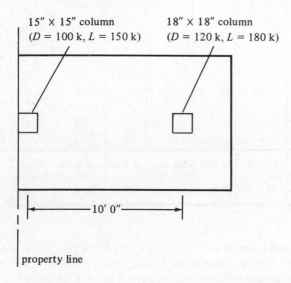

15″ × 15″ column
($D = 100$ k, $L = 150$ k)

18″ × 18″ column
($D = 120$ k, $L = 180$ k)

|← 10′ 0″ →|

property line

11.18 Determine the footing areas required for the loads given in the accompanying table so that the usual soil pressures are equal. Assume $q_e = 5$ ksf and a usual live load percentage of 30% for all of the footings.

Footing	Dead load	Live load
A	120 k	150 k
B	150 k	200 k
C	100 k	180 k
D	120 k	160 k
E	140 k	200 k
F	120 k	180 k

In Problems 11.19 and 11.20 determine the width required for the wall footings. Assume footings have total thicknesses of 24 in.

Problem	Reinforced concrete wall thickness	D	L	f'_c	f_y	q_a	Lateral load	Height of lateral load above top of footing	Distance from bottom of footing to final grade
11.19	12"	14 k/ft	22 k/ft	3 ksi	60 ksi	4 ksf	8 k/ft	4'	4'
11.20	14"	16 k/ft	20 k/ft	4 ksi	60 ksi	4 ksf	10 k/ft	5'	5'

(Answer to Problem 11.19: 15 ft 6 in.)

11.21 Repeat Problem 11.13 if a lateral force $V_u = 120$ k acts at the base of the column. Use the shear-friction concept. Assume that footing concrete has not been intentionally roughened before column concrete is placed. Thus $\mu = 0.6\lambda = (0.6)(1.0) = 0.6$. (*Ans.* 6 #8 dowels, 29 in. into footing, 22 in. into column)

In Problems 11.22 and 11.23, design plain concrete wall footings of uniform thickness.

Problem	Reinforced concrete wall thickness	D	L	f'_c	q_a	Distance from bottom of footing to final grade
11.22	14"	12 k/ft	14 k/ft	3 ksi	4 ksf	4'
11.23	12"	10 k/ft	16 k/ft	4 ksi	5 ksf	5'

(Answer to Problem 11.23: 6 ft wide, 28 in. thick)

11.24 Repeat Problem 11.1 as a plain concrete footing of uniform thickness if the live load is reduced to 15 k/ft.

In Problems 11.25 and 11.26 design square plain concrete column footings of uniform thickness.

Problem	Reinforced concrete column size	D	L	f_c'	q_a	Distance from bottom of footing to final grade
11.25	10″ × 10″	40 k	60 k	3 ksi	4 ksf	5′
11.26	12″ × 12″	70 k	90 k	3.5 ksi	4 ksf	5′

(Answer to Problem 11.25: 5 ft 6 in. × 5 ft 6 in., 25 in. thick)

For Problems 11.27 to 11.30 use the enclosed computer disk.

11.27 Repeat Prob. 11.2. (*Ans.* 27 in. footing, #8 @ $7\frac{1}{2}$ in. main steel)

11.28 Repeat Prob. 11.8.

11.29 Repeat Prob. 11.10 (*Ans.* 26 in. footing, 11 #10 bars both directions)

11.30 Repeat Prob. 11.16

PROBLEMS WITH SI UNITS

In Problems 11.31 and 11.32 design wall footings for the values given. The walls are to consist of reinforced concrete. Concrete weight = 23.5 kN/m³, soil weight = 16 kN/m³, $V_c = 0.166\sqrt{f_c'}\,bd$.

Problem	Wall thickness	D	L	f_c'	f_y	q_a	Distance from bottom of footing to final grade
11.31	300 mm	150 kN/m	200 kN/m	20.7 MPa	413.7 MPa	170 kN/m²	1.500 m
11.32	400 mm	180 kN/m	250 kN/m	27.6 MPa	413.7 MPa	210 kN/m²	1.200 m

(Answer to Problem 11.31: 2.50 m wide, 370-mm depth, #6 @ 170-mm main)

In Problems 11.33 to 11.35 design square single column footings for the values given. $V_c = 0.166\sqrt{f_c'}\,bd$. Concrete weight = 23.5 kN/m³, soil weight = 16 kN/m³.

Problem	Column size	D	L	f_c'	f_y	q_a	Distance from bottom of footing to final grade
11.33	350 mm × 350 mm	400 kN	500 kN	20.7 MPa	413.7 MPa	170 kN/m²	1.200 m
11.34	400 mm × 400 mm	650 kN	800 kN	27.6 MPa	413.7 MPa	170 kN/m²	1.200 m
11.35	450 mm × 450 mm	750 kN	1000 kN	27.6 MPa	413.7 MPa	210 kN/m²	1.600 m

(Answer to Problem 11.33: 2.5 m × 2.5 m, 480-mm depth, 9 #8 each direction)

(Answer to Problem 11.35: 3.2 m × 3.2 m, 600-mm depth, 10 #8 each direction)

11.36 Design a reinforced concrete rectangular combined footing for the two columns shown in the accompanying illustration. The bottom of the footing is to be 1.800 m below the final grade, $f_c' = 27.6$ MPa, $f_y = 344.8$ MPa. $V_c = 0.166\sqrt{f_c'}bd$, $q_a = 210$ kN/m², concrete weight = 23.5 kN/m³, and soil weight = 16 kN/m³.

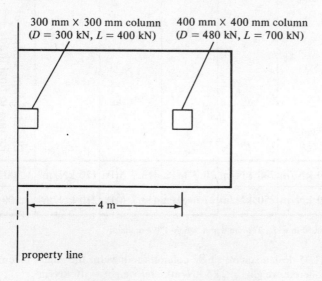

300 mm × 300 mm column
($D = 300$ kN, $L = 400$ kN)

400 mm × 400 mm column
($D = 480$ kN, $L = 700$ kN)

←——— 4 m ———→

property line

11.37 Design a plain concrete wall footing for a 300-mm-thick reinforced concrete wall that supports a 100-kN/m dead load (including its own weight) and a 120-kN/m live load. $f'_c = 20.7$ MPa, $f_y = 344.8$ MPa, and $q_a = 170$ kN/m². The base of the footing is to be 1.200 m below the final grade. Allowable $f_t = 0.415\phi\sqrt{f'_c}$, concrete weight = 22.7 kN/m³, and soil weight = 16 kN/m³. (*Ans.* 1.50 m width, 540 mm depth)

11.38 Design a square plain concrete column footing to support a 300-mm × 300-mm reinforced concrete column that in turn is supporting a 130-kN dead load and a 200-kN live load. $f'_c = 27.6$ MPa, $f_y = 344.8$ MPa, and $q_a = 210$ kN/m². The base of the footing is to be 1.500 m below the final grade. Allowable $f_t = 0.415\phi\sqrt{f'_c}$, concrete weight = 22.7 kN/m³, and soil weight = 16 kN/m³.

12

Retaining Walls

12.1 INTRODUCTION

A retaining wall is a structure built for the purpose of holding back or retaining or providing one-sided lateral confinement for soil or other loose material. The loose material being retained pushes against the wall, tending to overturn and slide it. Retaining walls are used in many design situations where there are abrupt changes in the ground slope. Perhaps the most obvious examples to the reader occur along highway or railroad cuts and fills. Often retaining walls are used in these locations to reduce the quantities of cut and fill as well as to reduce the right-of-way width required if the soils were allowed to assume their natural slopes. Retaining walls are used in many other locations as well, such as for bridge abutments, basement walls, and culverts.

Several different types of retaining walls are discussed in the next section, but whichever type is used, there will be three forces involved that must be brought into equilibrium. These include (1) the gravity loads of the concrete wall and any soil on top of the footing (the so-called *developed weight*), (2) the lateral pressure from the soil, and (3) the bearing resistance of the soil. In addition, the stresses within the structure have to be within permissible values and the loads must be supported in a manner such that undue settlements do not occur.

12.2 TYPES OF RETAINING WALLS

Retaining walls are generally classed as being gravity or cantilever types, with several variations possible. These are described in the paragraphs to follow, with reference being made to Figure 12.1.

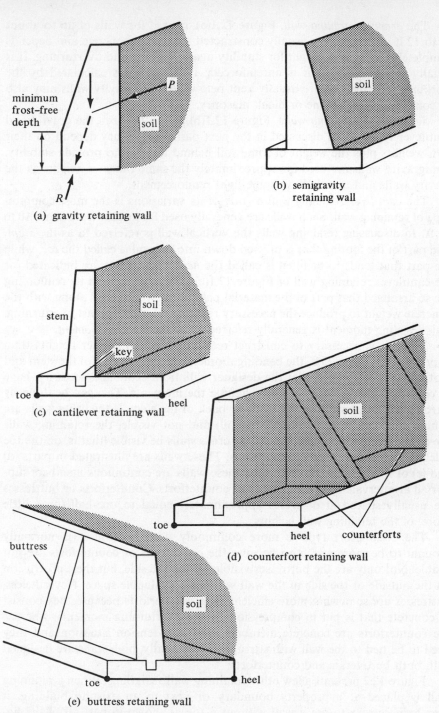

Figure 12.1 Retaining walls.

The *gravity retaining wall*, Figure 12.1(a), is used for walls of up to about 10 to 12 ft in height. It is usually constructed with plain concrete and depends completely on its own weight for stability against sliding and overturning. It is usually so massive that it is unreinforced. Tensile stresses calculated by the working stress method are usually kept below $1.6\sqrt{f'_c}$. Gravity walls may also be constructed with stone or block masonry.

Semigravity retaining walls, Figure 12.1(b), fall in between the gravity and cantilever types (to be discussed in the next paragraph). They depend on their own weights plus the weight of some soil behind the wall to provide stability. Semigravity walls are used for approximately the same range of heights as the gravity walls and usually have some light reinforcement.

The *cantilever retaining wall* or one of its variations is the most common type of retaining wall. Such walls are generally used for heights from about 10 to 25 ft. In discussing retaining walls the vertical wall is referred to as the *stem*. The part of the footing that is pressed down into the soil is called the *toe*, while the part that tends to be lifted is called the *heel*. These parts are indicated for the cantilever retaining wall of Figure 12.1(c). The concrete and its reinforcing are so arranged that part of the material behind the wall is used along with the concrete weight to produce the necessary resisting moment against overturning. This resisting moment is generally referred to as the *righting moment*.

When it is necessary to construct retaining walls of greater heights than approximately 20 to 25 ft, the bending moments at the junction of the stem and footing become so large that the designer will, from economic necessity, have to consider other types of walls to handle the moments. This can be done by introducing cross walls on the front or back of the stem. If the cross walls are behind the stem (that is, inside the soil) and not visible, the retaining walls are called *counterfort walls*. Should the cross walls be visible (that is, on the toe side), the walls are called *buttress walls*. These walls are illustrated in parts (d) and (e) of Figure 12.1. The stems for these walls are continuous members supported at intervals by the buttresses or counterforts. Counterforts or buttresses are usually spaced at distances approximately equal to one-half (or a little more) of the retaining wall heights.

The counterfort type is more commonly used because it is normally thought to be more attractive because the cross walls or counterforts are not visible. Not only are the buttresses visible on the toe side, but their protrusion on the outside or toe side of the wall will use up valuable space. Nevertheless, buttresses are somewhat more efficient than counterforts because they consist of concrete that is put in compression by the overturning moments, whereas the counterforts are concrete members used in a tension situation and they need to be tied to the wall with stirrups. Occasionally, high walls are designed with both butteresses and counterforts.

Figure 12.2 presents a few other retaining wall variations. When a retaining wall is placed at a property boundary or next to an existing building, it may be necessary to use a wall without a toe as shown in part (a) of the figure or without a heel as shown in part (b). Another type of retaining wall very often encountered is the bridge abutment shown in part (c) of the figure. Abut-

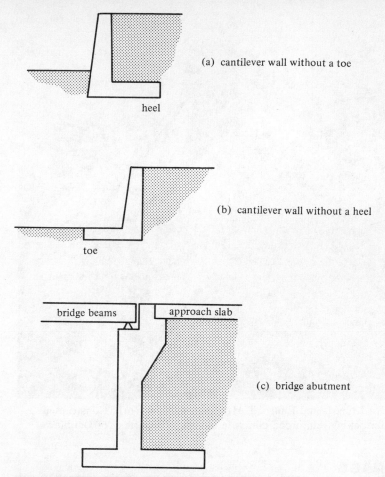

(a) cantilever wall without a toe

heel

(b) cantilever wall without a heel

toe

bridge beams   approach slab

(c) bridge abutment

Figure 12.2 More retaining walls.

ments may very well have wing wall extensions on the sides to retain the soil in the approach area. The abutment, in addition to other loads, will have to support the end reactions from the bridge.

The use of precast retaining walls is becoming more and more common each year. The walls are built with some type of precast units, and the footings are probably poured in place. The results are very attractive, and the units are high-quality concrete members made under "plant controlled" conditions. Less site preparation is required and the erection of the walls is much quicker than cast-in-place ones. The precast units can later be disassembled and the units used again. Other types of precast retaining walls consist of walls or sheetings actually driven into the ground before excavation. Also showing promise are the gabion or wire baskets of stone used in conjunction with geotextile reinforced embankments.

Retaining wall for Long Island Railroad, Huntington, New York. Constructed with precast interlocking reinforced concrete modules. (Courtesy of Doublewal Corporation.)

12.3 DRAINAGE

One of the most important items in designing and constructing successful retaining walls is the prevention of water accumulation behind the walls. If water is allowed to build up there, the result can be great lateral water pressures against the wall and perhaps an even worse situation in cold weather climates due to frost action.

The best possible backfill for a retaining wall is a well-drained and cohesionless soil. Furthermore, this is the condition for which the designer normally plans and designs. In addition to a granular backfill material, weep holes of 4 in. or more in diameter (the large sizes are used for easy cleaning) are placed in the walls approximately 5 ft on center, horizontally and vertically, as shown in Figure 12.3(a). If the backfill consists of a coarse sand, it is desirable to put a few shovels of pea gravel around the weep holes to try to prevent the sand from stopping up the holes.

Weep holes have the disadvantages that the water draining through the wall is somewhat unsightly and may very well cause a softening of the soil in the area of the highest soil pressure (under the footing toe). A better method

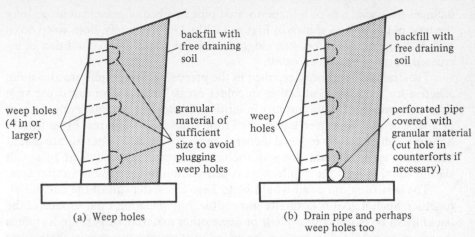

weep holes
(4 in or
larger)

backfill with
free draining
soil

granular
material of
sufficient
size to avoid
plugging
weep holes

(a) Weep holes

weep
holes

backfill with
free draining
soil

perforated pipe
covered with
granular material
(cut hole in
counterforts if
necessary)

(b) Drain pipe and perhaps
weep holes too

Figure 12.3 Retaining wall drainage.

Gang-formed panels. (Courtesy of Burke Concrete Accessories, Inc.)

includes the use of a 6- or 8-in. perforated pipe in a bed of gravel running along the base of the wall, as shown in Figure 12.3(b). Unfortunately, both weep holes and drainage pipes can become clogged, with the result that a condition of increased water pressure can occur.

The drainage methods described in the preceding paragraphs are also quite effective for reducing frost action in colder areas. Frost action can cause very large movements of walls, not just in terms of inches but perhaps even in terms of a foot or two, and over a period of time can lead to failures. Frost action, however, can be greatly reduced if coarse, properly drained materials are placed behind the walls. The thickness of the filled material perpendicular to a wall should equal at least the depth of frost penetration in the ground in that area.

The best situation of all would be to keep the water out of the backfill altogether. Such a goal is normally impossible, but sometimes the surface of the backfill can be paved with asphalt or some other material, or perhaps a surface drain can be provided to remove the water, or it may be possible in some other manner to interrupt the source of the water before it can get to the backfill.

12.4 FAILURES OF RETAINING WALLS

The number of failures or partial failures of retaining walls is rather alarming. The truth of the matter is that if large safety factors were not used, the situation would be even more severe. One reason for the large number of failures is the fact that designs are so often based on methods that are suitable for only certain special situations. For instance, if a wall that has a saturated clay behind it is designed by a method that is suitable for a dry granular material, future trouble will be the result.

12.5 LATERAL PRESSURES ON RETAINING WALLS

The actual pressures that occur behind retaining walls are quite difficult to estimate because of the large number of variables present. These include the kinds of backfill materials and their compactions and moisture contents, the types of materials beneath the footings, the presence or absence of surcharge, and other items. As a result, the detailed estimation of the lateral forces applied to various retaining walls is clearly a problem in theoretical soil mechanics. For this reason the discussion to follow is limited to a rather narrow range of cases.

If a retaining wall is constructed against a solid rock face, there will be no pressure applied to the wall by the rock. But if the wall is built to retain a body of water, hydrostatic pressure will be applied to the wall. At any point the pressure (p) will equal wh, where w is the unit weight of the water and h is the vertical distance from the surface of the water to the point in question.

If a wall is built to retain a soil, the soil's behavior will generally be some where in between that of rock and water (*but as we will learn the pressure caused by some soils is much higher than that caused by water*). The pressure exerted

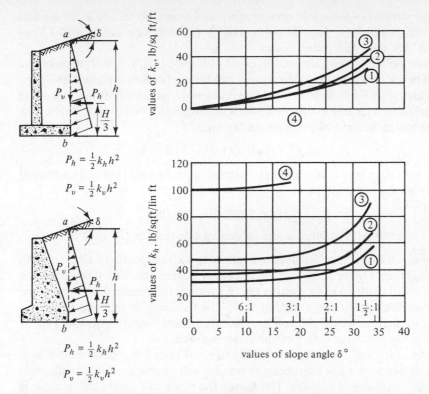

$$P_h = \tfrac{1}{2} k_h h^2$$

$$P_v = \tfrac{1}{2} k_v h^2$$

$$P_h = \tfrac{1}{2} k_h h^2$$

$$P_v = \tfrac{1}{2} k_v h^2$$

Figure 12.4 Chart for estimating pressure of backfill against retaining walls supporting backfills with plane surface. Use of this chart is limited to walls not over about 20 ft high. (1) Backfill of coarse-grained soil without admixture of fine particles, very permeable, as clean sand or gravel. (2) Backfill of coarse-grained soil of low permeability due to admixture of particles of silt size. (3) Backfill of fine silty sand, granular materials with conspicuous clay content, and residual soil with stones. (4) Backfill of very soft or soft clay, organic silt, or silty clay.[1]

against the wall will increase, as did the water pressure, with depth but not as rapidly. This pressure at any depth can be estimated with the following expression:

$$p = Cwh$$

In this equation w is the unit weight of the soil, h is the distance from the surface to the point in question, and C is a constant that is dependent on the characteristics of the backfill. Unfortunately, the value of C can vary quite a bit, being perhaps as low as 0.3 or 0.4 for loose granular soils and perhaps as high as 0.9 or even 1.0 or more for some clay soils. Figure 12.4 presents charts which are sometimes used for estimating the vertical and horizotal pressures applied by soil backfills of up to 20-ft heights. Several different types of backfill materials are considered in the figure.

[1] Peck, R. B., Hanson, and Thornburn, T. H., 1974, *Foundation Engineering*, 2nd ed. (New York: John Wiley & Sons), p. 425.

Unit weights of soils will vary roughly as follows: 90 to 100 lb/ft^3 for soft clays, 100 to 120 lb/ft^2 for stiff clays, 110 to 120 lb/ft^3 for sands, and 120 to 130 lb/ft^3 for sand and gravel mixes.

If you carefully study the second chart of Figure 12.4, you will probably be amazed to see how high lateral pressures can be, particularly for clays and silts. As an illustration a 1-ft-wide vertical strip is considered for a 15-ft-high retaining wall backfilled with soil number (④) with an assumed δ of 10° (6:1 slope). The total estimated horizontal pressure on the strip is

$$P_h = \tfrac{1}{2}k_h h^2 = (\tfrac{1}{2})(102)(15)^2 = 11{,}475 \text{ lb}$$

If a 15-ft-deep lake is assumed to be behind the same wall, the total horizontal pressure on the strip will be

$$P_h = (\tfrac{1}{2})(15)(15)(62.4) = 7020 \text{ lb}$$

(only 61% as large as the estimated pressure for the soil)

For this introductory discussion, a retaining wall supporting a sloping earth fill is shown in Figure 12.5. Part of the earth behind the wall (shown by the hatched area) tends to slide along a curved surface (represented by the dashed line) and push against the retaining wall. The tendency of this soil to slide is resisted by friction along the soil underneath (called *internal friction*) and by friction along the vertical face of the retaining wall.

Internal friction is greater for a cohesive soil than for a noncohesive one, but the wetter such a soil becomes, the smaller will be its cohesiveness and thus the flatter the plane of rupture. The flatter the plane of rupture, the greater is the volume of earth tending to slide and push against the wall. Once again it can be seen that good drainage is of the utmost importance. Usually the designer assumes that there is a cohesionless granular backfill behind the walls.

Due to lateral pressure the usual retaining wall will give or deflect a little because it is constructed of elastic materials. Furthermore, unless the wall rests

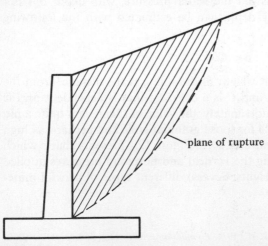

plane of rupture

Figure 12.5

on a rock foundation, it will tilt or lean a small distance away from the soil due to the compressible nature of the supporting soils. For these reasons, retaining walls are frequently constructed with a slight batter, or inclination, toward the backfill so that the deformations described are not obvious to the passerby.

Under the lateral pressures described, the usual retaining wall will move a little distance and *active soil pressure* will develop, as shown in Figure 12.6. Among the many factors that affect the pressure applied to a particular wall are the kind of backfill material used, the drainage situation, the level of the water table, the seasonal conditions such as dry or wet or frozen, the presence of trucks or other equipment on the backfill, and so on.

For design purposes it is usually satisfactory to assume that the active pressure varies linearly with the depth of the backfill. In other words, it is just as though (so far as lateral pressure is concerned) behind the wall there is a liquid of some weight that can vary from considerably less than the weight of water to considerably more. The chart of Figure 12.4 shows this large variation in possible lateral pressures. The assumed lateral pressures are often referred to as *equivalent fluid pressures*. Values from 30 to 50 pcf are normally assumed but may be much too low for clay and silt materials.

If the wall moves away from the backfill and against the soil at the toe, a passive soil pressure will be the result. Passive pressure, which is also assumed to vary linearly with depth, is illustrated in Figure 12.6. The inclusion or non-inclusion of passive pressure in the design calculations is a matter of judgment on the designer's part. For effective passive pressure to be developed at the toe, the toe concrete must be placed against undisturbed earth without the use of vertical forms. Even if this procedure is followed, the designer will probably reduce the height of the undisturbed soil (h' in Figure 12.6) used in the calculations to account for some disturbance of the earth during construction operations.

As long as the backfills are granular, noncohesive, and dry, the assumption of an equivalent liquid pressure is fairly satisfactory. Formulas based on an assumption of dry sand or gravel backfills are not satisfactory for soft clays or

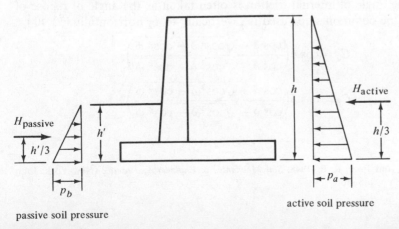

passive soil pressure

active soil pressure

Figure 12.6

saturated sands. Actually, clays should not be used for backfills because their shear characteristics change easily and they may tend to creep against the wall, increasing pressures as time goes by.

If a linear pressure variation is assumed, the active pressure at any depth can be determined as

$$p_a = C_a w h$$

or, for passive pressure,

$$p_p = C_p w h'$$

In these expressions, C_a and C_p are the approximate coefficients of active and passive pressures, respectively. These coefficients can be calculated by theoretical equations such as those of Rankine or Coulomb.[2] For a granular material, typical values of C_a and C_p are 0.3 and 3.3. The Rankine equation (published in 1857) neglects the friction of the soil on the wall, whereas the Coulomb formula (published in 1776) takes it into consideration. These two equations were developed for cohesionless soils. For cohesive soils containing clays and/or silts it is necessary to use empirical values determined from field measurements (such as those given in Fig. 12.4).

It has been estimated that the cost of constructing retaining walls varies directly with the square of their heights. Thus as retaining walls become higher, the accuracy of the computed lateral pressures becomes more and more important in providing economical designs. Since the Coulomb equation does take into account friction on the wall, it is thought to be the more accurate one and is often used for walls of over 20 ft. The Rankine equation is commonly used for ordinary retaining walls of 20 ft or less in height. It is interesting to note that the two methods give identical results if the friction of the soil on the wall is neglected.

The Rankine expressions for the active and passive pressure coefficients are given at the end of this paragraph, with reference being made to Figure 12.7. In these expressions ϕ is the angle the backfill makes with the horizontal, while α is the angle of internal friction of the soil. For well-drained sand or gravel backfills, the angle of internal friction is often taken as the angle of repose of the slope. One common slope used is 1 vertically to $1\frac{1}{2}$ horizontally ($33°40'$).

$$C_a = \cos \delta \left(\frac{\cos \delta - \sqrt{\cos^2 \delta - \cos^2 \phi}}{\cos \delta + \sqrt{\cos^2 \delta - \cos^2 \phi}} \right)$$

$$C_p = \cos \delta \left(\frac{\cos \delta + \sqrt{\cos^2 \delta - \cos^2 \phi}}{\cos \delta - \sqrt{\cos^2 \delta - \cos^2 \phi}} \right)$$

---

[2] Terzaghi, K., and Peck, R. B., 1948, *Soil Mechanics in Engineering Practice* (New York: John Wiley & Sons), pp. 138–166.

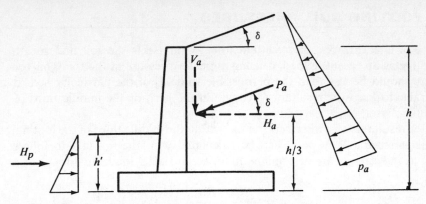

Figure 12.7

Should the backfill be horizontal—that is, should δ be equal to zero—the expressions become

$$C_a = \frac{1 - \sin \phi}{1 + \sin \phi}$$

$$C_p = \frac{1 + \sin \phi}{1 - \sin \phi}$$

One trouble with using these expressions is in the determinations of ϕ. It can be as small as $0°$ to $10°$ for soft clays and as high as $30°$ or $40°$ for some granular materials. As a result, the values of C_a can vary from perhaps 0.30 for some granular materials up to about 1.0 for some wet clays.

Once the values of C_a and C_p are determined, the total horizontal pressures, H_a and H_p, can be calculated as being equal to the areas of the respective triangular pressure diagrams. For instance, with reference made to Figure 12.7, the value of the active pressure is

$$H_a = (\tfrac{1}{2})(p_a)(h) = (\tfrac{1}{2})(C_a w h)(h)$$

$$H_a = \frac{C_a w h^2}{2}$$

and, similarly,

$$H_p = \frac{C_p w h'2}{2}$$

In addition to these lateral pressures applied to the retaining wall, it is considered necessary in many parts of the country to add the effect of frost action at the top of the stem perhaps as much as 600 or 700 lb per linear foot in areas experiencing extreme weather conditions.

12.6 FOOTING SOIL PRESSURES

Because of lateral forces, the resultant force R intersects the soil underneath the footing as an eccentric load, causing a greater pressure at the toe. This toe pressure should be less than the permissible value q_a of the particular soil. It is desirable to keep the resultant force within the kern or the middle third of the footing base.

If the resultant force intersects the soil within the middle third of the footing, the soil pressure at any point can be calculated with the formula to follow exactly as the stresses are determined in an eccentrically loaded column.

$$q = -\frac{R_v}{A} \pm \frac{R_v ec}{I}$$

In this expression, R_v is the vertical component of R or the total vertical load, e is the eccentricity of the load from the center of the footing, A is the area of a 1-ft-wide strip of soil of a length equal to the width of the footing base, and I is the moment of inertia of the same area about its centroid. This expression is correct only if R_v falls within the kern.

This expression can be reduced to the following expression, in which L is the width of the footing from heel to toe.

$$q = -\frac{R_v}{L} \pm \frac{R_v e(L/2)}{L^3/12} = -\frac{R_v}{L}\left(1 \pm \frac{6e}{L}\right)$$

If the resultant force falls outside of the middle third of the footing, the preceding expressions are not applicable because they indicate a tensile stress on one side of the footing—a stress the soil cannot supply. For such cases the soil pressures can be determined as previously described in Section 11.12 and Figure 11.23 of the preceding chapter. Such a situation should not be permitted in a retaining wall and is not considered further.

The values of soil pressure computed in this manner are only rough estimates of the real values and thus should not be valued too highly. The true pressures are appreciably affected by quite a few items other than the retaining wall weight. Included are drainage conditions, temperature, settlement, pore water, and so on.

12.7 DESIGN OF SEMIGRAVITY RETAINING WALLS

As previously mentioned, semigravity retaining walls are designed to resist earth pressure by means of their own weight plus some developed soil weight. Because

Retaining wall for United States Army Corps of Engineers, Colchester, Connecticut. Constructed with precast interlocking reinforced concrete modules. (Courtesy of Doublewal Corporation.)

they are normally constructed with plain concrete, stone, or perhaps some other type of masonry, their design is based on the assumption that only very little tension or none at all can be permitted in the structure. If the resultant of the earth pressure and the wall weight (including any developed soil weight) falls within the middle third of the wall base, tensile stresses will probably be negligible in the wall.

A wall size is assumed, safety factors against sliding and overturning are calculated, the point where the resultant force strikes the base is determined, and the soil pressures are calculated. It is normally felt that safety factors against sliding should be at least 1.5 for cohesionless backfills and 2.0 for cohesive ones, and safety factors of 2.0 for overturning are normally specified. A suitable wall is probably obtained after two or three trial sizes. Example 12.1 illustrates the calculations that need to be made for each trial.

Figure 12.8(a) shows a set of approximate dimensions that are often used for sizing semigravity walls. Dimensions may be assumed in the vicinity of the values given, and the safety factors against overturning and sliding computed. If the values are not suitable, the dimensions are adjusted and the safety factors are recalculated, and so on. Semigravity walls are normally trapezoidal in shape, as shown in Figure 12.8(a), but sometimes they may have broken backs, as illustrated in Figure 12.8(b).

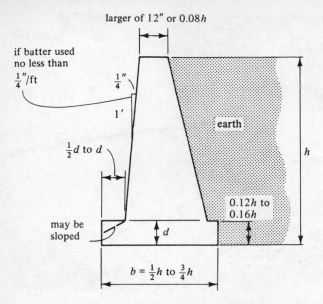

(a) some approximate dimensions for semigravity walls

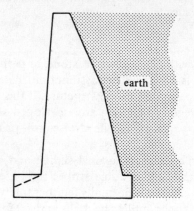

(b) broken-back semigravity wall

Figure 12.8

■ EXAMPLE 12.1

A semigravity retaining wall consisting of plain concrete (weight = 145 lb/ft³) is shown in Figure 12.9. The bank of supported earth is assumed to weigh 110 lb/ft³, to have a ϕ of 30°, and to have a coefficient of friction against sliding on soil of 0.5. Determine the safety factors against overturning and sliding and

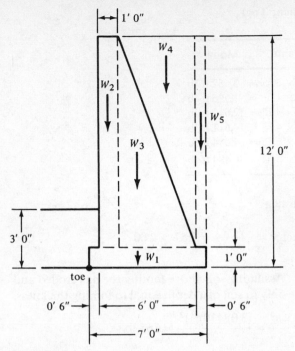

Figure 12.9

determine the bearing pressure underneath the toe of the footing. Use the Rankine expression for calculating the horizontal pressures.

SOLUTION

Computing the Soil Pressure Coefficients

$$C_a = \frac{1 - \sin \phi}{1 + \sin \phi} = \frac{1 - 0.5}{1 + 0.5} = 0.333$$

$$C_p = \frac{1 + \sin \phi}{1 - \sin \phi} = \frac{1 + 0.5}{1 - 0.5} = 3.00$$

The Value of H_a

$$H_a = \frac{C_a w h^2}{2} = \frac{(0.333)(110)(12)^2}{2} = 2637 \text{ lb}$$

Overturning Moment

$$\text{O.T.M.} = (2637)(\tfrac{12}{3}) = 10{,}548 \text{ ft-lb}$$

Righting Moments (Taken About Toe)

Force		Moment arm		Moment
$W_1 = (7)(1)(145)$	$=$	$1015\ lb \times 3.5'$	$=$	$3,552\ ft\text{-}lb$
$W_2 = (1)(11)(145)$	$=$	$1595\ lb \times 1.0$	$=$	$1,595\ ft\text{-}lb$
$W_3 = (\frac{1}{2})(5)(11)(145)$	$=$	$3988\ lb \times 3.17$	$=$	$12,642\ ft\text{-}lb$
$W_4 = (\frac{1}{2})(5)(11)(110)$	$=$	$3025\ lb \times 4.83$	$=$	$14,611\ ft\text{-}lb$
$W_5 = (0.5)(11)(110)$	$=$	$605\ lb \times 6.75$	$=$	$4,084\ ft\text{-}lb$
	$R_v =$	$10,228\ lb$	$M =$	$36,484\ ft\text{-}lb$

Safety Factor Against Overturning

$$\text{Safety factor} = \frac{36,484}{10,228} = 3.57 > 2.00 \qquad \underline{OK}$$

Safety Factor Against Sliding   Assuming soil above footing toe has eroded and thus the passive pressure is due only to soil of a depth equal to footing thickness,

$$H_p = \frac{C_p wh^2}{2} = \frac{(3.0)(110)(1)^2}{2} = 165\ lb$$

$$\text{Safety factor against sliding} = \frac{(0.5)(10,228) + 165}{2637} = 2.00 > 1.50 \qquad \underline{OK}$$

Distance of Resultant from Toe

$$\text{Distance} = \frac{36,484 - 10,228}{10,228} = 2.57' > 2.33' \qquad \underline{\text{inside middle third}}$$

Soil Pressure Under Heel and Toe

$$A = (1)(7.0) = 7.0\ ft^2$$

$$I = (\tfrac{1}{12})(1)(7)^3 = 28.58\ ft^4$$

$$f_{toe} = -\frac{R_v}{A} - \frac{R_v ec}{I} = -\frac{10,228}{7.0} - \frac{(10,228)(3.50 - 2.57)(3.50)}{28.58}$$

$$= -1461 - 1165 = -2826\ psf$$

$$f_{heel} = -\frac{R_v}{A} + \frac{R_v ec}{I} = -1461 + 1165 = -296\ psf \qquad \blacksquare$$

12.8 EFFECTS OF SURCHARGE

Should there be earth or other loads on the surface of the backfill, as shown in Figure 12.10, the horizontal pressure applied to the wall will be increased. If the surcharge is uniform over the sliding area behind the wall, the resulting

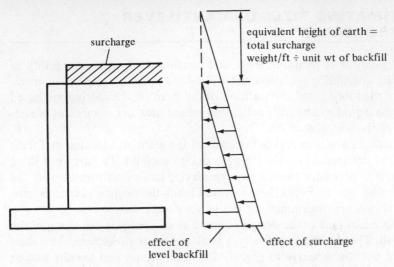

Figure 12.10

pressure is assumed to equal the pressure that would be caused by an increased backfill height having the same total weight as the surcharge. It is usually easy to handle this situation by adding a uniform pressure to the triangular soil pressure for a wall without surcharge, as shown in the figure.

If the surcharge does not cover the area entirely behind the wall, some rather complex soil theories are available to consider the resulting horizontal pressures developed. As a consequence, the designer usually uses a rule of thumb to cover the case, a procedure that works reasonably well. He or she may assume, as shown in Figure 12.11, that surcharge cannot affect the pressure above the intersection of a 45° line from the edge of the surcharge to the wall. The lateral pressure is increased, as by a full surcharge, below the intersection point. This is shown in the right side of the figure.

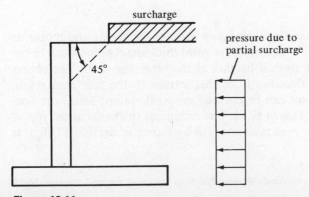

Figure 12.11

12.9 ESTIMATING SIZES OF CANTILEVER RETAINING WALLS

The statical analysis of retaining walls and consideration of their stability as to overturning and sliding are based on service-load conditions. In other words, the length of the footing and the position of the stem on the footing are based entirely on the actual soil backfill, estimated lateral pressure, coefficient of sliding friction of the soil, and so on.

On the other hand, the detailed designs of the stem and footing and their reinforcing are determined by the strength design method. To carry out these calculations, it is necessary to multiply the service loads and pressures by the appropriate load factors. From these factored loads the bearing pressures, moments, and shears are determined for use in the design.

Thus the initial part of the design consists of an approximate sizing of the retaining wall. Though this is actually a trial-and-error procedure, the values obtained are not too sensitive to slightly incorrect values and usually one or two trials are sufficient.

There are various rules of thumb available with which excellent initial size estimates can be made. In addition, various handbooks present the final sizes of retaining walls that have been designed for certain specific cases. This information will enable the designer to estimate very well the proportions of a wall to be designed. The *CRSI Handbook* is one such useful reference.[3] In the next few paragraphs, suggested methods are presented for estimating sizes without the use of a handbook. These approximate methods are very satisfactory as long as the conditions are not too much out of the ordinary.

Height of Wall

The necessary elevation at the top of the wall is normally obvious from the conditions of the problem. The elevation at the base of the footing should be selected so that it is below frost penetration in the particular area—about 3 to 6 ft below ground level in the northern part of the United States. From these elevations the overall height of the wall can be determined.

Stem Thickness

Stems are theoretically thickest at their bases because the shears and moments are greatest there. They will ordinarily have total thicknesses somewhere in the range of 7% to 12% of the overall heights of the retaining walls. The shears and moments in the stem decrease from the bottom to the top; as a result, thicknesses and reinforcement can be reduced proportionately. Stems are normally tapered, as shown in Figure 12.12. The minimum thickness at the top of the stem is 8 in., with 12 in. preferable. As will be shown in Section 12.10, it is

---

[3] *CRSI Handbook*, 1978 (Chicago: Concrete Reinforcing Steel Institute), pp. 14-3 through 14-25.

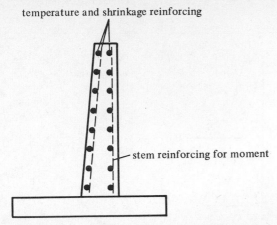

temperature and shrinkage reinforcing

stem reinforcing for moment

Figure 12.12

necessary to have a mat of reinforcing in the inside face of the stem and another mat in the outside face. To provide room for these two mats of reinforcing, for cover and spacing between the mats, a minimum total thickness of at least 8 in. is required.

If ρ in the stem is limited to a maximum value of approximately $\dfrac{0.18f'_c}{fy}$ the stem thickness required for moment will probably provide sufficient shear resistance without using stirrups. Furthermore it will probably be sufficiently thick to limit lateral deflections to reasonable values.

For heights up to about 12 ft, the stems of cantilever retaining walls are normally made of constant thickness because the extra cost of setting the tapered formwork is usually not offset by the saving in concrete. Above 12-ft heights, concrete savings are usually sufficiently large to make tapering economical.

Actually the sloping face of the wall can be either the front or the back, but if the outside face is tapered it will tend to counteract somewhat the deflection and tilting of the wall due to lateral pressures. A taper or batter of $\frac{1}{4}$ in. per foot of height is often recommended to offset deflection or the forward tilting of the wall.

Base Thickness

The final thickness of the base will be determined on the basis of shears and moments. For estimating, though, its total thickness will probably fall somewhere between 7% and 10% of the overall wall height. Minimum thicknesses of at least 10 to 12 in. are used.

Base Length

For preliminary estimates the base length can be taken to be about 40% to 60% of the overall wall height. A little better estimate, however, can be made

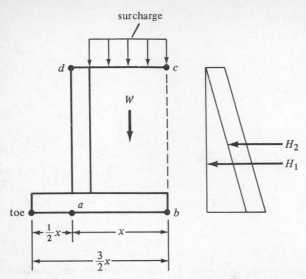

Figure 12.13

by using the method described by the late Professor Ferguson in his reinforced concrete text.[4] For this discussion reference is made to Figure 12.13. In this figure, W is assumed to equal the weight of the material within area *abcd*. This area contains both concrete and soil, but the author assumes here that it is all soil. This means a little larger safety factor will be developed against overturning than assumed. When surcharge is present, it will be included as an additional depth of soil, as shown in the figure.

If the sum of moments about point a due to W and the lateral forces H_1 and H_2 equals zero, the resultant force R will pass through point a. Such a moment equation can be written, equated to zero, and solved for x. Should the distance from the footing toe to point a be equal to one-half of the distance x in the figure and the resultant force R pass through point a, the footing pressure diagram will be triangular. In addition, if moments are taken about the toe of all the loads and forces for the conditions described, the safety factor against overturning will be approximately 2.

A summary of the preceding approximate first trial sizes for cantilever retaining walls is shown in Figure 12.14. These sizes are based on the dimensions of walls successfully constructed in the past. They often will be on the conservative side.

■ EXAMPLE 12.2
Using the approximate rules presented in this section, estimate the sizes of the parts of the retaining wall shown in Figure 12.15. The soil weighs 100 lb/ft³, and a surcharge of 300 psf is present. Assume $C_a = 0.32$. (For many practical soils such as clays or silts C_a will be two or more times this large.)

---

[4] Ferguson, P. M., 1979, *Reinforced Concrete Fundamentals*, 4th ed. (New York: John Wiley & Sons), p. 256.

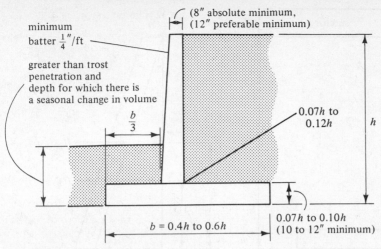

Figure 12.14

SOLUTION

Stem Thickness   Assume 12-in. thickness at top.

$$\text{Assume bottom thickness} = 0.07h = (0.07)(21) = 1.47' \quad \underline{\textbf{say } 1'6''}$$

Base Thickness   Assume base $t = 7\%$ to 10% of overall wall height:

$$t = (0.07)(21) = 1.47' \qquad \underline{\textbf{say } 1'6''}$$

$$\text{Height of stem} = 21'0'' \text{ minus } 1'6'' = \underline{19'6''}$$

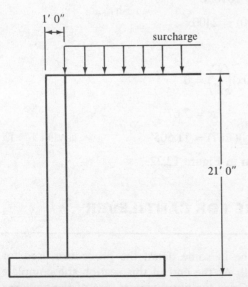

Figure 12.15

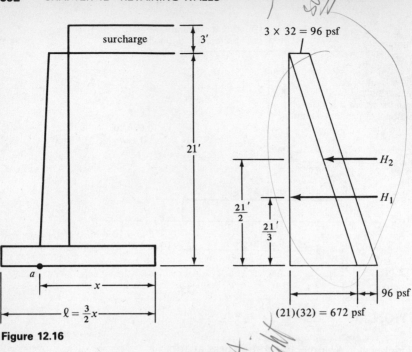

Figure 12.16

Base Length and Position of Stem   Calculating horizontal forces without load factors, as shown in Figure 12.16.

$$\rho_a = C_a wh = (0.32)(100)(21) = 672 \text{ lb}$$

$$H_1 = (\tfrac{1}{2})(21)(672) = 7056 \text{ lb}$$

$$H_2 = (21)(96) = 2016 \text{ lb}$$

$$W = (x)(24)(100) = 2400x$$

$$\Sigma M_a = 0$$

$$-(7056)(7.00) - (2016)(10.5) + (2400x)\left(\frac{x}{2}\right) = 0$$

$$x = 7.67'$$

$$b = (\tfrac{3}{2})(7.67) = 11.505' \qquad \underline{\text{say } 11'6''} \quad \blacksquare$$

The final trial dimensions are shown in Figure 12.22.

12.10 DESIGN PROCEDURE FOR CANTILEVER RETAINING WALLS

This section is presented to describe in some detail the procedure used for designing a cantilever retaining wall. At the end of this section the complete design of such a wall is presented. Once the approximate size of the wall has

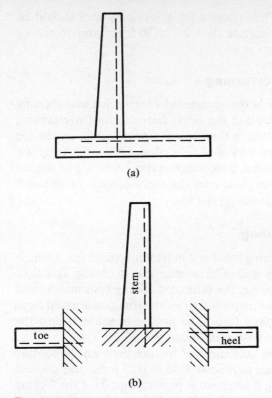

Figure 12.17

Figure 12.17

been established, the stem, toe, and heel can be designed in detail. Each of these parts will be designed individually as a cantilever sticking out of a central mass, as shown in Figure 12.17.

Stem

The values of shear and moment at the base of the stem due to lateral earth pressures are computed and used to determine the stem thickness and necessary reinforcing. Because the lateral pressures are considered to be live load forces, a load factor of 1.7 is used.

It will be noted that the bending moment requires the use of vertical reinforcing bars on the soil side of the stem. In addition, temperature and shrinkage reinforcing must be provided. In Section 14.3 of the ACI Code, a minimum value of horizontal reinforcing equal to 0.0025 of the area of the wall bt is required as well as a minimum amount of vertical reinforcing (0.0015). These values may be reduced to 0.0020 and 0.0012 if the reinforcing is $\frac{5}{8}$ in. or less in diameter and if it consists of bars or welded wire fabric with f_y equal to or greater than 60,000 psi.

The major changes in temperature occur on the front or exposed face of the stem, and for this reason most of the horizontal reinforcing (perhaps two-thirds) should be placed on that face with just enough vertical steel used to support the horizontal bars. The concrete for a retaining wall should be placed in fairly short lengths not greater than 20 or 30-ft sections to reduce shrinkage stresses.

Factor of Safety Against Overturning

Moments are taken about the toe of the unfactored overturning and righting forces. Traditionally it has been felt that the safety factor against overturning should be at least equal to 2. In making these calculations, backfill on the toe is usually neglected because it may very well be eroded. Of course, there are cases where there is a slab (for instance, a highway pavement on top of the toe backfill), which holds the backfill in place over the toe. For such situations it may be reasonable to include the loads on the toe.

Factor of Safety Against Sliding

A consideration of sliding for retaining walls is a most important topic because a very large percentage of retaining wall failures occur due to sliding. To calculate the factor of safety against sliding, the estimated sliding resistance (equal to the coefficient of friction for concrete on soil times the resultant vertical force μR_v) is divided by the total horizontal force. The passive pressure against the wall is probably neglected and the unfactored loads are used.

Typical design values of μ, the coefficient of friction between the footing concrete and the supporting soil, are as follows: 0.45 to 0.55 for coarse-grained soils, with the lower value applying if some silt is present, and 0.6 if the footing is supported on sound rock with a rough surface. Values of 0.3 to 0.35 are probably used if the supporting material is silt.

It is usually felt that the factor of safety against sliding should be at least equal to 1.5. If it is less than this value, a lug or key is normally provided, as shown in Figure 12.18, with the front face cast directly against undisturbed soil.

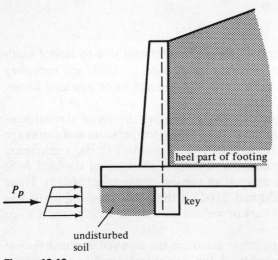

Figure 12.18

(Some designers feel that the construction of keys disturbs the soil so much that they are not worthwhile.) Keys are thought to be particularly necessary for moist clayey soils. The purpose of a key is to cause the development of passive pressure in front of and below the base of the footing, as shown by P_p in the figure. The actual theory involved, and thus the design of keys, is still a question among geotechnical engineers. As a result, many designers select the sizes of keys by rules of thumb. One common practice is to give them a depth between two-thirds and the full depth of the footing. They are usually made approximately square in cross section and have no reinforcing provided other than perhaps the dowels mentioned in the next paragraph.

Keys are often located below the stem so that some dowels or extended vertical reinforcing may be extended into the key. If this procedure is used, the front face of the key needs to be at least 5 or 6 in. in front of the back face of the stem to allow room for the dowels. From a soil mechanics view, keys may be a little more effective if they are placed a little further toward the heel.

If the key can be extended down into a very firm soil or even rock, the result will be a greatly increased sliding resistance—that resistance being equal to the force necessary to shear the key off from the footing, that is, a shear friction calculated as described in Sections 7.12 and 11.13 of this text.

Heel Design

The lateral earth pressure tends to cause the retaining wall to rotate about its toe. This action tends to pick up the heel into the backfill. The backfill pushes down on the heel cantilever, causing tension in its top. The major force applied to the heel of a retaining wall is the downward weight of the backfill behind the wall. Although it is true there is some upward soil pressure, many designers choose to neglect it because it is relatively small. The downward loads tend to push the heel of the footing down, and the necessary upward reaction to hold it attached to the stem is provided by the vertical tensile steel in the stem, which is extended down into the footing.

Because the reaction in the direction of the shear does not introduce compression into the heel part of the footing in the region of the stem, it is not permissible to determine V_u at a distance d from the face of the stem, as provided in Section 11.1.3.1 of the ACI Code. The value of V_u is determined instead at the face of the stem due to the downward loads. This shear is often of such magnitude as to control the thickness, but the moment at the face of the stem should be checked also. Because the load here consists of soil and concrete, a load factor of 1.4 is used for making the calculations.

It will be noted that the bars in the heel will be in the top of the footing. As a result, the required development length of these "top bars" is equal to 1.3 times the basic development value and may be rather large.

The percentage of flexural steel required for the heel will frequently be less than the ρ_{min} of $200/f_y$. Despite the fact that the Code (10.5.3) exempts slabs of uniform thickness from this ρ_{min} value, it is recommended that it be used because the retaining wall is a major "beamlike" structure.

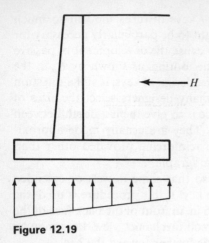

Figure 12.19

Toe Design

The toe is assumed to be a beam cantilevered from the front face of the stem. The loads it must support include the weight of the cantilever slab and the upward soil pressure beneath. Usually any earth fill on top of the toe is neglected (as though it has been eroded). Obviously, such a fill would increase the upward soil pressure beneath the footing, but because it acts downward and cancels out the upward pressure, it produces no appreciable changes in the shears and moments in the toe.

A study of Figure 12.19 shows that the upward soil pressure is the major force applied to the toe. Because this pressure is primarily caused by the lateral force H, a load factor of 1.7 is used for the calculations. The maximum moment for design is taken at the face of the stem, whereas the maximum shear for design is assumed to occur at a distance d from the face of the stem because the reaction in the direction of the shear does introduce compression into the toe of the footing. The average designer makes the thickness of the toe the same as the thickness of the heel, although such a practice is not essential.

It is a common practice in retaining wall construction to provide a shear keyway between the base of the stem and the footing. This practice, though definitely not detrimental, is of questionable value. The keyway is normally formed by pushing a beveled 2 × 4 or 2 × 6 into the top of the footing, as shown in Figure 12.20. After the concrete hardens, the wood member is removed, and when the stem is cast in place above, a keyway is formed. It is becoming more and more common simply to use a roughened surface on the top of the footing where the stem will be placed. This practice seems to be just as satisfactory as the use of a keyway.

In Example 12.3, #8 bars 6 in. on center are selected for the vertical steel at the base of the stem. Either these bars need to be embedded into the footing for development purposes or dowels equal to the stem steel need to be used for the transfer. This latter practice is quite common because it is rather difficult to hold the stem steel in position while the base concrete is placed.

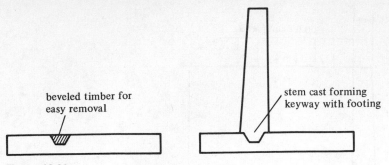

beveled timber for
easy removal

stem cast forming
keyway with footing

Figure 12.20

The required development length of the #8 bars down into the footing or for #8 dowels is 35 in. by Table A.15 (see Appendix) when $f_y = 60,000$ psi and $f'_c = 3000$ psi. This length cannot be obtained vertically in the 2-ft-0-in. footing used unless the bars or dowels are either bent as shown in Figure 12.21(a) or extended through the footing and into the base key as shown in Figure 12.21(b). Actually, the required development length can be reduced if more but smaller dowels are used. For #7 dowels ℓ_d is 26 in., and for #6 dowels it is 19 in.

If instead of dowels the vertical stem bars are embedded into the footing, they should not extend up into the wall more than 8 or 10 ft before they are spliced because they are difficult to handle in construction and may easily be bent out of place or even broken. Actually, you can see after examining Figure 12.21(a) that such an arrangement of stem steel can on some occasions be very advantageous economically.

The bending moment in the stem decreases rapidly above the base; as a result, the amount of reinforcing can be similarly reduced. It is to be remembered that these bars can be cut off only in accordance with the ACI Code development length requirements.

Example 12.3 illustrates the detailed design of a cantilever retaining wall. Several important descriptive remarks are presented in the solution, and these should be carefully read.

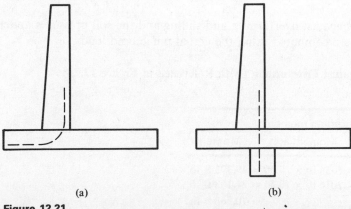

(a)

(b)

Figure 12.21

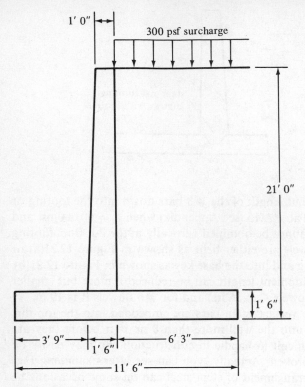

Figure 12.22

■ EXAMPLE 12.3

Complete the design of the cantilever retaining wall whose dimensions were estimated in Example 12.2 and are shown in Figure 12.22, if $f'_c = 3000$ psi, $f_y = 60,000$ psi, $q_a = 4000$ psf, and the coefficient of sliding friction equals 0.50 for concrete on soil. Use ρ approximately equal to $0.18f'_c/f_y$ to maintain reasonable deflection control.

SOLUTION

The safety factors against overturning and sliding and the soil pressures under the heel and toe are computed using the actual unfactored loads.

Safety Factor Against Overturning (with Reference to Figure 12.23)

	Overturning moment	
Force	Moment arm	Moment
$H_1 = (\frac{1}{2})(21)(672) = 7056$ lb $\times$ 7.00 = 49,392 ft-lb		
$H_2 = (21)(96)$ $= 2016$ lb $\times$ 10.50 = 21,168 ft-lb		
Total		70,560 ft-lb

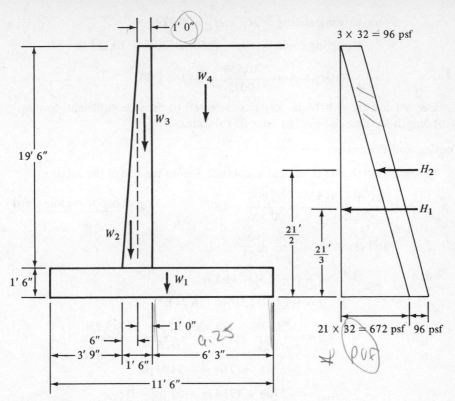

Figure 12.23

	Righting moment	
Force	Moment arm	Moment
$W_1 = (1.5)(11.5)(150) =$	2588 lb × 5.75 =	14,881 ft-lb
$W_2 = (\frac{1}{2})(19.5)(\frac{6}{12})(150) =$	731 lb × 4.08 =	2982 ft-lb
$W_3 = (19.5)(\frac{12}{12})(150) =$	2925 lb × 4.75 =	13,894 ft-lb
$W_4 = (22.5)(6.25)(100) =$	14,062 lb × 8.37 =	117,699 ft-lb[a]
$R_v = 20,306$ lb	$M = 149,456$ ft-lb	

[a] Includes surcharge.

$$\text{Safety factor against overturning} = \frac{149,456}{70,560} = 2.12 > 2.00 \quad \underline{\textbf{OK}}$$

Factor of Safety Against Sliding Here the passive pressure against the wall is neglected. Normally it is felt that the factor of safety should be at least 1.5. If it is not satisfactory, a shear key against sliding is normally used. The key should develop a passive pressure that is sufficient to resist the excess lateral force.

$$\text{Force causing sliding} = H_1 + H_2 = 9072 \text{ lb}$$

$$\text{Resisting force} = \mu R_v = (0.50)(20,306) = 10,153 \text{ lb}$$

$$\text{Safety factor} = \frac{10,153}{9072} = 1.12 < 1.50$$

Use a 1-ft-6-in. × 1-ft 6 in. key (size selected to provide sufficient development length for dowels selected later in solution).

Footing Soil Pressures

$R_v = 20,306$ lb and is located a distance $\bar{x}$ from the toe of the footing

$$\bar{x} = \frac{149,456 - 70,560}{20,306} = \frac{78,896}{20,306} = 3.89' \qquad \underline{\text{just inside middle third}}$$

$$\text{Soil pressure} = -\frac{R_v}{A} \pm \frac{Mc}{I}$$

$$A = (1)(11.5) = 11.5 \text{ ft}^2$$

$$I = (\tfrac{1}{12})(1)(11.5)^3 = 126.74 \text{ ft}^4$$

$$f_{toe} = -\frac{20,306}{11.5} - \frac{(20,306)(5.75 - 3.89)(5.75)}{126.74}$$

$$= -1766 - 1714 = -3480 \text{ psf}$$

$$f_{heel} = -1766 + 1714 = -52 \text{ psf}$$

Design of Stem   The lateral forces applied to the stem are calculated using a load factor of 1.7, as shown in Figure 12.24.

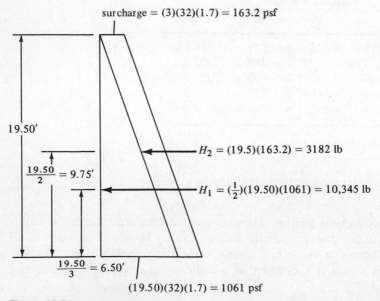

surcharge $= (3)(32)(1.7) = 163.2$ psf

19.50′

$\dfrac{19.50}{2} = 9.75'$

$H_2 = (19.5)(163.2) = 3182$ lb

$H_1 = (\tfrac{1}{2})(19.50)(1061) = 10,345$ lb

$\dfrac{19.50}{3} = 6.50'$

$(19.50)(32)(1.7) = 1061$ psf

Figure 12.24

Design of Stem for Moment

$$M_u = (H_1)(6.50) + (H_2)(9.75) = (10{,}345)(6.50) + (3182)(9.75)$$

$$M_u = 98{,}267 \text{ ft-lb}$$

Use

$$\rho = \text{approximately } \frac{0.18f_c'}{f_y} = \frac{(0.18)(3000)}{60{,}000} = 0.009$$

$$\frac{M_u}{\phi bd^2} \text{ (from Table A.13)} = 482.6$$

$$bd^2 = \frac{(12)(98{,}267)}{(0.9)(482.6)} = 2715$$

$$d = \sqrt{\frac{2715}{12}} = 15.04''$$

$$h = 15.04'' + 2'' + \frac{1''}{2} = 17.54'' \quad \underline{\text{say } 18'' \ (d = 15.50'')}$$

$$\frac{M_u}{\phi bd^2} = \frac{(12)(98{,}267)}{(0.90)(12)(15.5)^2} = 454.5$$

$$\rho = 0.00841 \text{ (from chart)}$$

$$A_s = (0.00841)(12)(15.5) = 1.56 \text{ in.}^2 \qquad \underline{\text{Use } \#8 \text{ at } 6''}$$

Minimum vertical ρ by ACI Section 14.3 = 0.0015 $\qquad \underline{\text{OK}}$

Minimum horizontal $A_s = (0.0025)(12)(\text{average stem } t)$

$$= (0.0025)(12)\left(\frac{12 + 18}{2}\right) = 0.450 \text{ in.}^2$$

(say one-third inside face and two-thirds outside face).

$$\underline{\text{Use } \#4 \text{ at } 7\tfrac{1}{2}'' \text{ outside face and } \#4 \text{ at } 15'' \text{ inside face}}$$

Checking Shear Stress in Stem   Actually V_u at a distance d from the top of the footing can be used, but for simplicity,

$$V_u = H_1 + H_2 = 10{,}345 + 3182 = 13{,}527 \text{ lb}$$

$$\phi V_c = \phi 2\sqrt{f_c'}bd = (0.85)(2\sqrt{3000})(12)(15.5)$$

$$= 17{,}319 > 13{,}527 \qquad \underline{\text{OK}}$$

Design of Heel   The upward soil pressure is conservatively neglected, and a load factor of 1.4 is used for calculating the shear and moment because soil and concrete make up the load.

(handwritten annotations: "heel base 1.0", "75000", "(S) heel", "concrete")

$$V_u = (22.5)(6.25)(100)(1.4) + (1.5)(6.25)(150)(1.4) = \boxed{21,655 \text{ lb}}$$

$$\phi V_c = (0.85)(2\sqrt{3000})(12)(14.5) = 16,202 < 21,655 \qquad \underline{\text{no good}}$$

Try 24-in. Depth ($d = 20.5$ in.)   Neglecting slight change in V_u with different depth

$$\phi V_c = (0.85)(2\sqrt{3000})(12)(20.5)$$
$$= 22,906 > 21,655 \qquad \underline{\text{OK}}$$

$$M_u \text{ at face of stem} = (21,655)\left(\frac{6.25}{2}\right) = 67,672 \text{ ft-lb}$$

$$\frac{M_u}{\phi bd^2} = \frac{(12)(67,672)}{(0.9)(12)(20.5)^2} = 178.9$$

$$\rho = 0.00310 < \frac{200}{f_y} = 0.00333$$

Using 0.00333,

$$A_s = (0.00333)(12)(20.5) = 0.82 \text{ in.}^2/\text{ft} \qquad \text{Use } \#8 \text{ @ } 11''$$

$$\ell_d \text{ required} = (32.86'')(1.3 \text{ for top bars})$$
$$= 55.86'' < 6'0'' \text{ available} \qquad \underline{\text{OK}}$$

Heel reinforcing is shown in Figure 12.25.

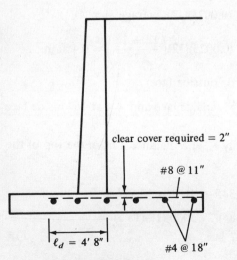

clear cover required = 2″

#8 @ 11″

$\ell_d = 4'\ 8''$

#4 @ 18″

Figure 12.25 Heel reinforcing.

Note: Temperature and shrinkage steel is normally considered to be unnecessary in the heel and toe. However, the author has placed #4 bars at 18 in. on center in the long direction, as shown in Figures 12.25 and 12.27, to serve as spacers for the flexural steel and to form mats out of the reinforcing.

Design of Toe   For service loads the soil pressures previously determined are multiplied by a load factor of 1.7 because they are primarily caused by the lateral forces, as shown in Figure 12.26.

$$V_u = 11,092 + 7529 = 18,621$$

(The shear can be calculated a distance d from the face of the stem because the reaction in the direction of the shear does introduce compression into the toe of the slab, but this advantage is neglected because 18,621 lb is already less than the 21,655 lb shear in the heel, which was satisfactory.)

$$M_u \text{ at face of stem} = (7529)\left(\frac{3.75}{3}\right) + (11,092)\left(\frac{2}{3} \times 3.75\right) = 37,141 \text{ ft-lb}$$

$$\frac{M_u}{\phi b d^2} = \frac{(12)(37,141)}{(0.9)(12)(20.5)^2} = 98.2$$

$$\rho = \text{less than } \rho_{min}$$

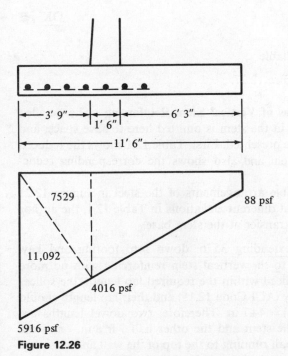

3' 9"
1' 6"
6' 3"
11' 6"

7529

88 psf

11,092

4016 psf

5916 psf

Figure 12.26

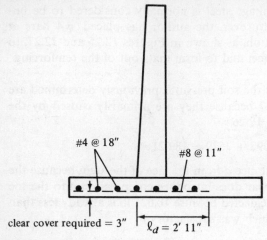

Figure 12.27 Toe reinforcing.

Therefore, use

$$\frac{200}{60,000} = 0.00333$$

$$A_s = (0.00333)(12)(20.5) = 0.82 \text{ in.2/ft}$$

ℓ_d required $= 32.86''$

Use #8 at 11"

OK ∎

$< 3''6''$ available

Toe reinforcing is shown in Figure 12.27.

Selection of Dowels and Lengths of Vertical Stem Reinforcing The detailed selection of vertical bar lengths in the stem is omitted here to save space, and only a few general comments are presented. First, Table 12.1 shows the reduced bending moments up in the stem and also shows the corresponding reductions in reinforcing required.

After considering the possible arrangements of the steel in Figure 12.21 and the required areas of steel at different elevations in Table 12.1, the author decided to use dowels for load transfer at the stem base.

Use #8 dowels at 6 in. extending 35 in. down into footing and key.

If these dowels are spliced to the vertical stem reinforcing with no more than one-half the bars being spliced within the required lap length, the splices will fall into the class B category (ACI Code 12.15) and their lap length should at least equal $1.3\ell_d = (1.3)(34.4) = 44.7$ in. Therefore, two dowel lengths are used—half 3 ft 10 in. up into the stem and the other half 7 ft 8 in.—and the #7 bars are lapped over them, half running to the top of the wall and the other half to middepth. Actually, a much more refined design can be made that involves more cutting of bars. For such a design a diagram comparing the theoretical steel area required at various elevations in the stem and the actual steel

TABLE 12.1

Distance from top of stem	M_u (ft-lb)	Effective stem d (in.)	ρ	A_s required (in.²/ft)	Bars needed
5'	3176	11.04	Use $\rho_{min} = 0.00333$	0.44	#8 at 21" #7 at 16"
10'	17,218	12.58	Use $\rho_{min} = 0.00333$	0.50	#8 at 18" or #7 at 14"
15'	48,960	14.12	0.00482	0.82	#8 at 11" or #7 at $8\frac{1}{2}$"
19.5'	98,267	15.50	0.00841	1.56	#8 at 6"

furnished is very useful. It is to be remembered (Code 12.10.3) that the bars cut off must run at least a distance d or 12 diameters beyond their theoretical cutoff points and must also meet the necessary development length requirements.

12.11 WALL JOINTS

Construction joints may be used both horizontally and vertically between successive pours of concrete. The surface of the hardened concrete can be cleaned and roughened, or keys may be used as shown in Figure 12.28(a) to form horizontal construction joints.

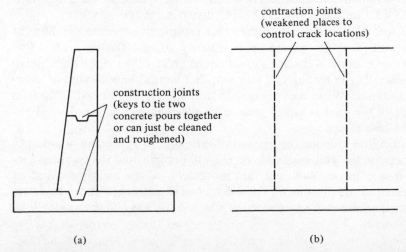

construction joints (keys to tie two concrete pours together or can just be cleaned and roughened)

contraction joints (weakened places to control crack locations)

(a)   (b)

Figure 12.28

Box culvert. (Courtesy of Economy Forms Corporation.)

If concrete is restrained from free movement when shrinking, as by being attached to more rigid parts of the structure, it will crack at points of weakness. Contraction joints are weakened places constructed so that shrinkage failures will occur at prepared locations. When the shrinkage tensile stresses become too large, they will pull these contraction joints apart and form rather neat cracks rather then the crooked unsightly ones that might otherwise occur. In addition to handling shrinkage problems, contraction joints are useful in handling differential settlements. They need to be spaced at intervals from about 25 to 40 ft on center (the AASHTO says not greater than 30 ft).

Expansion joints are vertical joints that completely separate the different parts of a wall. They are placed approximately 50 to 100 ft on centers (the AASHTO says maximum spacing should not be greater than 90 ft). Reinforcing bars are generally run through all joints so that vertical and horizontal alignment is maintained. When the bars do run through a joint, one end of the bars on one side of the joint is either greased or sheathed so that the desired expansion can take place.

It is difficult to estimate the amount of shrinkage or expansion of a particular wall because the wall must slide on the soil beneath, and the resulting frictional resistance may be sufficient that movement will be greatly reduced or even prevented. A rough value for the width of an expansion joint can be determined from the following expression, in which ΔL is the change in length, L

is the distance between joints, ΔT is the estimated temperature change, and 0.000005 is the estimated coefficient of contraction of the wall.

$$\Delta L = (0.000005L)(AT)$$

PROBLEMS

In Problems 12.1 to 12.4 use the Rankine equation to calculate the total horizontal active force and the overturning moment for the wall shown in the accompanying illustration. Assume that $\phi = 30°$ and the soil weighs 100 lb/ft³. Neglect the fill on the toe for each wall.

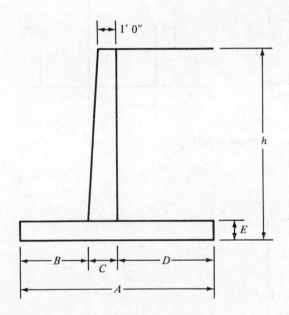

Problem	A	B	C	D	E	h
12.1	8'0"	2'0"	1'6"	4'6"	1'6"	14'0"
12.2	10'6"	2'6"	1'9"	6'3"	1'8"	18'0"
12.3	11'0"	3'6"	1'6"	6'0"	1'6"	20'0"
12.4	12'6"	4'0"	1'6"	7'0"	2'0"	22'0"

(Answer to Problem 12.1: 3266 lbs; 15,242 ft-lbs)

(Answer to Problem 12.3: 6666 lbs; 44,440 ft-lbs)

12.5 Repeat Problem 12.1 if δ is 20°. (*Ans.* 4059 lb; 18,943 ft-lbs)

12.6 Repeat Problem 12.3 if δ is 23°40′.

In Problems 12.7 to 12.9 determine the safety factors against overturning and sliding for the gravity and semigravity walls shown if $\phi = 30°$ and the coefficient of friction (concrete on soil) is 0.5. Compute also the soil pressure under the toe and heel of each footing. The soil weighs 100 lb/ft³, and the plain concrete used in the footing weighs 145 lb/ft³. Determine horizontal pressures using the Rankine equation.

Problem 12.7 (*Ans.* 5.69, 2.67, −2193 psf, −1015 psf)

Problem 12.8

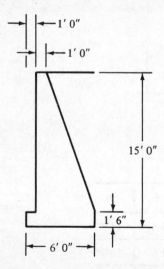

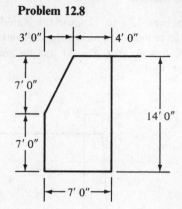

Problem 12.9 (*Ans.* 1.73, 1.32, −4739 psf, 0 psf)

In Problems 12.10 to 12.13, if Rankine's coefficient C_a is 0.60, the soil weight 110 lb/ft³, the concrete weight 150 lb/ft³, and the coefficient of friction (concrete on soil) is 0.45, determine the safety factors against overturning and sliding for the wall shown in the accompanying illustration.

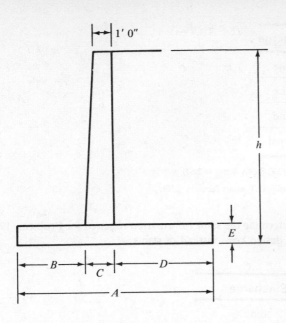

Problem	A	B	C	D	E	h
12.10	8'0"	2'0"	1'0"	5'0"	1'3"	14'0"
12.11	11'0"	2'6"	1'6"	7'0"	1'6"	15'0"
12.12	13'6"	4'0"	1'6"	8'0"	1'6"	18'0"
12.13	14'0"	3'6"	1'6"	9'0"	1'9"	20'0"

(Answer to Problem 12.11: 2.70, 0.93)

(Answer to Problem 12.13: 2.41, 0.86)

12.14 Repeat Problem 12.4 assuming a surcharge of 200 psf.

12.15 Repeat Problem 12.9 assuming a surcharge of 200 psf, 110 lb/ft³ soil weight and 145 lb/ft³ concrete. (*Ans.* 1.36, 1.12, −8089 psf, 0 psf.)

12.16 Repeat Problem 12.12 assuming a surcharge of 330 psf. Also determine toe and heel soil pressures.

In Problems 12.17 to 12.20 determine approximate dimensions of retaining walls, check safety factors against overturning and sliding, and calculate soil pressures for the wall shown. Also determine the required stem thickness at their bases and select vertical reinforcing there, using $f_y = 60,000$ psi, $f'_c = 3000$ psi, $q_a = 5000$ psf, $\rho =$ approximately $0.18 f'_c / f_y$, angle of internal friction = 33°40′, and coefficient of sliding friction (concrete on soil) = 0.45. Soil weight = 100 lb/ft³. Concrete weight = 150 lb/ft³.

Problem	h	Surcharge
12.17	12'0"	None
12.18	16'0"	None
12.19	18'0"	None
12.20	15'0"	200 psf

(Answer to Problem 12.17: 6 ft wide, O.T. Safety factor = 2.62)

(Answer to Problem 12.19: 8 ft 6 in. wide, O.T. safety factor = 2.30)

In Problems 12.21 to 12.23 determine the same information required for Problems 12.17 to 12.20 with same data, but design heels instead of stems.

Problem	h	Surcharge
12.21	14'0"	None
12.22	18'0"	300 psf
12.23	20'0"	300 psf starting 4' 0" from inside face of wall

(Answer to Problem 12.21: 6 ft 6 in. wide, O.T. safety factor = 2.24)

(Answer to Problem 12.23: 10 ft 3 in. wide, O.T. safety factor = 2.16)

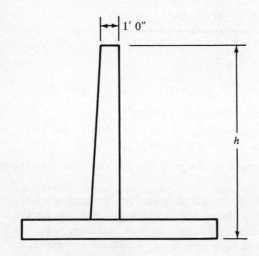

PROBLEMS WITH SI UNITS
In Problems 12.24 to 12.26 use the Rankine equation to calculate the total horizontal force and the overturning moment for the wall shown. Assume $\sin \phi = 0.5$ and the soil weighs 16 kN/m³.

Problem	A	B	C	D	E	h
12.24	2.400 m	600 mm	500 mm	1.300 mm	450 mm	4 m
12.25	2.700 m	700 mm	500 mm	1.500 mm	500 mm	6 m
12.26	3.150 m	800 mm	550 mm	1.800 mm	500 mm	8 m

(Answer to Problem 12.25: 95.904 kN, 191.908 kN·m)

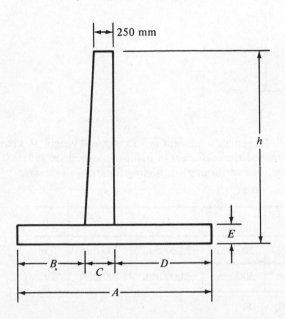

In Problems 12.27 and 12.28 determine the safety factors against overturning and sliding for the gravity and semigravity walls shown if $\phi = 30°$ and the coefficient of sliding (concrete on soil) is 0.45. Compute also the soil pressure under the toe and heel of each footing. The soil weighs 16 kN/m³ and the plain concrete used in the footing weighs 22.7 kN/m³.

Problem 12.27 (*Ans.* 3.06, 1.68, −141.87 kN/m², −17.03 kN/m²)

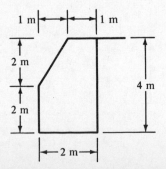

Problem 12.28

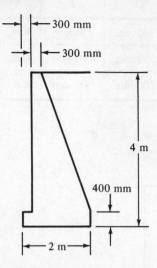

In Problems 12.29 and 12.30, if Rankine's coefficient is 0.35, the soil weight 16 kN/m³, the concrete weight 23.5 kN/m³, and the coefficient of friction (concrete on soil) is 0.50, determine the safety factors against overturning and sliding for the wall shown.

Problem	A	B	C	D	E	h
12.29	4 m	1.500 mm	300 mm	2.200 mm	700 mm	5 m
12.30	5 m	1.500 mm	500 mm	3.000 mm	800 mm	7 m

(Answer to Problem 12.29: 5.32, 1.77)

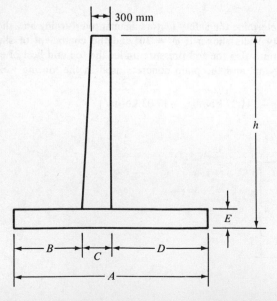

In Problems 12.31 to 12.33 select approximate dimensions for the cantilever retaining wall shown below and determine reinforcing required at base of stem using the following data: $f'_c = 20.7$ MPa, $f_y = 413.7$ MPa, $\rho =$ approximately $\frac{1}{2}\rho_{max}$, angle of internal friction $= 33°40'$, coefficient of sliding friction (concrete on soil) $= 0.50$. Soil weight $=$ 16 kN/m³ and reinforced concrete weight $= 23.5$ kN/m³.

Problem	h	Surcharge
12.31	4 m	None
12.32	6 m	None
12.33	7 m	4 kN/m

(Answer to Problem 12.31: use 320 mm stem @ base with $d = 250$ mm and #5 bars @ 240 mm)

(Answer to Problem 12.33: use 560 mm stem @ base with $d = 490$ mm and #7 bars @ 170 mm)

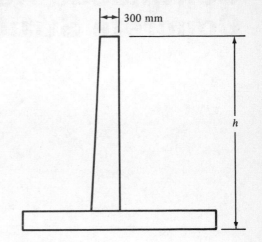

13

Continuous Reinforced Concrete Structures

13.1 INTRODUCTION

During the construction of reinforced concrete structures, as much concrete as possible is placed in each pour. For instance, the concrete for a whole concrete floor or for a large part of it, including the supporting beams and girders and parts of the columns, may be placed at the same time. The reinforcing bars extend from member to member, as from one span of a beam into the next. When there are construction joints, the reinforcing bars are left protruding from the older concrete so they may be lapped or spliced to the bars in the newer concrete. In addition, the old concrete is cleaned so that the newer concrete will bond to it as well as possible. The result of all these facts is that reinforced concrete structures are generally monolithic or continuous and thus statically indeterminate.

A load placed in one span of a continuous structure will cause shears, moments, and deflections in the other spans of that structure. Not only are the beams of a reinforced concrete structure continuous, but the entire structure is continuous. In other words, loads applied to a column affect the beams, slabs, and other columns, and vice versa.

The result is that more economical structures are obtained because the bending moments are smaller and thus member sizes are smaller. Although the analyses and designs of continuous structures are more complicated than they are for statically determinate structures, this fact becomes less and less important each year because of the constantly increasing availability of modern computers.

414

Confinazas Financial Center, Caracas, Venezuela. (Courtesy of Economy Forms Corporation.)

13.2 GENERAL DISCUSSION OF ANALYSIS METHODS

In reinforced concrete design today we use elastic methods to analyze structures loaded with factored or ultimate loads. Such a procedure probably doesn't seem quite correct to the reader, but it does yield satisfactory results. The reader might very well ask, "Why don't we use ultimate or inelastic analyses for reinforced concrete structures?" The answer is that our theory and tests are just not sufficiently advanced.

It is true that under certain circumstances some modifications of moments are permitted to recognize ultimate or inelastic behavior as described in Section 13.5 of this chapter. In general, however, we will discuss elastic analyses for reinforced concrete structures. Actually no method of analysis, elastic or inelastic, will give exact results because of the unknown effects of creep, settlement, shrinkage, workmanship, and so on.

13.3 QUALITATIVE INFLUENCE LINES

There are many possible methods that might be used to analyze continuous structures. The most common hand calculation method is moment distribution,

One Shell Plaza, Houston, Texas. (Courtesy of Master Builders.)

but other methods are frequently used, such as matrix methods, computer solutions, and others. Whichever method is used, you should understand that to determine maximum shears and moments at different sections in the structure, it is necessary to consider different positions of the live loads. As a background for this material, a brief review of *qualitative influence lines* is presented.

Qualitative influence lines are based on a principle introduced by the German professor Heinrich Müller-Breslau. This principle is as follows: *The deflected shape of a structure represents to some scale the influence line for a function such as reaction, shear, or moment if the function in question is allowed to act through a small distance.* In other words, the structure draws its own influence line when the proper displacement is made.

The shape of the usual influence line needed for continuous structures is so simple to obtain with the Müller-Breslau principle that in many situations it is unnecessary to compute the numerical values of the coordinates. It is possible to sketch the diagram roughly with sufficient accuracy to locate the critical positions for live loads for various functions of the structure. These diagrams are referred to as *qualitative* influence lines, whereas those with numerical values are referred to as *quantitative* influence lines.[1]

If the influence line is desired for the left reaction of the continuous beam of Figure 13.1(a), its general shape can be determined by letting the reaction act upward through a unit distance, as shown in Figure 13.1(b). If the left end of the beam is pushed up, the beam will take the shape shown. This distorted shape can be easily sketched by remembering that the other supports are considered to be unyielding. The influence line for V_c, drawn in a similar manner, is shown in Figure 13.1(c).

Figure 13.1(d) shows the influence line for positive moment at point x near the center of the left-hand span. The beam is assumed to have a pin or hinge inserted at x and a couple is applied adjacent to each side of the pin, which will cause compression in the top fibers. Bending the beam on each side of the pin causes the left span to take the shape indicated, and the deflected shape of the remainder of the beam may be roughly sketched. A similar procedure is used to draw the influence line for negative moment at point y in the third span, except that a moment couple is applied at the assumed pins, which will tend to cause compression in the bottom beam fibers, corresponding with negative moment.

Finally, qualitative influence lines are drawn for positive shear at points x and y. At point x the beam is assumed to be cut, and the two vertical forces of the nature required to give positive shear are applied to the beam on the sides of the cut section. The beam will take the shape shown in Figure 13.1(f). The same procedure is used in Figure 13.1(g) to draw a diagram for positive shear at point y. [Theoretically, for qualitative shear influence lines, it is necessary to have a moment on each side of the cut section sufficient to maintain equal

[1] McCormac, J. C., and Elling, R. E., 1988, *Structural Analysis: A Classical and Matrix Approach* (New York: HarperCollins), pp. 344–346.

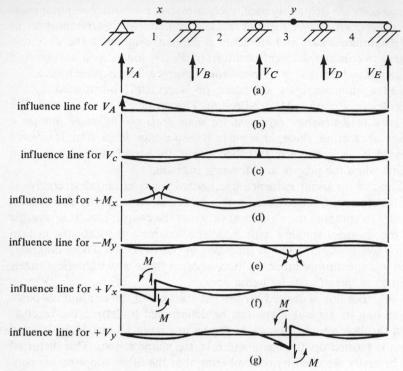

Figure 13.1 Influence lines.

slopes. Such moments are indicated in parts (f) and (g) of the figure by the letter M.]

From these diagrams considerable information is available concerning critical live-loading conditions. If a maximum positive value of V_A were desired for a uniform live load, the load would be placed in spans 1 and 3, where the diagram has positive ordinates; if maximum negative moment were required at point y, spans 2 and 4 would be loaded; and so on.

Qualitative influence lines are particularly valuable for determining critical load positions for buildings, as illustrated by the moment influence line for the building of Figure 13.2. In drawing diagrams for an entire frame, the joints are assumed to be free to rotate, but the members at each joint are assumed to be rigidly connected to each other so that the angles between them do not change during rotation. The influence line shown in the figure is for positive moment at the center of beam AB.

The spans that should be loaded to cause maximum positive moment are obvious from the diagram. It should be realized that loads on a member more than approximately three spans away have little effect on the function under consideration.

In the past few paragraphs, influence lines have been used to determine the critical positions for placing live loads to cause maximum moments. The same results can be obtained (and perhaps more easily) by considering the deflected

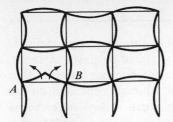

Figure 13.2

shape or curvature of a member under load. If the live loads are placed so that they cause the greatest curvature at a particular point, they will have bent the structure the greatest amount at that point, which means that the greatest moment will have been obtained.

For the continuous beam of Figure 13.3(a) it is desired to cause the maximum negative moment at support *B* by the proper placement of a uniform live load. In part (b) of the figure the deflected shape of the beam is sketched as it would be when a negative moment occurs at *B*, and the rest of the beam's deflected shape is drawn as shown by the dashed line. Then the live uniform load is placed in those locations which would exaggerate that deflected shape. This is done by placing the load in spans 1, 2, and 4.

A similar situation is shown in Figure 13.3(c), where it is desired to obtain maximum positive moment at the middle of the second span. The deflected shape of the beam is sketched as it would be when a positive moment occurs in that span, and the rest of the beam's deflected shape is drawn in. To exaggerate this positive or downward bending in the second span it can be seen that the live load should be placed in spans 2 and 4.

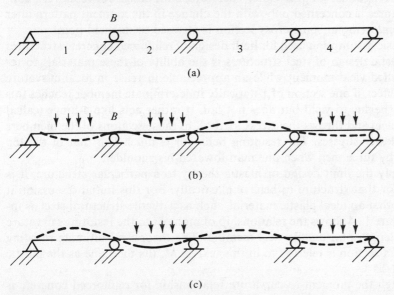

Figure 13.3

13.4 LIMIT DESIGN

Throughout the preceding ten chapters of this book, ultimate design methods have been used to proportion members that were analyzed by elastic methods. This procedure certainly must seem inconsistent to you because members that are subjected to loads greater than about half their ultimate capacities do not behave elastically. Although the method is inconsistent, it does nevertheless produce safe and conservative designs. Furthermore, the evaluation of the true collapse load or limit strength of an entire reinforced concrete frame requires a rather difficult analysis.

It can be clearly shown that a statically indeterminate beam or frame will normally not collapse when its ultimate moment capacity is reached at just one section. Instead, there is a redistribution of the moments in the structure. Its behavior is rather similar to the case where three men are walking along with a log on their shoulders and one of the men gets tired and lowers his shoulder just a little. The result is a redistribution of loads to the other men and thus changes in the shears and moments through the log.

It might be well at this point to attempt to distinguish between the terms *plastic design* as used in structural steel and *limit design* as used in reinforced concrete. In structural steel, plastic design involves both (a) the increased resisting moment of a member after the extreme fiber of the member is stressed to its yield point and (b) the redistribution or change in the moment pattern in the member. (Load and resistance factor design (LRFD) is a steel design method which incorporates most of the theory associated with plastic design.) In reinforced concrete the increase in resisting moment of a section after part of the section has been stressed to its yield point has already been accounted for in the strength design procedure. Therefore, limit design for reinforced concrete structures is concerned only with the change in the moment pattern after the steel reinforcing at some cross section is stressed to its yield point.

The basic assumption used for limit design of reinforced concrete structures and for plastic design of steel structures is the ability of these materials to resist a so-called yield moment while an appreciable increase in local curvature occurs. In effect, if one section of a statically indeterminate member reaches this moment, it begins to yield but does not fail. It rather acts like a hinge (called a *plastic hinge*) and throws the excess load off to those sections of the members that have lesser stresses. The resulting behavior is much like that of the log supported by three men when one man lowered his shoulder.

To apply the limit design or plastic theory to a particular structure, it is necessary for that structure to behave plastically. For this initial discussion it is assumed that an ideal plastic material, such as a ductile structural steel, is involved. Figure 13.4 shows the relationship of moment to the resulting curvature of a short length of a ductile steel member. The theoretical ultimate resisting moment of a section is referred to in this text as M_n (its the same as the plastic moment M_p).

Although the moment-to-curvature relationship for reinforced concrete is quite different from the ideal one pictured in Figure 13.4, the actual curve can

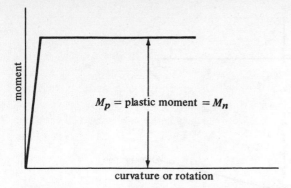

Figure 13.4 Moment–curvature relationship for an ideal plastic material.

be approximated reasonably well by the ideal one, as shown in Figure 13.5. The dashed line in the figure represents the ideal curve, while the solid line is a typical one for reinforced concrete. Tests have shown that the lower the reinforcing percentage in the concrete ρ or $\rho - \rho'$ (where ρ' is the percentage of compressive reinforcing), the closer will the concrete curve approach the ideal curve. This is particularly true when very ductile reinforcing steels such as grade 40 are used. Should a large percentage of steel be present in a reinforced concrete member, the yielding that actually occurs before failure will be so limited that the ultimate or limit behavior of the member will not be greatly affected by yielding.

The Collapse Mechanism

To consider moment redistribution in steel or reinforced concrete structures, it is felt necessary first to consider the location and number of plastic hinges re-

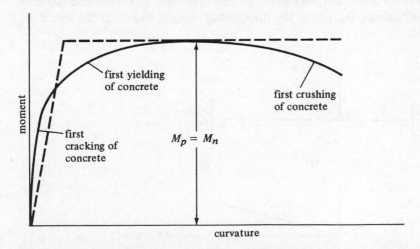

Figure 13.5 Typical moment–curvature relationship for a reinforced concrete member.

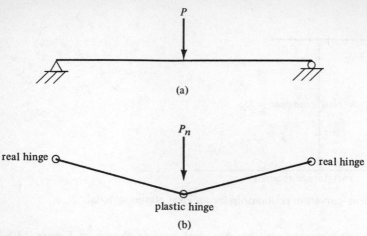

Figure 13.6

quired to cause a structure to collapse. A statically determinate beam will fail if one plastic hinge develops. To illustrate this fact, the simple beam of constant cross section loaded with a concentrated load at midspan shown in Figure 13.6(a) is considered. Should the load be increased until a plastic hinge is developed at the point of maximum moment (underneath the load in this case), an unstable structure will have been created, as shown in Figure 13.6(b). Any further increase in load will cause collapse.

The plastic theory is of little advantage for statically determinate beams and frames, but it may be of decided advantage for statically indeterminate beams and frames. For a statically indeterminate structure to fail, it is necessary for more than one plastic hinge to form. The number of plastic hinges required for failure of statically indeterminate structures will be shown to vary from structure to structure but may never be less than two. The fixed-end beam of Figure 13.7 cannot fail unless the three plastic hinges shown in the figure are developed.

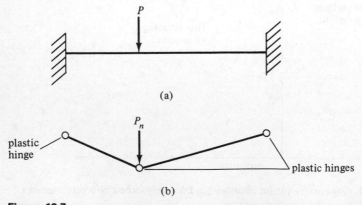

Figure 13.7

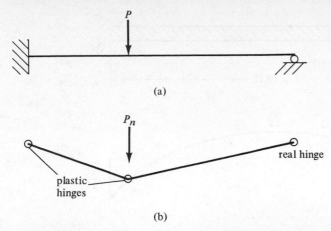

(a)

(b)

Figure 13.8

Although a plastic hinge may have formed in a statically indeterminate structure, the load can still be increased without causing failure if the geometry of the structure permits. The plastic hinge will act like a real hinge insofar as increased loading is concerned. As the load is increased, there is a redistribution of moment because the plastic hinge can resist no more moment. As more plastic hinges are formed in the structure, there will eventually be a sufficient number of them to cause collapse.

The propped beam of Figure 13.8 is an example of a structure that will fail after two plastic hinges develop. Three hinges are required for collapse, but there is a real hinge at the right end. In this beam the largest elastic moment caused by the design-concentrated load is at the fixed end. As the magnitude of the load is increased, a plastic hinge will form at that point.

The load may be further increased until the moment at some other point (here it will be at the concentrated load) reaches the plastic moment. Additional load will cause the beam to collapse. The arrangement of plastic hinges, and perhaps real hinges that permit collapse in a structure, is called the *mechanism*. Parts (b) of Figures 13.6, 13.7, and 13.8 show mechanisms for various beams.

Plastic Analysis by the Equilibrium Method

To plastically analyze a structure, it is necessary to compute the plastic or ultimate moments of the sections, to consider the moment redistribution after the ultimate moments develop, and finally to determine the ultimate loads that exist when the collapse mechanism is created. The method of plastic analysis known as the *equilibrium method* will be illustrated in this section.

As the first illustration the fixed-end beam of Figure 13.9 is considered. It is desired to determine the value of w_n, the theoretical ultimate load the beam can support. The maximum moments in a uniformly loaded fixed-end beam in the elastic range occur at the fixed ends, as shown in the figure.

If the magnitude of the uniform load is increased, the moments in the beam will be increased proportionately until a plastic moment is eventually developed at some point. Due to symmetry, plastic moments will be developed at the

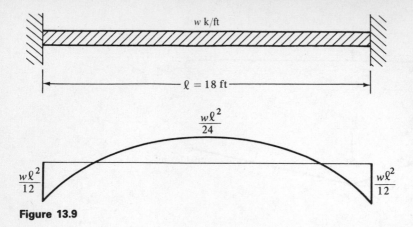

Figure 13.9

beam ends, as shown in Figure 13.10(b). Should the loads be further increased, the beam will be unable to resist moments larger than M_n at its ends. Those points will rotate through large angles, and thus the beam will be permitted to deflect more and permit the moments to increase out in the span. Although the plastic moment has been reached at the ends and plastic hinges are formed, the beam cannot fail because it has, in effect, become a simple end-supported beam for further load increases as shown in Figure 13.10(c).

The load can now be increased on this "simple" beam, and the moments at the ends will remain constant; however, the moment out in the span will increase as it would in a uniformly loaded simple beam. This increase is shown by the dashed line in Figure 13.11(b). The load may be increased until the moment at some other point (here the beam centerline) reaches the plastic moment. When this happens, a third plastic hinge will have developed and a mechanism will have been created, permitting collapse.

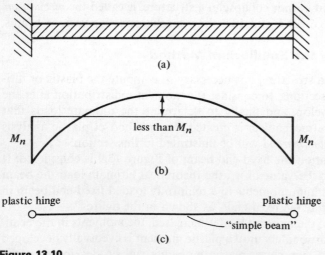

Figure 13.10

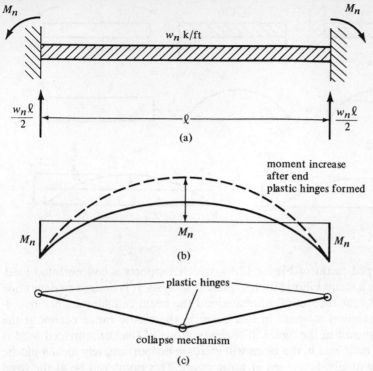

Figure 13.11

One method of determining the value of w_n is to take moments at the center line of the beam (knowing the moment there is M_n at collapse). Reference is made here to Figure 13.11(a) for the beam reactions.

$$M_n = -M_n + \left(w_n \frac{\ell}{2}\right)\left(\frac{\ell}{2} - \frac{\ell}{4}\right) = -M_n + \frac{w_n \ell^2}{8}$$

$$w_n = \frac{16M_n}{\ell^2}$$

The same value could be obtained by considering the diagrams shown in Figure 13.12. You will remember that a fixed-end beam can be replaced with a simply supported beam plus a beam with end moments. Thus the final moment diagram for the fixed-end beam equals the moment diagram if the beam had been simply supported plus the end moment diagram.

For the beam under consideration, the value of M_n can be calculated as follows (see Figure 13.13):

$$2M_n = \frac{w_n \ell^2}{8}$$

$$M_n = \frac{w_n \ell^2}{16}$$

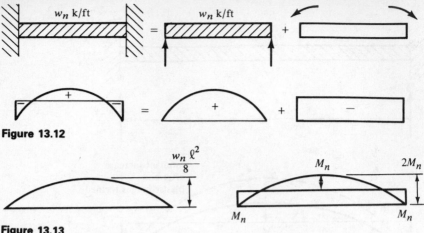

Figure 13.12

Figure 13.13

The propped beam of Figure 13.14, which supports a concentrated load, is presented as a second illustration of plastic analysis. It is desired to determine the value of P_n, the theoretical ultimate load the beam can support before collapse. The maximum moment in this beam in the elastic range occurs at the fixed end, as shown in the figure. If the magnitude of the concentrated load is increased, the moments in the beam will increase proportionately until a plastic moment is eventually developed at some point. This point will be at the fixed end, where the elastic moment diagram has its largest ordinate.

After this plastic hinge is formed, the beam will act as though it is simply supported insofar as load increases are concerned, because it will have a plastic hinge at the left end and a real hinge at the right end. An increase in the magnitude of the load P will not increase the moment at the left end but will increase the moment out in the beam, as it would in a simple beam. The increasing simple beam moment is indicated by the dashed line in Figure 13.14(c). Eventually the moment at the concentrated load will reach M_n and a mechanism will form consisting of two plastic hinges and one real hinge, as shown in Figure 13.14(d).

The value of the theoretical maximum concentrated load P_n that the beam can support can be determined by taking moments to the right or left of the load. Figure 13.14(e) shows the beam reactions for the conditions existing just before collapse. Moments are taken to the right of the load as follows:

$$M_n = \left(\frac{P_n}{2} - \frac{M_n}{20}\right)10$$

$$P_n = 0.3M_n$$

The subject of plastic analysis can be continued for different types of structures and loadings, as described in several textbooks on structural analysis or steel design.[2] The method has been proved to be satisfactory for ductile struc-

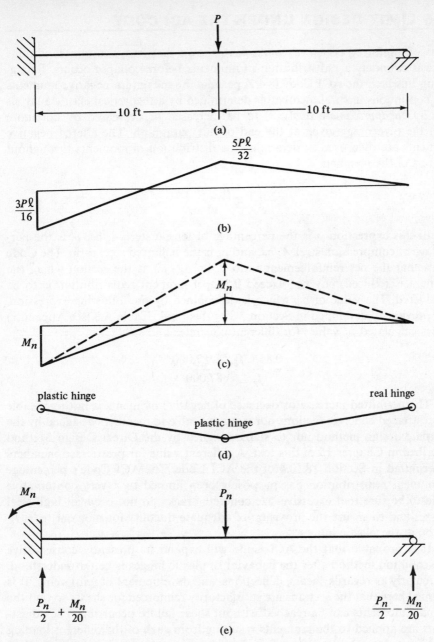

Figure 13.14

tural steels by many tests. Concrete, however, is a relatively brittle material, and the limit design theory has not been fully accepted by the ACI Code. The Code does recognize that there is some redistribution of moments and permits partial redistribution based on a rule of thumb that is presented in the next section of this chapter.

13.5 LIMIT DESIGN UNDER THE ACI CODE

Tests of reinforced concrete frames have shown that under certain conditions there is definitely a redistribution of moments before collapse occurs. Recognizing this fact, the ACI Code (8.4.1) permits the maximum negative moments in a continuous flexural member as determined by a theoretical elastic analysis (*not by an approximate analysis*) to be increased or decreased by not more than the percentage given at the end of this paragraph. The altered negative moments are then used to determine the distribution of moments throughout the rest of the member.

$$20\left(1 - \frac{\rho - \rho'}{\rho_b}\right)$$

In this expression ρ is the percentage of tensile steel A_s/bd, ρ' is the percentage of compression steel A_s'/bd, and ρ_b is the balanced steel ratio. The Code states that the net reinforcement ratio ρ, or $\rho - \rho'$, at the section where the moment is reduced may not exceed $0.50\rho_b$ if moment redistribution is to be considered. The value of ρ_b, which is determined by the following expression, was previously developed in Section 3.8 of this book. Table A.8 (see Appendix) shows calculated ρ_b values for different concretes and steels.

$$\rho_b = \frac{0.85\beta_1 f_c'}{f_y}\left(\frac{87,000}{87,000 + f_y}\right)$$

The permitted increase or decrease of negative moments is not applicable to prestressed concrete sections nor is it applicable to members designed by the alternate design method nor to slabs designed by the Direct Design Method described in Chapter 15 of this text. A different value for prestressed members is permitted in Section 18.10.4 of the ACI Code. The ACI Code's percentage of moment redistribution has purposely been limited to a very conservative value to be sure that excessive-size concrete cracks do not occur at high steel stresses and to ensure the provision of adequate ductility for moment redistribution at the plastic hinges.

It is probable that the ACI Code will expand its presently conservative redistribution method after the behavior of plastic hinges is better understood, particularly as regards shears, deflections, and development of reinforcing. It is assumed here that the sections are satisfactorily reinforced for shears so that the ultimate moments can be reached without shear failure occurring. The adjustments are applied to the moments resulting from each of the different loading conditions. The member in question will then be proportioned on the basis of the resulting moment envelope. Figures 13.15 through 13.18 illustrate the application of the moment redistribution permitted by the Code to a three-span continuous beam with a service dead load of 2 k/ft and a service live load of 3 k/ft. It will be noted in these figures that factored loads are used for all the calculations.

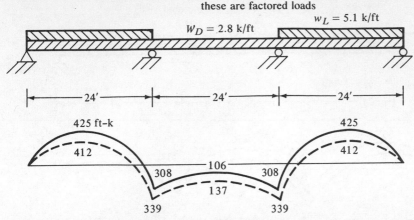

Figure 13.15 Maximum positive moment in end spans.

Three different live-load conditions are considered in these figures. To determine the maximum positive moment in span 1, live load is placed in spans 1 and 3 (Figure 13.15). Similarly, to produce maximum positive moment in span 2, the live load is placed in that span only (Figure 13.16). Finally, maximum negative moment at the first interior support from the left end is caused by placing the live load in spans 1 and 2 (Figure 13.17).

For this particular beam it is assumed that the Code permits a 10% up or down adjustment in the negative support moments. The result will be smaller design moments at the critical sections. Initially the loading for maximum positive moment in span 1 is considered as shown in Figure 13.15. If the maximum calculated negative moments of 308 ft-k at the interior supports are each in-

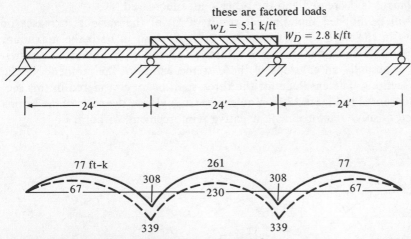

Figure 13.16 Maximum positive moment in span 2.

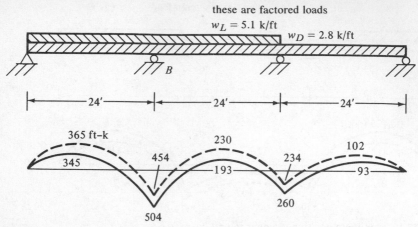

Figure 13.17 Maximum negative moment at support *B*.

creased by 10% to 339 ft-k, the maximum positive moments in spans 1 and 3 will be reduced proportionally to 412 ft-k.

In the same fashion in Figure 13.16, where the beam is loaded to produce maximum positive moment in span 2, an increase in the negative moments at the supports from 308 ft-k to 339 ft-k will reduce the maximum positive moment in span 2 from 261 ft-k to 230 ft-k.

Finally, in Figure 13.17 the live-load placement causes a maximum negative moment at the first interior support of 504 ft-k. If this value is reduced by 10%, the maximum moment there will be −454 ft-k. In this figure the author has reduced the negative moment at the other interior support by 10% also. Should it be of advantage, however, it can be assumed that one negative support moment is decreased and the other one is increased.

It will be noticed that the net result of all of the various increases or decreases in the negative moments is a net reduction in both the maximum positive and the maximum negative values. The result of these various redistributions is actually an envelope of the extreme values of the moments at the critical sections. The envelope for the three-span beam considered in this section is presented in Figure 13.18. You can see at a glance the parts of the beams that need positive reinforcement, negative reinforcement, or both.

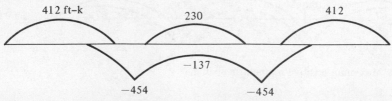

Figure 13.18 Moment envelope.

The reductions in bending moments due to moment redistribution as described here do not mean that the safety factors for continuous members will be less than those for simple spans. Rather, the excess strength that such members have due to this continuity is reduced so that the overall factors of safety are nearer but not less than those of simple spans.

Various studies have shown that cracking and deflection of members selected by the limit design process are no more severe than those for the same members designed without taking advantage of the permissible redistributions.[3,4]

13.6 PRELIMINARY DESIGN OF MEMBERS

Before an "exact" analysis of a building frame can be made, it is necessary to estimate the sizes of the members. Even if a computer design is used, it is economically advisable to make some preliminary estimates as to sizes. If an approximate analysis of the structure has been made, it will be possible to make very reasonable member size estimates. The result will be appreciable saving of both computer time and money.

An experienced designer can usually make very satisfactory preliminary size estimates based upon his or her previous experience. In the absence of such experience, however, the designer can still make quite reasonable size estimates based on his or her knowledge of structural analysis. For instance, to approximately size columns he or she can neglect moments, and assume an average axial stress or P_u/A_g value of about 0.4 to $0.6f'_c$. This rough value can be divided into the estimated column load to obtain its estimated area. If moments are large, lower values of average stress (0.4 to $0.5f'_c$) may be used; if moments are low, higher values (0.55 to $0.60f'_c$) may be used.

Preliminary beam sizes can be obtained by considering their approximate moments. A uniformly loaded simple beam will have a maximum bending moment equal to $w_u \ell^2/8$, whereas a uniformly loaded fixed-end beam will have a maximum moment of $w_u \ell^2/12$. For a continuous uniformly loaded beam, the designer might very well estimate a maximum moment somewhere in between the values given, perhaps $w_u \ell^2/10$, and use that value to estimate the beam size.

For many structures it is necessary to conduct at least two different analyses. One analysis is made to consider the effect of gravity loads as described in Section 13.7 of this chapter, while another might be made to consider the effect of lateral loads as discussed in Section 13.8. For the gravity loads only, $U = 1.4D + 1.7L$.

---

[3] Cohn, M. Z., 1964, "Rotational Compatibility in the Limit Design of Reinforced Concrete Continuous Beams." *Proceedings of the International Symposium on the Flexural Mechanics of Reinforced Concrete*, ASCE-ACI (Miami), pp. 359–382.

[4] Mattock. A. H., 1959, "Redistribution of Design Bending Moments in Reinforced Concrete Continuous Beams," *Proceedings of the Institution of Civil Engineers*, 113, pp. 35–46.

Because the gravity loads affect only the floor to which they are applied, each floor can probably be analyzed independently of the others. Such is not the case for lateral loads because lateral loads applied anywhere on the frame affect the lateral displacements throughout the frame and thus affect the forces in the frame below. For this case, $U = \frac{3}{4}(1.4D + 1.7L + 1.7W)$.

Sometimes a third analysis should be made—one that involves the possibility of force reversals on the windward side or even overturning of the structure. If overturning is being considered, the dead and live gravity loads should be reduced to their smallest possible values (that is, zero live load and $0.9D$ in case the dead loads have been overestimated a little) while the lateral loads are acting. For this case, $U = 0.9D + 1.3W$.

13.7 APPROXIMATE ANALYSIS OF CONTINUOUS FRAMES FOR VERTICAL LOADS

Statically indeterminate structures may be analyzed "exactly" or "approximately." Some approximate methods involving the use of simplifyiing assumptions are presented in this section. Despite the increased use of computers for making "exact" analyses, approximate methods are used about as much or more than ever for several reasons. These include the following.

1. The structure may be so complicated that no one who has the knowledge to make an "exact" analysis is available.
2. For some structures, either method may be subject to so many errors and imperfections that approximate methods may yield values as accurate as those obtained with an "exact" analysis. A specific example is the analysis of a building frame for wind loads where the walls, partitions, and floors contribute an indeterminate amount to wind resistance. Wind forces calculated in the frame by either method are not accurate.
3. To design the members of a statically indeterminate structure, it is necessary to make an estimate of their sizes before structural analysis can begin by an exact method. Approximate analysis of the structure will yield forces from which reasonably good initial estimates can be made as to member sizes.
4. Approximate analyses are quite useful in rough-checking exact solutions.

From the discussion of influence lines in Section 13.3 you can see that unless a computer is used (a rather practical alternative today) an exact analysis involving several different placements of the live loads would be a long and tedious affair. For this reason it is common when a computer is not readily available to use some approximate method of analysis, such as the ACI moment and shear coefficients, the equivalent rigid frame method, the assumed point-of-inflection-location method, and others discussed in the pages to follow.

ACI Coefficients for Continuous Beams and Slabs

A very common method used for the design of continuous reinforced concrete structures involves the use of the ACI coefficients given in Section 8.3.3 of the Code. These coefficients, which are reproduced in Table 13.1, provide estimated maximum shears and moments for buildings of normal proportions. The values calculated in this manner will usually be somewhat larger than those that would be obtained with an "exact" analysis. As a result, appreciable economy can normally be obtained by taking the time or effort to make such an analysis. In this regard it should be realized that these coefficients are considered to have their best application for continuous frames having more than three or four continuous spans.

In developing the coefficients the negative moment values were reduced to take into account the usual support widths and also some moment redistribution, as described in Section 13.5 of this text. In addition, the positive moment values have been increased somewhat to account for the moment redistribution. It will also be noted that the coefficients account for the fact that in monolithic construction the supports are not simple and moments are present at end supports, such as where those supports are beams or columns.

In applying the coefficients, w_u is the design load while ℓ_n *is the clear span for calculating positive moments and the average of the adjacent clear spans for calculating negative moments.* These values were developed for members with approximately equal spans (the larger of two adjacent spans not exceeding the smaller by more than 20%) and for cases where the ratio of the uniform service live load to the uniform service dead load is not greater than three. In

TABLE 13.1 ACI COEFFICIENTS

Positive moment	
End spans	
If discontinuous end is unrestrained	$\frac{1}{11}w_u\ell_n^2$
If discontinuous end is integral with the support	$\frac{1}{14}w_u\ell_n^2$
Interior spans	$\frac{1}{16}w_u\ell_n^2$
Negative moment at exterior face of first interior support	
Two spans	$\frac{1}{9}w_u\ell_n^2$
More than two spans	$\frac{1}{10}w_u\ell_n^2$
Negative moment at other faces of interior supports	$\frac{1}{11}w_u\ell_n^2$
Negative moment at face of all supports for (a) slabs with spans not exceeding 10 ft and (b) beams and girders where ratio of sum of column stiffnesses to beam stiffness exceeds eight at each end of the span	$\frac{1}{12}w_u\ell_n^2$
Negative moment at interior faces of exterior supports for members built integrally with their supports	
Where the support is a spandrel beam or girder	$\frac{1}{24}w_u\ell_n^2$
Where the support is a column	$\frac{1}{16}w_u\ell_n^2$
Shear in end members at face of first interior support	$1.15(w_u\ell_n/2)$
Shear at face of all other supports	$w_u\ell_n/2$

addition, the values are not applicable to prestressed concrete members. Should these limitations not be met, a more precise method of analysis must be used.

For the design of a continuous beam or slab, the moment coefficients provide in effect two sets of moment diagrams for each span of the structure. One diagram is the result of placing the live loads so that they will cause maximum positive moment out in the span, while the other is the result of placing the live loads so as to cause maximum negative moments at the supports. To be truthful, however, it is not possible to produce maximum negative moments at both ends of a span simultaneously. It takes one placement of the live loads to produce maximum negative moment at one end of the span and another placement to produce maximum negative moment at the other end. The assumption of both maximums occurring at the same time is on the safe side, however, because the resulting diagram will have greater critical values than are produced by either one of the two separate loading conditions.

The ACI coefficients give maximum points for a moment envelope for each span of a continuous frame. Typical envelopes are shown in Figure 13.19 for a continuous slab, which is assumed to be constructed integrally with its exterior supports, which are spandrel girders.

Example 13.1 presents the design of the slab of Figure 13.20 using the moment coefficients of the ACI Code. The calculations for this problem can be conveniently set up in some type of table such as the one shown in Figure 13.21. For this particular slab the author used an arrangement of reinforcement that included bent bars. It is quite common, however, in slabs, particularly those 5 in. or less in thickness, to use straight bars only in the top and bottom of the slab.

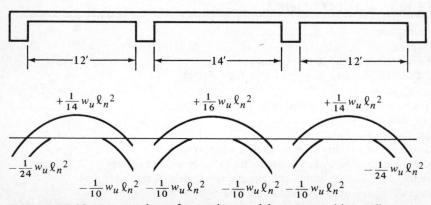

Figure 13.19 Moment envelopes for continuous slab constructed integrally.

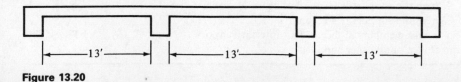

Figure 13.20

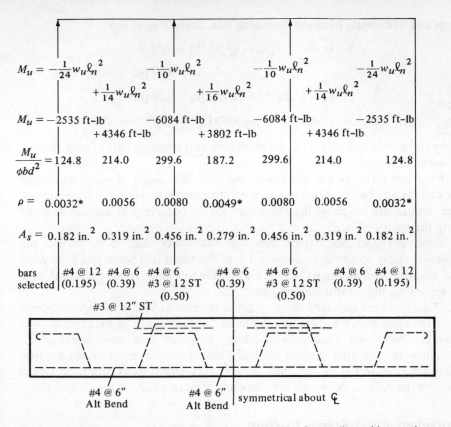

$$M_u = -\frac{1}{24}w_u\ell_n{}^2 \qquad -\frac{1}{10}w_u\ell_n{}^2 \qquad -\frac{1}{10}w_u\ell_n{}^2 \qquad -\frac{1}{24}w_u\ell_n{}^2$$
$$+\frac{1}{14}w_u\ell_n{}^2 \qquad +\frac{1}{16}w_u\ell_n{}^2 \qquad +\frac{1}{14}w_u\ell_n{}^2$$

$M_u = -2535$ ft-lb $\qquad -6084$ ft-lb $\qquad -6084$ ft-lb $\qquad -2535$ ft-lb
$\qquad +4346$ ft-lb $\qquad +3802$ ft-lb $\qquad +4346$ ft-lb

$\dfrac{M_u}{\phi bd^2} = 124.8 \qquad 214.0 \qquad 299.6 \qquad 187.2 \qquad 299.6 \qquad 214.0 \qquad 124.8$

$\rho = 0.0032^* \quad 0.0056 \quad 0.0080 \quad 0.0049^* \quad 0.0080 \quad 0.0056 \quad 0.0032^*$

$A_s = 0.182$ in.^{2} 0.319 in.^{2} 0.456 in.^{2} 0.279 in.^{2} 0.456 in.^{2} 0.319 in.^{2} 0.182 in.^{2}

bars   #4 @ 12   #4 @ 6   #4 @ 6   #4 @ 6   #4 @ 6   #4 @ 6   #4 @ 12
selected   (0.195)   (0.39)   #3 @ 12 ST   (0.39)   #3 @ 12 ST   (0.39)   (0.195)
                  (0.50)   (0.50)
       #3 @ 12" ST

#4 @ 6"   #4 @ 6"
Alt Bend   Alt Bend | symmetrical about $\mathcal{C}$

*Note: A minimum ρ is not specified in ACI Section 10.5.1 for tensile steel in negative moment regions.

Figure 13.21

■ EXAMPLE 13.1

Design the continuous slab of Figure 13.20 for moments calculated with the ACI coefficients. The slab is to support a service live load of 150 psf in addition to its own dead weight. $f_c' = 3000$ psi and $f_y = 40,000$ psi. The slab is to be constructed integrally with its spandrel girder supports.

SOLUTION

Minimum t for Deflection by ACI Code (9.5.2)

$$\text{Deflection multiplier for 40,000-psi steel} = 0.4 + \frac{40,000}{100,000} = 0.80$$

$$\text{Minimum } t \text{ for end span} = 0.8\,\frac{\ell}{24} = \frac{(0.8)(12 \times 13)}{24} = 5.2''$$

$$\text{Minimum } t \text{ for interior span} = 0.8\,\frac{\ell}{28} = \frac{(0.8)(12 \times 13)}{28} = 4.46''$$

Loads and Maximum Moment Assuming 6-in. Slab ($d = 4\frac{3}{4}$ in.)

$$w_D = \text{slab weight} = (\tfrac{6}{12})(150) = 75 \text{ psf}$$

$$w_L = 150 \text{ psf}$$

$$w_u = (1.4)(75) + (1.7)(150) = 360 \text{ psf}$$

$$M_{ax} \ M_u = \tfrac{1}{10}w_u\ell_n^2 = (\tfrac{1}{10})(360)(13)^2 = 6084 \text{ ft-lb} \qquad \blacksquare$$

Compute moments, ρ values, A_s requirements, and selecting bars at each section, as shown in Figure 13.21.

For floor slabs we are not concerned with the design of web reinforcing. For continuous beams, however, web reinforcing must be carefully designed. Such designs are based on the maximum shears occurring at various sections along the span.

From previous discussions you will remember that to determine the maximum shear occurring at Section 1-1 in the beam of Figure 13.22 the dead load would extend all across the span while the live load would be placed from the section to the most distant support.

If the live load is placed so as to cause maximum shears at various points along the span and the shear is calculated for each point, a maximum shear curve can be drawn. Practically speaking, however, it is unnecessary to go through such a lengthy process for buildings of normal proportions because the values that would be obtained do not vary significantly from the values given by the ACI Code, which are shown in Figure 13.23.

Equivalent Rigid-Frame Method

When continuous beams frame into and are supported by girders, the normal assumption made is that the girders provide only vertical support. Thus they are analyzed purely as continuous beams, as shown in Figure 13.24. The girders do provide some torsional stiffness, and if the calculated torsional moments exceed $\phi(0.5\sqrt{f_c'}\Sigma x^2 y)$ they must be considered, as will be described in Chapter 14.

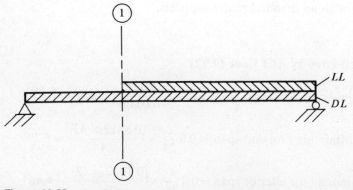

Figure 13.22

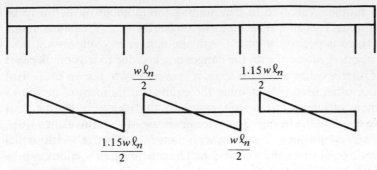

$$\frac{w\ell_n}{2} \qquad \frac{1.15\,w\ell_n}{2}$$

$$\frac{1.15\,w\ell_n}{2} \qquad \frac{w\ell_n}{2}$$

Figure 13.23

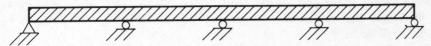

Figure 13.24

Where continuous beams frame into columns, the bending stiffnesses of the columns together with the torsional stiffnesses of the girders are of such magnitude that they need to be considered. An approximate method frequently used for analyzing such reinforced concrete members is the *equivalent rigid-frame method*. In this method, which is applicable only to gravity loads, the loads are assumed to be applied only to the floor or roof under consideration and the far ends of the columns are assumed to be fixed, as shown in Figure 13.25. The sizes of the members are estimated, and an analysis is made on the basis of moment distribution.

For this type of analysis it is necessary to estimate the sizes of the members and compute their relative stiffness or I/ℓ values. From these values, distribution factors can be computed and the method of moment distribution applied. The moments of inertia of both columns and beams are normally calculated on the basis of gross concrete sections, with no allowance made for reinforcing.

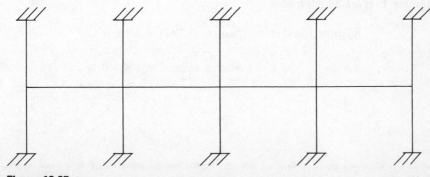

Figure 13.25

There is a problem involved in determining the moment of inertia to be used for continuous T beams. The moment of inertia of a T beam is much greater where there is positive moment with the flanges in compression than where there is negative moment with the flanges cracked due to tension. Because the moment of inertia varies along the span, it is necessary to use an equivalent value. A practice often used is to assume the equivalent moment of inertia as twice the moment of inertia of the web, assuming that the web depth equals the full effective depth of the beam.[5] Some designers use other equivalent values, such as assuming an equivalent T section with flanges of effective widths equal to so many (say 2 to 6) times the web width. These equivalent sections can be varied over a rather wide range without appreciably affecting the final moments.

The ACI Code (8.9.2) states that for such an approximate analysis, only two live-load combinations need to be considered. These are (1) live load placed on two adjacent spans and (2) live load placed on alternate spans. Example 13.2 illustrates the application of the equivalent rigid-frame method to a continuous T beam.

Computer results appear to indicate that the model shown in Figure 13.25 (as permitted by the ACI Code) may not be trustworthy for unsymmetrical loading. Differential column shortening can completely redistribute the moments obtained from the model (i.e., positive moments can become negative moments). As a result, designers should take into account possible axial deformations in their designs.

■ EXAMPLE 13.2

Using the equivalent rigid-frame method, draw the shear and moment diagrams for the continuous T beam of Figure 13.26. The beam is assumed to be framed into 16-in. × 16-in. columns and is to support a service dead load of 2 k/ft (including beam weight) and a service live load of 3 k/ft. Assume the live load is applied in the center span only. The girders are assumed to have a depth of 24 in. and a web width of 12 in. Assume that the I of the T beam equals two times the I of its web.

SOLUTION

Computing Fixed-End Moments

$$w_u \text{ in first and third spans} = (1.4)(2) = 2.8 \text{ k/ft}$$

$$M = \frac{(2.8)(24)^2}{12} = 134.4 \text{ ft-k}$$

---

[5] *Continuity in Concrete Building Frames*, 4th ed., 1959 (Chicago: Portland Cement Association), pp. 17–20.

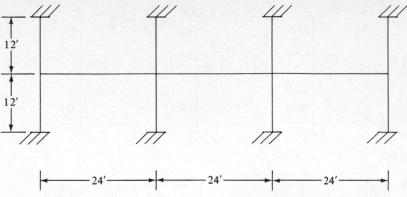

Figure 13.26

$$w_u \text{ in center span} = (1.4)(2) + (1.7)(3) = 7.9 \text{ k/ft}$$

$$M_u = \frac{(7.9)(24)^2}{12} = 379.2 \text{ ft-k}$$

Computing Stiffness Factors

$$I \text{ of columns} = (\tfrac{1}{12})(16)(16)^3 = 5461 \text{ in.}^4$$

$$k \text{ of columns} = \frac{I}{\ell} = \frac{5461}{12} = 455$$

$$\text{Equivalent } I \text{ of T beam} = (2)(\tfrac{1}{12}b_w h^3) = (2)(\tfrac{1}{12})(12)(24)^3 = 27,648 \text{ in.}^4$$

$$k \text{ of T beam} = \frac{27,648}{24} = 1152 \qquad\blacksquare$$

Record stiffness factors on the frame and compute distribution factors, as shown in Figure 13.27.

Balance fixed-end moments and draw shear and moment diagrams, as shown in Figure 13.28.

Assumed Points of Inflection

Another approximate method of analyzing statically indeterminate building frames is to assume the locations of the points of inflection in the members. Such assumptions have the effect of creating simple beams between the points of inflection in each span, and the positive moments in each span can be determined by statics. Negative moments occur in the girders between their ends and the points of inflection. They may be computed by considering the portion of each beam out to the point of inflection to be a cantilever. The shear at the end of each of the girders contributes to the axial forces in the columns. Similarly, the negative moments at the ends of the girders are transferred to the columns.

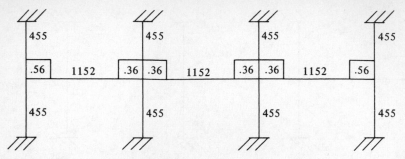

Figure 13.27

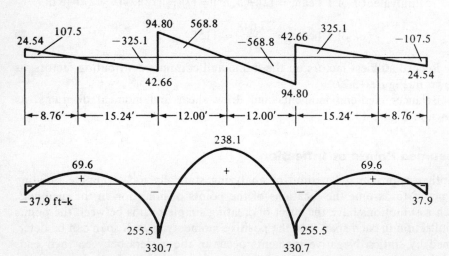

.56		.36	.36		.36	.36		.56
− 134.4		+ 134.4	− 379.2		+ 379.2	− 134.4		+ 134.4
{ + 75.3		+ 37.6				− 37.6		− 75.3
+ 37.3		+ 74.6	+ 74.6		+ 37.3			− 44.0
− 20.9		− 10.4	− 44.0		− 88.0	− 88.0		+ 24.6
+ 9.8		+ 19.6	+ 19.6		+ 9.8	+ 12.3		− 4.0
− 5.5		− 2.7	− 4.0		− 8.0	− 8.0		+ 2.2
+ 1.2		+ 2.4	+ 2.4		+ 1.2	+ 1.1		− 0.4
− 0.7		− 0.4	− 0.4		− 0.8	− 0.8		+ 0.2
− 37.9		+ 0.3	+ 0.3		+ 330.7	− 255.4		+ 37.7
		+ 255.4	− 330.7					
⌐33.60		33.60⌐	⌐94.80		94.80⌐	⌐33.60		33.60⌐
⌐9.06		9.06⌐				⌐9.06		9.06⌐
↑24.54		42.66⌐	⌐94.80		94.80⌐	⌐42.66		24.54↑

Figure 13.28

In Figure 13.29, beam *AB* of the building frame shown is analyzed by assuming points of inflection at the one-fifth points and assuming fixed supports at the beam ends.

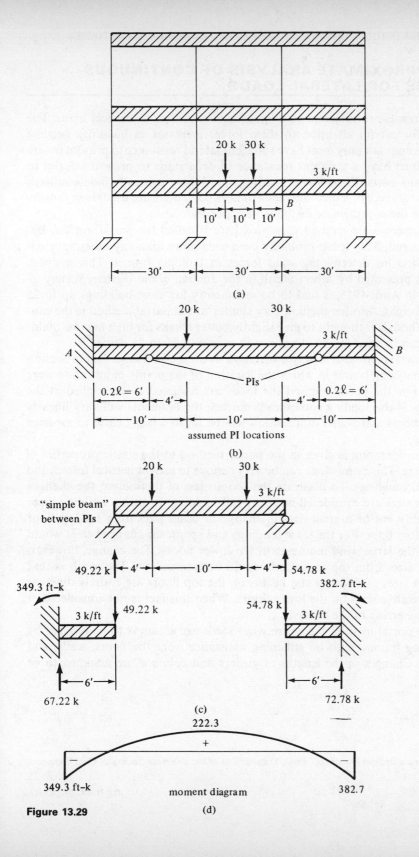

20 k 30 k

3 k/ft

A |← 10' →|← 10' →|← 10' →| B

|← 30' →|← 30' →|← 30' →|

(a)

20 k   30 k

3 k/ft

A ○————————————————————○ B

PIs

|← 0.2ℓ = 6' →|← 4' →| |← 4' →|← 0.2ℓ = 6' →|

|←————— 10' —————→|←————— 10' —————→|←————— 10' —————→|

assumed PI locations

(b)

20 k   30 k

3 k/ft

"simple beam" between PIs

49.22 k |← 4' →|←——— 10' ———→|← 4' →| 54.78 k

349.3 ft-k   382.7 ft-k

3 k/ft ↓49.22 k   54.78 k↓ 3 k/ft

|← 6' →| |← 6' →|

67.22 k   72.78 k

(c)

222.3

+

349.3 ft-k   moment diagram   382.7

(d)

Figure 13.29

13.8 APPROXIMATE ANALYSIS OF CONTINUOUS FRAMES FOR LATERAL LOADS

Building frames are subjected to lateral loads as well as to vertical loads. The necessity for careful attention to these forces increases as buildings become taller. Buildings not only must have sufficient lateral resistance to prevent failure but also must have a sufficient resistance to deflections to prevent injuries to their various parts. Rigid-frame buildings are highly statically indeterminate, and their analysis by "exact" methods (unless computers are used) is so lengthy as to make the approximate methods very popular.

The approximate method presented here is called the *portal method*. Because of its simplicity it has probably been used more than any other approximate method for determining wind forces in building frames. This method, which was presented by Albert Smith in the *Journal of the Western Society of Engineers* in April 1915, is said to be satisfactory for most buildings up to 25 stories in height. Another method very similar to the portal method is the cantilever method. It is thought to give slightly better results for high narrow buildings and can be used for buildings not in excess of 25 to 35 stories.[6]

The portal method is merely a variation of the method described in Section 13.7 for analyzing beams in which the location of the points of inflection were assumed. For the portal method the loads are assumed to be applied at the joints only. If this loading condition is correct, the moments will vary linearly in the members and points of inflection will be located fairly close to member midpoints.

No consideration is given in the portal method to the elastic properties of the members. These omissions can be very serious in unsymmetrical frames and in very tall buildings. To illustrate the seriousness of the matter, the changes in member sizes are considered in a very tall building. In such a building there will probably not be a great deal of change in beam sizes from the top floor to the bottom floor. For the same loadings and spans, the changed sizes would be due to the large wind moments in the lower floors. The change, however, in column sizes from top to bottom would be tremendous. The result is that the relative sizes of columns and beams on the top floors are entirely different from the relative sizes on the lower floors. When this fact is not considered, it causes large errors in the analysis.

In the portal method the entire wind loads are assumed to be resisted by the building frames, with no stiffening assistance from the floors, walls, and partitions. Changes in the lengths of girders and columns are assumed to be

---

[6] "Wind Bracing in Steel Buildings." 1940, *Transactions of the American Society of Civil Engineers*, 105, p. 1723.

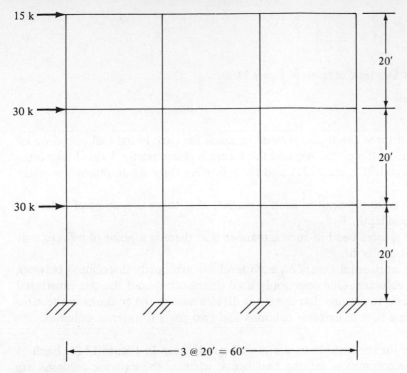

Figure 13.30

negligible. They are not negligible, however, in tall slender buildings, the height of which is five or more times the least horizontal dimension.

If the height of a building is roughly five or more times its least lateral dimension, it is generally felt that a more precise method of analysis should be used. There are several approximate methods that make use of the elastic properties of the structures and that give values closely approaching the results of the "exact" methods. These include the factor method,[7] the Witmer method of K percentages, and the Spurr method.[8]

The building frame shown in Figure 13.30 is analyzed by the portal method, as described in the following paragraphs.

---

[7] Norris, C. H., Wilbur, J. B., and Utku, S. 1976, *Elementary Structural Analysis*, 3rd ed. (New York: McGraw-Hill), pp. 207–212.

[8] "Wind Bracing in Steel Buildings," 1940, *Transactions of the American Society of Civil Engineers*, 105, pp. 1723–1727.

15 k

Figure 13.31 One level of frame of Figure 13.30.

At least three assumptions must be made for each individual portal or for each girder. In the portal method the frame is theoretically divided into independent portals (Figure 13.31), and the following three assumptions are made:

1. The columns bend in such a manner that there is a point of inflection at middepth.
2. The girders bend in such a manner that there is a point of inflection at their center lines.
3. The horizontal shears on each level are arbitrarily distributed between the columns. One commonly used distribution (and the one illustrated here) is to assume that the shear divides among the columns in the ratio of one part to exterior columns and two parts to interior columns.

The reason for the ratio in assumption 3 can be seen in Figure 13.31. Each of the interior columns is serving two bents, whereas the exterior columns are serving only one. Another common distribution is to assume that the shear V taken by each column is in proportion to the floor area it supports. The shear distribution by the two procedures would be the same for a building with equal bays, but for one with unequal bays the results would differ, with the floor area method probably giving more realistic results.

Frame Analysis by Portal Method

The frame of Figure 13.30 is analyzed in Figure 13.32 on the basis of these assumptions. The arrows shown on the figure give the direction of the girder shears and the column axial forces. You can visualize the stress condition of the frame if you assume that the wind is tending to push it over from left to right, stretching the left exterior columns and compressing the right exterior columns. Briefly the calculations were made as follows:

1. *Column shears.* The shears in each column on the various levels were first obtained. The total shear on the top level is 15 k. Because there are two exterior and two interior columns, the following expression may be written:

$$x + 2x + 2x + x = 15 \text{ k}$$

$$x = 2.5 \text{ k}$$

$$2x = 5.0 \text{ k}$$

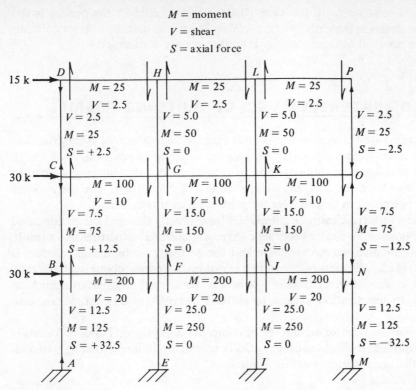

Figure 13.32

The shear in column *CD* is 2.5 k; in *GH* it is 5.0 k; and so on. Similarly, the member shears were determined for the columns on the first and second levels, where the total shears are 75 k and 45 k, respectively.

2. *Column moments.* The columns are assumed to have points of inflection at their middepths; therefore, their moments, top and bottom, equal the column shears times half the column heights.

3. *Girder moments and shears.* At any joint in the frame, the sum of the moments in the girders equals the sum of the moments in the columns. The column moments have been previously determined. By beginning at the upper left-hand corner of the frame and working across from left to right, adding or subtracting the moments as the case may be, the girder moments were found in this order: *DH*, *HL*, *LP*, *CG*, *GK*, and so on. It follows that with points of inflection at girder centerlines, the girder shears equal the girder moments divided by half-girder lengths.

4. *Column axial forces.* The axial forces in the columns may be directly obtained from the girder shears. Starting at the upper left-hand corner, the column axial force in *CD* is equal to the shear in girder *DH*. The axial force in column *GH* is equal to the difference between the two girder shears *DH* and

HL, which equals zero in this case. (If the width of each of the portals is the same, the shears in the girder on one level will be equal, and the interior columns will have no axial force, since only lateral loads are considered.)

13.9 COMPUTER ANALYSIS OF BUILDING FRAMES

All structures are three-dimensional, but theoretical analyses of such structures by hand calculation methods are so lengthy as to be impractical. As a result, such systems are normally assumed to consist of two-dimensional or planar systems and they are analyzed independently of each other. The methods of analysis presented in this chapter were handled in this manner.

Today, modern electronic computers have greatly changed the picture, and it is now possible to analyze complete three-dimensional structures. As a result, more realistic analyses are available and the necessity for high safety factors is reduced. The application of computers is not restricted merely to analysis; they are used in almost every phase of concrete work from analysis to design, to detailing, to specification writing, to material takeoffs, to cost estimating, and so on.

Another very major advantage of computer analysis for building frames is that the designer is able to consider quickly quite a few different loading patterns. The results are sometimes rather surprising.

13.10 LATERAL BRACING FOR BUILDINGS

For the usual building the designer will select relatively small columns. Although such a procedure results in more floor space, it also results in buildings with small lateral stiffnesses or resistance to wind and earthquake loads. Such buildings may have detrimental lateral deflections and vibrations during windstorms unless definite lateral stiffness or bracing is otherwise provided in the structure.

To provide lateral stiffness it will be necessary for the roof and floor slabs to be attached to rigid walls, stairwells, or elevator shafts. Sometimes structural walls, called *shear walls*, are added to a structure to provide the necessary lateral resistance. (The design of shear walls is considered in Chapter 17 of this text.) If it is not possible to provide such walls, stairwells or elevator shafts may be designed as large, box-shaped beams to transmit lateral loads to the supporting foundations. These members will behave as large cantilever beams. In designing such members the designer should try to keep resistance symmetrical so as to prevent uneven lateral twisting or torsion in the structures when lateral loads are applied.[9]

---

[9] Leet, K., 1991, *Reinforced Concrete Design*, 2nd ed. (New York: McGraw–Hill), pp. 453–454.

13.11 DEVELOPMENT LENGTH REQUIREMENTS FOR CONTINUOUS MEMBERS

In Chapter 6 a general introduction to the subject of development lengths for simple and cantilever beams was presented. In this section the ACI development length requirements for continuous members for both positive and negative reinforcing are presented. After studying this information the reader will be happy to read the end of this section where simplified design office practices for determining bar lengths are discussed.

Positive-Moment Reinforcement

Section 12.11 of the Code provides several detailed requirements for the lengths of positive-moment reinforcement. These are briefly summarized in the paragraphs to follow:

1. At least one-third of the positive steel in simple beams and one-fourth of the positive steel in continuous members must extend uninterrupted along the same face of the member at least 6 in. into the support (12.11.1). The purpose of this requirement is to make sure that the moment resistance of a beam will not be reduced excessively in parts of beams where the moments may be changed due to settlements, lateral loads, etc.

2. The positive reinforcement required in the preceding paragraph must, *if the member is part of a primary lateral load resisting system,* be extended into the support a sufficient distance to develop the yield stress of the bars at the face of the support. This requirement is included by the Code (12.11.2) to assure a ductile response to severe overstress as might occur with moment reversal during an earthquake or explosion. As a result of this requirement, it is necessary to have bottom bars lapped at interior supports and to use additional embedment lengths and hooks at exterior supports.

3. Section 12.11.3 of the Code says that at simple supports and at points of inflection (P.I.) the positive-moment tension bars must have their diameters limited to certain maximum sizes. The purpose of the limitation is to keep bond stresses within reason at these points of low moments and large shears. (It is to be remembered that when bar diameters are smaller the bars have greater surface area in proportion to their cross-sectional areas. Thus for bonding to concrete, the larger the bar diameter the larger must be their development lengths. This fact is reflected in the expressions for ℓ_d.) It has not been shown that long anchorage lengths are fully effective in developing bars in a short distance between a P.I. and a point of maximum bar stress, a condition that might occur in heavily loaded short beams with large bottom bars. It is specified that ℓ_d as computed by the requirements presented in Chapter 6 may not exceed the following:

$$\ell_d \leq \frac{M_n}{V_u} + \ell_a \qquad \text{(ACI Equation 12-2)}$$

In this expression M_n is the computed theoretical flexural strength of the member if all reinforcing in that part of the beam is assumed stressed to f_y and

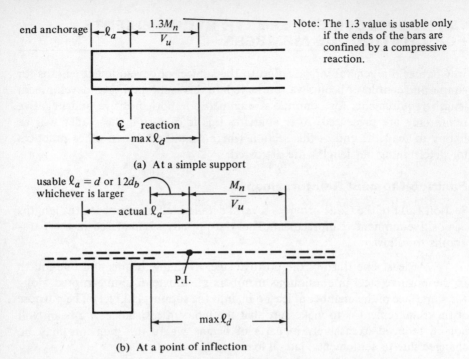

Note: The 1.3 value is usable only if the ends of the bars are confined by a compressive reaction.

(a) At a simple support

(b) At a point of inflection

Figure 13.33 Development length requirements for positive-moment reinforcing.

V_u is the maximum applied shear at the section. At a support, ℓ_a is equal to the sum of the embedment length beyond the ℄ of the support and the equivalent embedment length of any furnished hooks or mechanical anchorage. (Mechanical anchorage, which is permitted in Section 12.6.3 of the Code, consists of bars or plates or angles or other pieces welded or otherwise attached transversely to the flexural bars in locations where sufficient anchorage length is not available.) At a point of inflection, ℓ_a is equal to the larger of the effective depth of the member or $12d_b$. When the ends of the reinforcement are confined by a compression reaction such as where there is a column below but not when a beam frames into a girder, the value $1.3M_n/V_u$ is allowed. The values described here are summarized in Figure 13.33, and a brief numerical illustration for a simple end-supported beam is provided in Example 13.3.

■ EXAMPLE 13.3

At the simple support shown in Figure 13.34, two #9 bars have been extended from the maximum moment area and into the support. Are the bar sizes satisfactory if $f_y = 60$ psi, $f_c' = 3$ ksi, $b = 12$ in., $d = 24$ in., and $V_u = 65$ k, if normal sand-gravel concrete is used, and if the reaction is compressive? Assume cover for bars $>2d_b$ and clear spacing $>3d_b$.

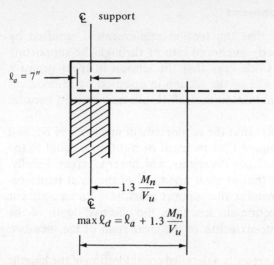

Figure 13.34

SOLUTION

$$a = \frac{A_s f_y}{0.85 f'_c b} = \frac{(2.00)(60)}{(0.85)(3)(12)} = 3.92''$$

$$M_n = A_s f_y \left(d - \frac{a}{2} \right) = (2.00)(60)\left(24 - \frac{3.92}{2} \right) = 2644.8 \text{ in.-k}$$

$$\ell_{db} = \frac{(0.04)(1.0)(60,000)}{\sqrt{3000}} = 43.8''$$

Modification factor from Table 6.1 Condition 1(d) = 1.0

$$\ell'_{db} = (1.0)(43.8) = 43.8''$$

No applicable modification from Table 6.2

$$\therefore \ell_d = \ell'_{db} = 43.8'' \qquad \qquad \underline{\text{Say } 44''}$$

Maximum permissible $\ell_d = \dfrac{M_n}{V_u} + \ell_a$

$$= (1.3)\left(\frac{2644.8}{65} \right) + 7 = 59.90'' > 44'' \quad \underline{\text{OK}} \qquad \blacksquare$$

Note: If this condition had not been satisfied, the permissible value of ℓ_d could have been increased by using smaller bars or by increasing the end anchorage ℓ_a as by the use of hooks.

Negative-Moment Reinforcement

Section 12.12 of the Code says that the tension reinforcement required by negative moments must be properly anchored into or through the supporting member. Section 12.12.1 of the Code says that the tension in these negative bars shall be developed on each side of the section in question by embedment length or end anchorage or a combination thereof. Hooks may be used because these are tension bars.

Section 12.10.3 of the Code says that the reinforcement must extend beyond the point where it is no longer required for moment by a distance equal to the effective depth of the member or 12 bar diameters, whichever is larger. Finally, Section 12.12.3 of the Code says that at least one-third of the total reinforcement provided for negative moment at the support must have an embedment length beyond the point of inflection no less than the effective depth of the member, 12 bar diameters, or one-sixteenth of the clear span of the member, whichever is greatest.

Example 13.4, which follows, presents a detailed consideration of the lengths required by the Code for a set of straight negative-moment bars at an interior support of a continuous beam.

■ EXAMPLE 13.4

At the first interior support of the continuous beam shown in Figure 13.35, six #8 straight bars (4.71 in.^{2}) are used to resist a moment M_u of 390 ft-k for which the calculated A_s required is 4.24 in.^{2} If $f_y = 60$ ksi, $f'_c = 3$ ksi, and normal sand–gravel concrete is used, determine the length of the bars as required by the ACI Code. The dimensions of the beam are $b_w = 14$ in. and $d = 24$ in. Assume bar cover $> 2d_b$ and clear spacing $> 3d_b$.

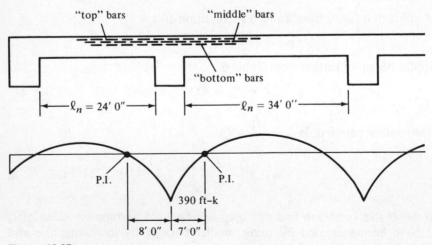

Figure 13.35

Note: The six negative-tension bars are actually placed in one layer in the top of the beam, but they are shown for illustration purposes in Figures 13.35 and 13.36 as though they are arranged in three layers of two bars each. In the solution that follows, the "bottom" two bars are cut off first, at the point where the moment plus the required development length is furnished; the "middle" two are cut when the moment plus the needed development length permits; and the "top" two are cut off at the required distance beyond the P.I.

SOLUTION

1. Basic development lengths required:

$$\ell_d = \frac{(0.04)(0.79)(60,000)}{\sqrt{3000}} = 34.61''$$

Table 6.1

Modification factor $= 1.0$ for Condition 1(d)

$$\ell'_{db} = (1.0)(34.61) = 34.61'' > \frac{(0.03)(1.0)(60,000)}{\sqrt{3000}} = 32.86''$$

Table 6.2

Modification factor for top bars $= 1.3$

$$\ell_d = (1.3)(34.61) = 44.99'' \qquad \underline{\text{say } 45''}$$

2. Section 12.12.3 of the Code requires one-third of the bars to extend beyond P.I. ($1/3 \times 6 = 2$ bars) for a distance equal to the largest of the following values (applies to "top" bars in figure).
   (a) d of web $= 24''$
   (b) $12d_b = (12)(1) = 12''$
   (c) $(\frac{1}{16})(24 \times 12)$ for 24'' span $= 18''$
       $(\frac{1}{16})(34 \times 12)$ for 34'' span $= 25\frac{1}{2}''$

   <u>Use 24'' for short span and $25\frac{1}{2}''$ for long span</u>

3. Section 12.10.3 of the Code requires that the bars shall extend beyond the point where they are no longer required by moment for a distance equal to the greater of:
   (a) $d = 24''$
   (b) $12d_b = (12)(1) = 12''$

4. Section 12.2.3.4 of the Code says that if #11 or smaller bars have clear spacings of at least $5d_b$ and have a $2.5d_b$ clear edge distance from the side face of the member, ℓ_{db} can be multiplied by 0.8 (it is assumed that such is not the case here).

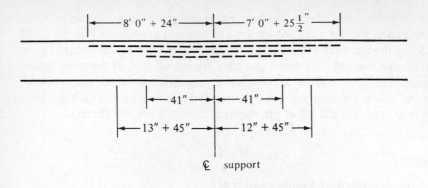

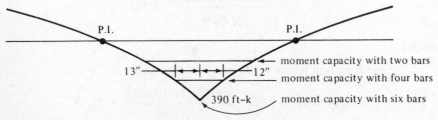

Figure 13.36

5. Section 12.3.3 says required ℓ_{db} can be multiplied by $A_{s\,reqd}/A_{s\,furn}$. This only applies to the first bars cut off ("bottom" bars here).

$$\text{Reduced } \ell_d = (48)\left(\frac{4.24}{4.71}\right) = 43''$$

6. In Figure 13.36 these values are applied to the bars in question and the results are given. ∎

The structural designer is seldom involved with a fixed moment diagram—the loads move and the moment diagram changes. Therefore, the Code (12.10.3) says that reinforcing bars should be continued for a distance of 12 bar diameters or the effective depth of the member, whichever is greater (except at the supports of simple spans and the free ends of cantilevers), beyond their theoretical cutoff points.

As previously mentioned, the bars must be embedded a distance ℓ_d from their point of maximum stress.

Next, the Code (12.11.1) says that at least one-third of the positive steel in simple spans and one-fourth of the positive steel in continuous spans must be continued along the same face of the beam at least 6 in. into the support.

Somewhat similar rules are provided by the Code (12.12.3) for negative steel. At least one-third of the negative steel provided at a support must be extended beyond its point of inflection a distance equal to one-sixteenth of the clear span or 12 bar diameters or the effective depth of the member, whichever

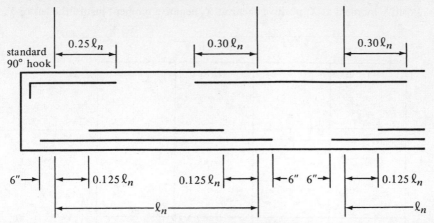

Figure 13.37 Recommended bar details for continuous beams.

is greater. Other negative bars must be extended beyond their theoretical point of cutoff by the effective depth, 12 bar diameters and at least ℓ_d from the face of the support.

Trying to go through these various calculations for cutoff or bend points for all of the bars in even a modest-sized structure can be a very large job. Therefore, the average designer or perhaps the structural draftsman will cut off or bend bars by certain rules of thumb, which have been developed to meet the code rules described here. In Figure 13.37 a sample set of such rules are given for continuous beams. In the *CRSI Handbook*[10] such rules are provided for several different types of structural members, such as solid one-way slabs, one-way concrete joists, two way slabs, and so forth.

PROBLEMS

In Problems 13.1 to 13.3 draw qualitative influence lines for the functions indicated in the structures shown.

13.1 Reactions at A and C, positive moment and positive shear at X.

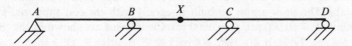

13.2 Reaction at D, negative moment at X, negative moment at D.

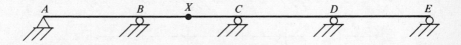

---

[10] Concrete Reinforcing Steel Institute (Chicago, 1984).

13.3 Positive moment at X, positive shear at X, negative moment just to the left of Y.

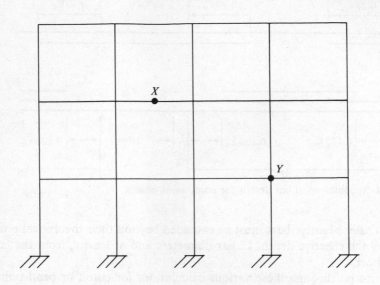

13.4 For the continuous beam shown and for a service dead load of 1.5 k/ft and a service live load of 2.4 k/ft, draw the moment envelope using factored loads assuming a permissible 10% up-or-down redistribution of the maximum negative moment.

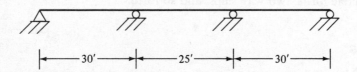

|←————— 30' —————|←——— 25' ———|←———— 30' ————→|

In Problems 13.5 and 13.6, design the continuous slabs shown using the ACI moment coefficients, assuming a service live load of 200 psf is to be supported in addition to the weight of the slabs. The slabs are to be built integrally with the end supports, which are spandrel beams. $f_y = 60{,}000$ psi, $f'_c = 3000$ psi. Clear spans are shown in the figure.
Problem 13.5 (*One ans.* $7\frac{1}{2}''$ slab, $d = 6\frac{1}{4}''$, #4 bars @ 7" positive A_s all 3 spans)

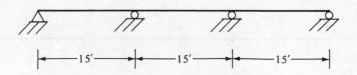

|←———— 15' ————|←——— 15' ———|←——— 15' ———→|

Problem 13.6

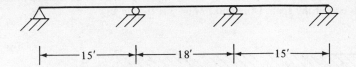

13.7 With the equivalent rigid-frame method, draw the shear and moment diagrams for the continuous beam shown using factored loads. Service dead load including beam weight is 1.5 k/ft and service live load is 2 k/ft. Place the live load in the center span only. Assume the I_s of the T beams equal two times the I_s of their webs. (*Ans.* max $-M = 160.3$ ft-k, max $+M = 114.7$ ft-k)

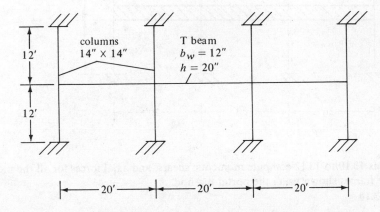

13.8 With the equivalent rigid-frame method, draw the shear and moment diagrams for the continuous beam shown using factored loads. Service dead load including beam weight is 2.4 k/ft and service live load is 3.2 k/ft. Place the live load in spans 1 and 2 only. Assume the I_s of the T beams equals two times the I_s of their webs.

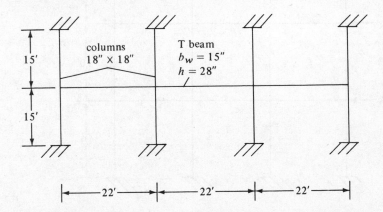

13.9   Draw the shear and moment diagrams for member AB of the frame shown in the accompanying illustration if points of inflection are assumed to be located at $0.15L$ from each end of the span. (*Ans* max. $M = 304$ ft-k @ ℄)

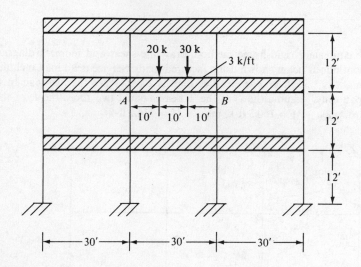

In Problems 13.10 to 13.12 compute moments, shears, and axial forces for all the members of the frames shown using the portal method.

Problem 13.10

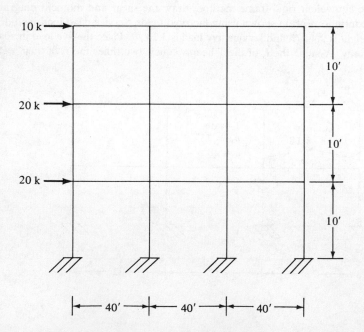

Problem 13.11 (*Ans.* for lower left column $V = 12.5$ k, $M = 75$ ft-k and $S = 6.6$ k)

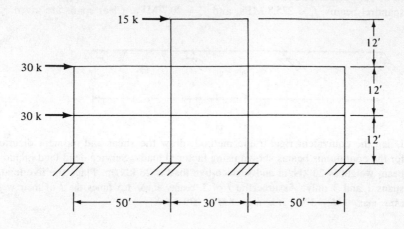

Problem 13.12

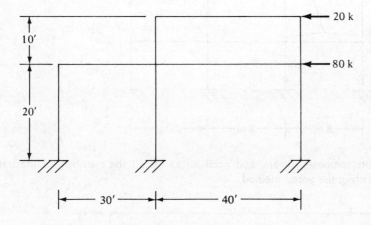

PROBLEMS WITH SI UNITS

13.13 For the continuous beam shown and for a service dead load of 26 kN/m and a service live load of 32 kN/m, draw the moment envelope assuming a permissible 10% up-or-down redistribution of the maximum negative moment. (*Ans.* max $-M = 1077$ ft-k, max $+M = 789$ ft-k)

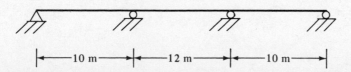

13.14 Design the continuous slab shown using the ACI moment coefficients, assuming the concrete weighs 23.5 kN/m³ and assuming a service live load of 9 kN/m is to be supported. The slabs are built integrally with the end supports, which are spandrel beams, $f_y = 275.8$ MPa, and $f'_c = 20.7$ MPa. Clear spans are given.

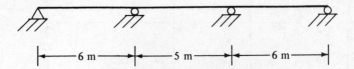

13.15 Using the equivalent rigid-frame method, draw the shear and moment diagrams for the continuous beams shown using factored loads. Service dead load including beam weight is 20 kN/m and service live load is 28 kN/m. Place the live load in spans 1 and 3 only. Assume the I of T beams equal 1.5 times the I of their webs. (*Ans.* max $-M = 379.7$ ft-k, max $+M = 279.2$ ft-k)

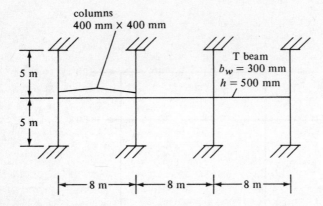

13.16 Compute moments, shears, and axial forces for all the members of the frame shown using the portal method.

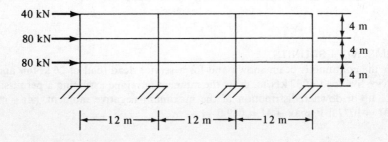

14

Torsion

14.1 INTRODUCTION

The average designer probably does not worry about torsion very much. He or she thinks almost exclusively of axial forces, shears, and bending moments, and yet most reinforced concrete structures are subject to some degree of torsion. Until recent years the safety factors required by codes for the design of reinforced concrete members for shear, moment, and so forth were so large that the effects of torsion could be safely neglected in all but the most extreme cases. Today, however, overall safety factors are less than they used to be and members are smaller, with the result that torsion is a more common problem.

Appreciable torsion does occur in many structures, such as in the main girders of bridges, which are twisted by transverse beams or slabs. It occurs in buildings where the edge of a floor slab and its beams are supported by a spandrel beam running between the exterior columns. This situation is illustrated in Figure 14.1, where the floor beams tend to twist the spandrel beam laterally. Other cases where torsion may be particularly significant occur in curved bridge girders, spiral stairways, balcony girders, and whenever large loads are applied to any beam "off center." An illustration of an off-center case where torsional stress can be very large is illustrated in Figure 14.2. It should be realized that if the supporting member is able to rotate, the resulting torsional stresses will be fairly small. If, however, the member is restrained, the torsional stresses can be quite large.

Christmas Common Bridge over the M40 in England. (Courtesy of Cement and Concrete Association.)

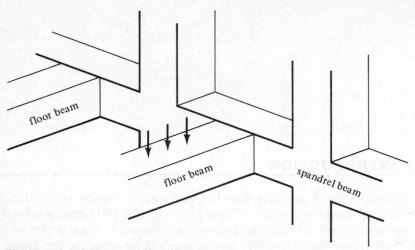

Figure 14.1 Torsion in spandrel beams.

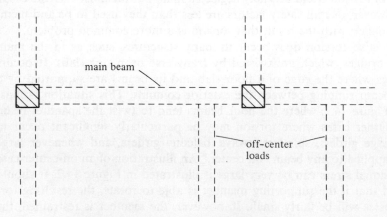

Figure 14.2 Off-center loads causing torsion in main beam.

Should a plain concrete member be subjected to pure torsion, it will crack and fail along 45° spiral lines due to the diagonal tension corresponding to the torsional stresses. For a very effective demonstration of this type of failure, you can take a piece of chalk in your hands and twist it until it breaks. Although the diagonal tension stresses produced by twisting are very similar to those caused by shear, they will occur on all faces of a member. As a result, they add to the stresses caused by shear on one side and subtract from them on the other.[1]

In recent years there have been more reports of structural failures attributed to torsion. As a result, a rather large amount of research has been devoted to the subject, and thus there is a much improved understanding of the behavior of structural members subjected to torsion. On the basis of this rather extensive experimental work, the ACI Code includes very specific requirements for the design of reinforced concrete members subjected to torsion or to torsion combined with shear and bending. It should be realized that maximum shears and torsional forces may occur in those areas where bending moments are small. For such cases the interaction of shear and torsion can be particularly important as it relates to design.

14.2 TORSIONAL REINFORCING

Reinforced concrete members subjected to large torsional forces may fail quite suddenly if they are not specially provided with torsional reinforcing. The addition of torsional reinforcing does not change the magnitude of the torsion that will cause diagonal tension cracks, but it does prevent the members from tearing apart. As a result, they will be able to resist substantial torsional moments without failure. Tests have shown that both longitudinal bars and closed stirrups (or spirals) are necessary to intercept the numerous diagonal tension cracks that occur on all surfaces of beams subject to appreciable torsional forces.

The normal ⊔-shaped stirrups are not satisfactory. They must be closed either by welding their ends together to form a continuous loop as illustrated in Figure 14.3(a) or by bending their ends around a longitudinal bar as shown in part (b) of the same figure. If one-piece stirrups such as these are used, the entire beam cage may have to be prefabricated and placed as a unit (and that may not be feasible if the longitudinal bars have to be passed between column bars) or the longitudinal bars will have to be threaded one by one through the closed stirrups and perhaps the column bars. It is easy to see that some other arrangement is usually desirable.

In the recent past it was rather common to use two overlapping ⊔ stirrups arranged as shown in Figure 14.4. Though this arrangement simplifies the placement of longitudinal bars, it has not proved to be very satisfactory. As described

---

[1] White, R. N., Gergely, P., and Sexsmith, R. G., 1974, *Structural Engineering*, vol. 3 (New York: John Wiley & Sons), pp. 423–424.

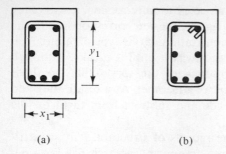

(a)   (b)

Figure 14.3 Closed stirrups (these types frequently impractical).

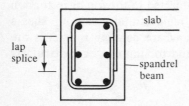

Figure 14.4 Overlapping ⊔ stirrups used as torsion reinforcing *but not desirable*.

in the ACI Commentary (R11.6.7.3), members which are primarily subjected to torsion lose their concrete side cover by spalling at high torques. Should this happen, the lapped spliced ⊔ stirrups of Figure 14.4 will prove to be ineffective and premature torsion failure may occur.

A much better type of torsion reinforcement consists of ⊔ stirrups, each with a properly anchored top bar such as the ones shown in Figure 14.5. It has been proved by testing that the use of torsional stirrups with 90° hooks results in spalling of the concrete outside the hooks. The use of 135° hooks for both the ⊔ stirrups and the top bars is very helpful in reducing this spalling.

Should lateral confinement of the stirrups be provided as shown in parts (b) and (c) of Figure 14.5, the 90° hooks will be satisfactory for the top bars.

Should beams with wide webs (say 2 ft or more) be used, it is often common to use multiple-leg stirrups. For such situations the outer stirrup legs will be proportioned for both shear and torsion while the interior legs will be designed to take only vertical shear.

The strength of closed stirrups cannot be developed unless additional longitudinal reinforcing is supplied. Longitudinal bars no smaller than #3 should be spaced uniformly around the insides of the stirrups, not more than 12 in. apart. There must be at least one bar in each corner of the stirrups to provide anchorage for the stirrup legs (Code 11.6.8.2). Otherwise, if the concrete inside the corners were to crush, the stirrups would slip and the result would be even larger torsional cracks.

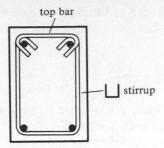

(a) No confining concrete either side—thus 135° hooks
     needed for both ends of top bar.

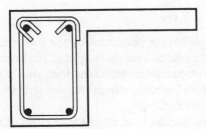

(b) Lateral confinement provided by slab on right-hand side—
     thus 90° hook permissible for top bar on that side.

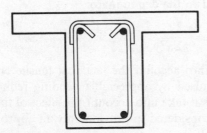

(c) Lateral confinement provided on both sides by
     concrete slab—thus 90° hooks permissible at
     both ends of top bar.

Figure 14.5 Recommended torsion reinforcement.

14.3 CALCULATION OF TORSIONAL STRESSES

The paragraphs to follow present information concerning the calculation of
torsional stresses for different kinds of beams.

Rectangular Cross Sections

The maximum torsional stresses caused in members with rectangular cross sections can be theoretically calculated with the expression to follow, in which T

is the torsional moment, x is the length of the short side of the cross section, y is the length of the long side, and k is a constant that depends on the magnitude of the y/x ratio.

$$v_t = \frac{kT}{\Sigma x^2 y}$$

The value of k varies from 3 (when y/x is large) to almost 5 (when y/x is 1.0). For purposes of design, the ACI uses a value of $k = 3$. If this value is substituted into the preceding equation, the following expression will be obtained for computing the approximate maximum torsional stress in a beam with a solid cross section composed of rectangles:

$$v_{tu} = \frac{3T_u}{\Sigma x^2 y}$$

In this equation, T_u is the ultimate torque produced by the factored loads and $\Sigma x^2 y$ is a value calculated from the component rectangles of the beam cross section. The short dimension of each rectangle is x and the long one is y. This elastic theory expression is only accurate for computing torsional stresses where y is considerably larger than x for each rectangle.

Despite the fact that results can be decidedly in error for sections of certain proportions, the ACI Code does not specify any limits on the use of this equation. The equation is written as shown at the end of this paragraph and is the same as the preceding equation except that ϕ, the capacity reduction factor with a value of 0.85, has been added to the denominator.

$$v_{tu} = \frac{3T_u}{\phi \Sigma x^2 y}$$

This expression does not take into account the fact that tensile cracking of a reinforced concrete member caused by appreciable bending reduces the overall cross-sectional area, nor does it take into account the effect of the cross section's properties. Thus it can be considered to provide only an approximate value of stress.

Compound Sections Consisting of Rectangles

The calculation of torsional stresses for compound sections such as T and L shapes is more complex. For such cases, however, a satisfactory approximation can be made by subdividing the sections into their component rectangles and assuming that each rectangle resists a portion of the total torque in proportion to its torsional rigidity.[2] The cross section is considered to consist of a set of rectangles that will give the greatest value of $\Sigma x^2 y$. The total figure is broken

---

[2] Bach, C., 1911, *Elastizität und Festigkeit* (Berlin: Springer).

down into rectangles so that the rectangles with the greatest width have the greatest length.

The flanges on the sides of a beam do add to the torsional strength of the member, but tests (Code 11.6.1.1) show that only the length of an overhanging flange extending out a distance equal to 3 times its thickness is effective in resisting torque.

Figure 14.6 shows the determination of $\Sigma x^2 y$ for several different beam cross sections. Part (a) of the figure shows the calculations for a rectangular section, whereas part (b) shows the calculations for a T beam that has torsional reinforcing in the web and in the flange directly above the web. For this case, $\Sigma x^2 y$ is calculated for the web rectangle extending to the top of the flange plus the flange rectangles on either side whose widths may not exceed $3h_f$.

If for T- and L-shaped beams the flanges are counted in calculating torsional strength, it will be necessary to provide torsional reinforcing in the flanges.

Figure 14.6(c) shows a T beam that has such reinforcement provided for both the web and the overhanging flanges. For such a case, $\Sigma x^2 y$ is calculated as being the larger of (1) the value obtained for the web rectangle extending up through the flange plus the value obtained for the overhanging rectangles or (2) the value obtained for the web rectangle below the flange plus the value obtained for the whole flange rectangle.[3]

Hollow Sections

Torsional stresses are quite low near the centers of beam cross sections and thus hollow sections have almost the same torsional strengths as do solid beams with the same outside dimensions. As a result the Code (11.6.1.2) states that the same equation used for solid sections may be used for hollow sections, provided that the wall thickness h is at least equal to $x/4$ (refer to Figure 14.7). If the wall thickness is less than $x/4$, the torsional strength of the hollow section is appreciably less than it is for a comparable solid section. If the wall thickness is less than $x/4$ but greater than $x/10$, the section can still be taken as a solid one, but the expression for v_{tu} must be modified as shown. Torsional test results of hollow sections show this reduction to be rather conservative, but such conservatism seems wise because hollow beams with thin walls subjected to torsion fail in a brittle manner, whereas solid sections fail in a ductile manner.[4]

$$v_{tu} = \frac{3T_u}{\Sigma x^2 y} \qquad \text{if } h \geq \frac{x}{4}$$

$$v_{tu} = \left(\frac{x}{4h}\right)\left(\frac{3T_u}{\Sigma x^2 y}\right) \qquad \text{if } h < \frac{x}{4} > \frac{h}{10}$$

---

[3] ACI Committee 318, 1989, *Commentary on Building Code Requirements for Reinforced Concrete* (ACI 318-89) (Detroit: American Concrete Institute), p. 152.

[4] ACI Committee 318, 1989, *Commentary on Building Code Requirements for Reinforced Concrete* (ACI 318-89) (Detroit: American Concrete Institute), p. 152.

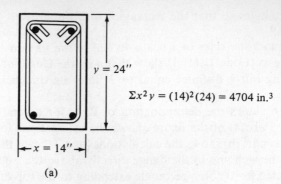

$$y = 24''$$

$$\Sigma x^2 y = (14)^2(24) = 4704 \text{ in.}^3$$

$$x = 14''$$

(a)

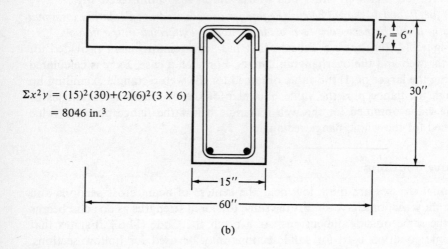

$$h_f = 6''$$

$$\Sigma x^2 y = (15)^2(30)+(2)(6)^2(3 \times 6)$$
$$= 8046 \text{ in.}^3$$

$$30''$$

$$15''$$

$$60''$$

(b)

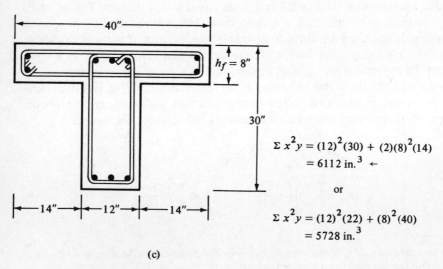

$$40''$$

$$h_f = 8''$$

$$30''$$

$$14'' \qquad 12'' \qquad 14''$$

$$\Sigma x^2 y = (12)^2(30) + (2)(8)^2(14)$$
$$= 6112 \text{ in.}^3 \leftarrow$$

or

$$\Sigma x^2 y = (12)^2(22) + (8)^2(40)$$
$$= 5728 \text{ in.}^3$$

(c)

Figure 14.6 Calculation of $\Sigma x^2 y$ values.

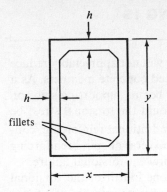

Figure 14.7 A hollow box section.

Should h be less than $x/10$, the stiffness of the wall must be considered. Actually such walls are usually avoided because of their greater flexibility and susceptibility to buckling.

As shown in Figure 14.7, fillets are needed at the corners of all box sections. They reduce the chances that the stirrup will tear out at the corners of the beam and they reduce the stress concentrations at the corners. The presence of fillets also makes it easier to get the concrete down in the form, with the result that the chance of honeycombing is reduced. The Commentary (R11.6.1.2) recommends certain limiting dimensions for these fillets. The minimum fillet leg sizes are $x/6$ if the longitudinal torsional reinforcing consists of less than 8 bars distributed around the section perimeter and $x/12$ but not necessarily larger than 4 in. if 8 or more bars are used and distributed around the section perimeter.

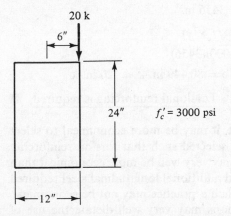

Figure 14.8

14.4 WHEN TORSIONAL REINFORCING IS REQUIRED BY THE ACI

If torsional stresses are less than about $1.5\sqrt{f'_c}$, they will not appreciably reduce either the shear or the flexural strengths of reinforced concrete members. As a result, torsion design is rarely necessary for interior beams supporting slabs on each side that have approximately equal spans. Should the torsional stress be higher than $1.5\sqrt{f'_c}$, it will be necessary to provide additional reinforcing, consisting of additional longitudinal reinforcing and transverse or spiral reinforcing used to restrict the development of cracks and to prevent torsional failure.

The $1.5\sqrt{f'_c}$ value is roughly equal to about one-fourth of the torsional stress required to produce diagonal cracks in a rectangular member. If $1.5\sqrt{f'_c}$ is substituted into the equation for v_{tu}, the limiting torsional moment T_u (above which torsional reinforcing is required) is obtained as given in Section 11.6.1 of the Code. Torques of this size or smaller will not reduce the vertical shearing strength or the bending strength of a member.

$$1.5\sqrt{f'_c} = \frac{3T_u}{\phi\Sigma x^2 y}$$

$$T_u = \phi(0.5\sqrt{f'_c}\,\Sigma c^2 y)$$

The following example illustrates the calculations involved in determining whether torsional reinforcing is required for a particular beam.

■ EXAMPLE 14.1

It is estimated that the rectangular cantilever beam shown in Figure 14.8 is going to be subjected to a torsional moment of 120 in.-k. Is torsional reinforcing required by the ACI Code?

SOLUTION

$$\Sigma x^2 y = (12)^2(24) = 3456 \text{ in.}^3$$

$$\text{Limiting torque} = T_u = \phi(0.5\sqrt{f'_c}x^2 y)$$
$$= 0.85(0.5\sqrt{3000})(3456)$$
$$= 80,449 \text{ in.-lb} = 80.449 \text{ in.-k} < 120 \text{ in.-k}$$

$$\therefore \text{ Torsional reinforcing is required} \quad ■$$

When appreciable torsion is present, it may be more economical to select a larger beam than would be normally selected such that torsion reinforcing does not have to be used. Such a beam may very well be more economical than a smaller one with the closed stirrups and additional longitudinal steel required for torsion design. On other occasions such a practice may not be economical, and sometimes architectural considerations may very well dictate the use of smaller sections.

The maximum calculated torque T_u can be set equal to the maximum value when torsion reinforcing is not provided and then the expression solved for $\Sigma x^2 y$. Then beams can be selected to have a $\Sigma x^2 y$ this large or larger.

$$\Sigma x^2 y = \frac{T_u}{\phi(0.5\sqrt{f'_c})}$$

For Example 14.1 we have

$$\Sigma x^2 y = \frac{(20,000)(6)}{(0.85)(0.5\sqrt{3000})} = 5155$$

Some possible beam dimensions for which torsion reinforcing is unnecessary are 12×36, 14×27, and 16×20.

14.5 TORSIONAL STRENGTH OF CONCRETE MEMBERS

The torsional strengths of sections made with homogeneous materials can be estimated quite accurately, but the torsional strength of nonhomogeneous reinforced concrete sections is another matter entirely. Although many investigations have been made into the subject, there is as yet no clear-cut theory available to handle torsion in reinforced concrete members. To further complicate the matter, pure torsion is seldom found in such members because it is usually combined with bending and shear and perhaps axial forces.

Several excellent textbooks contain a great deal of theoretical information concerning the present theories used for estimating the torsional strengths of reinforced concrete members.[5,6] In these books, torsional strength is considered for several different types of situations. These include torsion in plain concrete, torsion and shear in reinforced concrete beams without web reinforcement, torsion and shear in members with web reinforcement, and so on. The next few paragraphs summarize a little of the information obtained from these studies.

Torsion in Plain Concrete

Initially, plain concrete members subject to torsion are considered. The reason for considering such members is that they behave *almost* identically with longitudinally *or* transversely reinforced beams before cracking begins because the contribution of their reinforcing to torsional behavior is negligible. Actually if

[5] Bresler, B., ed., 1974, *Reinforced Concrete Engineering*, vol. 1 (New York: John Wiley & Sons), pp. 248–272.

[6] Park, R., and Paulay, T., 1975, *Reinforced Concrete Structures* (New York: John Wiley & Sons), chap. 8

both transverse and longitudinal steel are present, their effect on torsional behavior is appreciable.

A failure in torsion occurs by bending in a plane inclined at about 45° with the longitudinal axis of the beam. Quite a few theories have been presented for estimating the strength of plain concrete members subject to torsion, including the plastic theory, the membrane analogy, and others.[7]

The magnitude of the torsional resistance of a plain concrete member cannot be precisely related to a particular strength of the concrete such as its f_c' value. A large number of tests have been made on such members, however, and they seem to indicate that the torsional cracking stress will generally be between $4.0\sqrt{f_c'}$ and $7.0\sqrt{f_c'}$. The size of the specimens used in the tests affected the results somewhat. It is thought that torsional stresses can go about this high in reinforced concrete members before torsional cracks begin to develop. As a result, ACI Committee 438 uses $6.0\sqrt{f_c'}$ as the estimated cracking stress.

Reinforced Concrete Beams Without Web Reinforcement Subjected to Flexure and Torsion

As a result of flexure and torsion, some diagonal cracks are produced much as in beams subjected to flexure and shear. If the members are in flexure, there will be a compression zone on one side of the beam that restricts the cracks. As a result, though diagonal torsion and flexure cracks have developed, the other part of the beam is still capable of resisting some torsion. Again, the manner in which the torsion is resisted is not clearly understood. The compression side of the beam can resist some torsion, and additional torsional resistance is provided due to the dowel action of the reinforcing bars across the cracks on the tensile side of the member.

When the factored torsional moment $T_u = \phi T_n$ exceeds $\phi(0.5\sqrt{f_c'}\Sigma x^2 y)$, it is necessary to include torsion effects with those of moment and shear when the beam is designed. The nominal torsional strength of a member is furnished by the concrete and by the torsional reinforcing.

$$T_n = T_c + T_s$$

When torsional reinforcing is needed, the Code permits the torsional stress resisted by the concrete to go to $2.4\sqrt{f_c'}$ which is approximately 40% of its torsional cracking stress. This reduced value is used to estimate the torsional moment that can be resisted by the concrete when bending moment is simultaneously applied. The Code (11.6.6.1) says the torsional moment strength provided by the concrete is to be calculated by

$$T_c = \frac{0.8\sqrt{f_c'}\Sigma x^2 y}{\sqrt{1 + \left(\dfrac{0.4V_u}{C_t T_u}\right)^2}} \qquad \text{(ACI Equation 11-22)}$$

---

[7] Bresler B., ed., 1974, *Reinforced Concrete Engineering*, vol. 1 (New York, John Wiley & Sons), pp. 249–252.

In the preceding expression, V_u is the factored shear force at the section in question and T_u is the factored torsional moment at the same section. In developing the equation the value $b_w d/\Sigma x^2 y$ is replaced with C_t. Actually the effect of bending is not directly considered, but because the permissible torsional stress is less than half of the torsional cracking stress, the resulting T_c value is rather conservative for combinations of torsion, shear, and bending.

If T_u is greater than $\phi(0.5\sqrt{f_c'}\Sigma x^2 y)$ but less than the value of T_c calculated by this expression, it would seem that no torsional reinforcing is required. However, this is not correct because for such members (other than certain slabs, footing, joists, and beams described in Section 11.5.5.1 of the Code) a certain minimum amount of shear and torsional reinforcing is specified by the Code (11.5.5.5), as will be described.

Combined Shear and Torsion in Members with Web Reinforcement

To simplify the work of the reinforced concrete designer, the ACI has attempted to make the design procedure for torsional reinforcing as close as possible to the procedure used for shear reinforcing. If the factored torsional moment T_u exceeds ϕT_c, it will be necessary to provide torsional reinforcing. The nominal torsional moment strength T_s that must be supplied can be determined as follows:

$$T_u = \phi T_c + \phi T_s$$

$$T_s = \frac{T_u - \phi T_c}{\phi}$$

The nominal strength of torsion reinforcing can be determined with the following expression given in Section 11.6.9.1 of the Code.

$$T_s = \frac{A_t \alpha_t x_1 y_1 f_y}{s} \qquad \text{(ACI Equation 11-23)}$$

In this expression, α_t is a coefficient equal to $[0.66 + 0.33 (y_1/x_1)]$ but not greater than 1.50. The values of x_1 and y_1 are illustrated in Figure 14.3. The design yield strength of the stirrups is limited to a maximum of 60,000 psi in order to control crack sizes. Members with smaller cracks look better, and smaller cracks mean that greater amounts of torque can be transferred by shear friction. Higher-strength steels can be brittle near sharp bends as at stirrup corners, thus providing another reason for holding down yield strengths. (Cold working of steel bars during bending tends to cause brittleness in the higher-strength steels.)

Where torsional reinforcing is required, it is to be provided in addition to the reinforcing supplied for shear, flexure, and axial forces. The purpose of web reinforcement in members subject to torsion is quite similar to the purpose of web reinforcement in members subject to shear forces. Once the diagonal cracks begin to occur, the torsion cannot be resisted unless a different mechanism other than just the concrete is available. Such a mechanism is furnished by forming a type of space truss usually consisting of closed stirrups together with additional longitudinal steel in the corners of the stirrups and perhaps in between.

It is necessary to have both longitudinal and closed transverse reinforcement to resist the diagonal tension stresses caused by torsion. If one of these two types of torsional reinforcing is omitted, the other type is relatively ineffective. The stirrups must be closed because of the fact that torsional cracks can occur on all faces of a member. It is thus necessary for both the torsional and shear reinforcing to be adequately anchored at both ends so that they will be fully effective on either side of a crack.

The area of longitudinal bars required for flexural reinforcing must be increased by the larger of the following two values (Code 11.6.9.3).

$$A_\ell = 2A_t \left(\frac{x_1 + y_1}{s} \right) \qquad \text{ACI Equation 11-24}$$

$$A_\ell = \left[\frac{400xs}{f_y} \left(\frac{T_u}{T_u + (V_u/3C_t)} \right) - 2A_t \right] \left(\frac{x_1 + y_1}{s} \right) \qquad \text{ACI Equation 11-25}$$

In the second expression the value of A_ℓ need not exceed the value obtained if $50b_w s/f_y$ is substituted in place of $2A_t$.

The additional longitudinal bars must be at least #3 in size, they must be spread around the perimeter of the section with at least one bar placed in each corner of the closed stirrups, and their center-to-center spacing must not exceed 12 in. (Code 11.6.8.2). These bars are placed in the stirrup corners to provide anchorage for the stirrup legs and also to aid in fabricating the reinforcing cage. They have proved to be very effective in controlling cracks and developing torsional strength.

14.6 TORSION DESIGN PROCEDURE

Torsional stresses occur at the same time as flexural and shear stresses, with the result that it is necessary to consider the interaction of all three. Though several methods have been proposed to do just this, there is presently no rigorous method available that has been universally accepted by the design community.

In this section the approximate and conservative method required by Chapter 11 of the Code for combined shear and torsion design is presented. The procedure involves the determination of the web reinforcing required by shear, the determination of the web reinforcing required by torsion, the addition of the two, and finally the determination of the additional longitudinal steel needed. To simplify the calculations, it is convenient to calculate each of the two sets of web reinforcing in terms of required steel areas per inch of beam length. Although the design process is simple and straightforward, it is rather tedious and the following detailed list of steps may be helpful to you.

1. The beam dimensions (b and d) are selected for M_u. (They may have to be later revised if shear and/or torsional values are too large.) Then calculations are made to see whether shear and torsional reinforcing is required. The factored shear V_u and the factored torsion T_u are computed. As for shear reinforcing, those sections located less than a distance d from the face of the support may be designed for T_u at a distance d (Code 11.6.4). Shear reinforcing is needed if V_u is greater than $\phi V_c = \phi 2\sqrt{f'_c}b_w d$, whereas torsional reinforcing is required if T_u is greater than $\phi T_c = \phi(0.5\sqrt{f'_c}\Sigma x^2 y)$, with $\phi = 0.85$ in both cases.

2. The nominal shearing strength provided by the concrete V_c is reduced to the following value (Code 11.3.1.4) when the factored torsion T_u exceeds $\phi(0.5\sqrt{f'_c}\Sigma x^2 y)$.

$$V_c = \frac{2\sqrt{f'_c}b_w d}{\sqrt{1 + (2.5C_t T_u/V_u)^2}} \qquad \text{ACI Equation 11-5.}$$

The value of $V_s = (V_u - \phi V_c)/\phi$ is determined. Its maximum permissible value is $8\sqrt{f'_c}b_w d$ as stated by the Code (11.5.6.8). The area of the shear reinforcing required can be calculated with the following expression.

$$V_s = \frac{A_v f_y d}{s} \qquad \text{ACI Equation 11-17}$$

It is, however, convenient to compute the shear stirrup area required per inch of beam as follows:

$$\frac{A_v}{s} = \frac{V_s}{f_y d} = \frac{V_u - \phi V_c}{\phi f_y d}$$

You will remember that the area so obtained is for both of the legs of the stirrup. In Section 11.5.5.3 of the Code a minimum value for $A_v/s = 50b_w/f_y$ is required if $T_u \leq \phi(0.5)\sqrt{f'_c}\Sigma x^2 y)$.

3. The torsional moment strength contributed by the concrete is computed with the following expression from Section 11.6.6.1 of the Code.

$$T_c = \frac{0.8\sqrt{f'_c}\Sigma x^2 y}{\sqrt{1 + (0.4V_u/C_t T_u)^2}} \qquad \text{ACI Equation 11-22}$$

The value of the nominal torsional moment strength T_s that must be provided by the torsion reinforcement and that Section 11.6.9.4 of the Code says cannot exceed $4T_c$ is as follows:

$$T_s = \frac{T_u - \phi T_c}{\phi}$$

As for shear reinforcing, it is convenient to obtain the stirrup area required per inch of beam. You will note that the area obtained is for one leg of the stirrup (11.6.9.1).

$$T_s = \frac{A_t \alpha_t x_1 y_1 f_y}{s} \qquad \text{ACI Equation 11-23}$$

$$\frac{A_t}{s} = \frac{T_s}{\alpha_t x_1 y_1 f_y} = \frac{T_u - \phi T_c}{\phi f_y \alpha_t x_1 y_1}$$

4. The total area of stirrups per inch of beam for both shear and torsion can be determined by adding A_t/s and A_v/s, taking into account the fact that A_t is for one leg and A_v is for two. The area obtained may be no less than the value computed from the following formula given in Section 11.5.5.5 of the Code:

$$A_v + 2A_t = 50b_w s/f_y \qquad \text{(ACI Equation 11-16)}$$

The spacing of the closed stirrups selected may be not greater than $(x_1 + y_1)/4$ or 12 in., or the maximum spacings previously required for shear stirrups (11.6.8.1).

Torsional reinforcing (vertical stirrups and longitudinal bars) should be provided for a distance $b_t + d$ beyond the points where they are theoretically required. (The width of that part of the cross section containing the closed stirrups resisting torsion is b_t.) The Commentary R11.6.7.6 says that this distance, which is larger than that commonly required for shear and flexural bars, is desirable because of the long helical form taken by the torsional diagonal tension cracks. The longitudinal bars must also be effectively anchored so that their yield strengths can be developed as assumed in design. Thus they must be run full development lengths beyond the $b_t + d$ distances.

It should be noted that if an interior beam is loaded to produce maximum shear and moment (that is, with full load on the slab on both sides of the beam), the torsion will be reduced, or if it is loaded to produce maximum torsion (that is, with live load placed on the slab on one side of the beam only), the maximum shears and moments will be reduced. This is not the case for exterior beams, and hence it is necessary to add more longitudinal reinforcing. For interior beams, extra bars are not usually added except for corner bars added on the compression sides to form the steel cages with the closed stirrups.

Example 14.2 illustrates the design of torsional reinforcing for a rectangular beam.

■ EXAMPLE 14.2

In addition to the service live and dead uniform loads shown, the beam of Figure 14.9 is subjected to a 10 ft-k service dead-load torsion and a 12 ft-k

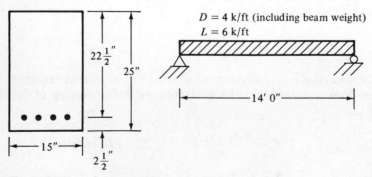

Figure 14.9

service live-load torsion at the face of the support. It is assumed that the magnitude of these torsional moments decreases uniformly from the beam end of the L. Design the shear and torsion reinforcement for this beam if $f'_c = 4000$ psi and $f_y = 60,000$ psi.

SOLUTION

Calculating Shear and Torsion Values

$$w_u = (1.4)(4) + (1.7)(6) = 15.8 \text{ k/ft}$$

$$V_u = (7)(15.8) = 110.6 \text{ k}$$

$$V_u \text{ @ a distance } d \text{ from face of support} = \left(\frac{61.5}{84}\right)(110.6) = 80.975 \text{ k}$$

$$= 80,975 \text{ lb}$$

$$T_u = (1.4)(10) + (1.7)(2) = 34.4 \text{ ft-k}$$

$$T_u \text{ @ a distance } d \text{ from face of support} = \left(\frac{61.5}{84}\right)(34.4) = 25.186 \text{ ft-k}$$

$$= 302,232 \text{ in.-lb}$$

This is a conservative value which does not account for any rotation of the beam.

Checking to See Whether Torsion Reinforcement is Required

$$\phi(0.5\sqrt{f'_c}\Sigma x^2 y) = (0.85)(0.5\sqrt{4000})(15)^2(25)$$

$$= 151,196 \text{ in.-lb} < T_u \text{ of } 302,232 \text{ in.-lb}$$

$$\therefore \text{ Torsion effects must be considered}$$

Design of Shear Reinforcement

$$C_t = \frac{b_w d}{\Sigma x^2 y} = \frac{(15)(22.5)}{(15)^2(25)} = 0.06$$

$$V_c = \frac{2\sqrt{f'_c}b_w d}{\sqrt{1 + [2.5C_t(T_u/V_u)]^2}} \quad \text{when } T_u > \phi(0.5\sqrt{f'_c}\Sigma x^2 y)$$

$$= \frac{(2\sqrt{4000})(15)(22.5)}{\sqrt{1 + [2.5 \times 0.06 \times (302,232/80,975)]^2}} = 37,250 \text{ lb}$$

$$V_u = \phi V_s + \phi V_c$$

$$V_s = \frac{V_u - \phi V_c}{\phi} = \frac{80,975 - (0.85)(37,250)}{0.85} = 58,015 \text{ lb}$$

Maximum V_s permitted by Code (11.5.6.8)

$$= 8\sqrt{f'_c}b_wd = (8\sqrt{4000})(15)(22.5) = 170,763 \text{ lb} > 58,015 \text{ lb} \qquad \text{OK}$$

$$\frac{A_v}{s} = \frac{V_u - \phi V_c}{\phi f_y d} = \frac{80,975 - (0.85)(37,250)}{(0.85)(60,000)(22.5)} = 0.0430 \text{ in.}^2/\text{in.}$$

$$\frac{A_v}{s} \text{ minimum by Code (11.5.5.3)} = \frac{50b_w}{f_y} = \frac{(50)(15)}{60,000}$$

$$= 0.0125 \text{ in.}^2/\text{in.} < 0.0430 \text{ in.}^2/\text{in.} \qquad \text{OK}$$

Torsional Moment Strength Contributed by Concrete

$$T_c = \frac{0.8\sqrt{f'_c}\Sigma x^2 y}{\sqrt{1 + \left(\dfrac{0.4V_u}{C_t T_u}\right)^2}} = \frac{(0.8\sqrt{4000})(15)^2(25)}{\sqrt{1 + \left(\dfrac{0.4 \times 80,975}{0.06 \times 302,232}\right)^2}} = 139,033 \text{ in.-lb}$$

$$T_s = \frac{T_u - \phi T_c}{\phi} = \frac{302,232 - (0.85)(139,033)}{0.85} = 216,532 \text{ in.-lb} < 4T_c \qquad \text{OK}$$

Determining Torsion Reinforcing Required Per Inch   (See Figure 14.10 for assumed dimensions.)

$$\alpha_t = \left[0.66 + (0.33)\left(\frac{y_1}{x_1}\right)\right] = \left[0.66 + (0.33)\left(\frac{21.5}{11.5}\right)\right] = 1.28 < 1.50$$

$$\frac{A_t}{s} = \frac{T_s}{\alpha_1 x_1 y_1 f_y} = \frac{216,532}{(1.28)(11.5)(21.5)(60,000)} = 0.0114 \text{ in.}^2/\text{in.}$$

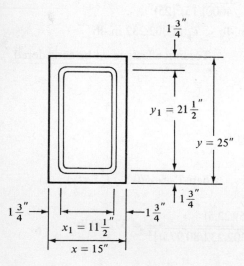

Figure 14.10

Selection of Web Reinforcing It will be noted that A_v in the expression to follow is the cross-sectional area of two legs for a closed stirrup while A_t is the area of one leg of the stirrup. Thus the equation to follow gives the total area of one leg of the stirrup.

$$\frac{A_t}{s} + \frac{A_v}{2s} = 0.0114 + \frac{0.0430}{2} = 0.0329 \text{ in.}^2/\text{in.}$$

For #3 stirrups,

$$s = \frac{0.11}{0.0329} = 3.34''$$

For #4 stirrups,

$$s = \frac{0.20}{0.0329} = 6.08''$$

For #5 stirrups,

$$s = \frac{0.31}{0.0329} = 9.42''$$

$$\text{Maximum permissible spacing} = \frac{x_1 + y_1}{4} = \frac{11.5 + 21.5}{4}$$

$$= 8.25'' \qquad \text{nor } 12'' \text{ as per Code 11.6.8.1}$$

<u>Use #5 closed stirrup at 8'' o.c.</u>

By proportions, the required spacing of the stirrups can be computed at other distances from the support. It will also be noted that farther out in the beam where T_u becomes less than $\phi(0.5\sqrt{f'_c}\Sigma x^2 y) = 151{,}196$ in.-lb the maximum spacing goes up to the maximum value for shear alone.

Check beam for minimum permissible area of stirrups by Section 11.5.5.5 of the Code:

$$A_v + 2A_t = 2A_{bar} = 50\frac{b_w s}{f_y}$$

$$2A_{bar} = (50)\left(\frac{15 \times 8}{60{,}000}\right) = 0.10 \text{ in.}^2$$

Minimum $A_{bar} = 0.05$ in.$^2 < 0.31$ in.^{2} furnished   OK

Increase in Longitudinal Steel Area

$$A_\ell = 2A_t\left(\frac{x_1 + y_1}{s}\right) = (2)(0.31)\left(\frac{11.5 + 21.5}{8.00}\right) = 2.56 \text{ in.}^2 \qquad \blacksquare$$

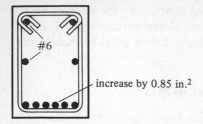

increase by 0.85 in.²

Figure 14.11 Additional longitudinal
reinforcing required for tension.

The second equation for A_ℓ (that is, ACI Equation 11-25) should also be checked, but in this case it did not control.

Add $2.56/3 = 0.853$ in.² (say 2 #6 = 0.88 in.²) at the top and middepth of the member and increase the required longitudinal reinforcing in the bottom by 0.853 in.², as shown in Figure 14.11.

14.7 THE TORSIONAL MOMENTS THAT HAVE TO BE CONSIDERED IN DESIGN

The reader is well aware from his or her studies of structural analysis that if one part of a statically indeterminate structure "gives" when a particular force is applied to that part, the amount of force that the part will have to resist will be appreciably reduced. For instance, if three men are walking along with a log on their shoulders (a statically indeterminate situation) and one of them lowers his shoulder a little under the load, there will be a major redistribution of the internal forces in the "structure" and a great deal less load for him to support. On the other hand, if two men are walking along with a log on their shoulders (a statically determinate situation) and one of them lowers his shoulder slightly, there will be little change in force distribution in the structure. These are similar to the situations that occur in statically determinate and indeterminate structures subject to torsional moments.

For a statically determinate structure there is only one path along which a torsional moment can be transmitted to the supports. This type of torsional moment, which is referred to as *equilibrium torsion* or *statically determinate torsion*, cannot be reduced by a redistribution of internal forces or by a rotation of the member. This kind of torsion was illustrated by the statically determinate cantilever beam of Example 14.1. The member must be designed to resist the full calculated torsional moment.

The torsional moment in a particular part of a statically indeterminate structure may be substantially reduced if that part of the structure cracks under the torsion and "gives" or rotates. The result will be a redistribution of forces in the structure. This type of torsion is referred to as *statically indeterminate torsion*, or *compatibility torsion* in the sense that the part of the structure in question twists in order to keep the deformations of the structure compatible.

Cracking occurs in reinforced concrete members at torques equal to about $T_u = \phi(4\sqrt{f'_c}\Sigma x^2 y/3)$. If this torque is reached in a statically indeterminate structure, there can be a large twist or rotation and large redistributions in forces in the structure. If the amount of torsional reinforcing is increased in the member, the ultimate torsional strength of the member will be increased, but this may not be particularly useful because that ultimate torque cannot be reached unless a larger rotation occurs.

The Code (11.6.3) states what can be done in statically indeterminate structures where a reduction of the torsional moment can occur as a result of a redistribution of internal forces. If the torsional moment exceeds the cracking torque $T_u = \phi(4\sqrt{f'_c}\Sigma x^2 y/3)$, torsional cracking is assumed to have occurred. (*To be sure that sufficient ductility and crack width control is provided, the member must be designed with the torsional reinforcing necessary to produce an ultimate torque equal to the cracking torque.*) A maximum factored torsional moment equal to the cracking torque at the section is assumed to occur at the critical section. Then, as illustrated in Example 14.3, this torsional moment is used to calculate new or adjusted values of shears and moments in the adjoining members. These values will be used for design of those members.

Should the torsional moment T_u calculated before cracking by an elastic analysis based on an uncracked section be between $\phi(0.5\sqrt{f'_c}x^2 y)$ and $\phi(4\sqrt{f'_c}x^2 y/3)$, the member may be designed using the actual computed torsional moments states the Commentary (R11.6.2 and R.11.6.3). In these same sections the Commentary warns the designer concerning certain situations where more exact analyses should be used. Such cases might occur where there are large torsional loadings near very stiff columns or near columns that rotate in reverse directions because of other loadings.

Example 14.3 is concerned with the statically indeterminate beam and floor slab structure with spandrel beams shown in Figure 14.12. It is desired to determine the design torsional moment (or cracking torque) of the spandrel beam.[8] This is done in step 1 of the solution.

Then in steps 2 and 3 the spandrel beam is assumed to be subjected to the cracking torque and is assumed to rotate and crack. The D end of beam DE is assumed to be capable of resisting an end moment equal to two times the cracking torque $= 2 \times 43.1 = 86.2$ ft-k because there is a spandrel beam on each side of the D end of the beam. Then the fixed end moments for beam DE are computed and balanced, with the moment at the D end being balanced to 86.2 ft-k.

In step 4 the reactions for beam DE are determined and then its shear and moment diagrams are drawn. These values are to be used in the design of the beam.

---

[8] Neville, Gerald B., ed., 1984, *Notes on ACI 318-83 Building Code Requirements for Reinforced Concrete with Design Applications*, 4th ed. (Skokie, Ill.: Portland Cement Association), pp. 14-11 through 14-13.

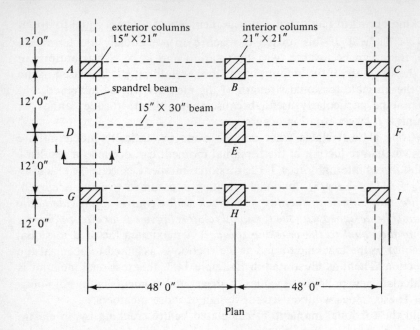

Plan

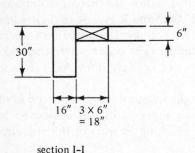

section I–I

Figure 14.12

Finally, in step 5 the shear in the spandrel beam is computed. The shear and torsional reinforcing for the spandrel beam could now be designed as they were for the statically determinate beam of Example 14.2, but space is not taken to show these calculations.

Instead of making the analysis of a statically indeterminate structure based on the idea that a member subject to torsion "gives" or rotates at a certain torque, Section 8.6.1 of the Code provides an alternative. It states that any reasonable assumption may be used for computing the relative flexural and torsional stiffnesses of the members of a statically indeterminate structure. Then a statically indeterminate analysis is made using the factored loads and the members are designed on a basis of these values.

■ EXAMPLE 14.3

Determine the cracking moment for the spandrel beam labeled AG in Figure 14.12. Assume that this torque is reached and determine the revised values of shear and moment for beam DE if $f'_c = 3000$ psi, $f_y = 60,000$ psi, and live load = 80 psf.

SOLUTION

Step 1. Compute the Maximum Torsional Moment for the Spandrel Beam

$$\Sigma x^2 y = (16)^2(30) + (6)^2(18) = 8328 \text{ in.}^3$$

$$T_u = \phi(4\sqrt{f'_c} x^2 y/3) = \frac{(0.85)(4\sqrt{3000})(8328)}{3}$$

$$= 516,962 \text{ in.-lb} = 43.1 \text{ ft-k}$$

Step 2. Determine Fixed-End Moments in Beam DE

$$\text{Clear span length} = 48.0 - \frac{8}{12} - \frac{10.5}{12} = 46.46'$$

$$DL = \frac{(15)(30)}{144}(150) + \left(12.0 - \frac{15}{12}\right)\left(\frac{6}{12}\right)(150) = 1275 \text{ lb/ft}$$

$$LL = (12)(80) = 960 \text{ lb/ft}$$

$$w_u = (1.4)(1.275) + (1.7)(0.960) = 3.417 \text{ k/ft}$$

$$\text{Fixed-end moments} = \frac{w_u \ell^2}{12} = \frac{(3.417)(46.46)^2}{12} = 614.6 \text{ ft-k}$$

Step 3. Balance the Moments in Beam DE Noting that spandrel beam AD will take a moment or torque of 43.1 ft-k at D as will spandrel beam DG, balance the D end of the beam to $(2)(43.1) = 86.2$ ft-k and make carryover to E:

-614.6	$+614.6$
$+528.4 \rightarrow$	$+264.2$
-86.2	$+878.8$

Step 4. Use the Moments from Step 3 to Calculate the Reactions in Beam *DE*
by Statics and Draw the Shear and Moment Diagrams

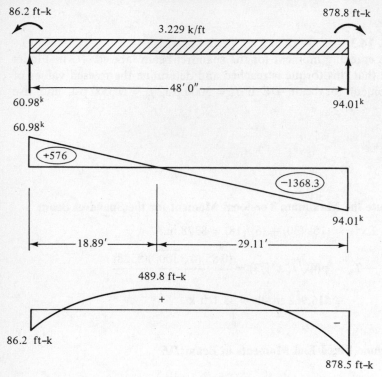

86.2 ft–k

878.8 ft–k

3.229 k/ft

← 48′ 0″ →

60.98^k

94.01^k

60.98^k

(+576)

(−1368.3)

94.01^k

← 18.89′ → ← 29.11′ →

489.8 ft–k

+

86.2 ft–k

−

878.5 ft–k

Step 5. Calculate Maximum V_u **for the Spandrel Beam**

0.7 k/ft

← clear span = $24.0 - \dfrac{(2)(7.5)}{12} = 22.75'$ →

Assume $d = 30 - 3 = 27''$.

V_u at a distance d from face of column

$$= \frac{60.98}{2} + \left(\frac{22.75}{2}\right)(0.7) - \left(\frac{27}{12}\right)(0.7) = 36.9 \text{ k}$$

∴ Design spandrel beam for $T_u = 43.1$ ft-k and $V_u = 36.9$ k ∎

14.8 COMPUTER EXAMPLE

Example 14.4 which follows illustrates the use of the application of the computer program CONCRETE to select torsion reinforcing for a beam. As with the other applications of the program the user can in a few seconds solve the problem using data by changing the input values shown on the screen.

■ EXAMPLE 14.4
Repeat Example 14.2 using the enclosed computer disk.

SOLUTION

DESIGN OF RECTANGULAR BEAMS FOR COMBINED SHEAR AND TORSION
- -

Beam Width	bw (in) =	15.0
Depth to Bars	d (in) =	22.5
Total Depth of Member	h (in) =	25.0
Stirrup Bar Number	=	#5
Cover	(in) =	1.50
Vu at Section	(lbs) =	80975
Tu at Section	(in-lbs) =	302232
Concrete Strength	f'c (psi) =	4000
Steel Yield Strength	fy (psi) =	60000

RESULTS
= = = = = =
Theoretical closed # 3 stirrup spacing = 3.358   ■

PROBLEMS

14.1 The beam shown in the accompanying illustration is subjected to a 12 ft-k service dead-load torsion and a 16 ft-k service live-load torsion at the face of the support. Assuming that torsion decreases uniformly from the beam end to the beam $\mathcal{C}$, design shear and torsion reinforcement if f'_c = 3000 psi and f_y = 50,000 psi. (*Ans.* Use #5 closed stirrups 1 @ 2″, 7 @ 7″, and 9 @ 11″ O.C.)

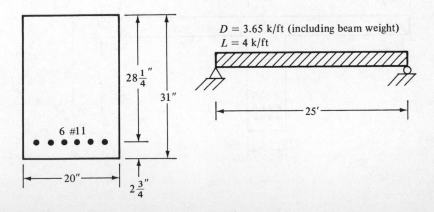

14.2   What minimum total theoretical depth is needed for the 20 in. wide beam shown if no torsional reinforcing is to be used? The concentrated load is located at the end of the cantilever 8 in. to one side of the beam $\mathcal{C}$, $f_y = 60,000$ psi, $f'_c = 4000$ psi.

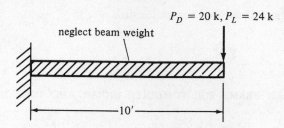

$P_D = 20$ k, $P_L = 24$ k

neglect beam weight

|← 10' →|

14.3   The reinforced concrete spandrel beam shown in the accompanying illustration is integrally connected with the 5-in. slab. If $f'_c = 4000$ psi and $f_y = 60,000$ psi, determine the spacing required for #3 stirrups at a distance d from the face of the support where $V_u = 32$ k and $T_u = 10$ ft-k. (*Ans.* #3 closed stirrups 6" O.C.)

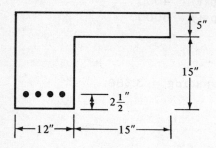

5"

15"

$2\frac{1}{2}''$

|← 12" →|← 15" →|

14.4   Design the torsion reinforcement for the beam shown in the accompanying figure at a section a distance d from the face of the support for a torsional moment of 24 ft-k. $V_u = 90$ k, $f'_c = 3000$ psi, and $f_y = 60,000$ psi. Cover = 1.50 in.

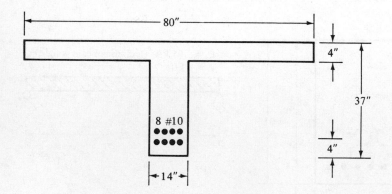

|← 80" →|

4"

37"

8 #10

4"

|← 14" →|

14.5 Select the torsional reinforcing required for the edge beam shown in the accompanying illustration if $f'_c = 4000$ psi and $f_y = 50,000$ psi. T_u equals 30 ft-k at the face of the support and is assumed to vary along the beam in proportion to the shear. (*Ans. #5 closed stirrups 8" O.C.*)

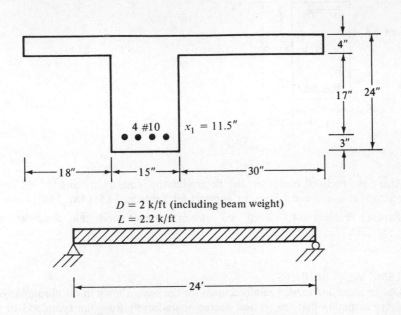

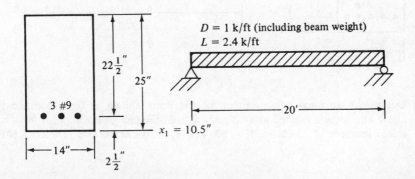

14.6 A 12-in × 22-in. spandrel beam ($d = 19$ in.) with a 20-ft simple span has a 4-in. slab on one side acting as a flange. It must carry a maximum V_u of 60 k and a maximum T_u of 20 ft-k at the face of the support. Assuming these values are zero at the beam $\mathcal{C}$, design torsional reinforcing, if $f'_c = 3000$ psi and $f_y = 60,000$ psi.

14.7 Design the web reinforcing for the beam shown if the load is acting 3 in. off center of the beam. Assume that the torsion equals the uniform load times the 3 in., $f'_c = 3000$ psi, and $f_y = 60,000$ psi. Use #4 stirrups. Assume the torsion value varies from a maximum at the support to 0 at the beam $\mathcal{C}$, as does the shear. (*Ans. #4 closed stirrups 8" O.C.*)

14.8   Is the beam shown satisfactory to resist a T_u of 20 ft-k and a V_u of 50 k if $f'_c = $ 3000 psi and $f_y = 60,000$ psi? The bars shown are used in addition to those provided for bending moment.

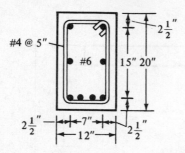

14.9   Using the enclosed computer disk determine the required spacing for #5 closed stirrups at a distance d from the support for Problem 14.1. (*Ans.* 7.47″)

14.10 Repeat Problem 14.7 using #3 closed stirrups and the computer disk CONCRETE.

PROBLEMS WITH SI UNITS

14.11 Design shear and torsion reinforcement for the beam shown in the accompanying figure assuming that the torsion decreases uniformly from the beam end to the beam ₵. The member is subjected to a 34-kN·m service dead-load torsion and a 40 kN·m service live-load torsion at the face of the support, $f_y = 344.8$ MPa, and $f'_c = 20.7$ MPa. (*Ans.* Use #4 closed stirrups 130 mm O.C.)

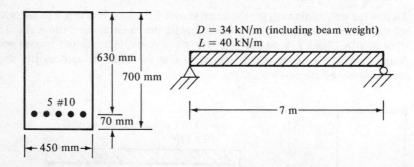

14.12 Design the torsional reinforcement for the beam shown in the accompanying figure at a section located at a distance d from the face of the support for a torsional moment of 40 kN·m, $V_u = 500$ kN, $f'_c = 20.7$ MPa, and $f_y = 275.8$ MPa.

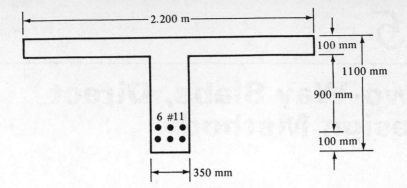

14.13 Design the web reinforcing for the beam shown if the load is acting 100 mm off center of the beam. Assume the torsion equals the uniform load times 100 mm, $f'_c = 17.2$ MPa, and $f_y = 275.8$ MPa. Use #4 stirrups and assume that the torsion and shear vary from a maximum at the support to 0 at the beam ℄. (*Ans.* #4 closed stirrups 140 mm O.C. at a distance d from face of support)

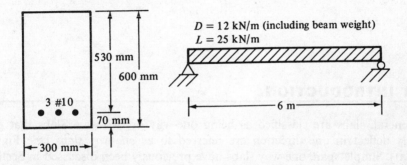

15

Two-Way Slabs, Direct Design Method

15.1 INTRODUCTION

In general, slabs are classified as being one-way or two-way. Slabs that primarily deflect in one direction are referred to as *one-way slabs* [see Figure 15.1(a)]. Simple-span, one-way slabs have previously been discussed in Section 3.16 of this text, while the design of continuous one-way slabs was considered in Section 13.7. When slabs are supported by columns arranged generally in rows so that the slabs can deflect in two directions, they are usually referred to as *two-way slabs*.

Two-way slabs may be strengthened by the addition of beams between the columns, by thickening the slabs around the columns (*drop panels*), and by flaring the columns under the slabs (*column capitals*). These situations are shown in Figure 15.1 and discussed in the next several paragraphs.

Flat plates [Figure 15.1(b)] are solid concrete slabs of uniform depths that transfer loads directly to the supporting columns without the aid of beams or capitals or drop panels. Flat plates can be quickly constructed due to their simple formwork and reinforcing bar arrangements. They need the smallest overall story heights to provide specified headroom requirements, and they also give the most flexibility in the arrangement of columns and partitions. They also provide little obstruction to light and have high fire resistance because there are few sharp corners where spalling of the concrete might occur.

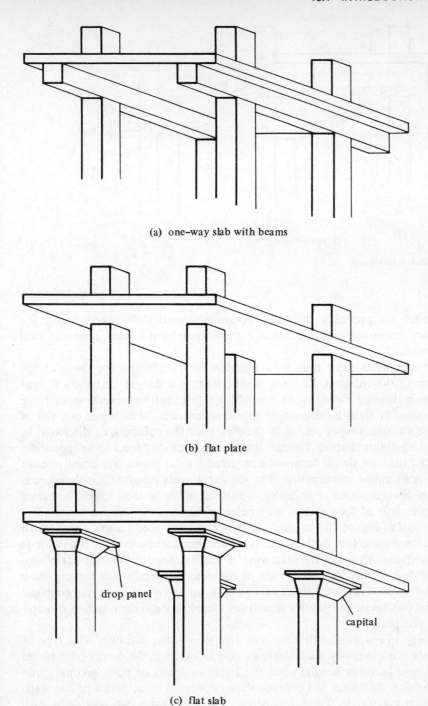

(a) one–way slab with beams

(b) flat plate

drop panel

capital

(c) flat slab

Figure 15.1 Slabs.

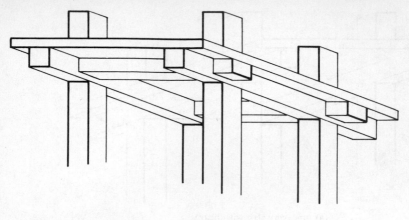

(d) two–way slab with beams

Figure 15.1 (*continued*).

Flat plates are probably the most commonly used slab system today for multistory reinforced concrete hotels, motels, apartment houses, hospitals, and dormitories.

For flat plates there may be a problem in transferring the shear at the perimeter of the columns. In other words, there is a danger that the columns may punch through the slabs. As a result, it is frequently necessary to increase column sizes or slab thicknesses or to use *shearheads*. Shearheads consists of steel I or channel shapes placed in the slab over the columns, as discussed in Section 15.5 of this chapter. Though such procedures may seem to be expensive, it is noted that the simple formwork required for flat plates will usually result in such economical construction that the extra costs required for shearheads are more than canceled. For heavy industrial loads or long spans, however, some other type of floor system may be required.

Flat slabs [Figure 15.1(c)] include two-way reinforced concrete slabs with capitals, drop panels, or both. These slabs are very satisfactory for heavy loads and long spans. Although the formwork is more expensive than for flat plates. flat slabs will require lesser amounts of concrete and reinforcing for the same loads and spans. They are particularly economical for warehouses, parking and industrial buildings, and similar structures where exposed drop panels or capitals are acceptable.

Finally, in Figure 15.1(d) a *two-way slab with beams* is shown. This type of floor system is obviously used where its cost is less than the costs of flat plates or flat slabs. In other words, when the loads or spans or both become quite large, the slab thickness and column sizes required for flat plates or flat slabs are of such magnitude that it is more economical to use two-way slabs with beams, despite the higher formwork costs.

Round columns. (Courtesy of Economy Forms Corporation.)

15.2 ANALYSIS OF TWO-WAY SLABS

Two-way slabs bend under load into dish-shaped surfaces, so there is bending in both principal directions. As a result, they must be reinforced in both directions by layers of bars that are perpendicular to each other. A theoretical elastic analysis for such slabs is a very complex problem due to their highly indeterminate nature. Numerical techniques such as finite difference and finite elements are required, but such methods are not really practical for routine design.

Actually, the fact that a great deal of stress redistribution can occur in such slabs at high loads makes it unnecessary to make designs based on theoretical analyses. As a result, the design of two-way slabs is generally based on empirical moment coefficients, which, though they might not accurately predict stress variations, result in slabs with satisfactory overall safety factors. In other words, if too much reinforcing is placed in one part of a slab and too little somewhere else, the resulting slab behavior will probably still be satisfactory. *The total amount of reinforcement in a slab seems more important than its exact placement.*

You should clearly understand that though this chapter and the next are devoted to two-way slab design based on approximate methods of analysis, there is no intent to prevent the designer from using more exact methods. He or she may design slabs on the basis of numerical solutions, yield-line analysis, or other theoretical methods, provided that it can be clearly demonstrated that he or she has met all the necessary safety and serviceability criteria required by the ACI Code.

Although it has been the practice of designers for many years to use approximate analyses for design and to use average moments rather than maximum ones, two-way slabs so designed have proved to be very satisfactory under service loads. Furthermore, they have proved to have appreciable overload capacity.

15.3 DESIGN OF TWO-WAY SLABS BY THE ACI CODE

The 1989 ACI Code (13.3.1.1) specifies in detail two methods for designing two-way slabs. These are the direct design method and the equivalent frame method.

Direct Design Method

The Code (13.6) provides a procedure with which a set of moment coefficients can be determined. The method, in effect, involves a single-cycle moment distribution analysis of the structure based on (a) the estimated flexural stiffnesses of the slabs, beams (if any), and columns and (b) the torsional stiffnesses of the slabs and beams (if any) transverse to the direction in which flexural moments are being determined. Some type of moment coefficients have been used satisfactorily for many years for slab design. They do not, however, give very satisfactory results for slabs with unsymmetrical dimensions and loading patterns.

Equivalent Frame Method

In this method a portion of the structure is taken out by itself, as shown in Figure 15.2, and analyzed much as a portion of a building frame was handled in Example 13.2. The same stiffness values used for the direct design method

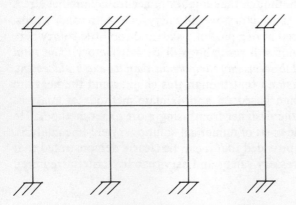

Figure 15.2 Equivalent frame method.

are used for the equivalent frame method. This latter method, which is very satisfactory for symmetrical frames as well as for those with unusual dimensions or loadings, is presented in Chapter 16 of this text.

15.4 COLUMN AND MIDDLE STRIPS

After the design moments have been determined by either the direct design method or the equivalent frame method, they are distributed across each panel. The panels are divided into column and middle strips, as shown in Figure 15.3, and positive and negative moments are estimated in each strip. The *column strip* is a slab with a width on each side of the column center line equal to one-fourth the smaller of the panel dimensions ℓ_1 and ℓ_2. It includes beams if they are present. The middle strip is the part of the slab in between the two column strips.

The part of the moments assigned to the column and middle strips may be assumed to be uniformly spread over the strips. As will be described later in this chapter, the percentage of the moment assigned to a column strip depends on the effective stiffness of that strip and on its *aspect ratio* ℓ_2/ℓ_1 (where ℓ_1 is the length of span, center-to-center, of supports in the direction in which moments are being determined and ℓ_2 is the span length, center-to-center, of supports in the direction transverse to ℓ_1).

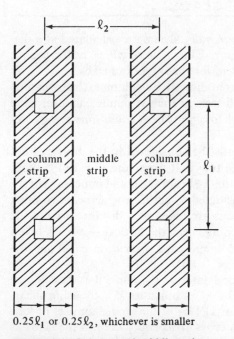

0.25ℓ_1 or 0.25ℓ_2, whichever is smaller

Figure 15.3 Column and middle strips.

Police Station, Philadelphia, Pennsylvania. (Courtesy of Portland Cement Association.)

15.5 SHEAR RESISTANCE OF SLABS

For two-way slabs supported by beams or walls, shears are calculated at a distance d from the face of the walls or beams. The value of ϕV_c is, as for beams, $\phi 2\sqrt{f_c'}\,b_w d$. Shear is not usually a problem for these types of slabs.

For flat slabs and flat plates supported directly by columns, shear may be the critical factor in design. In almost all tests of such structures, failures have been due to shear or perhaps shear and torsion. These conditions are particularly serious around exterior columns.

There are two kinds of shear that must be considered in the design of flat slabs and flat plates. These are the same two that were considered in column footings—one-way and two-way shears (that is, beam shear and punching shear). For beam shear analysis the slab is considered to act as a wide beam running between the supports. The critical sections are taken at a distance d from the face of the column or capital. For punching shear the critical section is taken at a distance $d/2$ from the face of the column, capital, or drop panel and the shear strength, as usually used in footings, is $\phi 4\sqrt{f_c'}\,b_w d$.

If shear stresses are too large around interior columns, it is possible to increase the shearing strength of the slabs by as much as 75% by using shearheads. A shearhead, as defined in Section 11.12.4 of the Code, consists of four steel I or channel shapes fabricated into cross arms and placed in the slabs as shown in Figure 15.4(a). The Code states that shearhead designs of this type

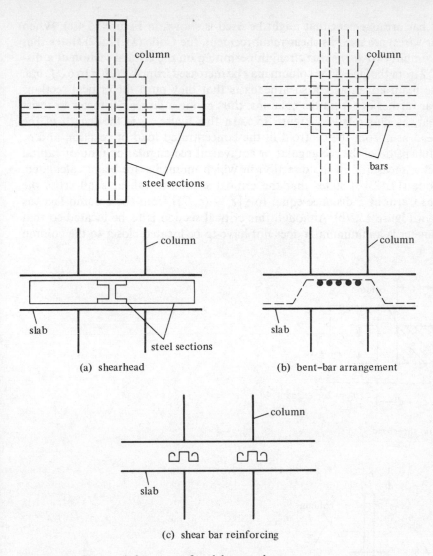

(a) shearhead

(b) bent–bar arrangement

(c) shear bar reinforcing

Figure 15.4 Shear reinforcement for slabs at columns.

do not apply at exterior columns. Thus special designs are required and the Code does not provide specific requirements. Shearheads increase the effective b_o for two-way shear and they also increase the negative moment resistance of the slab, as described in the Code (11.12.4.9). The negative moment reinforcing bars in the slab are usually run over the top of the steel shapes, while the positive reinforcing is normally stopped short of the shapes.

Another type of shear reinforcement permitted in slabs by the Code (11.12.3) involves the use of groups of bent bars or wires. One possible arrangement of such bars is shown in Figure 15.4(b). The bars are bent across the potential diagonal tension cracks at 45° angles and they are run along the bottoms of the slabs for the distances needed to fully develop the bar strengths. Another type of bar arrangement that might be used is shown in Figure 15.4(c). When bars (or wires) are used as shear reinforcement, the Code (11.12.3.2) states that the maximum two-way shear strength permitted on the critical section at a distance $d/2$ from the face of the column may be increased from $4\sqrt{f'_c}\,b_o d$ to $6\sqrt{f'_c}\,b_o d$.

The main advantage of shearheads is that they push the critical sections for shear further out from the columns, thus giving a larger perimeter to resist the shear, as illustrated in Figure 15.5. In this figure, ℓ_v is the length of the shearhead arm from the centroid of the concentrated load or reaction and c_1 is the dimension of the rectangular or equivalent rectangular column or capital or bracket measured in the direction in which moments are being calculated. The Code (11.12.4.7) states that the critical section for shear shall cross the shearhead arm at a distance equal to $\frac{3}{4}[\ell_v - (c_1/2)]$ from the column face, as shown in Figure 15.5(b). Although this critical section is to be located so that its perimeter is a minimum, it does not have to be located closer to the column

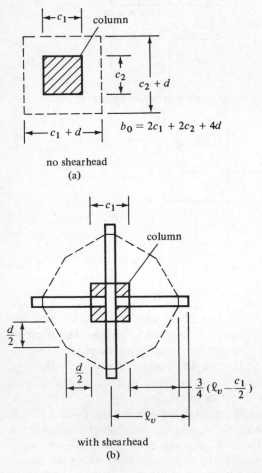

no shearhead
(a)

with shearhead
(b)

Figure 15.5

face than $d/2$ at any point. When steel I or channel shapes are used as shear-heads, the maximum shear strength can be increased to $7\sqrt{f'_c}\,b_o d$ at a distance $d/2$ from the column. According to the Code (11.12.4.8), this is only permissible if the maximum computed shear does not exceed $4\sqrt{f'_c}\,b_o d$ along the dashed critical section for shear shown in Figure 15.5(b).

In Section 15.12 the subject of shear stresses is continued with a consideration of the transfer of moments and shears between slabs and columns. The maximum load which a two-way slab can support is often controlled by this transfer strength.

15.6 DEPTH LIMITATIONS AND STIFFNESS REQUIREMENTS

It is obviously very important to keep the various panels of a two-way slab relatively level (that is, with reasonably small deflections). Thin reinforced two-way slabs have quite a bit of moment resistance, but deflections are often large. As a consequence, their depths are very carefully controlled by the ACI Code so as to limit these deflections. This is accomplished by requiring the designer to either (a) compute deflections and make sure they are within certain limitations or (b) use certain minimum thicknesses as specified in Section 9.5.2.1 of the Code.

Because deflection computations for two-way slabs are so very complicated, the average designer uses the minimum ACI thickness values which are given by three empirical equations. Involved in these equations are spans, panel shapes, flexural stiffness of beams if they are used, steel yield stresses, and so on. In these equations the following terms are used:

ℓ_n = the clear span in the long direction, measured face to face, of (a) columns for slabs without beams and (b) beams for slabs with beams

β = the ratio of the long to the short clear span

α_m = the average value of the ratios of beam-to-slab stiffness on all sides of a panel.

Throughout this chapter the letter α is used to represent the ratio of the flexural stiffness $(E_{cb}I_b)$ of a beam section to the flexural stiffness of the slab $(E_{cs}I_s)$ whose width equals the distance between the centerlines of the panels on each side of the beam; that is

$$\alpha = \frac{E_{cb}I_b}{E_{cs}I_s}$$

where E_{cb} = the modulus of elasticity of the beam concrete
   E_{cs} = the modulus of elasticity of the column concrete
   I_b = the gross moment of inertia about the centroidal axis of a section made up of the beam and the slab on each side of the beam extending a distance equal to the projection of the beam above

or below the slab (whichever is greater) but not exceeding four times the slab thickness (ACI 13.2.4)

I_s = the moment of inertia of the gross section of the slab taken about the centroidal axis and equal to $h^3/12$ times the slab width, where the width is the same as for x

The minimum thickness of slabs or other two-way construction may be obtained by substituting into the three equations to follow, which are given in Section 9.5.3.3 of the Code. In the equations the quantity β (ratio of clear span in long direction to clear span in short direction) is used to take into account the effect of the shape of the panel on its deflection, while the effect of beams (if any) is represented by α_m. If there are no beams present (as is the case for flat slabs), α_m will equal 0.

$$h = \frac{\ell_n\left(0.8 + \dfrac{f_y}{200,000}\right)}{36 + 5\beta\left[\alpha_m - 0.12\left(1 + \dfrac{1}{\beta}\right)\right]} \qquad \text{(ACI Equation 9-11)}$$

but should not be less than

$$h = \frac{\ell_n\left(0.8 + \dfrac{f_y}{200,000}\right)}{36 + 9\beta} \qquad \text{(ACI Equation 9-12)}$$

However, the thickness determined by the two preceding equations need not be greater than

$$h = \frac{\ell_n\left(0.8 + \dfrac{f_y}{200,000}\right)}{36} \qquad \text{(ACI Equation 9-13)}$$

Equations 9-11 and 9-13 are included primarily for slabs which have edge beams on all four sides of each panel. Regardless of the value obtained for a particular slab, the thickness used may not be less than the following:

(a) For slabs with $\alpha_m < 2.0$:   5 in.
(b) For slabs with $\alpha_m \geq 2.0$: $3\frac{1}{2}$ in.

Should we have slabs without beams but with drop panels extending for at least one-sixth of the span lengths in each direction measured from column center lines and with projections below the slabs of at least one-fourth of the slab thickness beyond the drop, the thickness required by the three equations may be reduced by 10% (ACI Section 9.5.3.4).

For panels with discontinuous edges, the Code (9.5.3.5) requires that edge beams be used, which have a minimum stiffness ratio α equal to 0.8, or else that the minimum slab thicknesses, as determined by ACI Equations 9-11 through 9-13, must be increased by 10%.

TABLE 15.1   MINIMUM THICKNESS OF SLABS WITHOUT INTERIOR BEAMS

Yield stress f_y psi[a]	Without drop panels[b]			With drop panels[b]		
	Exterior panels		Interior panels	Exterior panels		Interior panels
	Without edge beams	With edge beams[c]		Without edge beams	With edge beams[c]	
40,000	$\dfrac{\ell_n}{33}$	$\dfrac{\ell_n}{36}$	$\dfrac{\ell_n}{36}$	$\dfrac{\ell_n}{36}$	$\dfrac{\ell_n}{40}$	$\dfrac{\ell_n}{40}$
60,000	$\dfrac{\ell_n}{30}$	$\dfrac{\ell_n}{33}$	$\dfrac{\ell_n}{33}$	$\dfrac{\ell_n}{33}$	$\dfrac{\ell_n}{36}$	$\dfrac{\ell_n}{36}$

[a] For values of reinforcement, yield stress between 40,000 and 60,000 psi minimum thickness shall be obtained by linear interpolation.

[b] Drop panel is defined in ACI Sections 13.4.7.1 and 13.4.7.2.

[c] Slabs with beams between columns along exterior edges. The value of α for the edge beam shall not be less than 0.8.

For flat plates and flat slabs, minimum thicknesses may be determined directly by referring to Table 15.1, which is Table 9.5(c) of the ACI Code.

The designer may use slabs of lesser thicknesses than those required by the ACI Code as described in the preceding spans if deflections are computed and found to be equal to or less than the limiting values given in Table 9.5(b) of the ACI Code (Table 5.2 in this text).

Should the various rules for minimum thickness be followed but the resulting slab be insufficient to provide the shear capacity required for the particular column size, column capitals will probably be required. Beams running between the columns may be used for some slabs where partitions or heavy equipment loads are placed near column lines. A very common case of this type occurs where exterior beams are used when the exterior walls are supported directly by the slab. Another situation where beams may be used occurs where there is concern about the magnitude of slab vibrations.[1] Example 15.1 illustrates the application of the minimum slab thickness rules for a flat plate, while Example 15.2 does the same for a two-way slab with beams.

■ EXAMPLE 15.1

Using the ACI Code, determine the minimum thickness required for the corner panel of the flat plate floor shown in Figure 15.6. Edge beams are not used along the panel edges, $f'_c = 3000$ psi, and $f_y = 60,000$ psi.

---

[1] Fintel, M., ed., 1974, *Handbook of Concrete Engineering* (New York: Van Nostrand Reinhold), p. 58.

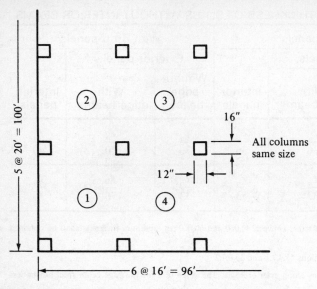

Figure 15.6 A flat-plate floor slab.

SOLUTION

$$\beta = \frac{20 - \frac{16}{12}}{16 - \frac{12}{12}} = 1.245$$

$$\alpha_m = 0$$

Panel 1   Substituting into ACI Equations 9-11 through 9-13, we use

$$\ell_n = 20 - \frac{16}{12} = 18.67 \text{ ft}$$

Because these panels have discontinuous edges, the values obtained by the three equations are increased by 10% (thus multiplying them by 1.10 as per ACI Section 9.5.3.5).

$$h = \frac{(18.67)\left(0.8 + \dfrac{60,000}{200,000}\right)(1.10)}{36 + (5)(1.245)\left[0 - 0.12\left(1 + \dfrac{1}{1.245}\right)\right]}$$

$$= 0.652 \text{ ft} = 7.82 \text{ in.}$$

but should not be less than

$$h = \frac{(18.67)\left(0.8 + \dfrac{60,000}{200,000}\right)(1.10)}{36 + (9)(1.245)}$$

$$= 0.479 \text{ ft} = 5.74 \text{ in.}$$

However, it need not be greater than

$$h = \frac{(18.67)\left(0.8 + \dfrac{60,000}{200,000}\right)(1.10)}{36}$$

$$= 0.628 \text{ ft} = 7.53 \text{ in.} \quad \leftarrow$$

Noting that minimum h for slabs without drop panels (Code 9.5.3.2) is 5 in., the controlling h is 7.53 in.

<u>Use $7\frac{1}{2}$ in. slab</u>

Instead of substituting into these three equations, we can (because there are no interior beams) go to Table 15.1 and pick out h as follows:

$$h = \frac{\ell_n}{30} \text{ for slabs without interior edge beams}$$

$$h = \frac{18.67}{30} = 0.622 \text{ ft} = 7.47 \text{ in.} \qquad \blacksquare$$

■ EXAMPLE 15.2

The two-way slab shown in Figure 15.7 has been assumed to have a thickness of 7 in. Section A-A in the figure shows the beam cross section. Check the ACI equations to determine if the slab thickness is satisfactory for an interior panel. $f'_c = 3000$ psi, $f_y = 60,000$ psi.

SOLUTION

Computing α_1

$$I_s = \text{gross moment of inertia of slab 20 ft wide}$$
$$= (\tfrac{1}{12})(12 \times 20)(7)^3 = 6860 \text{ in.}^4$$

I_b = gross I of T-beam cross section shown in Figure 15.7 about centroidal axis = 18,060 in.^{4}

$$\alpha_1 = \frac{EI_b}{EI_s} = \frac{(E)(18,060)}{(E)(6,860)} = 2.63$$

Computing α_2 and α_m

$$I_s \text{ for 24-ft-wide slab} = (\tfrac{1}{12})(12 \times 24)(7)^3 = 8232 \text{ in.}^4$$

$$I_b = 18,060 \text{ in.}^4$$

$$\alpha_2 = \frac{(E)(18,060)}{(E)(8232)} = 2.19$$

$$\alpha_m = \frac{\alpha_1 + \alpha_2}{2} = \frac{2.63 + 2.19}{2} = 2.41$$

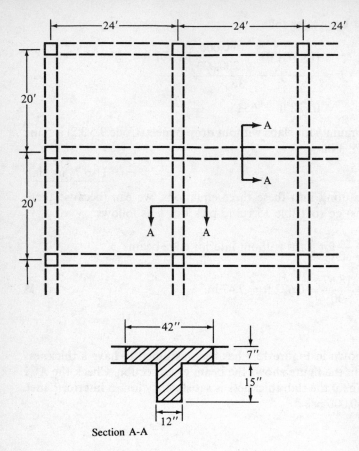

Section A-A

Figure 15.7 A two-way slab.

Determining slab thickness as per ACI 9.5.3

$$\beta = \frac{\ell_{n \text{ long}}}{\ell_{n \text{ short}}} = \frac{24 - \frac{12}{12}}{20 - \frac{12}{12}} = \frac{23}{19} = 1.21$$

$$h = \frac{(12)(23)\left(0.8 + \dfrac{60{,}000}{200{,}000}\right)}{36 + (5)(1.21)\left[2.41 - 0.12\left(1 + \dfrac{1}{1.21}\right)\right]} = 6.16 \text{ in.}$$

but h should not be less than

$$h = \frac{(12)(23)\left(0.8 + \dfrac{60{,}000}{200{,}000}\right)}{36 + (9)(1.21)} = 6.47 \text{ in.} \leftarrow$$

However, it need not be more than

$$h = \frac{(12)(23)\left(0.8 + \dfrac{60,000}{200,000}\right)}{36} = 8.43 \text{ in.}$$

Ans. The 7-in. thickness assumed is more than sufficient. ∎

15.7 LIMITATIONS OF DIRECT DESIGN METHOD

For the moment coefficients determined by the direct design method to be applicable, the Code (13.6.1) says that the following limitations must be met, unless a theoretical analysis shows that the strength furnished after the appropriate capacity reduction or ϕ factors are applied is sufficient to support the anticipated loads and provided that all serviceability conditions such as deflection limitations are met:

1. There must be at least three continuous spans in each direction.
2. The panels must be rectangular, with the length of the longer side of any panel not being more than 2.0 times the length of its shorter side.
3. Span lengths of successive spans in each direction may not differ in length by more than one-third of the longer span.
4. Columns may not be offset by more than 10% of the span length in the direction of the offset from either axis between center lines of successive columns.
5. The live load shall not be more than three times the dead load. All loads must be due to gravity and must be uniformly distributed over an entire panel.
6. If a panel is supported on all sides by beams, the relative stiffness of those beams in the two perpendicular directions, as measured by the following expression, shall not be less than 0.2 nor greater than 5.0.

$$\frac{\alpha_1 \ell_2^2}{\alpha_2 \ell_1^2}$$

The terms ℓ_1 and ℓ_2 were shown in Figure 15.3.

15.8 DISTRIBUTION OF MOMENTS IN SLABS

The total moment M_o that is resisted by a slab equals the sum of the maximum positive and negative moments in the span. It is the same as the total moment that occurs in a simply supported beam. For a uniform load it is as follows:

$$M_o = \frac{(w_u \ell_2)(\ell_1)^2}{8}$$

In this expression ℓ_1 is the span length, center to center, of supports in the direction in which moments are being taken and ℓ_2 is the length of the span transverse to ℓ_1, measured center to center of the supports.

The moment that actually occurs in such a slab has been shown by experience and tests to be somewhat less than the value determined by the above M_o expression. For this reason ℓ_1 is replaced with ℓ_n, the clear span measured face to face of the supports in the direction in which moments are taken. The Code (13.6.2.5) states that ℓ_n may not be taken to be less than 65% of the span center-to-center of supports. Using ℓ_n the total moment becomes

$$M_o = \frac{(w_u \ell_2)\ell_n)^2}{8} \qquad \text{(ACI Equation 13-3)}$$

When the moment is being calculated in the long direction, the total static moment is written here as $M_{o\ell}$ and in the short direction as M_{os}.

It is next necessary to know what proportion of these total moments are positive and what proportion are negative. If a slab was completely fixed at the end of each panel, the division would be as it is in a fixed-end beam, two-thirds negative and one-third positive, as shown in Figure 15.8.

This division is reasonably accurate for interior panels where the slab is continuous for several spans in each direction with equal span lengths and loads. In effect, the rotation of the interior columns is assumed to be small, and moment values of $0.65M_o$ for negative moment and $0.35M_o$ for positive moment are specified by the Code (13.6.3.2). For cases where the span lengths and loadings are different, the proportion of positive and negative moments may vary appreciably and the use of a more detailed method of analysis is desirable. The equivalent frame method (Chapter 16) will provide rather good approximations for such situations.

The relative stiffnesses of the columns and slabs of exterior panels are of far greater significance in their effect on the moments than is the case for interior panels. The magnitudes of the moments are very sensitive to the amount of torsional restraint supplied at the discontinuous edges. This restraint is provided both by the flexural stiffness of the slab and by the flexural stiffness of the exterior column.

Should the stiffness of an exterior column be quite small, the end negative

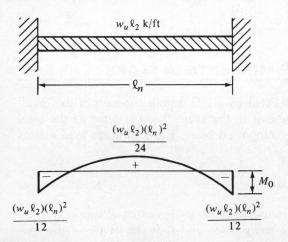

Figure 15.8

moment will be very close to zero. If the stiffness of the exterior column is very large, the positive and negative moments will still not be the same as those in an interior panel unless an edge beam with a very large torsional stiffness is provided that will substantially prevent rotation of the discontinuous edge of the slab.

If a 2-ft-wide beam were to be framed into a 2-ft-wide column of infinite flexural stiffness in the plane of the beam, the joint would behave as would a perfectly fixed end, and the negative beam moment would equal the fixed-end moment.

If a two-way slab 24 ft wide were to be framed into this same 2-ft-wide column of infinite stiffness, the situation of no rotation would occur only along the part of the slab at the column. For the remaining 11-ft widths of slab on each side of the column, there would be rotation varying from zero at the side face of the column to maximums 11 ft on each side of the column. As a result of this rotation, the negative moment at the face of the column would be less than the fixed-end moment. Thus the stiffness of the exterior column is reduced by the rotation of the attached transverse slab.

To take into account the fact that the rotation of the edge of the slab is different at different distances from the column, the exterior columns and slab edge beam are replaced with an equivalent column that has the same flexibility as the column plus the edge beam. It can be seen that this is quite an involved process; therefore, instead of requiring a complicated analysis, the Code (13.6.3.3) provides a set of percentages for dividing the total factored static moment into its positive and negative parts in an end span. These divisions, which are shown in Table 15.2, include values for unrestrained edges (as where

TABLE 15.2 DISTRIBUTION OF TOTAL SPAN MOMENT IN AN END SPAN (ACI CODE 13.6.3.3)

	(1)	(2)	(3)	(4)	(5)
			Slab without beams between interior supports		
	Exterior edge unrestrained	Slab with beams between all supports	Without edge beam[a]	With edge beam	Exterior edge fully restrained
Interior negative factored moment	0.75	0.70	0.70	0.70	0.65
Positive moment	0.63	0.57	0.52	0.50	0.35
Exterior negative factored moment	0	0.16	0.26	0.30	0.65

[a] See ACI Section 13.6.3.6.

the slab is simply supported on a masonry or concrete wall) and for restrained edges (as where the slab is constructed integrally with a very stiff reinforced concrete wall so that the little rotation occurs at the slab-to-wall connection).

In Figure 15.9 the distribution of the total factored moment for the interior and exterior spans of a flat-plate structure is shown. The plate is assumed to be constructed without beams between interior supports and without edge beams.

The next problem is to estimate what proportion of these moments are taken by the column strips and what proportions are taken by the middle strips. For this discussion a flat-plate structure is assumed and the moment resisted by the column strip is estimated by considering the tributary areas shown in Figure 15.10.

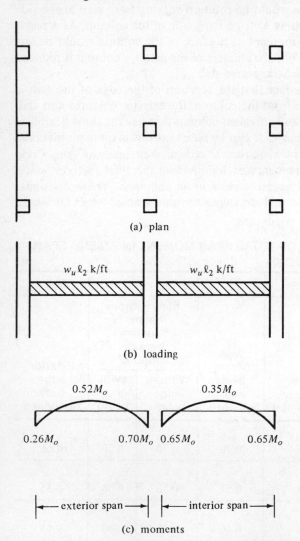

(a) plan

$w_u \ell_2$ k/ft $w_u \ell_2$ k/ft

(b) loading

$0.52M_o$ $0.35M_o$

$0.26M_o$ $0.70M_o$ $0.65M_o$ $0.65M_o$

|← exterior span →| |← interior span →|

(c) moments

Figure 15.9 Sample moments for a flat plate with no edge beams.

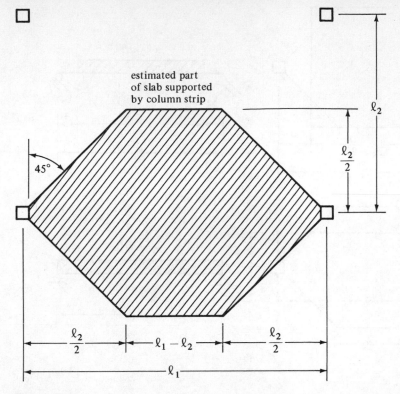

estimated part
of slab supported
by column strip

$45°$

ℓ_2

$\dfrac{\ell_2}{2}$

$\dfrac{\ell_2}{2}$ $\ell_1 - \ell_2$ $\dfrac{\ell_2}{2}$

ℓ_1

Figure 15.10

To simplify the mathematics, the load to be supported is assumed to fall within the dashed lines shown in either part (a) or (b) of Figure 15.11. The corresponding load is placed on the simple span, and its centerline moment is determined as an estimate of the portion of the static moment taken by the column strip.[2]

In Figure 15.11(a), the load is concentrated near the ℄ of the beam, thus causing the moment to be overestimated a little, while in Figure 15.11(b) the load is spread uniformly from end to end, causing the moment to be underestimated. Based on these approximations, the estimated moments in the column strips for square panels will vary from $0.5M_o$ to $0.75M_o$. As ℓ_1 becomes larger than ℓ_2, the column strip takes a larger proportion of the moment. For such cases about 60–70% of M_o will be resisted by the column strip.

If you sketch in the approximate deflected shape of a panel, you will see that a larger portion of the positive moment is carried by the middle strip than by the column strip, and vice versa for the negative moments. As a result about

[2] White, R. N., Gergely, P., and Sexsmith, R. G., 1974, *Structural Engineering, vol. 3: Behavior of Members and Systems* (New York: John Wiley & Sons), pp. 456–461.

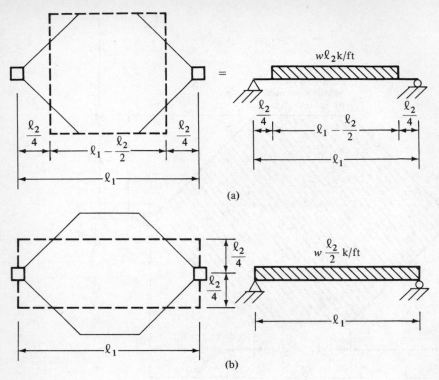

Figure 15.11

60% of the positive M_o and about 70% of the negative M_o is expected to be resisted by the column strip.[3]

Section 13.6.4.1 of the Code states that the column strip shall be proportioned to resist the percentages of the total interior negative design moment given in Table 15.3.

TABLE 15.3 PERCENTAGES OF INTERIOR NEGATIVE DESIGN MOMENTS TO BE RESISTED BY COLUMN STRIP

$\dfrac{\ell_2}{\ell_1}$	0.5	1.0	2.0
$\dfrac{\alpha_1 \ell_2}{\ell_1} = 0$	75	75	75
$\dfrac{\alpha_1 \ell_2}{\ell_1} \geq 1.0$	90	75	45

[3] White, R. N., Gergely, P., and Sexsmith, R. G., 1974, *Structural Engineering, vol. 3: Behavior of Members and Systems* (New York: John Wiley & Sons), pp. 459–461.

TABLE 15.4 PERCENTAGES OF EXTERIOR NEGATIVE DESIGN MOMENT TO BE RESISTED BY COLUMN STRIP

$\dfrac{\ell_2}{\ell_1}$		0.5	1.0	2.0
$\dfrac{\alpha_1 \ell_2}{\ell_2} = 0$	$\beta_t = 0$	100	100	100
	$\beta_t \geq 2.5$	75	75	75
$\dfrac{\alpha_1 \ell_2}{\ell_1} \geq 1.0$	$\beta_t = 0$	100	100	100
	$\beta_t \geq 2.5$	90	75	45

In the table, α_1 is again the ratio of the stiffness of a beam section to the stiffness of a width of slab bounded laterally by the ℄ of the adjacent panel, if any, on each side of the beam and equals $E_{cb}I_b/E_{cs}I_s$.

Section 13.6.4.2 of the Code states that the column strip is to be assumed to resist percentages of the exterior negative design moment, as given in Table 15.4.

The column strip (Section 13.6.4.4 of the Code) is to be proportioned to resist the portion of the positive moments given in Table 15.5.

In Section 13.6.5 the Code requires that the beam be allotted 85% of the column strip moment if $\alpha_1(\ell_2/\ell_1) \geq 1.0$. Should $\alpha_1(\ell_2/\ell_1)$ be between 1.0 and zero, the moment allotted to the beam is determined by linear interpolation from 85% to 0%. The part of the moment not given to the beam is allotted to the slab in the column strip.

Finally, the Code (13.6.6) requires that the portion of the design moments not resisted by the column strips as previously described is to be allotted to the

TABLE 15.5 PERCENTAGES OF POSITIVE DESIGN MOMENT TO BE RESISTED BY COLUMN STRIP

$\dfrac{\ell_2}{\ell_1}$	0.5	1.0	2.0
$\dfrac{\alpha_1 \ell_2}{\ell_1} = 0$	60	60	60
$\dfrac{\alpha_1 \ell_2}{\ell_1} \geq 1.0$	90	75	45

corresponding half middle strip. The middle strip will be designed to resist the total of the moments assigned to its two half middle strips.

15.9 DESIGN OF AN INTERIOR FLAT PLATE

In this section an interior flat plate is designed by the direct design method. The procedure specified in Chapter 13 of the ACI Code is applicable not only to flat plates but also to flat slabs, waffle flat slabs, and two-way slabs with beams. The steps necessary to perform the designs are briefly summarized at the end of this paragraph. The order of the steps may have to be varied somewhat for different types of slab designs. Either the direct design method or the equivalent frame method may be used to determine the design moments. The design steps are as follows:

1. Estimate the slab thickness to meet the Code requirements.
2. Determine the depth required for shear.
3. Calculate the total static moments to be resisted in the two directions.
4. Estimate the percentages of the static moments that are positive and negative, and proportion the resulting values between the column and middle strips.
5. Select the reinforcing.

Example 15.3 illustrates this method of design applied to a flat plate. In the solution it will be noted that the $M_u/\phi bd^2$ values are quite small and thus most of the ρ values do not fall within Table A.13 (see Appendix). For such cases the author assumes a value of $z = 0.95d$ and solves the following expression for A_s:

$$A_s = \frac{M_u}{\phi f_y z}$$

After this steel area is determined, it is theoretically necessary to compute the value z and recalculate the steel area and so forth. The author did not make these last calculations because he felt the values would change by negligible amounts. Such a trial-and-error procedure is illustrated, however, at the end of this paragraph for the positive moment in the long-span column strip of Example 15.3. In this example the width of the column strip is 8 ft, M_u is 42.5 ft-k, f_y is 60,000 psi, and $f'_c = 3000$ psi.

Assume $z = (0.95)(6.5) = 6.18$ in.

$$A_s = \frac{M_u}{\phi f_y z} = \frac{(12)(42.5)}{(0.9)(60)(6.18)} = 1.53 \text{ in.}^2$$

$$a = \frac{(1.53)(60)}{(0.85)(3)(12 \times 8)} = 0.375''$$

$$z = 6.50 - \frac{0.375}{2} = 6.31''$$

$$A_s = \frac{(12)(42.5)}{(0.9)(60)(6.31)} = 1.50 \text{ in.}^2$$

Therefore, negligible change from 1.53 in.^{2}

For several locations in Example 15.3 the steel areas calculated as described here for moment are less than the temperature and shrinkage minimum $0.0018bh$ for grade 60 bars. The larger value of the two must be used. Actually, the temperature percentage includes bars in the top and bottom of the slab. In the negative moment region, some of the positive steel bars have been extended into the support region and are also available for temperature and shrinkage steel. If desirable, these positive bars can be lapped instead of being stopped in the support.

■ EXAMPLE 15.3

Design an interior flat plate for the structure considered in Example 15.1. This plate is shown in Figure 15.12. Assume a service live load equal to 80 psf, a service dead load equal to 110 psf (including slab weight), $f_y = 60,000$ psi, $f'_c = 3000$ psi, and column heights of 12 ft.

SOLUTION

Estimate Slab Thickness Try a $7\frac{1}{2}$-in. slab as calculated in Example 15.1.

Determine Depth Required for Shear Using d for shear equal to the estimated average of the d values in the two directions, we obtain

$$d = 7.50'' - \tfrac{3}{4}'' \text{ cover} - 0.50'' = 6.25''$$

$$w_u = (1.4)(110) + (1.7)(80) = 290 \text{ psf}$$

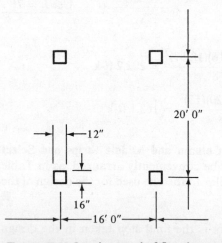

Figure 15.12 Interior panel of flat-plate structure of Example 15.12.

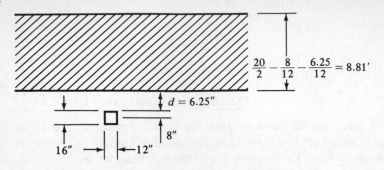

Figure 15.13

Checking One-Way or Beam Shear (Rarely Checked as It Seldom Controls)

Using the dimensions shown in Figure 15.13, we obtain

$$V_{u1} = (8.81)(290) = 2555 \text{ lb for a } 12'' \text{ width}$$

$$\phi V_c = \phi 2\sqrt{f_c'}b_o d$$
$$= (0.85)(2\sqrt{3000})(12)(6.25)$$
$$= 6983 \text{ lb} > 2555 \text{ lb} \qquad \underline{\text{OK}}$$

Checking Two-Way or Punching Shear Around the Column

$$b_o = (2)(16 + 6.25) + (2)(12 + 6.25) = 81''$$

$$V_{u2} = \left[(20)(16) - \left(\frac{16 + 6.25}{12}\right)\left(\frac{12 + 6.25}{12}\right)\right](0.290)$$

$$= 91.98 \text{ k} = 91{,}980 \text{ lb}$$

$$\phi V_c = (0.85)(4\sqrt{3000})(81)(6.25)$$
$$= 94{,}277 \text{ lb} > 91{,}980 \text{ lb} \qquad \underline{\text{OK}}$$

$$\text{Use } h = 7\tfrac{1}{2}''.$$

Calculate Static Moments

$$M_{o\ell} = \frac{w_u \ell_2 \ell_n^2}{8} = \frac{(0.290)(16)(18.67)^2}{8} = 202.2 \text{ ft-k}$$

$$M_{os} = \frac{w_u \ell_1 \ell_n^2}{8} = \frac{(0.290)(20)(15)^2}{8} = 163.1 \text{ ft-k}$$

Proportion the Static Moments to the Column and Middle Strips and Select the Reinforcing

These calculations can be conveniently arranged, as in Table 15.6 which follows. This table is very similar to the one used for the design of the continuous one-way slab in Chapter 13. ∎

The selection of the reinforcing bars is the final step taken in the design of this flat plate. The Code Figure 13.4.8 (given as Figure 15.14 here) shows the

TABLE 15.6

	Long span (estimate $d = 6.50''$)				Short span (estimate $d = 6.00''$)			
	Column strip (8')		Middle strip (8')		Column strip (8')		Middle strip (12')	
	−	+	−	+	−	+	−	+
M_u	(0.65)(0.75)(202.2) = −98.6 ft-k	(0.35)(0.60)(202.2) = +42.5 ft-k	(0.65)(202.2) − 98.6 = −32.8 ft-k	(0.35)(202.2) − 42.5 = +28.3 ft-k	(0.65)(0.75)(163.1) = −79.5 ft-k	(0.35)(0.60)(163.1) = +34.3 ft-k	(0.65)(163.1) − 79.5 = −26.5 ft-k	(0.35)(163.1) − 34.3 = +22.8 ft-k
$\dfrac{M_u}{\phi b d^2}$	324.1 psi	139.7 psi	107.8 psi	93.0 psi	306.7 psi	132.3 psi	68.2 psi	58.6 psi
ρ	0.00580	Not in table; assume z = 6.18"	Not in table; assume z = 6.18"	Not in table; assume z = 6.18"	0.00546	Not in table; assume z = 5.70"	Not in table; assume z = 5.70"	Not in table; assume z = 5.70"
A_s	3.62 in.²	1.53 in.²	Temperature controls 1.30 in.²	Temperature controls 1.30 in.²	3.14 in.²	1.34 in.²	Temperature controls 1.94 in.²	Temperature controls 1.94 in.²
Bars selected	19 #4	8 #4	7 #4	7 #4	16 #4	7 #4	10 #4	10 #4

513

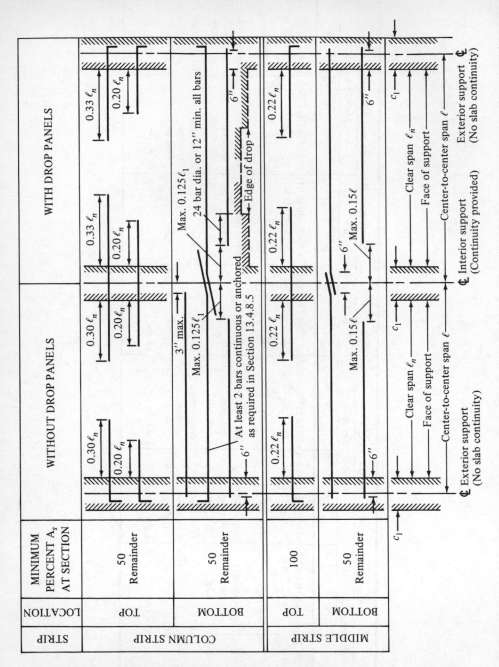

Figure 15.14 Minimum extensions for reinforcement in slabs without beams (see ACI Section 12.11.1 for reinforcement extension into supports).

minimum lengths of slab reinforcing bars for flat plates and for flat slabs with drop panels. This figure shows that some of the positive reinforcing must be run into the support area within 3 in. of the support

The bars selected for this flat plate are shown in Figure 15.15. Bent bars are used in this example, but straight bars could have been used just as well. There seems to be a trend among designers in the direction of using more straight bars in slabs and fewer bent bars.

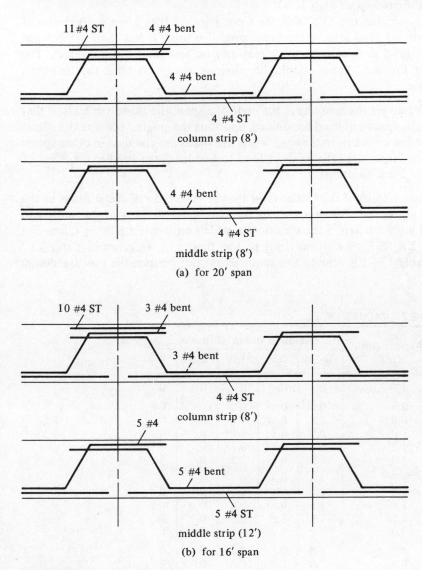

11 #4 ST   4 #4 bent

4 #4 bent

4 #4 ST

column strip (8′)

4 #4 bent

4 #4 ST

middle strip (8′)

(a) for 20′ span

10 #4 ST   3 #4 bent

3 #4 bent

4 #4 ST

5 #4   column strip (8′)

5 #4 bent

5 #4 ST

middle strip (12′)

(b) for 16′ span

Figure 15.15

15.10 MINIMUM COLUMN STIFFNESSES

The moments in a continuous floor slab are appreciably affected by different positions or patterns of the live loads. The usual procedure, however, is to calculate the total static moments assuming that all panels are subjected to full live load. When different loading patterns are used, the moments can be changed so much that overstressing may occur in the slab. For such situations the Code permits overstresses as high as 33%.

To stay within the 33% limit, the Code requires that when β_a (the ratio of the unfactored dead load to the unfactored live load) is less than 2.0, column stiffnesses (and thus column sizes) may not be less than certain values. The reasoning for such limits should be obvious after you read the following statements:

1. The larger the columns at the ends of a span, the closer the ends of that span approach fixed conditions, and thus the smaller become the effects on the moments in the span when live loads are changed in other spans.
2. The larger the ratio of dead load to live load, the smaller the effect of live load variations..

Section 13.6.10 of the Code states that α_c [the ratio of the stiffness of the columns above and below a slab to the combined stiffness of the slabs and beams at a joint taken in the direction in which moments are being taken and equal to $\Sigma K_c / \Sigma(K_s + K_b)$] may not be less than α_{min}, as given in Table 15.7 (Code Table 13.6.10). Should this requirement not be met, the positive design

TABLE 15.7 VALUES OF α_{min}

β_a	Aspect ratio ℓ_2/ℓ_1	Relative beam stiffness α				
		0	0.5	1.0	2.0	4.0
2.0	0.5–2.0	0	0	0	0	0
1.0	0.5	0.6	0	0	0	0
	0.8	0.7	0	0	0	0
	1.0	0.7	0.1	0	0	0
	1.25	0.8	0.4	0	0	0
	2.0	1.2	0.5	0.2	0	0
0.5	0.5	1.3	0.3	0	0	0
	0.8	1.5	0.5	0.2	0	0
	1.0	1.6	0.6	0.2	0	0
	1.25	1.9	1.0	0.5	0	0
	2.0	4.9	1.6	0.8	0.3	0
0.33	0.5	1.8	0.5	0.1	0	0
	0.8	2.0	0.9	0.3	0	0
	1.0	2.3	0.9	0.4	0	0
	1.25	2.8	1.5	0.8	0.2	0
	2.0	13.0	2.6	1.2	0.5	0.3

moment in the panels supported by the columns in question must be increased. This is done by multiplying them by the coefficient δ_s determined from the following expression:

$$\delta_s = 1 + \left(\frac{2 - \beta_a}{4 + \beta_a}\right)\left(1 - \frac{\alpha_c}{\alpha_{min}}\right)$$

It will be noted in Table 15.7 that if β_a is 2.0 or more, α_{min} is 0. As β_a falls below 2.0, the value of α_{min} increases. For the flat plate designed in Example 15.3 with $L = 80$ psf and $D = 110$ psf, $\beta_a = 110/80 = 1.375$, the aspect ratio $(\ell_2/\ell_1) = 20/16 = 1.25$, and $\alpha = 0$ (with no beams present). The values to follow can be determined from the table:

With $\beta_a = 2.0, \dfrac{\ell_2}{\ell_1}$ = from 0.5 to 2.0, $\alpha = 0$, $\alpha_{min} = 0$

With $\beta_a = 1.0, \dfrac{\ell_2}{\ell_1} = 1.25$, $\alpha = 0$, $\alpha_{min} = 0.8$

With $\beta_a = 1.375, \dfrac{\ell_2}{\ell_1} = 1.25$, $\alpha = 0$, $\alpha_{min} = 0.5$ (by interpolation
between the two preceding α_{min} values)

From structural analysis texts the stiffness of a prismatic member such as a flat plate may be taken as $K_s = 4EI_s/\ell$, where I_s is the moment of inertia of the gross section of the slab about the centroidal axis. Should beams or drop panels or column capitals be present, the slab will not be prismatic and thus I will vary and it will be necessary to calculate a more correct stiffness value or take it from tables for that purpose. For the slab of Example 15.3,

$$I_s = (\tfrac{1}{12})(12 \times 16)(7.5)^3 = 6750 \text{ in.}^4$$

$$K_s = \frac{4EI_c}{\ell} = \frac{(4E)(6750)}{(12)(20)} = 112.5E$$

$$K_b = 0 \text{ as there is no beam}$$

Referring back to α_{min},

$$\alpha_{min} = \frac{K_c}{K_s}$$

$$0.50 = \frac{K_c}{K_s} = \frac{K_c}{112.5E}$$

from which

$$K_c = 56.25E$$

Remembering that

$$K_c = \frac{4EI_s}{\ell} = \frac{4EI_c}{(12)(12)} = \frac{EI_c}{36}$$

$$56.25E = \frac{EI_c}{36}$$

$$I_c = 2025 \text{ in.}^4 < I \text{ of column used here}$$

$$= (\tfrac{1}{12})(16)(12)^3 = 2304 \text{ in.}^4 \qquad \underline{OK}$$

15.11 ANALYSIS OF TWO-WAY SLABS WITH BEAMS

In this section the moments are determined by the direct design method for an exterior panel of a two-way slab with beams. It is felt that if you can handle this problem, you can handle any other case that may arise in flat plates, flat slabs, or two-way slabs with beams using the direct design method.

The requirements of the Code are so lengthy and complex that in Example 15.4, which follows, the steps and appropriate code sections are spelled out in detail. The practicing designer should obtain a copy of the *CRSI Handbook*, because the tables therein will be of tremendous help in slab design.

■ EXAMPLE 15.4

Determine the negative and positive moments required for the design of the exterior panel of the two-way slab with beam structure as shown in Figure 15.16. The slab is to support a live load of 120 psf and a dead load of 100 psf including the slab weight. The columns are 15 in. × 15 in. and 12 ft long. The slab is supported by beams along the column line with the cross section shown. Determine the slab thickness and check the shear stress if $f'_c = 3000$ psi and $f_y = 60,000$ psi.

SOLUTION

1. Checking ACI Code limitations (13.6.1). These conditions are met.
2. Minimuum thickness as required by Code (9.5.3.1):
   (a) Assume $h = 6$ in.
   (b) Effective flange projection of column line beam as specified by Code (13.2.4)

   $$= 4h_f = (4)(6) = 24'' \quad \text{or} \quad h - h_f = 20 - 6 = \underline{14''}$$

   (c) Gross moments of inertia of T beams. The following values are the gross moments of inertia of the edge and interior beams computed respectively about their centroidal axes. Many designers use approximate values for these moments of inertias, I_s, with almost identical results for slab thicknesses. One common practice is to use 2 times the gross moment of inertia of the stem using a depth of stem running from top of slab to bottom of stem) for interior beams and $1\frac{1}{2}$ times the stem gross moment of inertia for edge beams.

   $$I \text{ for edge beams} = 13,476 \text{ in.}^4$$

   $$I \text{ for interior beams} = 25,570 \text{ in.}^4$$

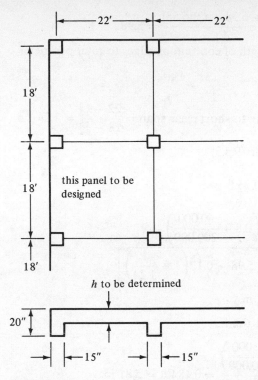

this panel to be designed

h to be determined

Figure 15.16

(d) Calculating α values (where α is the ratio of the stiffness of the beam section to the stiffness of a width of slab bounded laterally by the centerline of the adjacent panel, if any, on each side of the beam):

For edge beam $\left(\text{width} = \frac{1}{2} \times 22 + \dfrac{7.5}{12} = 11.625 \text{ ft}\right)$

$$I_s = (\tfrac{1}{12})(12 \times 11.625)(6)^3 = 2511 \text{ in.}^4$$

$$\alpha = \frac{13{,}467}{2511} = 5.36$$

For 18-ft interior beam (with 22-ft slab width)

$$I_s = (\tfrac{1}{12})(12 \times 22)(6)^3 = 4752 \text{ in.}^4$$

$$\alpha = \frac{25{,}570}{4752} = 5.38$$

For 22-ft interior beam (with 18-ft slab width)

$$I_s = (\tfrac{1}{12})(12 \times 18)(6)^3 = 3888 \text{ in.}^4$$

$$\alpha = \frac{25{,}570}{3888} = 6.58$$

$$\text{Avg. } \alpha = \alpha_m = \frac{5.36 + 5.38 + (2)(6.58)}{4} = 5.98$$

β_s = ratio of length of continuous edges to total perimeter

$$= \frac{62}{80} = 0.78$$

β = ratio of long to short clear span $= \dfrac{22 - \frac{15}{12}}{18 - \frac{15}{12}} = 1.24$

(e) Thickness limits by Code (9.5.3)

$$\ell_n = 22.0 - \frac{15}{12} = 20.75 \text{ ft}$$

$$h = \frac{20.75\left(0.8 + \dfrac{60,000}{200,000}\right)}{36 + (5)(1.24)\left[5.98 - 0.12\left(1 + \dfrac{1}{1.24}\right)\right]}$$

$$= 0.318 \text{ ft} = 3.82 \text{ in.}$$

but h should not be less than

$$h = \frac{20.75\left(0.8 + \dfrac{60,000}{200,000}\right)}{36 + (9)(1.24)} = 0.484 \text{ ft} = 5.81 \text{ in.}$$

and it need not be greater than

$$h = \frac{20.75\left(0.8 + \dfrac{60,000}{200,000}\right)}{36} = 0.634 \text{ ft} = 7.61 \text{ in.}$$

Also, h may not be less than $3\frac{1}{2}''$.

Use $h = 6''$ (will check shear later)

3. Moments for the short-span direction centered on interior column line:

$$w_u = (1.4)(100) + (1.7)(120) = 344 \text{ psf}$$

$$M_o = \frac{(w\ell_2)(\ell_n)^2}{8} = \frac{(0.344)(22)(16.75)^2}{8} = 265 \text{ ft-k}$$

(a) Dividing this static design moment into negative and positive portions, as per Section 13.6.3.2 of Code:

negative design moment $= (0.65)(265) = -172$ ft-k

positive design moment $= (0.35)(265) = +93$ ft-k

(b) Allotting these interior moments to beam and column strips, as per Section 13.6.4 of Code:

$$\frac{\ell_2}{\ell_1} = \frac{22}{18} = 1.22$$

$$\alpha_1 = \alpha \text{ in direction of short span} = 5.38$$

$$\alpha_1 \frac{\ell_2}{\ell_1} = (5.38)(1.22) = 6.56$$

The portion of the interior negative moment to be resisted by the column strip, as per Table 15.3 of this chapter, by interpolation is $(0.68)(-172) = 117$ ft-k. This -117 ft-k is allotted 85% to the beam (Code 13.6.5), or -99 ft-k, and 15% to the slab, or -18 ft-k. The remaining negative moment, $172 - 117 = 55$ ft-k, is allotted to the middle strip.

The portion of the interior positive moment to be resisted by the column strip, as per Table 15.5 of this chapter, by interpolation is $(0.68)(+93) = +63$ ft-k. This 63 ft-k is allotted 85% to the beam (Code 13.6.5), or $+54$ ft-k, and 15% to the slab, or $+9$ ft-k. The remaining positive moment, $93 - 63 = 30$ ft-k, goes to the middle strip.

4. Moments for the short-span direction centered on the edge beam:

$$M_o = \frac{(w\ell_2)(\ell_n)^2}{8} = \frac{(0.344)(11.62)(16.75)^2}{8} = 140 \text{ ft-k}$$

(a) Dividing this static design moment into negative and positive portions, as per Section 13.6.3.2 of the Code:

Negative design moment $= (0.65)(140) = -91$ ft-k

Positive design moment $= (0.35)(140) = +49$ ft-k

(b) Allotting these exterior moments to beam and column strips, as per Section 13.6.4 of the Code:

$$\frac{\ell_2}{\ell_1} = \frac{22}{18} = 1.22$$

$$\alpha_1 = \alpha \text{ for edge beam} = 5.36$$

$$\alpha_1 \frac{\ell_2}{\ell_1} = (5.36)(1.22) = 6.54$$

5. The portion of the exterior negative moment going to the column strip, from Table 15.4 of this chapter, by interpolation is $(0.68)(-91) = -62$ ft-k. This -62 ft-k is allotted 85% to the beam (Code 13.6.5), or -53 ft-k, and 15% to the slab, or -9 ft-k. The remaining negative moment, $91 - 62 = -29$ ft-k, is allotted to the middle strip.

The portion of the exterior positive moment to the resisted by the column strip, as per Table 15.5, is $(0.68)(+49) = +33$ ft-k. This 33 ft-k is allotted 85% to the beam, or $+28$ ft-k, and 15% to the slab, or $+5$ ft-k. The remaining positive moment, $49 - 33 = +16$ ft-k, goes to the middle strip.

A summary of the short-span moments is presented in Table 15.8.

TABLE 15.8   SHORT-SPAN MOMENTS (ft-k)

	Column strip moments		Middle strip slab moments
	Beam	Slab	
Interior slab-beam strip			
Negative	-99	-18	-55
Positive	$+54$	$+9$	$+30$
Exterior slab-beam strip			
Negative	-53	-9	-29
Positive	$+28$	$+5$	$+16$

6. Moments for the long-span direction:

$$M_o = \frac{(w\ell_2)(\ell_n)^2}{8} = \frac{(0.344)(18)(20.75)^2}{8} = 333.3 \text{ ft-k}$$

(a) From Table 15.2 of this Chapter (Code 13.6.3.3) for an end span with beams between interior supports but with no edge beams:

Interior negative factored moment

$$= 0.70M_o = -(0.70)(333.3) = -233 \text{ ft-k}$$

Positive factored moment

$$= 0.52M_o = +(0.52)(333.3) = +173 \text{ ft-k}$$

Exterior negative factored moment

$$= 0.26M_o = -(0.26)(333.3) = -87 \text{ ft-k}$$

These factored moments may be modified by 10% according to Section 13.6.7 of the Code, but this reduction is neglected here.

(b) Allotting these moments to beam and column strips:

$$\frac{\ell_2}{\ell_1} = \frac{18}{22} = 0.818$$

$$\alpha_1 = \alpha_1 \text{ for the 22' beam} = 6.58$$

$$\alpha_1 \frac{\ell_2}{\ell_1} = (6.58)(0.818) = 4.21$$

Next an expression is given for β_t. It is the ratio of the torsional stiffness of an edge beam section to the flexural stiffness of a width of slab equal to the span length of the beam measured center to center of supports.

$$\beta_t = \frac{E_{cb}C}{2E_{cs}I_s}$$

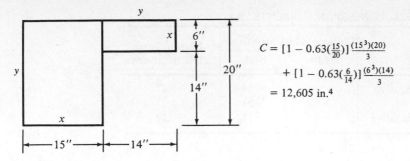

$$C = [1 - 0.63(\tfrac{15}{20})]\tfrac{(15^3)(20)}{3}$$
$$+ [1 - 0.63(\tfrac{6}{14})]\tfrac{(6^3)(14)}{3}$$
$$= 12{,}605 \text{ in.}^4$$

Figure 15.17 Evaluation of C.

Involved in the equation is a term C which is a property of the cross-sectional area of the torsion arm estimating the resistance to twist.

$$C = \Sigma\left(1 - 0.63\frac{x}{y}\right)\frac{x^3 y}{3}$$

where x is the length of the short side of each rectangle and y is the length of the long side of each rectangle. The exterior beam considered here is described in ACI Section 13.2.4 and is shown in Figure 15.17 together with the calculation of C.

$$\beta_t = \frac{(E_c)(12{,}605)}{(2)(E_c)(3888)} = 1.62 \qquad \blacksquare$$

The portion of the interior negative design moment allotted to the column strip, from Table 15.2, by interpolation is $(0.80)(-233) = -186$ ft-k. This -186 ft-k is allotted 85% to the beam (Code 13.6.5), or -158 ft-k, and 15% to the slab, or -28 ft-k. The remaining negative moment, $-233 + 186 = -47$ ft-k, is allotted to the middle strip.

The portion of the positive design moment to be resisted by the column strip, as per Table 15.5 is $(0.80)(173) = +138$ ft-k. This 138 ft-k is allotted 85% to the beam, or $(0.85)(138) = 117$ ft-k, and 15% to the slab, or $+21$ ft-k. The remaining positive moment, $173 - 138 = 35$ ft-k, goes to the middle strip.

The portion of the exterior negative moment allotted to the column strip, from Table 15.4 is $(0.86)(-87) = -75$ ft-k. This -75 ft-k is allotted 85% to the beam, or -64 ft-k, and 15% to the slab, or -11 ft-k. The remaining negative moment, -12 ft-k, is alloted to the middle strip.

A summary of the long-span moments is presented in Table 15.9.

6. Check shear strength in the slab at a distance d from the face of the beam. Shear is assumed to be produced by the load on the tributary area shown in Figure 15.18, working with a 12-in.-wide strip as shown.

$$V_u = (0.344)\left(9 - \frac{7.5}{12} - \frac{5}{12}\right) = 2.738 \text{ k} = 2738 \text{ lb}$$

TABLE 15.9   LONG-SPAN MOMENTS (ft-k)

| | Column strip moments | | Middle strip slab moments |
	Beam	Slab	
Interior negative	−158	−28	−47
Positive	+17	+21	+35
Exterior negative	−64	−11	−12

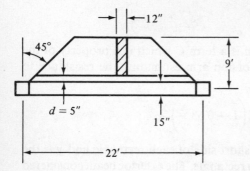

Figure 15.18

$$\phi V_c = (0.85)(2\sqrt{3000})(12)(5)$$
$$= 5587 \text{ lb} > 2738 \text{ lb} \qquad\qquad \underline{\text{OK}}$$

15.12   TRANSFER OF MOMENTS AND SHEARS BETWEEN SLABS AND COLUMNS

On many occasions the maximum load that a two-way slab can support is dependent upon the strength of the joint between the column and the slab. Not only is the load transferred by shear from the slab to the column along an area around the column, but also there may be moments that have to be transferred as well. The moment situation is usually most critical at the exterior columns.

If there are moments to be transferred, they will cause shear stresses of their own in the slabs as will be described in this section. Furthermore, shear forces resulting from moment transfer must be considered in the design of the lateral column reinforcement (that is, ties and spirals) as stated in Section 11.11.1 of the Code.

When columns are supporting slabs without beams (that is, flat plates or flat slabs), the load transfer situation between the slabs and columns is extremely *critical*. Perhaps if we don't have the flexural reinforcing designed just right throughout the slab as to quantities and positions, we can get by with it; however, if we handle the shear strength situation incorrectly, the results may very well be disastrous.

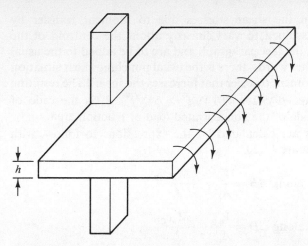

Figure 15.19

The serious nature of this problem is shown in Figure 15.19, where it can be seen that if there is no spandrel beam, all of the total exterior slab moment has to be transferred to the column. The transfer is made by both flexure and eccentric shear, the latter being located at a distance of about $d/2$ from the column face.

Section 13.6.3.6 of the Code states that for moment transfer between the slab and edge column the full column strip nominal moment strength provided, that is $M_n = M_u/\phi = A_s f_y(d - a/2)$, must be used as the transfer moment for gravity load.

In Section 11.12.6, a fraction of the unbalanced moment equal to γ_v shall be considered to be transferred by eccentricity of shear about the centroid of a critical section perpendicular to the plane of the slab and located so that its perimeter is a minimum, but it doesn't have to be closer than $d/2$ to the perimeter of the column; that is,

$$\gamma_v = 1 - \frac{1}{1 + \dfrac{2}{3}\sqrt{\dfrac{c_1 + d}{c_2 + d}}} = 1 - \frac{1}{1 + \dfrac{2}{3}\sqrt{\dfrac{b_1}{b}}} \qquad \text{ACI Equation 11-42}$$

where c_1 is the dimension of the column measured in the direction of the span for which moments are being determined and c_2 is the other column dimension.

The fraction of the unbalanced moment that is to be transferred by flexure over an effective slab width between lines that are one and one-half times the slab or drop panel thickness $(1.5h)$ outside opposite faces of the column or capital is given by the following expression (Code 13.3.3.2):

$$\gamma_f = \frac{1}{1 + \dfrac{2}{3}\sqrt{\dfrac{c_1 + d}{c_2 + d}}} = \frac{1}{1 + \dfrac{2}{3}\sqrt{\dfrac{b_1}{b_2}}} \qquad \text{ACI Equation 13-1}$$

From this information the shear stresses due to moment transfer by eccentricity of shear are assumed to vary linearly about the centroid of the critical section described in the last paragraph and are to be added to the usual factored shear forces. (In other words, there is the usual punching shear situation plus a twisting due to the moment transfer that increases the shear.) The resulting shear stresses may not exceed $\phi(2 + 4/\beta_c)\sqrt{f'_c} > \phi4\sqrt{f'_c}$. β_c is the ratio of the long side to the short side of the concentrated load or reaction area.

The combined stresses are calculated by the expressions to follow, with reference being made to Figure 15.20.

$$v_u \text{ along } AB = \frac{V_u}{A_c} + \frac{\gamma_v M_u c_{AB}}{J_c}$$

$$v_u \text{ along } CD = \frac{V_u}{A_c} - \frac{\gamma_v M_u c_{CD}}{J_c}$$

In these expressions, A_c is the area of the concrete along the assumed critical section. For instance, for the interior column of Figure 15.20(a) it would be equal to $(2a + 2b)d$, and for the edge column of Figure 15.20(b) it would equal $(2a + b)d$.

J_c is a property analogous to the polar moment of inertia about the z–z axis of the shear areas located around the periphery of the critical section. First

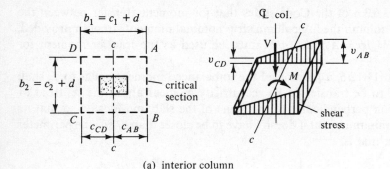

(a) interior column

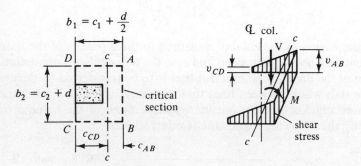

(b) edge column

Figure 15.20 Assumed distribution of shear stress.

the centroid of the shear area A_c is located by taking moments. The centroid is shown with the distances c_{AB} and c_{CD} in both parts (a) and (b) of Figure 15.20. Then the value of J_c is computed for the shear areas. For the interior column of part (c) it is

$$J_c = d\left[\frac{a^3}{6} + \frac{ba^2}{2}\right] + \frac{ad^3}{6}$$

and for the edge column of part (b) it is

$$J_c = d\left[\frac{2a^3}{3} - (2a + b)(c_{AB})^2\right] + \frac{ad^3}{6}$$

Example 15.5 shows the calculations involved for shear and moment transfer for an exterior column.

The Commentary (11.12.6.1) states that the critical section described for two-way action for slabs with a critical section located at $d/2$ from the perimeter of the column is appropriate for the calculation of shear stresses caused by moment transfer even when shearheads are used. Thus the critical sections for direct shear and shear due to moment transfer will be different from each other.

The total reinforcing that will be provided in the column strip must include additional reinforcing concentrated over the column to resist the part of the bending moment transferred by flexure $= \gamma_f M_u = \gamma_f(0.26M_o)$.

■ **EXAMPLE 15.5**
For the flat slab of Figure 15.21, compute the negative steel required in the column strip for the exterior edge indicated. Also check the slab for moment and shear transfer at the exterior column; $f'_c = 3000$ psi, $f_y = 60,000$ psi, and $LL = 100$ psf. An 8-in. slab has already been selected with $d = 6.75$ in.

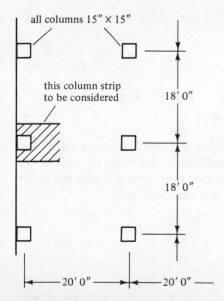

Figure 15.21

SOLUTION

1. Compute w_u and $M_{o\ell}$:

$$w_d = (\tfrac{8}{12})(150) = 100 \text{ psf}$$

$$w_\ell = 100 \text{ psf}$$

$$w_u = (1.4)(100) + (1.7)(100) = 310 \text{ psf}$$

$$M_{o\ell} = \frac{w_u \ell_2 \ell_n^2}{8} = \frac{(0.310)(18)(18.75)^2}{8} = 245.21 \text{ ft-k}$$

2. Determine the exterior negative moment as per Section 13.6.3.3:

$$-0.26M_{o\ell} = -(0.26)(245.21) = 63.75 \text{ ft-k}$$

Width of column strip $= (0.25)(18) = 4'6'' = 54''$

3. Section 13.6.4.2 shows that 100% of the exterior negative moment is to be resisted by the column strip.

4. Design of steel in column strip:

$$\frac{M_u}{\phi b d^2} = \frac{(12)(63,750)}{(0.9)(54)(6.75)^2} = 345$$

$$\rho = 0.0062$$

$$A_s = (0.0062)(54)(6.75) = 2.26 \text{ in.}^2 \quad \underline{\text{Use } \# 6 @ 10''}$$

5. Moment transfer design:
   (a) The Code (13.3.3.2) states that additional bars must be added over the column in a width = column diameter + (2)(1.5h) = 15 + (2)(1.5 × 8) = 39''.
   (b) The additional reinforcing needed over the columns is to be designed for a moment $= \gamma_f M_u$

$$\gamma_f = \cfrac{1}{1 + \cfrac{2}{3}\sqrt{\cfrac{c_1 + d}{c_2 + d}}} = \cfrac{1}{1 + \cfrac{2}{3}\sqrt{\cfrac{15 + 6.75}{15 + 6.75}}} = 0.60$$

$$\gamma_f M_u = (0.60)(63.75) = 38 \text{ ft-k}$$

   (c) Add two #6 bars in the 39'' width and check to see whether the moment transfer situation is satisfactory. To resist the 38 ft-k we now have the two #6 bars just added plus three #6 bars put in for the column strip design, or five #6 bars (2.21 in.^{2}).

$$a = \frac{A_s f_y}{0.85 f'_c b} = \frac{(2.21)(60)}{(0.85)(3)(39)} = 1.333''$$

$$\phi M_n = M_u = \phi A_s f_y \left(d - \frac{a}{2} \right)$$

$$= \frac{(0.9)(2.21)(60)\left(6.75 - \dfrac{1.333}{2}\right)}{12}$$

$$= 60.5 \text{ ft-k} > 38 \text{ ft-k} \qquad \text{OK}$$

6. Compute combined shear stress at exterior column due to shear and moment transfer:

   (a) Nominal moment strength of full column strip with eight #6 bars (3.53 in.²).

   $$a = \frac{(3.53)(60)}{(0.85)(3)(54)} = 1.538''$$

   $$M_n = \frac{(3.53)(60)\left(6.75 - \dfrac{1.538}{2}\right)}{12} = 105.56 \text{ ft-k}$$

   (b) Fraction of unbalanced moment carried by eccentricity of shear = $\gamma_v M_n$.

   $$\gamma_v = 1 - \frac{1}{1 + \dfrac{2}{3}\sqrt{\dfrac{15 + 6.75}{15 + 6.75}}} = 0.40$$

   $$\gamma_v M_n = (0.40)(105.60) = 42.22 \text{ ft-k}$$

   (c) Compute properties of critical section for shear (Figure 15.22).

   $$A_c = (2a + b)d = (2 \times 18.375 + 21.75)(6.75) = 394.875 \text{ in.}^2$$

   $$c_{AB} = \frac{(2)(18.375)(6.75)\left(\dfrac{18.375}{2}\right)}{394.875} = 5.77 \text{ in.}$$

   $$J_c = d\left[\frac{2a^3}{3} - (2a + b)(c_{AB})^2\right] + \frac{ad^3}{6}$$

   $$= 6.75\left[\frac{(2)(18.375)^3}{3} - (2 \times 18.375 + 21.75)(5.77)^2\right]$$

   $$+ \frac{(18.375)(6.75)^3}{6}$$

   $$= 15{,}714 \text{ in.}^4$$

   (d) Compute gravity load shear to be transferred at the exterior column.

   $$V_u = \frac{w_u \ell_1 \ell_2}{2} = \frac{(0.310)(18)(20)}{2} = 55.8 \text{ k}$$

$$a = 15 + \frac{6.75}{2} = 18.375''$$

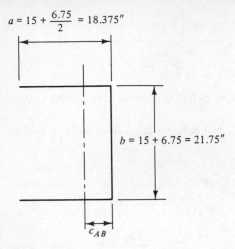

$b = 15 + 6.75 = 21.75''$

c_{AB}

Figure 15.22

(e) Combined stresses.

$$v_u = \frac{V_u}{A_c} + \frac{\gamma_v M_n c_{AB}}{J_c}$$

$$= \frac{55,800}{394.875} + \frac{(12 \times 42,220)(5.77)}{15,714}$$

$$= 141.3 + 186 = 327.3 \text{ psi}$$

$$> 4\sqrt{3000} = 219 \text{ psi} \qquad\qquad \text{N.G.} \quad\blacksquare$$

∴ It is necessary to do one or more of the following: increase depth of slab, use higher-strength concrete, use drop panel, install shear-head reinforcing.

Factored Moments in Columns and Walls

It is obvious that if there is an unbalanced loading of two adjoining spans, the result will be an additional moment at the connection of walls and columns to slabs. The Code (13.6.9.2) provides the approximate equation listed at the end of this paragraph to consider the effects of such situations. This particular equation was derived for two adjoining spans, one longer than the other. It was assumed that the longer span was loaded with dead load plus one-half live load and that only dead load was applied to the shorter span.

$$M = 0.07[(w_d + 0.5w_\ell)\ell_2\ell_n^2 - w_d'\ell_2'(\ell_n')^2] \qquad \text{(ACI Equation 13.4)}$$

In this expression w_d', ℓ_2', and ℓ_n' are for the shorter span. It will be noted that w_ℓ is a factored value. The resulting approximate value should be used for unbalanced moment transfer by gravity loading at interior columns unless a more theoretical analysis is used.

15.13 OPENINGS IN SLAB SYSTEMS

According to the Code (13.5), openings can be used in slab systems if adequte strength is provided and if all serviceability conditions of the ACI including deflections are met.

1. If openings are located in the area common to intersecting middle strips, it will be necessary to provide the same total amount of reinforcing in the slab that would have been there without the opening.
2. For openings in intersecting column strips, the width of the openings may not be more than one-eighth the width of the column strip in either span. An amount of reinforcing equal to that interrupted by the opening must be placed on the sides of the opening.
3. Openings in an area common to one column strip and one middle strip may not interrupt more than one-fourth of the reinforcing in either strip. An amount of reinforcing equal to that interrupted shall be placed around the sides of the opening.
4. The shear requirements of Section 11.1.2.5 of the Code must be met.

PROBLEMS

15.1. Using the ACI Code, determine the minimum thickness required for panels ① and ③ of the flat plate floor shown in the accompanying illustration. Edge beams are not used along the exterior floor edges $f'_c = 3000$ psi and $f_y = 60,000$ psi. (*Ans.* 9.14″ and 8.18″)

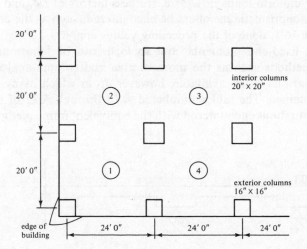

15.2. Assume that the floor system of Problem 15.1 is to support a service live load of 80 psf and a service dead load of 20 psf in addition to its own weight. If the columns are 10 ft long, design an interior flat plate for panel ③.

15.3. Design a flat plate for panel ① for the structure loads and material considered in Problems 15.1 and 15.2.

16

Two-Way Slabs, Equivalent Frame Method

16.1 MOMENT DISTRIBUTION FOR NONPRISMATIC MEMBERS

Most of the moment distribution problems which the student has previously faced have dealt with prismatic members for which carryover factors of $\frac{1}{2}$, fixed-end moments for uniform loads of $w_u \ell^2/8$, stiffness factors of I/ℓ, and so on were used. Should nonprismatic members be encountered, such as the continuous beam of Figure 16.1, none of the preceding values apply.

Carryover factors, fixed-end moments, and so forth can be laboriously obtained by various methods such as the moment-area and column-analogy methods.[1] There are various tables available, however, from which many of these values can be obtained. The tables numbered A.18 through A.22 of the Appendix cover most situations encountered with the equivalent frame method.

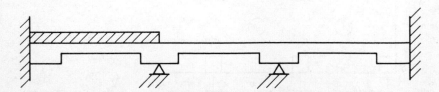

Figure 16.1

---

[1] McCormac, J. C., 1984, *Structural Analysis*, 4th ed. (New York: Harper & Row), pp. 333–334, 567–582.

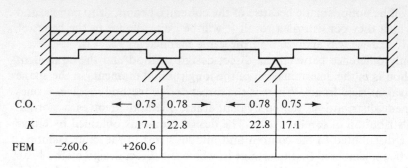

Figure 16.2

Before the equivalent frame method is discussed, an assumed set of fixed-end moments are balanced in Example 16.1 for the nonprismatic beam of Figure 16.1.

■ EXAMPLE 16.1

The carryover factors (C.O.), stiffness factors (K), and fixed-end moments (FEM) shown in Figure 16.2 have been calculated for the continuous nonprismatic member of Figure 16.1. Balance these moments by moment distribution.

SOLUTION

		0.43	0.57		0.57	0.43	
	−260.6	+260.6					
	− 84.1	−112.1	−148.5	−115.8			
			+ 51.5	+ 66.0	+49.8		+ 37.3
	− 16.6	− 22.1	− 29.4	− 22.9			
			+ 10.2	+ 13.1	+ 9.8		+ 7.4
	− 3.3	− 4.4	− 5.8	− 4.5			
			+ 2.0	+ 2.6	+ 1.9		+ 1.4
	− 0.7	− 0.9	− 1.1	− 0.9			
final				+ 0.5	+ 0.4		+ 0.3
moments	− 365.3	+121.1	−121.1	− 61.9	+61.9		+ 46.4

16.2 INTRODUCTION TO THE EQUIVALENT FRAME METHOD

With the preceding example the author hoped to provide the reader with a general idea of the kinds of calculations he or she will face with the equivalent frame method. A part of a two-way slab building will be taken out by itself and analyzed by moment distribution. The slab-beam members of this part of the structure will be nonprismatic because of the columns, beams, drop panels, and so on of which they consist. As a result it will be quite similar to the beam of Figure 16.1, and we will analyze it in the same manner.

The only difference between the direct design method and the equivalent frame method is in the determination of the longitudinal moments in the spans of the equivalent rigid frame. Whereas the direct design method involves a one-cycle moment distribution, the equivalent frame method involves a normal moment distribution of several cycles. The design moments obtained by either method are distributed to the column and middle strips in the same fashion.

It will be remembered that the range in which the direct design method can be applied is limited by a maximum 3-to-1 ratio of live-to-dead load and a maximum ratio of the longitudinal span length to the transverse span length of 2 to 1. In addition, the columns may not be offset by more than 10% of the span length in the direction of the offset from either axis between centerlines of successive columns. There are no such limitations on the equivalent frame method. This is a very important matter because so many floor systems do not meet the limitations specified for the direct design method.

Analysis by either method will yield almost the same moments for those slabs that meet the limitations required for application of the direct design method. For such cases it is then simpler to use the direct design method.

The equivalent frame method merely involves the elastic analysis of a structural frame consisting of a row of equivalent columns and horizontal slab members that are each one panel long and have a transverse width equal to the distance between centerlines of the panels on each side of the columns in question. For instance, the hatched strip of the floor system of Figure 16.3 can be taken out and analyzed. This hypothetical strip is called a *slab beam*.

For vertical loads each floor, together with the columns above and below, is analyzed separately. For such an analysis the far ends of the columns are considered to be fixed. Figure 16.4 shows a typical equivalent slab beam as described in this chapter.

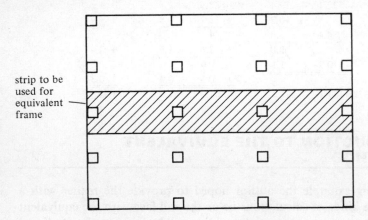

strip to be used for equivalent frame

Figure 16.3 Slab beam.

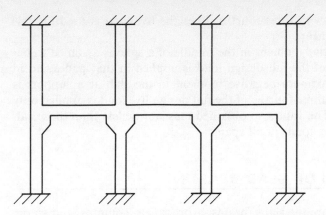

Figure 16.4 Equivalent frame for the crosshatched strip of Figure 16.3 for vertical loading.

Should there be a large number of panels, the moment at a particular joint in a slab beam can be satisfactorily obtained by assuming that the member is fixed two panels away. This simplification is permissible because vertical loads in one panel only appreciably affect the forces in that panel and in the one adjacent to it.

For lateral loads it is necessary to consider an equivalent frame that extends for the entire height of the building because the forces in a particular member are affected by the lateral forces on all the stories above the floor being considered. When lateral loads are considered, it will be necessary to analyze the frame for them and combine the results with analyses made for gravity loads.

The equivalent frame is made up of the horizontal slab, any beams spanning in the direction of the frame being considered, the columns or other members that provide vertical support above and below the slab, and any parts of the structure that provide moment transfer between the horizontal and vertical members. You can see there will be quite a difference in moment transfer from the case where a column provides this transfer and where there is a monolithic reinforced concrete wall extending over the full length of the frame. For cases in between, the stiffnesses of the torsional members such as edge beams will be estimated.

The same minimum ACI slab thicknesses must be met as in the direct design method. The depths should be checked for shear at columns and other supports as specified in Section 11.12 of the Code. Once the moments have been computed, it will also be necessary to check for moment shear transfer at the supports.

The analysis of the frame is made for the full design live load applied to all spans unless the actual design live load exceeds 0.75 times the actual dead load (ACI Code 13.7.6). When the live load is greater than 0.75 times the dead load, a pattern loading with three-fourths times the live load is used for calculating moments and shears.

The maximum positive moment in the middle of a span is assumed to occur when three-fourths of the full design load is applied in that panel and in alternate spans. The maximum negative moment in the slab at a support is assumed to occur when three-fourths of the full design live load is applied only to the adjacent panels. The values so obtained may not be less than those calculated assuming full live loads in all spans.

16.3 PROPERTIES OF SLAB BEAMS

The parts of the frame are the slabs, beams, drop panels, columns, and so on. Our first objective is to compute the properties of the slab beams and the columns (that is, the stiffness factors, distribution factors, carryover factors, and fixed-end moments). To simplify this work the properties of the members of the frame are based on their gross moments of inertia rather than their transformed or cracked sections. If the cracked sections were used, the computations would be much longer and the final results would probably be little better. Despite the use of the gross dimensions of members, the calculations involved in determining the properties of nonprismatic members is still a lengthy task and we will find the use of available tables very helpful.

Figures 16.5 and 16.6 present sketches of two-way slab structures together with the equivalent frames that will be used for their analysis. A flat slab with columns is shown in Figure 16.5(a). Cross sections of the structure are shown through the slab in part (b) of the figure and through the column in part (c). Then in part (d) the equivalent frame that will be used for the actual numerical calculations is shown. In this figure, E_{cs} is the modulus of elasticity of the concrete slab. With section 2-2 a fictitious section is shown which will have a stiffness approximately equivalent to that of the actual slab and column. An expression for I for the equivalent section is also given. In this expression, c_2 is the width of the column in a direction perpendicular to the direction of the span and ℓ_2 is the width of the slab beam. The gross moment of inertia at the face of the support is calculated and is divided by $(1 - c_2/\ell_2)^2$. This approximates the effect of the large increase in depth provided by the column for the distance in which the slab and column are in contact.

In Figure 16.6, similar sketches and I values are shown for a slab with drop panels. Figure 13.7.3 of the 1983 ACI Commentary shows such information for slab systems with beams and for slab systems with column capitals.

With the equivalent slab beam stiffness diagram it is possible using the conjugate beam method, column analogy, or some other method to compute stiffness factors, distribution factors, carryover factors, and fixed-end moments for use in moment distribution. Tables A.18 through A.22 of the Appendix of this text provide tabulated values of these properties for various slab systems. The numerical examples of this chapter make use of this information.

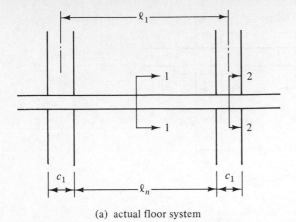

(a) actual floor system

$$I_1 = (\tfrac{1}{12})(\ell_2)(h)^3$$

(b) section 1-1

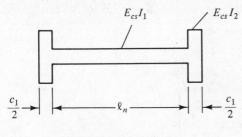

$$I_2 = \frac{I_1}{(1 - \dfrac{c_2}{\ell_2})^2}$$

(c) equivalent section 2-2

$E_{cs}I_1$ $E_{cs}I_2$

(d) equivalent slab beam
     stiffness diagram

Figure 16.5 Slab system without beams.

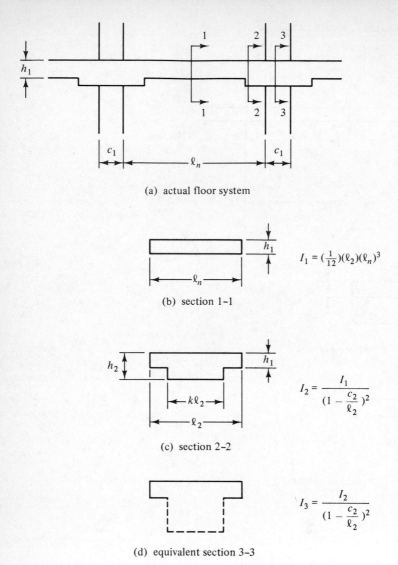

(a) actual floor system

$$I_1 = (\tfrac{1}{12})(\ell_2)(\ell_n)^3$$

(b) section 1–1

$$I_2 = \dfrac{I_1}{(1 - \dfrac{c_2}{\ell_2})^2}$$

(c) section 2–2

$$I_3 = \dfrac{I_2}{(1 - \dfrac{c_2}{\ell_2})^2}$$

(d) equivalent section 3–3

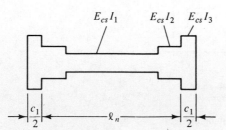

(e) equivalent slab beam stiffness diagram

Figure 16.6 Slab systems with drop panels.

16.4 PROPERTIES OF COLUMNS

The length of a column is assumed to run from the middepth of the slab on one floor to the middepth of the slab on the next floor. For stiffness calculations the moments of inertia of columns is based on their gross dimensions. Thus if capitals are present, the effect of their dimensions must be used for those parts of the columns. Columns are assumed to be infinitely stiff for the depth of the slabs.

Figure 16.7 shows a sample column, together with its column stiffness diagram. Similar diagrams are shown for other columns (as where there are drop panels, capitals, etc.) in Figure 13.7.4 of the 1983 ACI Commentary.

With a column stiffness diagram the column flexural stiffness K_c can be determined by the conjugate beam procedure or other methods. Tabulated values of K_c are given in Table A.22 of the Appendix of this text for typical column situations.

In applying moment distribution to a particular frame we need the stiffnesses of the slab beam, the torsional members, and the equivalent column so that the distribution factors can be calculated. For this purpose the equivalent column, the equivalent slab beam, and the torsional members are needed at a particular joint.

For this discussion, reference is made to Figure 16.8, where it is assumed that there is a column above and below the joint in question. Thus the column stiffness (K_c) here is assumed to include the stiffness of the column above (K_{ct}) and the one below (K_{cb}). Thus $\Sigma K_c = K_{ct} + K_{cb}$. In a similar fashion the total torsional stiffness is assumed to equal that of the torsional members on both sides of the joint $(\Sigma K_t = K_{t1} + K_{t2})$. For an exterior frame the torsional member will be located on one side only.

The stiffness of the torsional member (K_t) is to be computed with the following expression according to the Code (13.7.5.2).

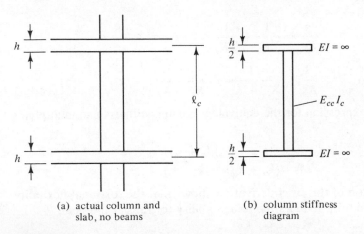

(a) actual column and
slab, no beams

(b) column stiffness
diagram

Figure 16.7

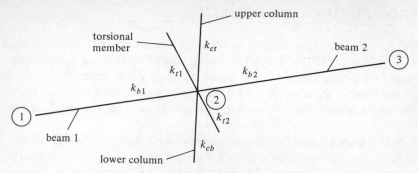

Figure 16.8

$$K_t = \Sigma \frac{9E_{cs}C}{\ell_2 \left(1 - \dfrac{c_2}{\ell_2}\right)^3}$$ (ACI Equation 13-6)

In this expression, C is to be determined with the following expression by dividing the cross section of the torsional member into rectangular parts and summing up the C values for the different parts.

$$C = \Sigma \left(1 - 0.63 \frac{x}{y}\right) \frac{x^3 y}{3}$$ (ACI Equation 13-7)

If there is no beam framing into the column in question, a part of the slab equal to the width of the column or capital shall be used as the effective beam. If a beam frames into the column, a T beam or L beam will be assumed with flanges of widths equal to the projection of the beam above or below the slab but not more than four times the slab thickness.

The flexibility of the equivalent column is equal to the reciprocal of its stiffness, as follows:

$$\frac{1}{K_{ec}} = \frac{1}{\Sigma K_c} + \frac{1}{\Sigma K_t}$$

$$\frac{1}{K_{ec}} = \frac{1}{K_{ct} + K_{cb}} + \frac{1}{K_c + K_t}$$

Solving this expression for the equivalent column stiffness and multiplying through by K_c:

$$K_{ec} = \frac{(K_{ct} + k_{cb})(k_t + k_t)}{(k_{ct} + k_{cb}) + (k_t + k_t)}$$

An examination of this brief derivation shows that the torsional flexibility of the slab column joint reduces the joint's ability to transfer moment.

After the value of K_{ec} is obtained, the distribution factors can be computed as follows, with reference again made to Figure 16.8.

$$\text{DF for beam 2-1} = \frac{K_{b1}}{K_{b1} + K_{b2} + K_{ec}}$$

$$\text{DF for beam 2-3} = \frac{K_{b2}}{K_{b1} + K_{b2} + K_{ec}}$$

$$\text{DF for column above} = \frac{K_{ec}/2}{K_{b1} + K_{b2} + K_{ec}}$$

16.5 EXAMPLE PROBLEMS

Example 16.2 illustrates the determination of the moments in a flat-plate structure by the equivalent frame method.

■ EXAMPLE 16.2

Using the equivalent frame method, determine the design moments for the hatched strip of the flat-plate structure shown in Figure 16.9 if $f'_c = 3000$ psi, $f_y = 60,000$ psi, $LL = 70$ psf, and $DL = 112$ psf (including slab weight). Column lengths = $9'6''$.

SOLUTION

1. Determine the depth required for ACI depth limitations (9.5.3.1). Assume that this has been done and that a preliminary slab $h = 8''$ ($d = 6.75''$) has been selected.

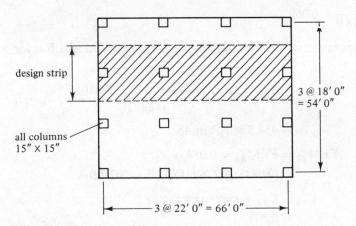

Figure 16.9

2. Check beam shear for exterior column:

$$w_u = (1.4)(112) + (1.7)(70) = 276 \text{ psf}$$

$$V_u \text{ for } 12'' \text{ width} = (0.276)\left(11.0 - \frac{7.5}{12} - \frac{6.75}{12}\right) = 2.708 \text{ k/ft}$$

$$\phi V_c = \frac{(0.85)(2\sqrt{3000})(12)(6.75)}{1000} = 7.542 \text{ k/ft} > 2.708 \text{ k/ft} \qquad \underline{\text{OK}}$$

3. Check two-way shear around interior columns:

$$V_u = \left[(18)(22) - \left(\frac{15 + 6.75}{12}\right)^2\right](0.276) = 108.39 \text{ k}$$

$$\phi V_c = \frac{(0.85)(4\sqrt{3000})(4)(15 + 6.75)(6.75)}{1000} = 109.36 \text{ k} > 108.39 \text{ k}$$

<u>OK</u> (close)

4. Using tables in the Appendix, determine stiffness factor and fixed-end moments for the 22-foot spans:

$$I_s = \frac{\ell_2 h^3}{12} = \frac{(12 \times 18)(8)^3}{12} = 9216 \text{ in.}^4$$

$E_{cs} = 3.16 \times 10^6$ psi from Table A.1 of Appendix

With reference to Table A.18.

$$C_{1A} = C_{2A} = C_{1B} = C_{2B} = 15'' = 1.25'$$

$$\frac{C_{1A}}{\ell_1} = \frac{1.25}{22.0} = 0.057$$

$$\frac{C_{1B}}{\ell_2} = \frac{1.25}{18.0} = 0.069$$

By interpolation in table (noting that A is for near end and B is for far end),

$$K_{AB} = \frac{4.12 E_{cs} I_s}{\ell_1} = \frac{(4.12)(3.16 \times 10^6)(9216)}{(12)(22)}$$

$$= 454.5 \times 10^6 \text{ in.-lb}$$

$$\text{FEM}_{AB} = \text{FEM}_{BA} = 0.0843 w_u \ell_2 \ell_1^2$$

$$= (0.0843)(0.276)(18)(22)^2 = 202.7 \text{ ft-k}$$

$$C_{AB} = C_{BA} = \text{carryover factor}$$

$$= 0.507$$

5. Determine column stiffness:

$$I_c = (\tfrac{1}{12})(15)(15)^3 = 4219 \text{ in.}^4$$

$$E_{cc} = 3.16 \times 10^6 \text{ psi}$$

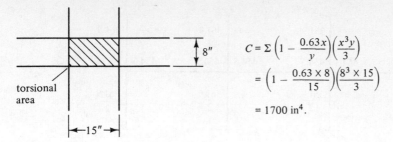

$$C = \Sigma \left(1 - \frac{0.63x}{y}\right)\left(\frac{x^3 y}{3}\right)$$

$$= \left(1 - \frac{0.63 \times 8}{15}\right)\left(\frac{8^3 \times 15}{3}\right)$$

$$= 1700 \text{ in}^4.$$

Figure 16.10

Using Table A.22 of Appendix,

$$\ell_n = 9' \, 6'' = 9.50 \text{ ft}$$

$$\ell_c = 9.50 - \tfrac{8}{12} = 8.833 \text{ ft}$$

$$\frac{\ell_n}{\ell_c} = \frac{8.833}{9.50} = 0.930$$

$$\frac{h_a}{h_b} = \frac{4}{4} = 1.0$$

By interpolation,

$$K_c = \frac{4.81 \, E_{cc}I_c}{H} = \frac{(4.81)(3.16 \times 10^6)(4219)}{114}$$

$$= 562.5 \times 10^6 \text{ in.-lb}$$

$$C_{AB} = 0.55$$

6. Determine torsional stiffness of slab section (see Figure 16.10):

$$C = \Sigma \left(1 - 0.63 \frac{x}{y}\right)\left(\frac{x^3 y}{3}\right)$$

$$= \left(1 - \frac{0.63 \times 8}{15}\right)\left(\frac{8^3 \times 15}{3}\right)$$

$$= 1700 \text{ in.}^4$$

$$K_t = \frac{9 E_{cs} C}{\left[\ell_2 \left(1 - \frac{c_2}{\ell_2}\right)^3\right]} = \frac{(9)(3.16 \times 10^6)(1700)}{\left[(12)(18)\left(1 - \frac{15}{(12)(22)}\right)^3\right]}$$

$$= 266.8 \times 10^6 \text{ psi}$$

7. Compute K_{ec}, the stiffness of the equivalent column:

$$K_{ec} = \frac{\Sigma K_c \Sigma K_t}{\Sigma K_c + \Sigma K_t} = \frac{(2 \times 562.5)(2 \times 266.8)}{2 \times 562.5 + 2 \times 266.8}$$

$$= 361.9 \times 10^6 \text{ in.-lb}$$

A summary of the stiffness values is shown in Figure 16.11.

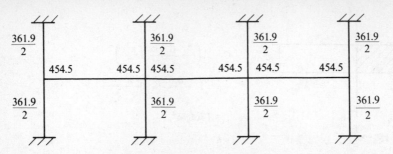

Figure 16.11

8. Computing distribution factors and balancing moments (see Figure 16.12):
   The author does not show moments at tops and bottoms of columns, but this could easily be done by multiplying the balanced column moments at the joints with the slabs by the carryover factor for the columns, which is 0.55.) ∎

A summary of the moment values for Example 16.2 is given in Figure 16.13. The positive moments shown in each span are assumed to equal the simple beam centerline moments plus the average of the end negative moments. This is correct if the end moments in a particular span are equal and is approximately correct if the end moments are unequal. For span 1,

$$+M = \frac{(0.276)(18)(22)^2}{8} - \left(\frac{96.0 + 237.0}{2}\right) = 134.1 \text{ ft-k}$$

The negative moments shown in Figure 16.12 were calculated at the centerlines of the supports. At these supports the cross section of the slab beam is very large due to the presence of the column. At the face of the column, however, the cross section is far smaller, and the Code (13.7.7) specifies that negative reinforcing be designed for the moment there. (If the column is not rectangular, it is replaced with a square column of the same total area and the moment is computed at the face of that fictitious column.)

The design moments shown in Figure 16.14 were determined by drawing the shear diagram and computing the area of that diagram between the centerline of each support and the face of the column.

For interior columns the critical section (for both column and middle strips) is to be taken at the face of the supports, but not at a distance greater than $0.175\ell_1$ from the center of the column. At exterior supports with brackets or capitals, the moment used in the span perpendicular to the edge shall be computed at a distance from the face of the support element not greater than one-half of the projection of the bracket or capital beyond the face of the supporting element.

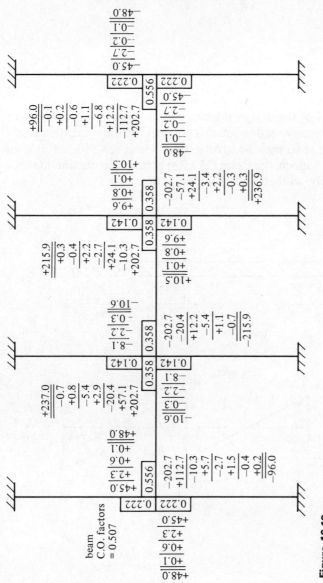

Figure 16.12

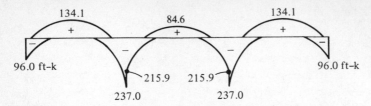

134.1   84.6   134.1

96.0 ft–k   96.0 ft–k

215.9   215.9

237.0   237.0

Figure 16.13

Sometimes the total of the design moments (that is, the positive moment plus the average of the negative end moments) obtained by the equivalent frame method for a particular span may be greater than $M_o = w_u \ell_2 \ell_n^2 / 8$, as used in Chapter 15. Should this happen, the Code (13.7.7.4) permits a reduction in those moments proportionately, so their sum does equal M_o.

$0.276 \times 18 = 4.968$ k/ft

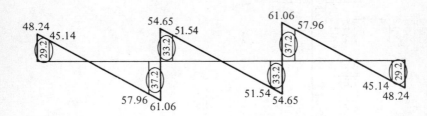

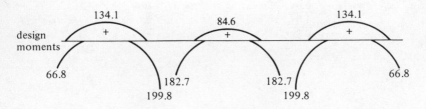

design moments

134.1   84.6   134.1

66.8   66.8

182.7   182.7

199.8   199.8

Figure 16.14

16.6 COMPUTER ANALYSIS

The *equivalent frame method* was developed with the intention that the moment distribution method was to be used for the structural analysis. Truthfully the method is so involved that it is not satisfactory for hand calculations. It is possible, however, to use computers and plane frame analysis programs if the structure is especially modeled. (In other words we must establish various nodal points in the structure so as to account for the changing moments of inertia along the member axes.) There are also some computer programs on the market especially written for these frames. The best known is called "ADOSS" (analysis and design of concrete floor systems) and was prepared by the Portland Cement Association.

PROBLEMS

16.1 Determine the end moments for the beams and columns of the frame shown for which the fixed-end moments, carryover factors, and distribution factors (circled) have been computed. Use the moment distribution method. (*Ans.* 44.9 ft-k and 21.0 ft-k at column bases)

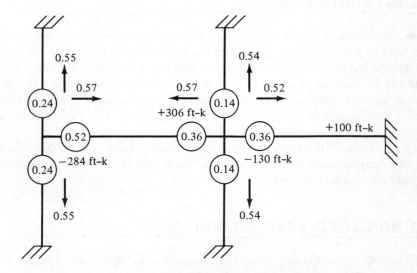

16.2 Using the equivalent frame method, repeat Example 16.1 if the left hand beam span is 40 ft and the right hand one is 30 ft. Assume no changes in distribution and carry over factors. The service live load is 90 psf and the service dead load (including slabweight) is 130 psf.

17

Walls

17.1 INTRODUCTION

Before the advent of frame construction during the nineteenth century, most walls were of a load-bearing type. Since the late 1800s, however, the non-load-bearing wall has become quite common because other members of the structural frame can be used to provide stability. As a result, we today have walls that serve all sorts of purposes, such as retaining walls, basement walls, partition walls, fire walls, and so on. These walls may or may not be of a load-bearing type.

In this chapter the following kinds of concrete walls will be considered: non-load-bearing, load-bearing, and shear walls (the latter being either load-bearing or non-load-bearing).

17.2 NON-LOAD-BEARING WALLS

Non-load-bearing walls are those that support only their own weights and perhaps some lateral loads. Falling into this class are retaining walls, basement walls, and facade-type walls. For non-load-bearing walls the ACI Code provides several specific limitations, and these are listed at the end of this paragraph. The values given for minimum reinforcing quantities and wall thicknesses do not have to be met if lesser values can be proved to be satisfactory by structural analysis (14.2.7). The numbers given in parentheses are ACI section numbers.

1. The thickness of a non-load-bearing wall cannot be less than 4 in. or 1/30 times the least distance between members that provide lateral support (14.6.1).

2. The minimum amount of vertical reinforcement as a percent of gross concrete area is 0.0012 for deformed bars #5 or smaller with $f_y =$ at least 60,000 psi, 0.0015 for other deformed bars, and 0.0012 for smooth or deformed welded wire fabric not larger than W31 or D31—that is, $\frac{5}{8}$ in. in diameter (14.3.2).

3. The vertical reinforcement does not have to be enclosed by ties unless the percent of vertical reinforcing is greater than 0.01 times the gross concrete area or where the vertical reinforcing is not required as compression reinforcing (14.3.6).

4. The minimum amount of horizontal reinforcing as a percent of gross concrete area is 0.0020 for deformed bars #5 or smaller with $f_y \geq$ 60,000 psi, 0.0025 for other deformed bars, and 0.0020 for smooth or deformed welded wire fabric not larger than W31 or D31 (14.3.3).

5. The spacing of vertical and horizontal reinforcement may not exceed three times the wall thickness, or 18 in. (14.3.5).

6. Reinforcing for walls more than 10 in. thick (not including basement walls) must be placed in two layers as follows: one layer containing from one-half to two-thirds of the total reinforcing placed in the exterior surface not less than 2 in. nor more than one-third times the wall thickness from the exterior surface; the other layer placed not less than $\frac{3}{4}$ in. nor more than one-third times the wall thickness from the interior surface (14.3.4).

7. For walls less than 10 in. thick the Code does not specify two layers of steel, but to control shrinkage it is probably a good practice to put one layer on the face of walls exposed to view and one on the nonstressed side of foundation walls 10 ft or more in height.

8. In addition to the reinforcing specified in the preceding paragraphs, at least two #5 bars must be provided around all window and door openings. These bars must be extended for full development lengths beyond the corners of the openings but not less than 24 in. (14.3.7).

9. For cast-in-place walls the area of reinforcing across the interface between a wall and footing must be no less than the minimum vertical wall reinforcing given in paragraph 2 of this list (15.8.2.2).

10. When precast bearing or nonbearing concrete wall panels are used to span horizontally to columns or isolated footings, the panels must conform to the requirements of Section 10.7 of the Code relating to deep flexural members (Code 16.3).

17.3 LOAD-BEARING CONCRETE WALLS— EMPIRICAL DESIGN METHOD

Most of the concrete walls in buildings are load-bearing walls that support not only vertical loads but also some lateral moments. As a result of their considerable in-plane stiffnesses, they are quite important in resisting wind and earthquake forces.

Load-bearing walls with solid rectangular cross sections may be designed as were columns subject to axial load and bending or they may be designed by an empirical method given in Section 14.5 of the Code. The empirical method may only be used if the resultant of all the factored loads falls within the middle third of the wall (that is, the eccentricity must be equal to or less than one-sixth the thickness of the wall). *Whichever of the two methods is used, the design must meet the minimum requirements given in the preceding section of this chapter for non-load-bearing walls.*

This section is devoted to the empirical design method, which is applicable to relatively short vertical walls with approximately concentric loads. The Code (14.5.2) provides an empirical formula for calculating the design axial load strength of solid rectangular cross-section walls with *e* less than one-sixth of wall thicknesses. Should walls have nonrectangular cross sections (such as ribbed wall panels) and/or should *e* be greater than one-sixth wall thicknesses, the regular design procedure for columns subject to axial load and bending (Code 14.4) must be followed.

The practical use of the empirical wall formula, which is given at the end of this paragraph, is for relatively short walls with small moments. When lateral loads are involved, *e* will quickly exceed one-sixth of wall thicknesses. The number 0.55 in the equation is an eccentricity factor that causes the equation to yield a strength approximately equal to that which would be obtained by the axial load and bending procedure of Chapter 10 of the Code if the eccentricity is $h/6$.

$$\phi P_{aw} = 0.55 \phi f'_c A_g \left[1 - \left(\frac{k\ell_c}{32h} \right)^2 \right] \quad \text{(ACI Equation 14-1)}$$

where $\phi = 0.70$
  A_g = gross area of the wall section (in.^{2})
  ℓ_c = vertical distance between supports (in.)
  h = overall thickness of member (in.)
  k = effective length factor determined in accordance with the values given in Table 17.1.

Other ACI requirements for load-bearing concrete walls designed by the empirical formula follow.

1. The thickness of the walls may not be less than 1/25 the supported height or length, whichever is smaller, or less than 4 in. (14.5.3.1).

TABLE 17.1 EFFECTIVE LENGTH FACTORS FOR
       LOAD-BEARING WALLS (14.5.2)

1. Walls braced top and bottom against lateral translation and	
(a) Restrained against rotation at one or both ends	
(top and/or bottom)	0.80
(b) Not restrained against rotation at either end	1.0
2. For walls not braced against lateral translation	2.0

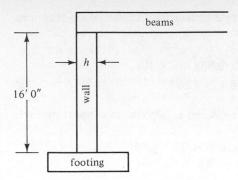

Figure 17.1

2. The thickness of exterior basement walls and foundation walls may not be less than $7\frac{1}{2}$ in. (14.5.3.2).
3. The horizontal length of a wall that can be considered as being effective for each concentrated load may not exceed the smaller of the center-to-center distance between loads or the bearing width plus four times the wall thickness. This provision may be waived if a larger value can be proved satisfactory by a detailed analysis (14.2.4).
4. Load-bearing walls should be anchored to intersecting elements such as floors or roofs, or they should be anchored to columns, pilasters, footings, buttresses, and walls. (14.2.6)

The empirical method is quite easy to apply because only one calculation has to be made to determine the design axial strength of a wall. Example 17.1, which follows, illustrates the design of a bearing wall with a small moment.[1]

■ EXAMPLE 17.1
Design a concrete bearing wall using the ACI empirical formula to support a set of precast concrete roof beams 7'0" on center as shown in Figure 17.1. The bearing width of each beam is 10". The wall is considered to be laterally restrained top and bottom and is further assumed to be restrained against rotation at the footing; thus $k = 0.8$. Neglect wall weight. Other data: $f'_c = 3000$ psi, $f_y = 60,000$ psi, beam reaction, $D = 30$ k, $L = 18$ k.

SOLUTION

1. Determine minimum wall thickness (14.5.3.1):
   (a) $h = (\frac{1}{25})(12 \times 16) = 7.68'' \leftarrow$ <u>Try 8"</u>
   (b) $h = 4''$
   Compute factored beam reactions:

$$P_u = (1.4)(30) + (1.7)(18) = 72.6 \text{ k}$$

---

[1] For the example problems presented in this chapter the author has followed the general procedures used in *Notes on ACI 318-83 Building Code Requirements for Reinforced Concrete with Design Applications*, 4th ed., 1984, Gerald B. Neville, ed. (Skokie, Ill.: Portland Cement Association), pp. 22-1 through 22-32.

2. Is bearing strength of wall concrete satisfactory under beam reactions (10.15)?

$$\phi(0.85f'_cA_1) = (0.70)(0.85)(3)(8 \times 10)$$
$$= 142.8 \text{ k} > 72.6 \text{ k} \qquad \text{OK}$$

3. Horizontal length of wall to be considered as effective in supporting each concentrated load (14.2.4)
   (a) Center-to-center spacing of beams = $7'0'' = 84''$
   (b) Width of bearing + $4h = 10 + (4)(8) = 42''$ ←
4. Design strength of wall:

$$\phi P_{nw} = 0.55\phi f'_c A_g \left[1 - \left(\frac{k\ell_c}{32h} \right)^2 \right]$$

$$= (0.55)(0.70)(3)(8 \times 42) \left[1 - \left(\frac{0.80 \times 12 \times 16}{32 \times 8} \right)^2 \right]$$

$$= 258.7 \text{ k} > 72.6 \text{ k} \qquad \text{OK}$$

5. Select reinforcing (14.3.2 and 14.3.3)

Maximum spacing = $(3)(8) = 24''$ or $\underline{18''}$

Vertical $A_s = (0.0012)(12)(8) = 0.115 \text{ in.}^2/\text{ft}$ #4 @ 18″

Horizontal $A_s = (0.0020)(12)(8) = 0.192 \text{ in.}^2/\text{ft}$ #4 @ 12″ ∎

17.4 LOAD-BEARING CONCRETE WALLS— RATIONAL DESIGN

Reinforced concrete-bearing walls may be designed as columns by the rational method of Chapter 10 of the Code whether the eccentricity is smaller or larger than $h/6$ (they must be designed as columns if $e > h/6$). Should $k\ell_u/r$ be larger than 100, however, the Code (10.11.4.3) states that a theoretical analysis involving axial loads, moments, deflections, creep, and so on must be used. In any case the minimum vertical and horizontal reinforcing requirements of Section 14.3 of the Code must be met.

The design of walls as columns is difficult unless design aids are available. There are various wall design aids available from the Portland Cement Association, but the designer can prepare his or her own aids such as axial load and bending interaction diagrams. The designs may be complicated by the fact that walls often will be classified as "long columns," with the result that the slenderness requirements of Section 10.11.5 of the Code will have to be met.

Walls with $k\ell_u/r$ greater than 100 (and for which a theoretical analysis is required) are rather common, particularly in tilt-up wall construction. The

Portland Cement Association has available a design aid that is particularly useful for such cases.[2]

17.5 SHEAR WALLS

As previously indicated, the in-plane stiffness of a reinforced concrete wall can be quite large. For this reason such walls are often used in design to provide resistance to wind and earthquake forces.

When earthquake-resistant construction is being considered, it is to be remembered that the relatively stiff parts of a structure will attract much larger forces than will the more flexible parts. A structure with reinforced concrete shear walls is going to be quite stiff and thus will attract large seismic forces. If the shear walls are brittle and fail, the rest of the structure may not be able to take the shock. But if the shear walls are ductile (and they will be if properly reinforced), they will be very effective in resisting seismic forces.

Figure 17.2 shows a shear wall subjected to a lateral force V_u. The wall is in actuality a cantilever beam of width h and overall depth ℓ_w. In part (a) of the figure the wall is being bent from left to right by V_u, with the result that tensile bars are needed on the left or tensile side. If V_u is applied from the right side as shown in part (b) of the figure, tensile bars will be needed on the right-hand end of the wall. Thus it can be seen that a shear wall needs tensile reinforcing on both sides because V_u can come from either direction. For flexural calculations the depth of the beam from the compression end of the wall to the center of gravity of the tensile bars is estimated to be about 0.8 times the wall length ℓ_w. (If a larger value of d is obtained by a proper strain compatibility analysis, it may be used.)

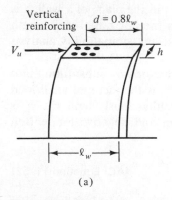
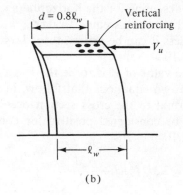

(a)   (b)

Figure 17.2 Shear wall.

---

[2] "Tilt-Up Load Bearing Walls—A Design Aid," EBO74D, 1980 (Skokie, Ill.: Portland Cement Association), 28 pp.

The shear wall acts as a vertical cantilever beam and in providing lateral support is subjected to both bending and shear forces. For such a wall the maximum shear V_u and the maximum moment M_u can be calculated at the base. If flexural stresses are calculated, their magnitude will be affected by the design axial load N_u, and thus its effect should be included in the analysis. If N_u is a compressive load, the permissible shear in the wall ϕV_c will be larger; if N_u is a tensile load, the opposite is true.

Shear is more important in walls with small height-to-length ratios. Moments will be more important for higher walls, particularly those with uniformly distributed reinforcing.

It is necessary to provide both horizontal and vertical shear reinforcing for shear walls. The Commentary (11.10.9) says that in low walls the horizontal shear reinforcing is less effective and the vertical shear reinforcing is more effective. For high walls the situation is reversed. The vertical shear reinforcing contributes to the shear strength of a wall by shear friction.

17.6 ACI PROVISIONS FOR SHEAR WALLS

1. The factored beam shear must be equal to or less than the permissible shear strength of the wall.

$$V_u \leq \phi V_n$$

2. The design shear strength of a wall is equal to the design shear strength of the concrete plus that of the shear reinforcing.

$$V_u \leq \phi V_c + \phi V_s$$

3. The shear strength V_n at any horizontal section in the plane of the wall may not be taken greater than $10\sqrt{f'_c}\,hd$ (11.10.3).

4. In designing for the horizontal shear forces in the plane of a wall, d is to be taken as equal to $0.8\ell_w$, where ℓ_w is the horizontal wall length unless it can be proved to be larger by a strain compatibility analysis (11.10.4).

5. The value of V_c is to be the lesser value obtained by substituting into the two equations that follow, in which N_u is the factored axial load normal to the cross section occurring simultaneously with V_u. N_u is to be considered positive for compression and negative for tension (11.10.6).

$$V_c = 3.3\sqrt{f'_c}\,hd + \frac{N_u d}{4\ell_w} \qquad \text{(ACI Equation 11-32)}$$

or

$$V_c = \left[0.6\sqrt{f'_c} + \frac{\ell_w(1.25\sqrt{f'_c} + 0.2N_u/\ell_w h)}{\dfrac{M_u}{V_u} - \dfrac{\ell_w}{2}} \right] hd \qquad \text{(ACI Equation 11-33)}$$

The first of these equations was developed to predict the inclined cracking strength at any section through a shear wall corresponding to a principal tensile stress of about $4\sqrt{f'_c}$ at the centroid of the wall cross section. The second equation was developed to correspond to an occurrence of a flexural tensile stress of $6\sqrt{f'_c}$ at a section $\ell_w/2$ above the section being investigated. Should $M_u/V_u - \ell_w/2$ be negative, the second equation will have no significance and will not be used.

6. The values of V_c computed by the two preceding equations at a distance from the base equal to $\ell_w/2$ or $h_w/2$ (whichever is less) are applicable for all sections between this section and one at the wall base (11.10.7).

7. Should the factored shear V_u be less than $\phi V_c/2$ computed as described in the preceding two paragraphs, it will be necessary to provide a minimum amount of both horizontal and vertical reinforcing as described in Section 11.10.9.2 through 11.10.9.5 or Chapter 14 of the Code.

8. Should V_u be greater than $\phi V_c/2$, shear wall reinforcing must be designed as described in Section 11.10.9 of the Code.

9. If the factored shear force V_u exceeds the shear strength ϕV_c, the value of V_s is to be determined from the following expression, in which A_v is the area of the horizontal shear reinforcement and s_2 is the spacing of the shear or torsional reinforcing in a direction perpendicular to the horizontal reinforcing (11.10.9.1).

$$V_s = \frac{A_v f_y d}{s_2} \qquad \text{(ACI Equation 11-34)}$$

10. The amount of horizontal shear reinforcing ρ_h (as a percent of the gross vertical concrete area) shall not be less than 0.0025 (11.10.9.2).

11. The maximum spacing of horizontal shear reinforcing s_2 shall not be greater than $\ell_u/5$, $3h$, or 18 in. (11.10.9.3).

12. The amount of vertical shear reinforcing ρ_n (as a percent of the gross horizontal concrete area) shall not be less than the value given by the following equation, in which h_w is the total height of the wall (11.10.9.4).

$$\rho_n = 0.0025 + 0.5\left(2.5 - \frac{h_w}{\ell_w}\right)(\rho_h - 0.0025) \qquad \text{(ACI Equation 11-35)}$$

However, it need not be greater than 0.0025 or the required horizontal shear reinforcing ρ_h.

For high walls the vertical reinforcing is much less effective than it is in low walls. This fact is reflected in the preceding equation where for walls with a height/length ratio less than 0.5 the amount of vertical reinforcing required equals the horizontal reinforcing required. If the ratio is larger than 2.5, only a minimum amount of vertical reinforcing is required.

13. The maximum spacing of vertical shear reinforcing s_1 shall not be greater than $\ell_w/3$, $3h$, or 18 in. (11.10.9.5).

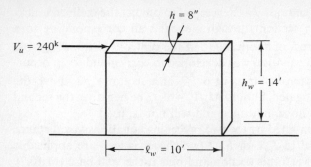

Figure 17.3

■ EXAMPLE 17.2

Design the reinforced concrete wall shown in Figure 17.3 if $f'_c = 3000$ psi and $f_y = 60,000$ psi.

SOLUTION

1. Is the wall thickness satisfactory?

$$V_u = \phi 10\sqrt{f'_c}\,hd$$

$$d = 0.8\ell_w = (0.8)(12 \times 10) = 96''$$

$$V_u = (0.85)(10)(\sqrt{3000})(8)(96)$$

$$V_u = 357{,}553 \text{ lb} = 357.6 \text{ k} > 240 \text{ k} \qquad\qquad \text{OK}$$

2. Compute V_c for wall (lesser of two values):

(a) $V_c = 3.3\sqrt{f'_c}\,hd + \dfrac{N_u d}{4h_w} = (3.3\sqrt{3000})(8)(96) + 0$

$\qquad = 138{,}815 \text{ lb} = 138.8 \text{ k} \leftarrow$

(b) $V_c = \left[0.6\sqrt{f'_c} + \dfrac{\ell_w(1.25\sqrt{f'_c} + 0.2N_u/\ell_w h)}{\dfrac{M_u}{V_u} - \dfrac{\ell_w}{2}} \right] hd$

Computing V_u and M_u at the lesser of $\ell_w/2 = 10/2 = 5'$ or $h_w/2 = 14/2 = 7'$ from base:

$$V_u = 240 \text{ k}$$

$$M_u = 240(14 - 5) = 2160 \text{ ft-k} = 25{,}920 \text{ in.-k}$$

$$V_c = \left[0.6\sqrt{3000} + \dfrac{(12 \times 10)(1.25)(\sqrt{3000}) + 0}{\dfrac{25{,}920}{240} - \dfrac{(12)(10)}{2}} \right](8)(96)$$

$\qquad = 156{,}692 \text{ lb} = 156.7 \text{ k}$

3. Is shear reinforcing needed?

$$\frac{\phi V_c}{2} = \frac{(0.85)(138.8)}{2} = 69.4 \text{ k} < 240 \text{ k} \qquad \underline{\text{yes}}$$

4. Select horizontal shear reinforcing:

$$V_u = \phi V_c + \phi V_s$$

$$V_u = \phi V_c + \phi \frac{A_v f_y d}{s_2}$$

$$\frac{A_v}{s_2} = \frac{V_u - \phi V_c}{\phi f_y d} = \frac{240 - (0.85)(138.8)}{(0.85)(60)(96)} = 0.0249$$

Try different-size horizontal stirrups □ with A_v = two-bar cross-sectional areas. Compute s_2 = vertical spacing of horizontal stirrups.

<u>Try #3</u>

$$s_2 = \frac{(2)(0.11)}{0.0249} = 8.84''$$

<u>Try #4</u>

$$s_2 = \frac{(2)(0.20)}{0.0249} = 16.06''$$

Maximum vertical spacing of horizontal stirrups:

$$\frac{\ell_w}{5} = \frac{(12)(10)}{5} = 24''$$

$$3h = (3)(8) = 24''$$

$$18'' = 18'' \leftarrow$$

<u>Try #4 @ 16''</u>

$$\rho_h = \frac{A_v}{A_g}$$

where A_g = wall thickness times the vertical spacing of the horizontal stirrups.

$$\rho_h = \frac{(2)(0.20)}{(8)(16)} = 0.003125$$

which is greater than the minimum ρ_n of 0.0025 permitted by Code.

<u>Use #4 horizontal stirrups 16'' o.c. vertically.</u>

5. Design vertical shear reinforcing:

$$\text{min. } \rho_n = 0.0025 + 0.5\left(2.5 - \frac{h_w}{\ell_w}\right)(\rho_h - 0.0025)$$

$$= 0.0025 + 0.5\left(2.5 - \frac{12 \times 14}{12 \times 10}\right)(0.003125 - 0.0025)$$

$$= 0.00284$$

Assume vertical stirrups $\square$ with A_v = two-bar cross-sectional areas and with s_1 = horizontal spacing of vertical stirrups.

Try #4

$$s_1 = \frac{(2)(0.20)}{(8)(0.00284)} = 17.60''$$

Maximum horizontal spacing of vertical stirrups:

$$\frac{\ell_w}{3} = \frac{(12)(10)}{3} = 40''$$

$$3h = (3)(8) = 24''$$

$$18'' = 18'' \leftarrow$$

Use #4 vertical stirrups 16" o.c. horizontally.

6. Design vertical flexural reinforcing:

$$M_u = (240)(14) = 3360 \text{ ft-k @ base of wall}$$

$$\frac{M_u}{\phi b d^2} = \frac{(12)(3360)(1000)}{(0.9)(8)(96)^2} = 607.6$$

$$\rho = 0.0118 \text{ from Appendix Table A.13}$$

$$A_s = \rho b d$$

where b is wall thickness and d is again $0.80\ell_w = (0.8)(12 \times 10) = 96''$.

$$A_s = (0.0118)(8)(96) = 9.06 \text{ in.}^2$$

Use 10 #9 bars each end (assuming V_u could come from either direction).

7. A sketch of the wall cross section is given in Figure 17.4.   ■

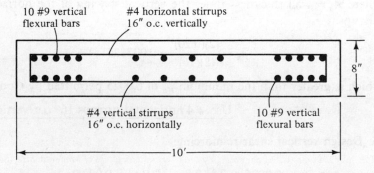

10 #9 vertical
flexural bars

#4 horizontal stirrups
16" o.c. vertically

8"

#4 vertical stirrups
16" o.c. horizontally

10 #9 vertical
flexural bars

10'

Figure 17.4

17.7 ECONOMY IN WALL CONSTRUCTION

To achieve economical reinforced concrete walls, it is necessary to consider such items as wall thicknesses, openings, footing elevations, and so on.

The thicknesses of walls should be sufficient to permit the proper placement and vibration of the concrete. All of the walls in a building should have the same thickness if practical. Such a practice will permit the reuse of forms, ties, and other items. Furthermore, it will reduce the possibilities of field mistakes.

As few openings as possible should be placed in concrete walls. Where openings are necessary it is desirable to repeat the sizes and positions of openings in different walls rather than using different sizes and positions. Furthermore, a few large openings are more economical than a larger number of small ones.

Much money can be saved if a footing elevation can be kept constant for any given wall. Such a practice will appreciably simplify the use of wall forms. If steps are required in a footing, their number should be kept to the minimum possible.[3]

PROBLEMS

17.1 Design a reinforced concrete bearing wall using the ACI empirical formula to support a set of precast concrete roof beams 6'0" on center as shown in the accompanying illustration. The bearing width of each beam is 8". The wall is considered to be laterally restrained top and bottom and is further assumed to be restrained against rotation at the footing. Neglect wall weight. Other data: $f'_c = 3000$ psi, $f_y = 60,000$ psi, beam reaction, $D = 25$ k, $L = 20$ k. (*Ans.* 7" wall with #3 @ 12" vertical steel)

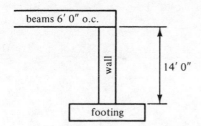

17.2 Repeat Problem 17.1 if the wall is not restrained against rotation at top or bottom, is 8'0" in height, and has an $f'_c = 4000$ psi.

---

[3] Neville, G. B., ed., 1984, *Simplified Design Reinforced Concrete Buildings of Moderate Size*, (Skokie, Ill.: Portland Cement Association), pp. 9-12 and 9-13.

17.3 Design the reinforced concrete wall shown if $f'_c = 4000$ psi and $f_y = 60,000$ psi. (*One ans.* 8 # 9 vertical flexural bars each end)

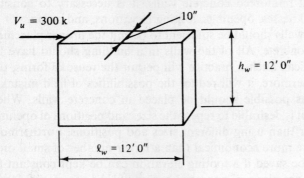

17.4 Repeat Problem 17.3 if $h_w = 15'0''$ and if $f'_c = 3000$ psi.

18

Prestressed Concrete

18.1 INTRODUCTION

Prestressing can be defined as the imposition of internal stresses into a structure that are of opposite character to those that will be caused by the service or working loads. A rather common method used to describe prestressing is shown in Figure 18.1, where a row of books has been squeezed together by a person's hands. The resulting "beam" can carry a downward load as long as the compressive stress due to squeezing at the bottom of the "beam" is greater than the tensile stress there due to the moment produced by the weight of the books and the superimposed loads. Such a beam has no tensile strength and thus no moment resistance until it is squeezed together or prestressed. You might very logically now expand your thoughts to a beam consisting of a row of concrete blocks squeezed together and then to a plain concrete beam with its negligible tensile strength similarly prestressed.

The theory of prestressing is quite simple and has been used for many years in various kinds of structures. For instance, wooden barrels have long been made by putting tightened metal bands around them, thus compressing the

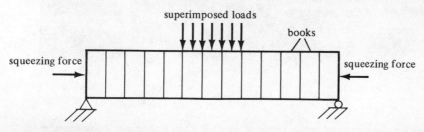

Figure 18.1 Prestressing.

staves together and making a tight container with resistance to the outward pressures of the enclosed liquids. Prestressing is primarily used for concrete beams to counteract tension stresses caused by the weight of the members and the superimposed loads. Should these loads cause a positive moment in a beam, it is possible by prestressing to introduce a negative moment that can counteract part or all of the positive moment. An ordinary beam has to have sufficient strength to support itself as well as the other loads, but it is possible with prestressing to produce a negative loading that will eliminate the effect of the beam's weight, thus producing a "weightless beam."

From the preceding discussion it is easy to see why prestressing has captured the imagination of so many persons and why it has all sorts of possibilities now and in the future.

In the earlier chapters of this book, only a portion of the cross sections of members in bending could be considered effective in resisting loads because a large part of those cross sections were in tension and thus cracked. If, however, flexural members can be prestressed so that their entire cross sections are kept in compression, then the properties of the entire sections are available to resist the applied forces.

For a little more detailed illustration of prestressing, reference is made to Figure 18.2. It is assumed that the following steps have been taken with regard to this beam:

Prestressed concrete channels, John A. Denies Sons Company Warehouse #4, Memphis, Tennessee. (Courtesy of Master Builders.)

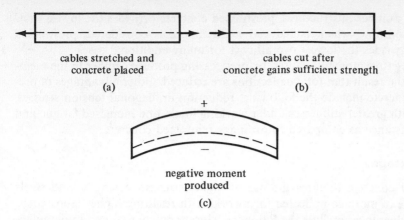

cables stretched and
concrete placed

(a)

cables cut after
concrete gains sufficient strength

(b)

negative moment
produced

(c)

Figure 18.2

1. Steel strands (represented by the dashed lines) were placed in the lower part of the beam form.
2. The strands were tensioned to a very high stress.
3. The concrete was placed in the form and allowed to gain sufficient strength for the prestressed strands to be cut.
4. The strands were cut.

The cut strands tend to resume their original length, thus compressing the lower part of the beam and causing a negative bending moment. The positive moment caused by the beam weight and any superimposed gravity loads is directly opposed by the negative moment. Another way of explaining this is to say that a compression stress has been produced in the bottom of the beam opposite in character to the tensile stress that is caused there by the working loads.

18.2 ADVANTAGES AND DISADVANTAGES OF PRESTRESSED CONCRETE

Advantages

As described in the preceding section, it is possible with prestressing to utilize the entire cross sections of members to resist loads. Thus smaller members can be used to support the same loads, or the same-size members can be used for longer spans. This is a particularly important advantage because member weights make up a substantial part of the total design loads of concrete structures.

Prestressed members are crack-free under working loads and, as a result, look better and are more watertight, thus providing better corrosion protection for the steel. Furthermore, crack-free prestressed members require less maintenance and last longer than cracked reinforced concrete members. Therefore, for a large number of structures, prestressed concrete provides the lowest first-cost solution, and when its reduced maintenance is considered, prestressed concrete provides the lowest overall cost for many additional cases.

The negative moments caused by prestressing produce camber in the members, with the result that total deflections are reduced. Other advantages of prestressed concrete include the following: reduction in diagonal tension stresses, sections with greater stiffnesses under working loads, and increased fatigue and impact resistance as compared to ordinary reinforced concrete.

Disadvantages

Prestressed concrete requires the use of higher-strength concretes and steels and the use of more complicated formwork, with resulting higher labor costs. Other disadvantages include the following: closer control required in manufacture, losses in the initial prestressing forces, additional stress conditions that must be checked in design, and end anchorages and end-bearing plates that may be required.

18.3 PRETENSIONING AND POSTTENSIONING

The two general methods of prestressing are pretensioning and posttensioning. *Pretensioning* was illustrated in Section 18.1, where the prestress tendons were tensioned before the concrete was placed. After the concrete had hardened sufficiently, the tendons were cut and the prestress force was transmitted to the concrete by bond. This method is particularly well suited for mass production because the casting beds can be constructed several hundred feet long. The tendons can be run for the entire bed lengths and used for casting several beams in a line at the same time, as shown in Figure 18.3.

In *posttensioned* construction (see Figure 18.4) the tendons are stressed after the concrete is placed and has gained the desired strength. Plastic or metal tubes or conduits or sleeves or similar devices with unstressed tendons inside (or later inserted) are located in the form and the concrete is placed. After the concrete

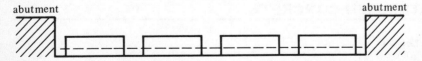

Figure 18.3 Prestress bed.

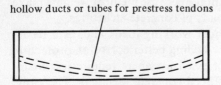

Figure 18.4 Posttensioned beam.

Prestresssed concrete segmental bridge over the River Trent near Scunthorpe, Lincolnshire, England. (Courtesy of Cement and Concrete Association.)

has sufficiently hardened, the tendons are stretched and mechanically attached to end anchorage devices to keep the tendons in their stretched positions. Thus in posttensioning, the prestress forces are transferred to the concrete not by bond but by end bearing.

It is actually possible in posttensioning to have either bonded or unbonded tendons. If bonded, the conduits are often made of aluminum, steel, or other metal sheathing. In addition, it is possible to use steel tubing or rods or rubber cores that are cast in the concrete and removed a few hours after the concrete is placed. After the steel is tensioned, cement grout is injected into the duct for bonding. The grout is also useful in protecting the steel from corrosion. If the tendons are to be unbonded, they should be greased to facilitate tensioning and to protect them from corrosion.[1]

18.4 MATERIALS USED FOR PRESTRESSED CONCRETE

The materials ordinarily used for prestressed concrete are concrete and high-strength steels. The concrete used is probably of a higher strength than that used for reinforced concrete members for several reasons. These include the following:

1. The modulus of elasticity of such concretes is higher, with the result that the elastic strains in the concretes are smaller when the tendons are cut. Thus the relaxations or losses in the tendon stresses are smaller.
2. In prestressed concrete the entire members are kept in compression and thus all the concrete is effective in resisting forces. Hence it is reasonable to pay for a more expensive but stronger concrete if all of it is going to

[1] Lin, T. Y. and Burns, N. H., *Design of Prestressed Concrete Structures*, 3rd ed., 1981 (New York: John Wiley & Sons), pp. 59–61.

be used. (In ordinary reinforced concrete members, more than half of the cross sections are in tension and thus assumed to be cracked. As a result, more than half of a higher-strength concrete used there would be wasted.)

3. Most prestressed work in the United States is of the precast, pretensioned type done at the prestress yard where the work can be carefully controlled; consequently, dependable higher-strength concrete can readily be obtained.

4. For pretensioned work the higher-strength concretes permit the use of higher bond stresses between the cables and the concrete.

High-strength steels are necessary to produce and keep satisfactory prestress forces in members. The strains that occur in these steels during stressing are much greater than those that can be obtained with ordinary reinforcing steels. As a result, when the concrete elastically shortens in compression and also shortens due to creep and shrinkage, the losses in strain in the steel (and thus stress) represent a smaller percentage of the total stress. Another reason for using high-strength steels is that a large prestress force can be developed in a small area.

Early work with prestressed concrete using ordinary-strength bars to induce the prestressing forces in the concrete resulted in failure because the low stresses that could be put into the bars were completely lost due to the concrete's shrinkage and creep. Should a prestress of 20,000 psi be put into such rods, the resulting strains would be equal to $20,000/(29 \times 10^6) = 0.00069$. This value is less than the long-term creep and shrinkage strain normally occurring in concrete, roughly 0.0008, which would completely relieve the stress in the steel. Should a high-strength steel be stressed to about 150,000 psi and have the same creep and shrinkage, the stress reduction will be of the order of $(0.0008)(29 \times 10^6) = 23,200$ psi, leaving $150,000 - 23,200 = 126,800$ psi in the steel (a loss of only 15.47% of the steel stress).[2]

Three forms of prestressing steel are used: single wires, wire strands, and bars. The greater the diameter of the wires, the smaller become their strengths and bond to the concrete. As a result, wires are manufactured with diameters from 0.192 in. up to a maximum of 0.276 in. (about $\frac{9}{32}$ in.). In posttensioning work, large numbers of wires are grouped in parallel into tendons. Strands that are made by twisting wires together are used for most pretensioned work. They are of the seven-wire type, where a center wire is tightly surrounded by twisting the other six wires helically around it. Strands are manufactured with diameters from $\frac{1}{4}$ to $\frac{1}{2}$ in. Sometimes large-size, high-strength, heat-treated alloy steel bars are used for posttensioned sections. They are available with diameters running from $\frac{3}{4}$ to $1\frac{3}{8}$ in.

High-strength prestressing steels do not have distinct yield points (see Figure 18.10) as do the structural carbon reinforcing steels. The practice of con-

[2] Winter, G., and Nilson, A. H., 1991, *Design of Concrete Structures*, 11th ed. (New York: McGraw-Hill), pp. 759–760.

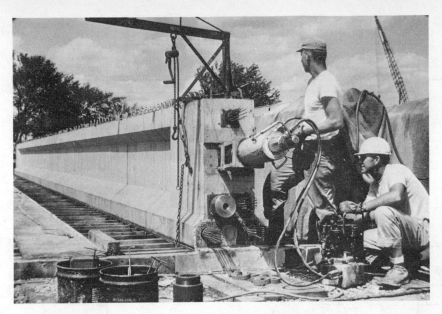

Prestressing girders for bridge in Butler, Pennsylvania. (Courtesy of Portland Cement Association.)

sidering yield points, however, is so firmly embedded in the average designer's mind that high-strength steels are probably given an arbitrary yield point anyway. The yield stress for wires and strands is usually assumed to be the stress that causes a total elongation of 1% to occur in the steel. For high-strength bars the yield stress is assumed to occur when a 0.2% permanent strain occurs.

18.5 STRESS CALCULATIONS

For a consideration of stresses in a prestressed rectangular beam, reference is made to Figure 18.5. For this example the prestress tendons are assumed to be straight, although it will later be shown that a curved shape is more practical for most beams. The tendons are assumed to be located an eccentric distance e below the centroidal axis of the beam. As a result, the beam is subjected to a combination of direct compression and a moment due to the eccentricity of the prestress. In addition, there will be a moment due to the external load including the beam's own weight. The resulting stress at any point in the beam caused by these three factors can be written as

$$f = -\frac{P}{A} \pm \frac{Pec}{I} \pm \frac{Mc}{I}$$

In Figure 18.5 a stress diagram is drawn for each of these three items and all three are combined to give the final stress diagram.

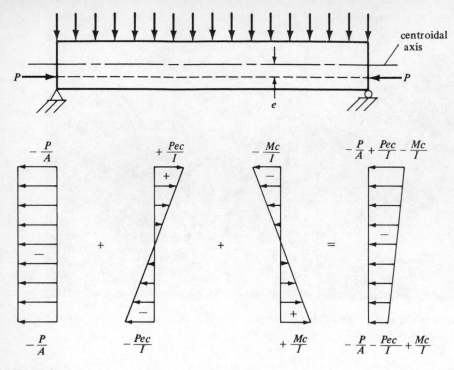

Figure 18.5

The usual practice is to base the stress calculations in the elastic range on the properties of the gross concrete section. The gross section consists of the concrete external dimensions with no additions made for the transformed area of the steel tendons nor subtractions made for the duct areas in posttensioning. This method is considered to give satisfactory results because the changes in stresses obtained if net or transformed properties are used are usually not significant.

Example 18.1 illustrates the calculations needed to determine the stresses at various points in a simple-span prestressed rectangular beam. It will be noted that, as there are no moments at the ends of a simple beam due to the external loads or to the beam's own weight, the Mc/I part of the stress equation is zero there and the equation reduces to

$$f = -\frac{P}{A} \pm \frac{Pec}{I}$$

■ **EXAMPLE 18.1**
Calculate the stresses in the top and bottom fibers at the center line and ends of the beam shown in Figure 18.6.

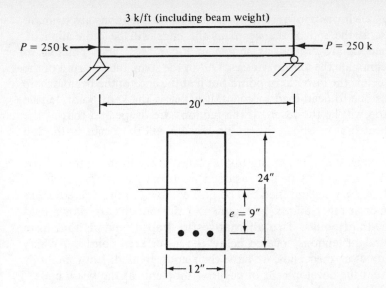

Figure 18.6

SOLUTION

Section Properties

$$I = (\tfrac{1}{12})(12)(24)^3 = 13,824 \text{ in.}^4$$

$$A = (12)(24) = 288 \text{ in.}^2$$

$$M = \frac{(3)(20)^2}{8} = 150 \text{ ft-k}$$

Stresses at Beam Centerline

$$f_{\text{top}} = -\frac{P}{A} + \frac{Pec}{I} - \frac{Mc}{I} = -\frac{250}{288} + \frac{(250)(9)(12)}{13,824} - \frac{(12)(150)(12)}{13,824}$$

$$= -0.868 + 1.953 - 1.562 = -0.477 \text{ ksi}$$

$$f_{\text{bottom}} = -\frac{P}{A} - \frac{Pec}{I} + \frac{Mc}{I} = -0.868 - 1.953 + 1.562 = -1.259 \text{ ksi}$$

Stresses at Beam Ends

$$f_{\text{top}} = -\frac{P}{A} + \frac{Pec}{I} = -0.868 + 1.953 = +1.085 \text{ ksi}$$

$$f_{\text{bottom}} = -\frac{P}{A} - \frac{Pec}{I} = -0.868 - 1.953 = -2.821 \text{ ksi}$$

■

In Example 18.1 it was shown that when the prestress tendons are straight, the tensile stress at the top of the beam at the ends will be quite high. If, however, the tendons are draped, as shown in Figure 18.7, it is possible to reduce or even eliminate the tensile stresses. Out in the span, the centroid of the strands may be below the lower kern point, but if at the ends of the beam, where there is no stress due to dead-load moment, it is below the kern point, tensile stresses in the top will be the result. If the tendons are draped so that at the ends they are located at or above this point, tension will not occur in the top of the beam.

In posttensioning, the sleeve or conduit is placed in the forms in the curved position desired. The tendons in pretensioned members can be placed at or above the lower kern points and then can be pushed down to the desired depth at the centerline or at other points. In Figure 18.7 the tendons are shown held down at the one-third points. Two alternatives to draped tendons that have been used are straight tendons, located below the lower kern point but which are encased in tubes at their ends, or have their ends greased. Both methods are used to prevent the development of negative moments at the beam ends.

Example 18.2 shows the calculations necessary to locate the kern point for the beam of Example 18.1. In addition, the stresses at the top and bottom of the beam ends are computed. It will be noted that according to these calculations the kern point is 4 in. below the middepth of the beam, and it would thus appear that the prestress tendons should be located at the kern point at the beam ends and pushed down to the desired depth further out in the beam. Actually, however, the tendons do not have to be as high as the kern points because the ACI Code (18.4.1) permits some tension in the top of the beam when the tendons are cut. This value is $3\sqrt{f'_{ci}}$, where f'_{ci} is the strength of the concrete at the time the tendons are cut, as determined by testing concrete cylinders. This permissible value equals about 40% of the cracking strength or modulus of rupture of the concrete ($7.5\sqrt{f'_{ci}}$) at that time. The stress at the bottom of the beam, which is compressive, is permitted to go as high as $0.60f'_{ci}$.

The Code actually permits tensile stresses at the ends of simple beams to go as high as $6\sqrt{f'_{ci}}$. These allowable tensile values are applicable to the stresses that occur immediately after the transfer of the prestressing forces and after the losses occur due to elastic shortening of the concrete and relaxation of the tendons and anchorage seats. It is further assumed that the time-dependent losses of creep and shrinkage have not occurred. A discussion of these various losses is presented in Section 18.7 of this chapter. If the calculated tensile stresses are greater than the permissible values, it is necessary to use some additional bonded reinforcing (prestressed or unprestressed) to resist the total tensile force in the concrete computed on the basis of an uncracked section.

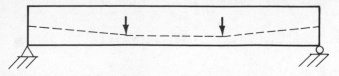

Figure 18.7 Draped tendons.

Section 18.4.2 of the Code gives allowable stresses at service loads after all prestress losses have occurred. The compression in the concrete may be $0.45f_c'$ while the allowable tensile stress in the precompressed tensile zone for ordinary prestressed beams is $6\sqrt{f_c'}$. If, however, certain conditions are met such as additional cover and satisfactory deflections, a tensile stress of $12\sqrt{f_c'}$ is permitted. This higher value, which is not applicable to two-way slabs, provides improved performance under service loads particularly where live loads are of a transient nature.

Only compressive stresses should be allowed in prestressed sections which are to be used in severe corrosive conditions. If tension cracks occur, the results may very well be increased cable corrosion.

■ EXAMPLE 18.2

Determine the location of the lower kern point at the ends of the beam of Example 18.1. Calculate the stresses at the top and bottom of the beam ends assuming the tendons are placed at the kern point.

SOLUTION

Locating the Kern Point

$$f_{\text{top}} = -\frac{P}{A} + \frac{Pec}{I} = 0$$

$$-\frac{250}{288} + \frac{(250)(e)(12)}{13,824} = 0$$

$$-0.868 + 0.217e = 0$$

$$e = 4''$$

Computing Stresses

$$f_{\text{top}} = -\frac{P}{A} - \frac{Pec}{I} = \frac{250}{288} + \frac{(250)(4)(12)}{13,824}$$

$$= -0.868 + 0.868 = 0$$

$$f_{\text{bottom}} = -\frac{P}{A} - \frac{Pec}{I} = -0.868 - 0.868 = 1.736 \text{ ksi} \qquad ■$$

18.6 SHAPES OF PRESTRESSED SECTIONS

For simplicity in introducing prestressing theory, rectangular sections are used for most of the examples of this chapter. From the viewpoint of formwork alone, rectangular sections are the most economical, but more complicated shapes, such as I's and T's, will require smaller quantities of concrete and prestressing steel to carry the same loads and, as a result, they frequently have the lowest overall costs.

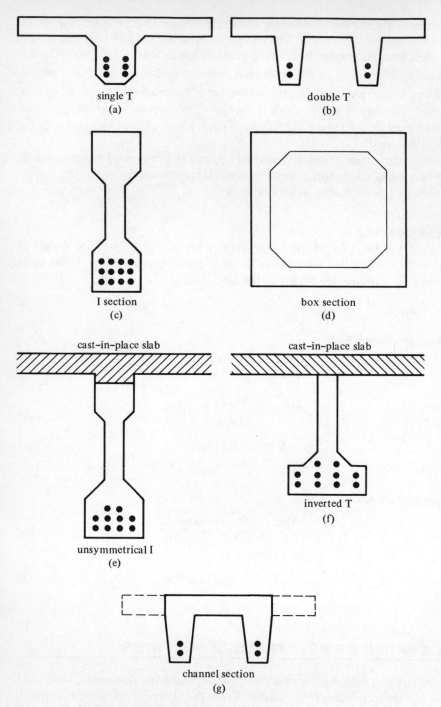

Figure 18.8 Commonly used prestressed sections.

If a member is to be made only one time, a cross section requiring simple formwork (thus often rectangular) will probably be used. For instance, simple formwork is essential for most cast-in-place work. Should, however, the forms be used a large number of times to make many identical members, more complicated cross sections, such as I's and T's, channels, or boxes, will be used. For such sections the cost of the formwork as a percentage of each member's total cost will be much reduced. Several types of commonly used prestressed sections are shown in Figure 18.8. The same general theory used for the determination of stresses and flexural strengths applies to shapes such as these as it does to rectangular sections.

The usefulness of a particular section depends on the simplicity and reusability of the formwork, the appearance of the sections, the degree of difficulty of placing the concrete, and the theoretical properties of the cross section. The greater the amount of concrete located near the extreme fibers of a beam, the greater will be the lever arm between the C and T forces and thus the greater the resisting moment. Of course, there are some limitations on the widths and thicknesses of the flanges. In addition, the webs must be sufficiently large to resist shear and to allow the proper placement of the concrete and at the same time be sufficiently thick to avoid buckling.

A prestressed T such as the one shown in Figure 18.8(a) is often a very economical section because a large proportion of the concrete is placed in the compression flange, where it is quite effective in resisting compressive forces. The double T shown in Figure 18.8(b) is used for schools, office buildings, stores, and so on and is probably the most used prestressed section in the United States today. The total width of the flange provided by a double T is in the range of about 5 to 8 ft and spans of 30 to 50 ft are common. You can see that a floor or roof system can easily and quickly be erected by placing a series of precast double T's side by side, ⊓⊓ ⊓⊓ ⊓⊓, the sections providing both the beams and slabs for the floor or roof system. Single T's are normally used for heavier loads and longer spans up to as high as 100 or 120 ft. Double T's for such spans would be very heavy and difficult to handle.

The I and box sections, shown in parts (c) and (d) of Figure 18.8, have a larger proportion of their concrete placed in their flanges, with the result that larger moments of inertia are possible (as compared to rectangular sections with the same amounts of concrete and prestressing tendons). The formwork, however, is complicated and the placing of concrete is difficult. Box girders are frequently used for bridge spans. Their properties are the same as for I sections. Unsymmetrical I's [Figure 18.8(e)], with large bottom flanges to contain the tendons and small top flanges, may be economical for certain composite sections where they are used together with a slab poured in place to provide the compression flange. A similar situation is shown in Figure 18.8(f), where an inverted T is used with a cast-in-place slab.

Many variations of these sections are used, such as the channel section shown in Figure 18.8(g). Such a section might be made by blocking out the flanges of a double-T form as shown, and the resulting members might be used for stadium seats or similar applications.

Posttensioned segmental precast concrete for East Moors Viaduct, Lanbury Way, Cardiff, South Wales. (Courtesy of Cement and Concrete Association.)

18.7 PRESTRESS LOSSES

The flexural stresses calculated for the beams of Examples 18.1 and 18.2 were based on initial stresses in the prestress tendons. The stresses, however, become smaller with time due to several factors. These factors, which are discussed in the paragraphs to follow, include

1. elastic shortening of the concrete
2. shrinkage and creep of the concrete
3. relaxation or creep in the tendons
4. slippage in posttensioning end anchorage systems
5. friction along the ducts used in posttensioning.

The ultimate strength of a prestressed member is almost completely controlled by the cross-sectional area of the cables. As a result, losses in prestress will have very little effect on its ultimate flexural strength. However, losses in prestress will cause more cracking to occur under working loads, with the result that deflections will be larger. Furthermore the member's shear and fatigue strength will be somewhat reduced.

Elastic Shortening of the Concrete

When the tendons are cut for a pretensioned member, the prestress force is transferred to the concrete, with the result that the concrete is put in compression and shortens, thus permitting some relaxation or shortening of the tendons. The stress in the concrete adjacent to the tendons can be computed as described in the preceding examples. The strain in the concrete ε_c, which equals f_c/E_c, is assumed due to bond to equal the steel strain ε_s. Thus the loss in prestress can be computed as $\varepsilon_s E_s$. An average value of prestress loss in pretensioned members due to elastic shortening is about 3% of the initial value.

An expression for the loss of prestress due to elastic shortening of the concrete can be derived as shown in the paragraphs to follow:

It can be seen that the compressive strain in the concrete due to prestress must equal the lessening of the steel strain

$$\varepsilon_c = \Delta \varepsilon_s$$

These values can be written in terms of stresses as follows:

$$\frac{f_c}{E_c} = \frac{\Delta f_s}{E_s}$$

Thus we can write

$$\Delta f_s = \frac{E_s}{E_c} f_c = n f_c$$

where f_c is the stress in the concrete after transfer of stresses from the cables.

If we express Δf_s as being the initial tendon stress f_{si} minus the tendon stress after transfer, we can write

$$f_{si} - f_s = n f_c$$

Then letting P_0 be the initial total cable stress and P_f the stress afterward, we obtain

$$P_0 - P_f = n \frac{P_f}{A_c} A_{ps}$$

$$P_0 = n \frac{P_f}{A_c} A_{ps} + P_f$$

$$P_0 = P_f \left(\frac{n A_{ps}}{A_c} + 1 \right) = \frac{P_f}{A_c} (n A_{ps} + A_c)$$

Then

$$f_c = \frac{P_0}{A_c + n A_{ps}} = \text{approximately } \frac{P_0}{A_g}$$

and finally

$$\Delta f_s = nf_c = \frac{nP_0}{A_g}$$

a value which can easily be calculated.

For posttensioned members the situation is a little more involved because it is rather common to stress a few of the strands at a time and connect them to the end plates. As a result, the losses vary, with the greatest losses occurring in the first strands stressed and the least losses occurring in the last strands stressed. For this reason an average loss may be calculated for the different strands. Losses due to elastic shortening average about $1\frac{1}{2}\%$ for posttensioned members. It is, by the way, often possible to calculate the expected losses in each set of tendons and overstress them by that amount so the net losses will be close to zero.

Shrinkage and Creep of the Concrete

The losses in prestressing due to the shrinkage and creep in the concrete are quite variable. For one thing the amount of shrinkage that occurs in concrete varies from almost zero up to about 0.0005 in./in. (depending on dampness and on the age of the concrete when it is loaded), with an average value of about 0.0003 in./in. being the usual approximation. The shrinkage loss is approximately 7% in pretensioned sections and 6% in posttensioned ones.

The loss in prestress due to shrinkage can be said to equal $\varepsilon_{sh}E_s$, where ε_{sh} is the shrinkage strain of the concrete. In reference 3 of this chapter, a recommended value of ε_{sh} is given which is to be determined by taking the basic shrinkage strain times a correction factor based on the volume (V) to surface (S) ratio times a relative humidity correction (H).

$$\varepsilon_{sh} = (0.00055)\left(1 - 0.06\frac{V}{S}\right)(1.5 - 0.15H)$$

Should the member be posttensioned, an additional multiplier is provided in reference 3 to take into account the time between the end of the moist curing until the prestressing forces are applied.

The amount of creep in the concrete depends on several factors, which have been previously discussed in this text and can vary from 1 to 5 times the instantaneous elastic shortening. Prestress forces are usually applied to pretensioned members much earlier in the age of the concrete than for posttensioned members. Pretensioned members are normally cast in a bed at the prestress yard where the speed of production of members is an important economic matter. The owner wants to tension the steel, place the concrete, and take the members out of the prestress bed as quickly as the concrete gains sufficient strength so that work can start on the next set of members. As a result, creep and shrinkage are larger, as are the resulting losses. Average losses used for pretensioned members are about 6% and about 5% for posttensioned members.

The losses in cable stresses due to concrete creep strain can be determined by multiplying an experimentally determined creep coefficient C_t by nf_c.

$$\Delta f_s = C_t n f_c$$

In reference 3 a value of $C_t = 2.0$ is recommended for pretensioned sections while 1.6 is recommended for posttensioned ones. The value f_c is defined as the stress in the concrete adjacent to the centroid of the tendons due to the initial prestress $(-P/A)$ and due to the permanent dead loads which are applied to the member after prestressing $(-Pec/I)$, where c is measured from the centroid of the section to the centroid of the tendons.

Relaxation or Creep in the Tendons

The plastic flow or relaxation of steel tendons is quite small when the stresses are low, but the percentage of relaxation increases as stresses become higher. In general, the estimated losses run from about 2% to 3% of the initial stresses. The amount of these losses actually varies quite a bit for different steels and should be determined from test data available from the steel manufacturer in question. A formula is available with which this loss can be computed.[3]

Slippage in Posttensioning End Anchorage Systems

When the jacks are released and the prestress forces are transferred to the end anchorage system, a little slippage of the tendons occurs. The amount of the slippage depends on the system used and tends to vary from about 0.10 in. to 0.20 in. Such deformations are quite important if the members and thus the tendons are short, but if they are long, the percentage is much less important.

Friction Along the Ducts Used in Posttensioning

There are losses in posttensioning due to friction between the tendons and the surrounding ducts. In other words, the stress in the tendons gradually falls off as the distance from the tension points increases due to friction between the tendons and the surrounding material. These losses are due to the so-called length and curvature effects.

The *length effect* is the friction that would have existed if the cable had been straight and not curved. Actually it is impossible to have a perfectly straight duct in posttensioned construction, and the result is friction, called the length effect or sometimes the *wobble effect*. The magnitude of this fraction is dependent on the stress in the tendons, their length, the workmanship for the particular member in question, and the coefficient of friction between the materials.

The *curvature effect* is the amount of friction that occurs in addition to the unplanned wobble effect. The resulting loss is due to the coefficient of friction between the materials caused by the pressure on the concrete from the tendons, which is dependent on the stress and the angle change in the curved tendons.

---

[3] Zia, P., Preston, H. K., Scott, N. L., and Workman, E. B., June 1979, "Estimating Prestress Losses," *Concrete International: Design & Construction*, vol. 1, no. 6 (Detroit: American Concrete Institute), pp. 32–38.

It is possible to substantially reduce frictional losses in prestressing by several methods. These include jacking from both ends, overstressing the tendons initially, and lubricating unbonded cables.

The ACI Code (18.6.2) requires that frictional losses for posttensioned members be computed with wobble and curvature coefficients experimentally obtained and verified during the prestressing operation. Furthermore, the Code gives equations for making the calculations.

In the past it was quite common to estimate total prestress losses rather than calculating each one separately, as described in the preceding paragraphs. Values in the range of 15–20% of the initial steel stresses were frequently used. The usual practice was to assume total losses in the tendons, not including friction, as 35,000 psi for pretensioned members and 25,000 psi for posttensioned ones. These same values were also recommended by previous ACI Commentaries.

The practice of using lump sum values for prestress losses is now considered to be obsolete because more accurate values can be obtained by calculating losses for each item as described in reference 3.

18.8 ULTIMATE STRENGTH OF PRESTRESSED SECTIONS

Considerable emphasis is given to the ultimate strength of prestressed sections, the objective being to obtain a satisfactory factor of safety against collapse. You might wonder why it is necessary in prestress work to consider *both* working-stress and ultimate-strength situations. The answer lies in the tremendous change that occurs in a prestressed member's behavior after tensile cracks occur. Before the cracks begin to form, the entire cross section of a prestressed member is effective in resisting forces, but after the tensile cracks begin to develop, the cracked part is not effective in resisting tensile forces. Cracking is usually assumed to occur when calculated tensile stresses equal the modulus of rupture of the concrete (about $7.5\sqrt{f_c'}$).

Another question that might enter your mind at this time is this: "What effect do the prestress forces have on the ultimate strength of a section?" The answer to the question is quite simple. An ultimate-strength analysis is based on the assumption that the prestressing strands are stressed above their yield point. If the strands have yielded, the tensile side of the section has cracked and the theoretical ultimate resisting moment is the same as for a nonprestressed beam constructed with the same concrete and reinforcing.

The theoretical calculation of ultimate capacities for prestressed sections is not such a routine thing as it is for ordinary reinforced concrete members. The high-strength steels from which prestress tendons are manufactured do not have distinct yield points. Despite this fact, the strength method for determining the ultimate moment capacities of sections checks rather well with load tests as long as the steel percentage is sufficiently small as to ensure a tensile failure and as long as bonded strands are being considered.

In the expressions used here, f_{ps} is the average stress in the prestressing steel at the design load. This stress is used in the calculations because the prestressing steels usually used in prestressed beams do not have well-defined yield points (that is, the flat portions that are common to stress-strain curves for ordinary structural steels). Unless the yield points of these steels are determined from detailed studies, their values are normally specified. For instance, the ACI Code (18.7.2) states that the following approximate expression may be used for calculating f_{ps}. In this expression f_{pu} is the ultimate strength of the prestressing steel, ρ_p is the percentage of prestress reinforcing A_{ps}/bd, and f_{se} is the effective stress in the prestressing steel after losses. If more accurate stress values are available, they may be used instead of the specified values. In no case may the resulting values be taken as more than the specified yield strength f_{py}, or $f_{se} + 60,000$. For bonded members,

$$f_{ps} = f_{pu}\left(1 - \frac{\gamma_p}{\beta_1}\left[\rho_p \frac{f_{pu}}{f'_c} + \frac{d}{d_p}(\omega - \omega')\right]\right) \quad \text{if } f_{se} \geq 0.5f_{pu}$$

(ACI Equation 18-3)

where γ_p is a factor for the type of prestress tendon ($\gamma_p = 0.55$ for f_{py}/f_{pu} not less than 0.80, 0.40 for f_{py}/f_{pu} not less than 0.85, and 0.28 for f_{py}/f_{pu} not less than 0.90), $d_p =$ distance from the extreme compression fiber to the centroid of the prestress reinforcement, $\omega = \rho f_y/f'_c$, and $\omega' = \rho' f_y/f'_c$.

If any compression reinforcing is considered in calculating f_{ps}, the terms in brackets may not be taken less than 0.17 (see Commentary R18.7.2) and d' (distance from extreme compression fiber to centroid of compression reinforcement) may not be taken less than $0.15d_p$.

For unbonded members with span to depth ≤ 35,

$$f_{ps} = f_{se} + 10,000 + \frac{f'_c}{100\rho_p} \quad \text{but not greater than } f_{py} \text{ nor } (f_{se} + 60,000)$$

(ACI Equation 18-4)

For unbonded members with span to depth > 35,

$$f_{ps} = f_{se} + 10,000 + \frac{f'_c}{300\rho_p} \quad \text{(ACI Equation 18-5)}$$

However, f_{ps} may not exceed f_{py}, or $f_{se} + 30,000$.

As in reinforced concrete members, the amount of steel in prestressed sections is limited to ensure tensile failures. The limitation rarely presents a problem except in members with very small amounts of prestressing or in members that have not only prestress strands but also some regular reinforcing bars.

In Section 18.8.1 of the Code ω, the ratio of prestressed and nonprestressed reinforcement used for calculating moment strength, shall be such that ω_p, $[\omega_p + (d/d_p)(\omega - \omega')]$, or $[\omega_{pw} + (d/d_p)(\omega_w - \omega'_w)]$ is not greater than $0.36\beta_1$. If these values are exceeded, the theoretical ultimate-strength equations do not give results that correspond well with test results. For such overreinforced members the design moment may not exceed the moment resistance calculated using the compression part of the internal resisting couple. For rectangular and flanged sections with $\omega > 0.3$, the 1983 ACI Commentary (18.8.2) recommends the following values.

For rectangular sections, or flanged sections in which the neutral axis falls in the flange,

$$\phi M_n = \phi[f'_c b d_p^2 (0.36\beta_1 - 0.08\beta_1^2)]$$

For flanged sections for which the neutral axis falls outside the flange,

$$\phi M_n = \phi[f'_c b_w d_p^2 (0.36\beta_1 - 0.08\beta_1^2) + 0.85f'_c (b - b_w) h_f (d_p - 0.5h_f)]$$

Example 18.3 illustrates the calculations involved in determining the permissible ultimate capacity of a rectangular prestressed beam. Some important comments about the solution and about ultimate-moment calculations in general are made at the end of the problem.

■ EXAMPLE 18.3

Determine the permissible ultimate moment capacity of the prestressed bonded beam of Figure 18.9 if f_{pu} is 275,000 psi and f'_c is 5000 psi.

SOLUTION

Approximate Value of f_{ps} from ACI Code

$$\rho_p = \frac{A_{ps}}{bd} = \frac{1.40}{(12)(21.5)} = 0.00543$$

$$\gamma_p = 0.40 \qquad (\text{for} f'_c = 5000 \text{ psi, } \beta_1 = 0.80)$$

$$f_{ps} = 275\left(1 - \frac{0.40}{0.80}\left[0.00543\left(\frac{275}{5}\right) + 0\right]\right) = 233.9 \text{ ksi}$$

Computing Value of Reinforcing Ratio (Code 18.8.1)

$$\omega_p = \frac{\rho_p f_{ps}}{f'_c} = \frac{(0.00543)(233.9)}{5} = 0.254 < 0.36\beta_1 = 0.288$$

$$\therefore \text{ Tensile failure occurs}$$

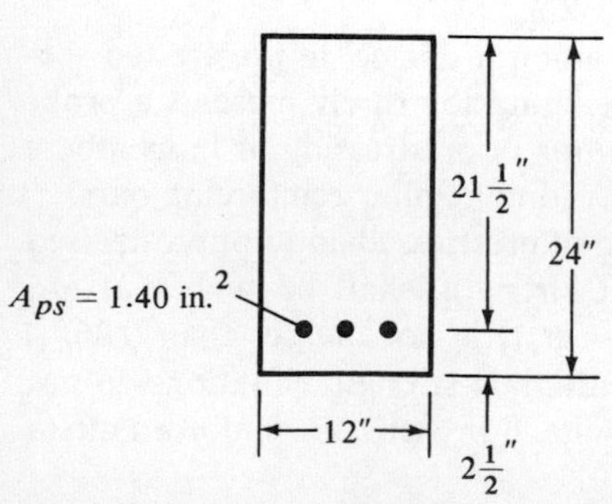

Figure 18.9

Moment Capacity

$$a = \frac{A_{ps}f_{ps}}{0.85f'_cb} = \frac{(1.40)(233.9)}{(0.85)(5)(12)} = \underline{6.42''}$$

$$M_u = \varphi A_{ps}f_{ps}\left(d - \frac{a}{2}\right) = (0.9)(1.40)(233.9)\left(21.5 - \frac{6.42}{2}\right)$$

$$= 5390 \text{ in.-k} = \underline{449.2 \text{ ft-k}}$$

Discussion

The approximate value of f_{ps} obtained by the ACI formula is very satisfactory for all practical purposes. Actually a slightly more accurate value of f_{ps} and thus of the moment capacity of the section can be obtained by calculating the strain in the prestress strands due to the prestress and adding to it the strain due to the ultimate moment. This latter strain can be determined from the values of a and the strain diagram as frequently used in earlier chapters for checking to see if tensile failures control in reinforced concrete beams. With the total strain a more accurate cable stress can be obtained by referring to the stress–strain curve for the prestressing steel being used. Such a curve is shown in Figure 18.10.

The analysis described herein is satisfactory for pretensioned beams or for bonded posttensioned beams but is not so good for unbonded posttensioned members. In these latter beams the steel can slip with respect to the concrete; as a result, the steel stress is almost constant throughout the member. The calculations for M_u for such members are less accurate than for bonded members. Unless some ordinary reinforcing bars are added to these members, large cracks may form which are not attractive and which can lead to some corrosion of the prestress strands.

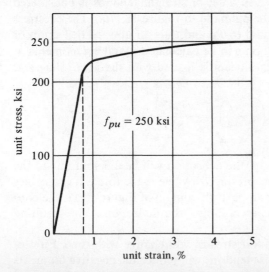

Figure 18.10 Typical stress–strain curve for high-tensile steel wire.

Should a prestressed beam be satisfactorily designed based on service loads, then checked by strength methods and found to have insufficient strength to resist the factored loads ($M_u = 1.4M_D + 1.7M_L$), nonprestressed reinforcement may be added to increase the factor of safety. The increase in T due to these bars is assumed to equal $A_s f_y$ (Code 18.7.3). The Code (18.8.3) further states that the total amount of prestressed and nonprestressed reinforcement shall be sufficient to develop an ultimate moment equal to at least 1.2 times the cracking moment of the section calculated with the modulus of rupture of the concrete. This additional steel also will serve to reduce cracks.

18.9 DEFLECTIONS

The deflections of prestressed concrete beams must be calculated very carefully. Some members that are completely satisfactory in all other respects are not satisfactory for practical use because of the magnitudes of their deflections.

In previous chapters, one method used for limiting deflections was to specify minimum depths for various types of members (as in Table 3.1 of this textbook). These minimum depths, however, are applicable only to nonprestressed sections. The actual deflection calculations are made as they are for members made of other materials, such as structural steel, reinforced concrete, and so on. There is, however, the same problem that exists for reinforced concrete members, and that is the difficulty of determining the modulus of elasticity to be used in the calculations. The modulus varies with age, with different stress levels, and with other factors. Usually the gross moments of inertia are used for immediate deflection calculations (ACI Code 9.5.4.1), although the transformed I may on occasion be used if there is a large percentage of reinforcing steel in the members or if the members have a relatively high tensile stress as permitted in Section 18.4.2(c) of the Code.

The deflection due to the force in a set of straight tendons is considered first in this section, with reference being made to Figure 18.11(a). The prestress forces cause a negative moment equal to Pe and thus an upward deflection or camber of the beam. This ℄ deflection can be calculated by taking moments at the point desired when the conjugate beam is loaded with the M/EI diagram. At the ℄ the deflection equals

$$-\left(\frac{Pe\ell}{2EI}\right)\left(\frac{\ell}{2}-\frac{\ell}{4}\right) = -\frac{Pe\ell^2}{8EI}$$

upwards for the straight cables down.

Should the cables not be straight, the deflection will be different due to the different negative moment diagram produced by the cable force. If the cables are bent down or curved, as shown in parts (b) and (c) of Figure 18.11, the conjugate beam can again be applied to compute the deflections. The resulting values are shown in the figure.

The deflections due to the tendon stresses will change with time. First of all, the losses in stress in the prestress tendons will reduce the negative moments

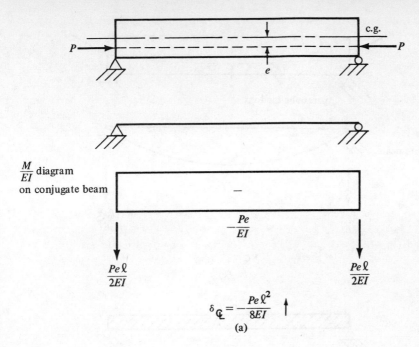

$$\delta_{\mathbb{C}} = -\frac{Pe\ell^2}{8EI} \uparrow$$
(a)

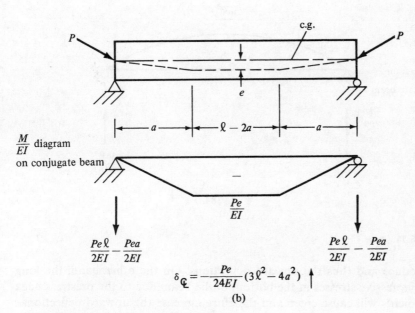

$$\delta_{\mathbb{C}} = -\frac{Pe}{24EI}(3\ell^2 - 4a^2) \uparrow$$
(b)

Figure 18.11 Deflections in prestressed beams.

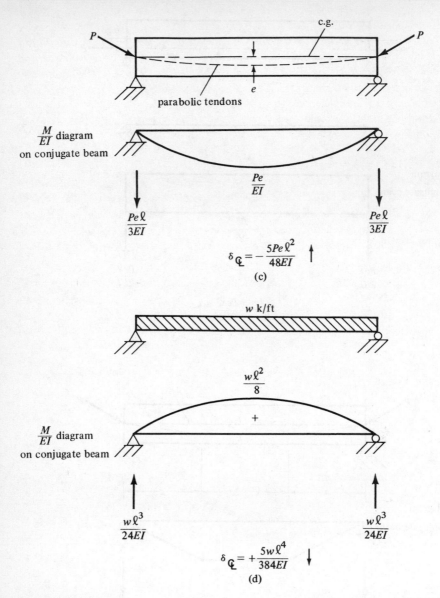

Figure 18.11 (*cont.*)

they produce and thus the upward deflections. On the other hand, the long-term compressive stresses in the bottom of the beam due to the prestress negative moments will cause creep and therefore increase the upward deflections.

In addition to the deflections caused by the tendon stresses, there are deflections due to the beam's own weight and due to the additional dead and live loads subsequently applied to the beam. These deflections can be computed and superimposed on the ones caused by the tendons. Figure 18.11(d) shows the ℄ deflection of a uniformly loaded simple beam obtained by taking moments at the ℄ when the beam is loaded with the M/EI diagram.

The ACI Code (9.5.4.1) permits the calculation of deflections for uncracked prestressed sections using gross moments of inertia. If, however, the calculated tensile stress is greater than the modulus of rupture ($7.5\sqrt{f'_c}$), the transformed moment of inertia must be used. Section R18.4.2(c) of the ACI Commentary requires the use of the transformed I for members designed for a tensile stress in the precompressed tensile zone equal to $12\sqrt{f'_c}$.

Example 18.4 shows the initial and long-term deflection calculations for a rectangular pretensioned beam.

■ EXAMPLE 18.4

The pretensioned rectangular beam shown in Figure 18.12 has straight cables with initial stresses of 175 ksi and final stresses after losses of 140 ksi. Determine the deflection at the beam ℄ immediately after the cables are cut and six months later under full dead and live loads. Assume a creep coefficient of 2.0 for this latter case. $E = 4 \times 10^6$ psi. Assume concrete uncracked.

SOLUTION

$$I_g = (\tfrac{1}{12})(12)(20)^3 = 8000 \text{ in.}^4$$

$$e = 6''$$

$$Bm \text{ weight} = \frac{(12)(20)}{144}(150) = 250 \text{ lb/ft}$$

Deflection Immediately After Cables Are Cut

$$\delta \text{ due to cable} = -\frac{Pe\ell^2}{8EI} = \frac{(1.2 \times 175,000)(6)(12 \times 30)^2}{(8)(4 \times 10^6)(8000)} = -0.638''\uparrow$$

$$\delta \text{ due to beam weight} = +\frac{5w\ell^4}{384EI} = \frac{(5)\left(\dfrac{250}{12}\right)(12 \times 30)^4}{(384)(4 \times 10^6)(8000)} = +0.142''\downarrow$$

$$\text{Total deflection} = \overline{-0.496''\uparrow}$$

Deflection at Six Months Under Full Load

$$\delta \text{ due to cable} = \left(\frac{140,000}{175,000}\right)(-0.638)(2)$$

$$= -1.021''$$

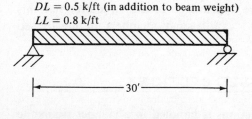

DL = 0.5 k/ft (in addition to beam weight)
LL = 0.8 k/ft

30'

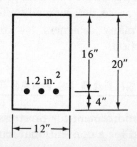

16"
20"
1.2 in.²
4"
12"

Figure 18.12

$$\delta \text{ due to beam weight} + \text{additional } DL = +\frac{5w\ell^4}{384EI}\,C_c$$

$$= \frac{(5)\left(\dfrac{750}{12}\right)(12 \times 30)^4}{(384)(4 \times 10^6)(8000)}(2.0)$$

$$= +0.852''\downarrow$$

$$\delta \text{ due to } LL = \left(\frac{800}{250}\right)(+0.142)$$

$$= +0.454''\downarrow$$

$$\text{total deflection} = \overline{+0.285''\downarrow} \qquad \blacksquare$$

Additional Deflection Comments

From the preceding example it can be seen that, not counting external loads, the beam is initially cambered upwards by 0.496 in.; as time goes by, this camber increases due to creep in the concrete. Such a camber is often advantageous in offsetting deflections caused by the superimposed loads. In some members, however, the camber can be quite large, particularly for long spans and where lightweight aggregates are used. If this camber is too large, the results can be quite detrimental to the structure (warping of floors, damage to roofing, cracking and warping of partitions, and so on).

To illustrate one problem that can occur, it is assumed that the roof of a school is being constructed by placing 50-ft double T's made with a lightweight aggregate, side by side over a classroom. The resulting cambers may be rather large and, worse, they may not be equal in the different sections. It then becomes necessary to force the different sections to the same deflection and tie them together in some fashion so that a smooth surface is provided for roofing. Once the surface is even, the members may be connected by welding together metal inserts, such as angles that were cast in the edges of the different sections for this purpose.

Both reinforced concrete members and prestressed members with overhanging or cantilevered ends will often have rather large deflections. The total deflections at the free ends of these members are due to the sum of the normal deflections plus the effect of support rotations. This latter effect may frequently be the larger of the two and, as a result, the sum of the two deflections may be so large as to detrimentally affect the appearance of the structure. For this reason, many designers try to avoid cantilevered members in prestressed construction.

18.10 SHEAR IN PRESTRESSED SECTIONS

Web reinforcement for prestressed sections is handled in a manner similar to that used for a conventional reinforced concrete beam. In the expressions that follow, b_w is the web width or the diameter of a circular section and d is the distance from the extreme fiber in compression to the centroid of the tensile reinforcement. Should the reaction introduce compression into the end region of a prestressed member, sections of the beam located at distances less than $h/2$ from the face of the support may be designed for the shear computed at $h/2$ where h is the overall thickness of the member.

$$v_u = \frac{V_u}{\phi b_w d}$$

The Code (11.4.1) provides two methods for estimating the shear strength that the concrete of a prestressed section can resist. There is an approximate method, which can only be used when the effective prestress force is equal to at least 40% of the tensile strength of the flexural reinforcement f_{pu}, and a more detailed analysis, which can be used regardless of the magnitude of the effective prestress force. These methods are discussed in the paragraphs to follow.

Approximate Method

With this method the nominal shear capacity of a prestressed section can be taken as

$$V_c = \left(0.6\sqrt{f_c'} + 700\,\frac{V_u d}{M_u}\right)b_w d \qquad \text{(ACI Equation 11-10)}$$

The Code (11.4.1) states that regardless of the value given by this equation, V_c need not be taken as less than $2\sqrt{f_c'}b_w d$ nor may it be larger than $5\sqrt{f_c'}b_w d$. In this expression, V_u is the maximum design shear at the section being considered and M_u is the design moment at the same section occurring simultaneously with V_u and d is the distance from the extreme compression fiber to the centroid of the prestressed tendons. The value of $V_u d/M_u$ is limited to a maximum value of 1.0.

More Detailed Analysis

If a more detailed analysis is desired (it will have to be used if the effective prestressing force is less than 40% of the tensile strength of the flexural reinforcement), the nominal shear force carried by the concrete is considered to equal the smaller of V_{ci} and V_{cw}, to be defined here. The shear force at diagonal cracking due to all design loads when the cracking is due to flexural shear is called V_{ci}. The shear force when cracking is due to the principal tensile stress is referred to as V_{cw}. In both expressions to follow, d is the distance from the extreme compression fiber to the centroid of the prestressed tendons or is $0.8h$, whichever is greater (Code 11.4.2.3).

The estimated shear capacity V_{ci} can be computed by the following expression, given by the ACI Code (11.4.2):

$$V_{ci} = 0.6\sqrt{f_c'}b_w d + V_d + \frac{V_i M_{cr}}{M_{max}} \qquad \text{but need not be taken less than } 1.7\sqrt{f_c'}b_w d$$

(ACI Equation 11-11)

In this expression, V_d is the shear at the section in question due to service dead load, M_{max} is the maximum bending moment at the section due to externally applied design loads, V_i is the shear that occurs simultaneously with M_{max}, and M_{cr} is the cracking moment, which is to be determined as follows:

$$M_{cr} = \left(\frac{I}{y_t}\right)(6\sqrt{f'_c} + f_{pe} - f_d) \qquad \text{(ACI Equation 11-12)}$$

where I = the moment of inertia of the section that resists the externally
applied loads

   y_t = the distance from the centroidal axis of the gross section
(neglecting the reinforcing) to the extreme fiber in tension

   f_{pe} = the compressive stress in the concrete due to prestress after
all losses at the extreme fiber of the section where the applied
loads cause tension

   f_d = the stress due to dead load at the extreme fiber where the applied
loads cause tension

From a somewhat simplified principal tension theory, the shear capacity of a beam is equal to the value given by the following expression but need not be less than $1.7\sqrt{f'_c}b_w d$.

$$V_{cw} = (3.5\sqrt{f'_c} + 0.3f_{pc})b_w d + V_p \qquad \text{(ACI Equation 11-13)}$$

In this expression, f_{pc} is the calculated compressive stress (in pounds per square inch) in the concrete at the centroid of the section resisting the applied loads due to the effective prestress after all losses have occurred. (Should the centroid be in the flange, f_{pc} is to be computed at the junction of the web and flange.) V_p is the vertical component of the effective prestress at the section under consideration. Alternately, the Code states that V_{cw} may be taken as the shear force that corresponds to a multiple of dead load plus live load, which results in a calculated principal tensile stress equal to $4\sqrt{f'_c}$ at the centroid of the member or at the intersection of the flange and web if the centroid falls in the web.

A further comment should be made here about the computation of f_{pc} for pretensioned members, since it is affected by the transfer length. The Code (11.4.3) states that the transfer length can be taken as 50 diameters for strand tendons and 100 diameters for wire tendons. The prestress force may be assumed to vary linearly from zero at the end of the tendon to a maximum at the aforesaid transfer distance. If the value of $h/2$ is less than the transfer length, it is necessary to consider the produced prestress when V_{cw} is calculated.

18.11 DESIGN OF SHEAR REINFORCEMENT

Should the computed value of V_u exceed ϕV_c, the area of vertical stirrups (the Code not permitting inclined stirrups or bent-up bars in prestressed members) must not be less than A_v as determined by the following expression from the Code (11.5.6.2):

$$V_s = \frac{A_v f_y d}{s} \qquad \text{(ACI Equation 11-17)}$$

As in conventional reinforced concrete design, a minimum area of shear reinforcing is required at all points where V_u is greater than $\frac{1}{2}\phi V_c$. This minimum area is to be determined from the expression to follow if the effective prestress is less than 40% of the tensile strength of the flexural reinforcement (ACI Code 11.5.5.3):

$$A_v = 50 \frac{b_w s}{f_y} \qquad \text{(ACI Equation 11-14)}$$

If the effective prestress is equal to or greater than 40% of the tensile strength of the flexural reinforcement, the following expression, in which A_{ps} is the area of prestressed reinforcement in the tensile zone, is to be used to calculate A_v:

$$A_v = \left(\frac{A_{ps}}{80}\right)\left(\frac{f_{pu}}{f_y}\right)\left(\frac{s}{d}\right)\sqrt{\left(\frac{d}{b_w}\right)} \qquad \text{(ACI Equation 11-15)}$$

Section 11.5.4.1 of the ACI Code states that in no case may the maximum spacing exceed $0.75h$ or 24 in. Examples 18.5 and 18.6 illustrate the calculations necessary for determining the shear strength and for selecting the stirrups for a prestressed beam.

Prestressed concrete Runnymede Bridge over Thames River near Egham, Surrey, England. (Courtesy of Cement and Concrete Association.)

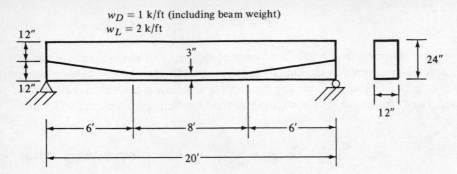

Figure 18.13

■ EXAMPLE 18.5

Calculate the shearing strength of the section shown in Figure 18.13 at 4 ft from the supports using both the approximate method and the more detailed method allowed by the ACI Code. Assume the area of the prestressing steel is 1.0 in.^{2}, the effective prestress force is 250 k, and $f'_c = 4000$ psi.

SOLUTION

Approximate Method

$$w_u = (1.4)(1.0) + (1.7)(2.0) = 4.8 \text{ k/ft}$$

$$V_u = (10)(4.8) - (4)(4.8) = 28.8 \text{ k}$$

$$M_u = (10)(4.8)(4) - (4)(4.8)(2) = 153.6 \text{ ft-k}$$

$$\frac{V_u d}{M_u} = \frac{(28.8)(24 - 3 - 3)}{(12)(153.6)} = 0.281 < 1.0 \qquad \underline{\text{OK}}$$

$$V_c = \left(0.6\sqrt{f'_c} + 700\frac{V_u d}{M_u}\right)b_w d$$

$$= [0.6\sqrt{4000} + (700)(0.281)](12)(18) = 50{,}684 \text{ lb}$$

$$\text{Maximum } V_c = (2\sqrt{4000})(12)(18) = 27{,}322 \text{ lb} < 50{,}684 \text{ lb}$$

$$\text{Maximum } V_c = (5\sqrt{4000})(12)(18) = 68{,}305 \text{ lb} > 50{,}684 \text{ lb}$$

$$V_c = 50{,}684 \text{ lb}$$

More Detailed Method

$$I = (\tfrac{1}{12})(12)(24)^3 = 13{,}824 \text{ in.}^4$$

$$y_t = 12''$$

f_{pe} = compressive stress in concrete due to prestress after all losses

$$= \frac{P}{A} + \frac{Pec}{I}$$

$$f_{pe} = \frac{250,000}{(12)(24)} + \frac{(250,000)(6)(12)}{13,824} = 2170 \text{ psi}$$

M_d = dead load moment at 4' point = $(10)(1)(4) - (4)(1)(2)$

  = 32 ft-k

f_d = stress due to the dead load moment = $\dfrac{(12)(32,000)(12)}{13,824}$

  = 333 psi

M_{cr} = cracking moment = $\left(\dfrac{I}{y_t}\right)(6\sqrt{f'_c} + f_{pe} - f_d)$

$$= \left(\frac{13,824}{12}\right)(6\sqrt{4000} + 2170 - 333) = 2,553,373 \text{ in.-lb}$$

  = 212,780 ft-lb

Beam weight = $\dfrac{(12)(24)}{144}(150) = 300 \text{ lb/ft}$

w_u not counting beam weight = $(1.4)(1.0 - 0.3) + (1.7)(2) = 4.38 \text{ k/ft}$

$M_{\text{max}} = (10)(4.38)(4) - (4)(4.38)(2) = 140.2 \text{ ft-k} = 140,200 \text{ ft-lb}$

V_i due to w_u occuring same time as $M_{\text{max}} = (10)(4.38) - (4)(4.38)$

$$= 26.28 \text{ k} = 26,280 \text{ lb}$$

V_d = dead load shear = $(10)(1.0) - (4)(1.0) = 6 \text{ k} = 6000 \text{ lb}$

$d = 24 - 3 - 3 = 18''$ or $(0.8)(24) = \underline{19.2''}$

$V_{ci} = 0.6\sqrt{f'_c}b_w d + V_d + \dfrac{V_i M_{cr}}{M_{\text{max}}}$

$$= (0.6\sqrt{4000})(12)(19.2) + 6000 + \frac{(26,280)(212,780)}{140,200} = 54,628 \text{ lb}$$

$$> (1.7\sqrt{4000})(12)(19.2) = 24,772 \text{ lb}$$

Computing V_{cw}

f_{pc} = calculated compressive stress in psi at the centroid of the concrete
due to the effective prestress

$$= \frac{250,000}{(12)(24)} = 868 \text{ psi}$$

V_p = vertical component of effective prestress at section

$$= \left(\frac{9}{72.56}\right)(250{,}000) = 31{,}009 \text{ lb}$$

$$V_{cw} = (3.5\sqrt{f_c'} + 0.3f_{pc})b_w d + V_p$$

$$= (3.5\sqrt{4000} + 0.3 \times 868)(12)(19.2) + 31{,}009 = 142{,}006 \text{ lb}$$

Using Lesser of V_{ci} or V_{cw}

$$V_c = 54{,}628 \text{ lb} \qquad \blacksquare$$

■ **EXAMPLE 18.6**

Determine the spacing of #3 ⊔ stirrups required for the beam of Example 18.5 at 4 ft from the end support if f_{pu} is 250 ksi for the prestressing steel and f_y for the stirrups is 40 ksi. Use the value of V_c obtained by the approximate method, 50,684 lb.

SOLUTION

$$w_u = (1.4)(1.0) + (1.7)(2.0) = 4.8 \text{ k/ft}$$

$$V_u = (10)(4.8) - (4)(4.8) = 28.8 \text{ k}$$

$$\phi V_c = (0.85)(50{,}684) = 43{,}081 \text{ lb}$$

$$> V_u = 28{,}800 \text{ lb}$$

$$V_u > \frac{\phi V_c}{2} = 21{,}541 \text{ lb} < \phi V_c$$

A minimum amount of reinforcement is needed.

Since effective prestress is greater than 40% of tensile strength of reinforcing,

$$A_v = \left(\frac{A_{ps}}{80}\right)\left(\frac{f_{pu}}{f_y}\right)\left(\frac{s}{d}\right)\sqrt{\left(\frac{d}{b_w}\right)}$$

$$(2)(0.11) = \left(\frac{1.0}{80}\right)\left(\frac{250}{40}\right)\left(\frac{s}{18}\right)\sqrt{\frac{18}{12}}$$

$$s = 41.38'', \text{ but maximum } s = (\tfrac{3}{4})(24) = 18'' \qquad \underline{\text{Use } 18''} \quad \blacksquare$$

18.12 ADDITIONAL TOPICS

This chapter has presented a brief discussion of prestressed concrete. There are quite a few other important topics that have been omitted from this introductory material. Several of these items are briefly mentioned in the paragraphs to follow.

Stresses in End Blocks

The part of a prestressed member around the end anchorages of the steel tendons is called the *end block*. In this region the prestress forces are transferred from very concentrated areas out into the whole beam cross section. It has been found that the length of transfer for posttensioned members is less than the height of the beam and in fact is probably much less.

For posttensioned members there is direct-bearing compression at the end anchorage; as a result, solid end blocks are usually used there to spread out the concentrated prestress forces. To prevent bursting of the block, either wire mesh or a grid of vertical and horizontal reinforcing bars is placed near the end face of the beam. In addition, both vertical and horizontal reinforcing is placed throughout the block.

For pretensioned members where the prestress is transferred to the concrete by bond over a distance approximately equal to the beam depth, a solid end block is probably not necessary, but spaced stirrups are needed. A great deal of information on the subject of end block stresses for posttensioned and pretensioned members is available.[4]

Composite Construction

Precast prestressed sections are frequently used in buildings and bridges in combination with cast-in-place concrete. Should such members be properly designed for shear transfer so the two parts will act together as a unit, they are called *composite sections*. Examples of such members were previously shown in parts (e) and (f) of Figure 18.8. In composite construction the parts that are difficult to form and that contain most of the reinforcing are precast, whereas the slabs and perhaps the top of the beams, which are relatively easy to form, are cast in place.

The precast sections are normally designed to support their own weights plus the green cast-in-place concrete in the slabs plus any other loads applied during construction. The dead and live loads applied after the slab hardens are supported by the composite section. The combination of the two parts will yield a composite section that has a very large moment of inertia and thus a very large resisting moment. It is usually quite economical to use (a) a precast prestressed beam made with a high-strength concrete and (b) a slab made with an ordinary grade of concrete. If this practice is followed, it will be necessary to account for the different moduli of elasticity of the two materials in calculating the composite properties (thus it becomes a transformed area problem).

Continuous Members

Continuous prestressed sections may be cast in place completely with their tendons running continuously from one end to the other. It should be realized for such members that where the service loads tend to cause positive moments,

---

[4] Guyon, Y., 1960, *Prestressed Concrete* (New York: John Wiley & Sons), pp. 127–174.

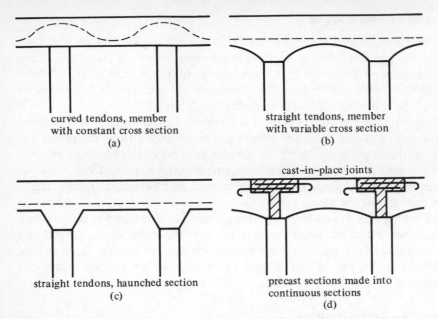

curved tendons, member
with constant cross section
(a)

straight tendons, member
with variable cross section
(b)

straight tendons, haunched section
(c)

cast–in–place joints

precast sections made into
continuous sections
(d)

Figure 18.14 Continuous beams.

the tendons should produce negative moments and vice versa. This means that the tendons should be below the member's center of gravity in normally positive moment regions and above the center of gravity in normally negative moment regions. To produce the desired stress distributions, it is possible to use curved tendons and members of constant cross section or straight tendons with members of variable cross section. In Figure 18.14 several continuous beams of these types are shown.

Another type of continuous section that has been very successfully used in the United States, particularly for bridge construction, involves the use of precast prestressed members made into continuous sections with cast-in-place concrete and regular reinforcing steel. Figure 18.14(d) shows such a case. For such construction the precast section resists a portion of the dead load, while the live load and the dead load that is applied after the cast-in-place concrete hardens are resisted by the continuous member.

Partial Prestressing

During the early days of prestressed concrete, the objective of the designer was to proportion members that could never be subject to tension when service loads were applied. Such members are said to be *fully prestressed*. Subsequent investigations of fully prestressed members have shown that they often have an appreciable amount of extra strength. As a result, many designers now believe that certain amounts of tensile stresses can be permitted under service loads. Members that are permitted to have some tensile stresses are said to be *partially prestressed*.

A major advantage of a partially prestressed beam is a decrease in camber. This is particularly important when the beam load or the dead load is quite low when compared to the total design load.

To provide additional safety for partially prestressed beams, it is common practice to add some conventional reinforcement. This reinforcement will increase the ultimate flexural strength of the members as well as help to carry the tensile stresses in the beam.[5]

PROBLEMS

18.1 The beam shown in the accompanying illustration has an effective total prestress of 240 k. Calculate the fiber stresses in the top and bottom of the beam at the ends and ℄. The tendons are assumed to be straight. (*Ans.* $f_{top} = -1.302$ ksi, $f_{bott} = -0.364$ ksi at the ℄.)

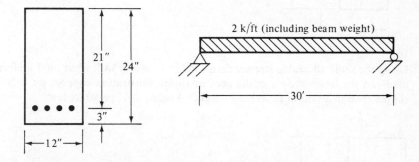

18.2 Compute the stresses in the top and bottom of the beam shown at the ends and center line immediately after the cables are cut. Assume straight cables. Initial prestress is 170 ksi.

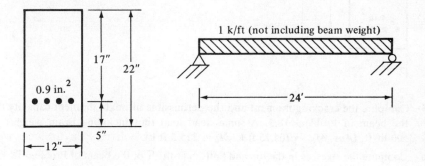

[5] Lin, T. Y. and Burns, N. H., 1981, *Design of Prestressed Concrete Structures*, 3rd ed. (New York: John Wiley & Sons), pp. 325–344.

18.3   The beam shown has a 30-ft simple span: $f'_c = 5000$ psi, $f_{pu} = 250,000$ psi, and the initial prestress is 160,000 psi.

   (a) Calculate the concrete stresses in the top and bottom of the beam at midspan immediately after the tendons are cut. (*Ans.* $f_{top} = +0.370$ ksi, $f_{bott} = -1.810$ ksi)

   (b) Recalculate the stresses at midspan after assumed prestress losses in the tendons of 20%. (*Ans.* $f_{top} = +0.212$ ksi, $f_{bott} = -1.364$ ksi)

   (c) What maximum service live load can this beam support if allowable stresses of $0.45f'_c$ in compression and $6\sqrt{f'_c}$ in tension are permitted? (*Ans.* 0.884 k/ft)

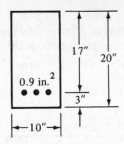

18.4   Using the same allowable stresses permitted in Problem 18.3, what total uniform load can the beam shown in the accompanying illustration support for a 50-ft simple span in addition to its own weight? Assume 20% prestress loss.

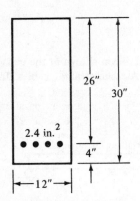

18.5   Compute the cracking moment and the permissible ultimate moment capacity of the beam of Problem 18.3. Assume dead load (not including beam weight) is 500 lb/ft. (*Ans.* $M_{cr} = 105.25$ ft-k, $M_u = 215.3$ ft-k)

18.6   Compute the stresses in the top and bottom at the ℄ of the beam of Problem 18.1 if it is picked up at the ℄. Assume 100% impact. Concrete weighs 150 lb-ft³.

18.7   Determine the stresses at the one-third points of the beam of Problem 18.6 if the beam is picked up at those points. (*Ans.* $f_{top} = +1.354$ ksi, $f_{bott} = -3.02$ ksi)

18.8 Calculate the design moment capacity of a 12-in. × 20-in. pretensioned beam that is prestressed with 1.2 in.^{2} of steel tendons stressed to an initial stress of 160 ksi. The center of gravity of the tendons is 3 in. above the bottom of the beam $f'_c = 5000$ psi and $f_{pu} = 250,000$ psi.

18.9 Compute the design moment capacity of the pretensioned beam of Problem 18.2 if $f'_c = 5000$ psi and $f_{pu} = 225$ ksi. (*Ans.* 208.1 ft-k)

18.10 Compute the design moment capacity of the bonded T beam shown in the accompanying illustration if $f'_c = 5000$ psi, $f_{pu}{'} = 250,000$ psi, and the initial stress in the cables is 160,000 psi.

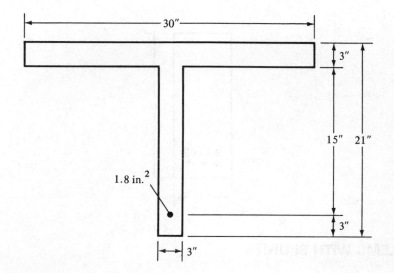

18.11 Calculate the deflection at the ℄ of the beam of Problem 18.3 immediately after the cables are cut, assuming cable stress $= 160,000$ psi. Calculate the same deflection two years later if a live load of 2 k/ft is considered and a creep coefficient of 1.8 is used. Assume a 20% prestress loss and straight cables. (*Ans.* $-0.464''$ and $+0.733''$)

18.12 Calculate the deflection at the ℄ of the beam of Problem 18.1 immediately after the cables are cut assuming P initial $= 240$ k and P after losses $= 190$ k. Repeat the calculation two years later if 20-k concentrated live loads are located at the one-third points of the beam and a creep coefficient of 2.0 is used. $f'_c = 5000$ psi. Assume that cables are straight and that no other loads are present other than the beam weight and the two 20-k loads. Use I_g for all calculations.

18.13 Determine the shearing strength of the beam shown in the accompanying illustration 3 ft from the supports using the approximate method allowed by the ACI Code. Determine the required spacing of #3 ⊔ stirrups at the same section, if $f'_c = 5000$ psi, $f_{pu} = 250,000$ psi, f_y for stirrups $= 50,000$ psi, and $f_{se} = 200,000$ psi. (*Ans.* $s = 21.73''$)

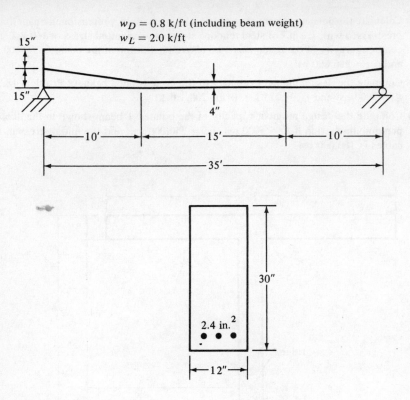

PROBLEMS WITH SI UNITS

18.14 Immediately after cutting the cables in the beam shown, they have an effective prestress of 1.260 GPa. Determine the stresses at the top and bottom of the beam at the ends and centerline. The concrete weighs 23.5 kN/m³. Cables are straight. $E_c = 27,924$ MPa.

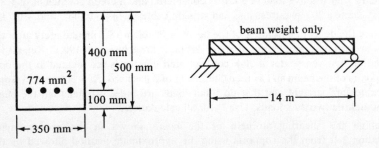

18.15 The beam shown has a 12-m simple span; $f'_c = 34.5$ MPa, $f_{pu} = 1.725$ GPa, and the initial prestress is 1.10 GPa.

(a) Calculate the concrete stresses in the top and bottom of the beam at mid-span immediately after the tendons are cut. (*Ans.* $f_{top} = +1.766$ MPa, $f_{bott} = -12.086$ MPa)

(b) Recalculate the stresses after assumed losses in the tendons of 18%. (*Ans.* $f_{top} = 0.618$ MPa, $f_{bott} = -9.08$ MPa)

(c) What maximum service uniform live load can the beam support in addition to its own weight if allowable stresses of $0.45f'_c$ in compression and $0.5\sqrt{f'_c}$ in tension are permitted? (*Ans.* 10.095 kN/m)

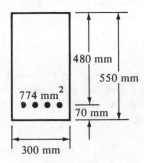

18.16 Using the same allowable stresses permitted and cable stresses in Problem 18.15, what total uniform load including beam weight can the beam shown in the accompanying illustration support for a 15-m simple span?

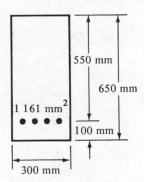

18.17 Compute the cracking moment and the design moment capacity of the bonded beam of Problem 18.15. (*Ans.* 181.68 kN·m, 438.07 kN·m)

18.18 Compute the stresses in the top and bottom of the beam of Problem 18.14 if it is picked up at its one-third points. Assume an impact of 100%.

18.19 Compute the design moment capacity of the bonded T beam shown in the accompanying illustration. $f'_c = 34.5$ MPa, $f_{pu} = 1.725$ GPa, and the initial stress in the cables is 1.100 GPa. (*Ans.* 843 kN·m)

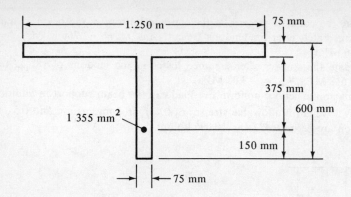

18.20 Calculate the deflection at the ℄ of the beam of Problem 18.14 immediately after the cables are cut. Repeat the calculation for a time two years later if 80 kN concentrated live loads are placed at the one-third points of the beam and a creep coefficient of 2.0 is used. Assume 18% loss in cable stresses. Use I_g.

19

Formwork

19.1 INTRODUCTION

Concrete forms are molds into which semiliquid concrete is placed. The molds need to be sufficiently strong to hold the concrete in the desired size and shape until the concrete hardens. Since forms are structures, they should be carefully and economically designed to support the imposed loads using the methods required for the design of other engineering structures.

Safety is a major concern in formwork because a rather large percentage of the accidents that occur during the construction of concrete structures is due to formwork failures. Normally formwork failures are not caused by the application of excessive gravity loading. Although such failures do occasionally occur, the usual failures are due to lateral forces that cause the supporting members to be displaced. These lateral forces may be caused by wind, by moving equipment on the forms, by vibration from passing traffic, or by the lateral pressure of freshly placed and vibrated concrete. A rather sad thing is that most of these failures could have been prevented if only a little additional lateral bracing had been used. There are, of course, other causes of failure, such as stripping the forms too early and improper control of the placement rate of the concrete.

Though you might think that shape, finish, and safety are the most important items in concrete formwork, you should realize that economy is also a major consideration. The cost of the formwork, which can range from one-third to almost two-thirds of the total cost of a concrete structure, is often more than the cost of both the concrete and the reinforcing steel. For the average concrete structure, the formwork is considered to represent about 50% of the total cost. *From this discussion it is obvious that any efforts made to improve the economy of concrete structures should be primarily concentrated on reducing formwork costs.* Formwork must be treated as an integral part of the overall job plan, and the lowest bidder will be the contractor who has planned the most economical forming job.

When designers are considering costs, they have a tendency to think only of quantities of materials. As a result, they will sometimes carefully design a structure with the lightest possible members and end up with some complicated and expensive formwork.

19.2 RESPONSIBILITY FOR FORMWORK DESIGN

Normally the structural designer is responsible for the design of reinforced concrete structures, while the design of the formwork and its construction is the responsibility of the contractor. For some structures, however, where the formwork is very complex, such as for shells, folded plates, and arches, it is desir-

Wastewater treatment plant, Santa Rosa, California. (Courtesy of Simpson Timber Company.)

able for the designer to take responsibility for the formwork design. There does seem to be an increasing trend, particularly on larger jobs, to employ structural designers to design and detail the formwork.

The contract documents should clearly specify the responsibility of each party for the work. Normally it is wise to allow the contractor as much room as possible to plan his or her own formwork and construction details, because more economical bids will often be the result.

It is to be remembered in contract law that if the contract documents describe exactly how something is to be done and at the same time spell out the results to be accomplished, it is not generally enforceable. In other words, if the exact form of the construction details are specified in the contract and followed by the contractor, and the formwork fails, the contractor cannot be held responsible.

19.3 MATERIALS USED FOR FORMWORK

Several decades ago, lumber was the universal material used for constructing concrete forms. The forms were constructed, used one time, and torn down, and there was little salvage other than perhaps a board or timber here or there. Such a practice still exists in some parts of the world where labor costs are very low. In a high-labor-cost area, such as the United States, the trend for several decades has been toward reusable forms made from many different materials. Wood is still the most widely used material, however, and no matter what types of material are used, some wood will still probably be required, whether it be lumber or plywood.

The lumber usually comes from the softwood varieties, with pine, fir, and western hemlock being the most common varieties. These woods are relatively light and are commonly available. Form lumber should be only partly seasoned because if it is too green, it shrinks and warps in hot dry weather. If it is too dry, it swells a great deal when it becomes wet. The lumber should be planed if it is going to be in contact with the concrete, but it is possible to use rough lumber for braces and shoring.

The introduction of plywood made great advances in formwork possible. Large sheets of plywood save a great deal of labor in the construction of the forms and at the same time result in large areas of joint-free concrete surfaces with corresponding reductions in costs of finishing and rubbing of the exposed concrete. Plywood also has considerable resistance to changes in shape when it becomes wet, can withstand rather rough usage, and can be bent to a certain degree, making it possible to construct curved surfaces.

Steel forms are extremely important for today's concrete construction. Not only are steel panel systems important for buildings, but steel bracing and framing are important where wood or plywood forms may be used. If steel forms are properly maintained, they may be reused many times. In addition, steel forms with their great strength may be used in places where other materials are not feasible, as in forming long spans. Forms for some other types of structures are frequently made with steel simply as a matter of expediency. Falling into this class are the forms for round columns, tunnels, and so on.

The Sonotube paper-fiber forms are another type of form often used for round columns. These patented forms are lightweight one-piece units that can be sawed to fit beams, utility outlets, and other structures. They have built-in moisture barriers that aid in curing. Because they are disposable, there are no cleaning, reshipping, or inventory costs. They can be quickly stripped with electric saws or hand tools.

Two other materials often used for concrete forms are glass-fiber-reinforced plastic and insulation boards. The glass-fiber-reinforced plastic can be sprayed over wooden base forms or can be used to fabricate special forms such as the pans for the waffle-type floors. Quite a few types of insulation board used as form liners are on the market. They are fastened to the forms, and when the forms are removed, the boards are left in place either bonded to the concrete or held in place by some type of clips.

19.4 FURNISHING OF FORMWORK

There are companies that specialize in formwork and from whom forms, shoring, and needed accessories can be rented or purchased. The contractor may very well take bids from these types of companies for forms for columns, floors, and walls. If structures are designed with standard dimensions, the use of rented forms may provide the most economical solution. Such a process makes the contractor's job a little easier. He or she can regulate and smooth out the work load of his company by assigning the responsibility and risk involved in the formwork to a company that specializes in that particular area. Renting formwork permits the contractor to reduce his investment in the many stock items, such as spacers, fasteners, and so forth, that would be necessary if he did all of the formwork himself.

If forms of the desired sizes are not available for rent or purchase, it is possible for the contractor to make his own forms or to have them made by one of the companies that specialize in form manufacture. If this latter course is chosen, the result is usually very high quality formwork.

Finally, the forms can be constructed in place. Though the quality may not be as high as for the last two methods mentioned, it may be the only option available for complicated structures for which the forms cannot be reused. Of course, for many jobs a combination of rented or built-in-place or shop-built forms may be used for best economy.

19.5 ECONOMY IN FORMWORK

The first opportunities for form economy occur during the design of the structure. Concrete designers can often save the owner a great deal of money if they will take into account those factors that permit formwork economy in the design of the structure. Among the items that can be done in this regard are the following:

1. Attempt to coordinate the design with the architectural design, such as varying room sizes a little to accommodate standard forms.
2. Keep story heights as well as the sizes of beams, slabs, and columns the same for as many floors as possible to permit the reuse of forms without change from floor to floor. Without a doubt the most important item affecting the total cost of formwork is the number of times the forms can be used.
3. Beams should be made as wide or wider than columns so as to make connections simpler. When beams are narrower than columns, formwork is complicated and expensive.
4. Remember in the design of columns for multistory buildings that their sizes may be kept constant for a good many floors by changing the percentage of steel used from floor to floor, starting at the top with approximately 1% steel and increasing it floor by floor until the maximum percentage is reached.
5. The more we keep columns the same size, the smaller our costs will be. Should it be necessary to change the size of a particular column as we move vertically in a building, it is desirable to do it in 2-in. increments—one on one side of a column on one floor and then one on the other side on another floor.
6. Design structures that permit the use of commercial forms, such as metal pans or corrugated metal deck sheets for floor or roof slabs.
7. It is desirable to use the same thickness for all of the reinforced concrete walls in a particular project. Wall openings should be kept to a minimum and should if possible be placed in the same locations in each wall. It is further desirable to keep footing elevations constant under a particular wall. Where this is not possible, as few footing steps as possible should be used.
8. Use beam and column shapes that do not have complicated haunches, offsets, and cutouts.
9. Usually wide flat beams are more economical than are deep narrow beams because it may be difficult to place the concrete and the reinforcing in the latter types.
10. For reinforced concrete frames the floor slabs and their required formwork are usually the most expensive items. From the standpoint of formwork, one-way flat slabs and two-way flat plates are the most economical floor systems. On the other extreme, the two-way beam and slab floors and the one-way beam and slab floors require the most expensive formwork. In between these two extremes fall the one-way and two-way pan floors.
11. Allow the contractor to use his or her own methods in constructing the forms, holding the contractor responsible only for their adequacy.

19.6 FORM MAINTENANCE

To be able to reuse forms many times, it is obvious that they should be properly removed, maintained, and stored. One particular thing the contractor can do to appreciably lengthen form lives is to require the same crews to do both erection and stripping. If while stripping a set of forms, the workers know that they are going to have to erect those same forms for the next job, they will be much more careful in their handling and maintenance. (Using the same crews for both jobs may not be possible in some areas because of labor contracts.)

Metal bars or pries should not be used for stripping plywood forms because they may easily cause splitting. Wooden wedges, which are gradually tapped to separate the forms from the hardened concrete, are preferable. After the forms are removed, they should be thoroughly cleaned and oiled. Any concrete that has adhered to metal parts of the forms should be thoroughly scraped off. The holes in wooden form faces caused by nails, ties (to be described in Section 19.7), and other items should be carefully filled with metal plates, cork, or plastic materials. After the cleaning and repairs are completed, the forms should be coated with oil or other preservative. Steel forms should be coated both front and back to keep them from rusting and to prevent spilled concrete from sticking to them.

Plywood surfaces should be lightly oiled, and the forms should be stored in a flat position with the oiled faces together. The stored panels should be kept out of the sun and rain or should be lightly covered so as to permit air circulation and prevent heat buildup. In addition, wood strips may be placed between forms to increase this ventilation. Although some plywood forms have reportedly been used successfully for 200 or more times, it is normal with proper maintenance to be able to use them approximately 30 or 40 times. After plywood panels are deemed to be no longer suitable for formwork, they can often be used by the wise contractor for subflooring or wall or roof sheathing.

Before forms are used, their surfaces should be wetted and oiled or coated with some type of material that will not stain or soften the concrete. The coating is primarily used to keep the concrete from sticking to the forms. There are a large number of compounds on the market that will reduce sticking and will at the same time serve as sealers or protective coatings to substantially reduce the absorption of water by the forms from the concrete.

Various types of oil or oil compounds have been the most commonly used form-release agents used in the past for both wood and metal surfaces. There are many form-release and sealer agents available, but before trying a new one on a large job, the contractor would be wise to experiment with a small application. No material should be used that will form a coating on the concrete which will, after stripping, interfere with the wetting of the concrete for curing purposes. In the same manner, if the concrete is later to be plastered or painted, the coating should not be one that will leave a waxy or oily surface that will interfere with the sticking of paint or plaster.

There are various kinds of products used to coat the plywood during its manufacture. Some manufacturers of these products say that the oiling of forms

Formwork for Water Tower Place, Chicago, Illinois. (Courtesy of Symons Corporation.)

coated with their products is not necessary in the field. Glass-fiber-reinforced plastic forms, which are used a great deal for precast concrete and for architectural concrete because of the very excellent surfaces produced, need to be oiled only very lightly before use. In fact, they have been satisfactorily used by some contractors with no oiling whatsoever.

19.7 DEFINITIONS

In this section several terms relating to formwork, with which you need to be familiar, are introduced. They are not listed alphabetically but rather are given in the order in which the author thinks they will help the reader understand the material in subsequent sections of this chapter.

Double-headed nails: A tremendous number of devices have been developed to simplify the erection and stripping of forms. For instance, there are the very simple double-headed nails. These nails can be easily removed, permitting the quick removal of braces and dismantling of forms.

Double-headed nail.

Sheathing: The material that forms the contact face of the forms to the concrete is called sheathing or lagging or sheeting. (See Figure 19.1.)

Joists: The usually horizontal beams that support the sheathing for floor and roof slabs are called joists. (See Figure 19.1.)

Stringers: The usually horizontal beams that support the joists and that rest on the vertical supports are called stringers. (See Figure 19.1.)

Shores: The temporary members (probably wood or metal) that are used for vertical support for formwork holding up fresh concrete are called shores. (See Figure 19.1.) Other names given to these members are struts, posts, or props. The whole system of vertical supports for a particular structure is called the shoring.

Ties: Devices used to support both sides of wall forms against the lateral pressure of the concrete are called ties. They pass through the concrete and are fastened on each side. (See Figure 19.2.) The lateral pressure caused by the fresh concrete, which tends to push out against the forms, is resisted in tension by the ties. They can, when correctly used, eliminate most of the external braces otherwise required for wall forms. They may or may not have a provision for holding the forms the desired distance apart.

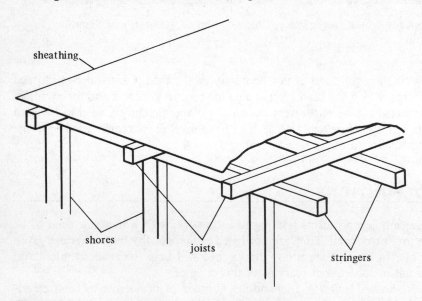

Figure 19.1

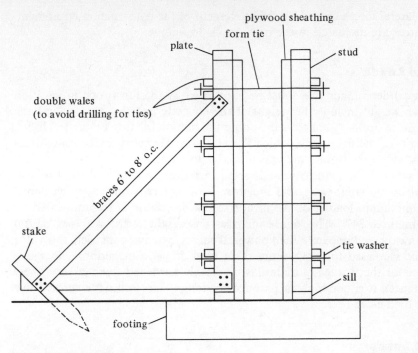

Figure 19.2 Wood formwork for a wall.

Spreaders: Braces, usually made of wood, that are placed in a form to keep its two sides from being drawn together are called spreaders. They should be removed when the concrete is placed so they are not cast in the concrete.

Snap ties: Rather than using spreaders separate from the ties, it is more common to use a type of spreader called a snap tie, which is combined with the spreader. The ends of these patented devices can be twisted or snapped off a little distance from the face of the concrete. After the forms are removed, the small holes resulting can easily be repaired with cement mortar.

Studs: The vertical members that support sheathing for wall forms are called studs. (See Figure 19.2.)

Wales: The long horizontal members that support the studs are called wales. (See Figure 19.2.) Notice that each wale usually consists of two timbers with the ties passing in between so that holes do not have to be drilled.

19.8 FORCES APPLIED TO CONCRETE FORMS

Formwork must be designed to resist all vertical and lateral loads applied to it until the concrete gains sufficient strength to do the job itself. The vertical loads applied to the forms include the weights of the concrete, the reinforcing steel, and the forms themselves, as well as the construction live loads. The lateral loads include the liquid pressure of freshly placed concrete, wind forces, and the lateral forces caused by the movement of the construction equipment. These forces are discussed in the paragraphs to follow.

Vertical Loads

The vertical dead loads that must be supported by the formwork in addition to its own weight include the weight of the concrete and the reinforcing bars. The weight of ordinary concrete including the reinforcing is taken as 150 lb/ft^3. The weight of the formwork, which is frequently neglected in the design calculations, may vary from 3 or 4 psf up to 12 to 15 psf.

The vertical live load to be supported includes the weight of the workers, the construction equipment, and the storage of materials on freshly hardened slabs. A minimum load of 50 psf of horizontal projection is recommended by ACI Committee 347.[1] This figure includes the weight of the workers, equipment, runways, and impact. When powered concrete buggies are used, this value should be increased to a minimum of at least 75 psf. Furthermore, to make allowance for the storage of materials on freshly hardened slabs (a very likely circumstance), it is common in practice to design slab forms for vertical live loads as high as 150 psf.

Lateral Loads

For walls and columns the loads are different from those of slabs. The semiliquid concrete is placed in the form and exerts lateral pressures against the form as does any liquid. The amount of this pressure is dependent on the concrete weight per cubic foot, the rate of placing, the temperature of the concrete, and the method of placing.

The exact pressures applied to forms by freshly placed concrete are difficult to estimate due to several factors. These factors include the following:

1. The rate of placing the concrete. Obviously, the faster the concrete is placed, the higher the pressure will be.
2. The temperature of the concrete. The liquid pressure on the forms varies quite a bit with different temperatures. For instance, the pressure when the concrete is at 50°F is appreciably higher than when the concrete is at 70°F. The reason for this is that concrete placed at 50°F hardens at a slower rate than does concrete deposited at 70°F. As a result, the colder concrete remains in a semiliquid state for a longer time and thus to a greater depth, and pressures may be as much as 25% or more higher. Because of this fact, forms for winter-placed concrete should be designed

[1] ACI, Committee 301, 1972, *Specifications for Structural Concrete for Buildings* (ACI 301-72) (Detroit: American Concrete Institute), p. 156.

Pulp mill kiln near Castlegar, British Columbia. (Courtesy of
Economy Forms Corporation.)

for much higher lateral pressures than those designed for summer-placed
concrete.

3. The method of placing the concrete. If high-frequency vibration is used,
the concrete is kept in a liquid state to a fairly high depth, and it acts
very much like a liquid with a weight per cubic foot equal to that of the
concrete. Vibration may increase lateral pressures by as much as 20%
over pressures caused by spading.

4. Size and shape of the forms and the consistency and proportions of the
concrete. These items do affect to some degree lateral pressures but for
the usual building are felt to be negligible.

The pressures that occur in formwork due to the semiliquid concrete are often of such magnitude that other people do not believe it when the designer says the lateral pressure is 1500 psf or 1800 psf or more. ACI Committee 347 has published recommended formulas for calculating lateral concrete pressures for different temperatures and rates of placing the concrete. In these expressions, which follow, p is the maximum equivalent liquid pressure (in pounds per square foot) at any elevation in the form, R is the vertical rate of concrete placement (in feet per hour), T is the temperature of the concrete in the forms, and h is the maximum height of fresh concrete above the point being considered. These expressions are to be used for walls for which concrete with a slump no greater than 4 in. is placed at a rate not exceeding 7 ft/hr. Vibration depth is limited to 4 ft below the concrete surface.[2]

In Walls with R not Exceeding 7 ft/hr

$$p = 150 + \frac{9000R}{T} \quad \text{(maximum 2000 psf or } 150h, \text{ whichever is less)}$$

In Walls with R Greater Than 7 ft/hr but not Greater Than 10 ft/hr

$$p = 150 + \frac{43{,}400}{T} + \frac{2800R}{T} \quad \text{(maximum 2000 psf or } 150h, \text{ whichever is less)}$$

Column forms are often filled very quickly. They may in fact be completely filled in less time than is required for the bottom concrete to set. Furthermore, vibration will frequently extend for the full depth of the form. As a result, greater lateral pressures are produced than would be expected for the wall conditions just considered. ACI Committee 347 presents the following equation for estimating the maximum pressures for column form design.

Columns with Maximum Horizontal Dimension of 6 ft or Less

$$p = 150 + \frac{9000R}{T} \quad \text{(maximum 3000 psf or } 150h, \text{ whichever is less)}$$

In Figure 19.3 a comparison of the pressures calculated with the preceding expressions is presented for temperatures ranging from 30°F to 100°F.

In addition to the lateral pressures on the forms caused by the fresh concrete, it is necessary for the braces and shoring to be designed to resist all other possible lateral forces, such as wind, dumping of concrete, movement of equipment, and other forces.

---

2 ACI Committee 301, 1972, *Specifications for Structural Concrete for Buildings* (ACI 301-72) (Detroit: American Concrete Institute), pp. 156–157.

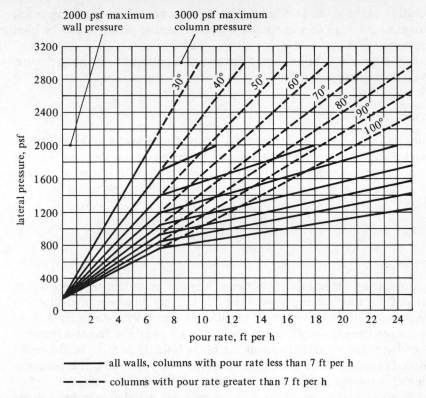

2000 psf maximum wall pressure

3000 psf maximum column pressure

——— all walls, columns with pour rate less than 7 ft per h

— — — columns with pour rate greater than 7 ft per h

Figure 19.3 Lateral pressure of fresh concrete in columns and walls. (Courtesy of American Plywood Association.)

19.9 ANALYSIS OF FORMWORK FOR FLOOR AND ROOF SLABS

This section is devoted to the calculations needed for analyzing formwork for concrete floor and roof slabs; its design is presented in the following section. Because wood formwork is the most common type faced in practice, the discussion and examples of this chapter are primarily concerned with wood.

Wood members have a very useful property with which you should be familiar, and that is their ability to support excessive loads for short periods of time. As a result of this characteristic, it is a common practice to allow them a 25% increase over their normal allowable stresses if the applied loads are of short duration.

Formwork is usually considered to be a temporary type of structure because it remains in place only a short time. Furthermore, the loads supported by the forms reach a peak during pouring activity and then rapidly fall off as the concrete hardens around the reinforcing and begins to support the load. As a result, the formwork designer will want to use increased allowable stresses where possible. The ability of wood to take overloads is based on the total time of application of the loads. In other words, if forms are to be used many times, the amount of overload they can resist is not as large as if they are to be used only one time.

In the paragraphs to follow, some discussion is presented regarding flexure, shear, and deflections in formwork.

Flexure

Normally sheathing, joists, and stringers are continuous over several spans. For such cases it seems reasonable to assume that the maximum moment is equal to the maximum moment that would occur in a uniformly loaded span continuous over three or more spans. This value is approximately equal to

$$M = \frac{w\ell^2}{10}$$

Shear

The horizontal and vertical shear stresses at any one point in a beam are equal. For materials for which strengths are the same in every direction, no distinction is made between shear values acting in different directions. Materials such as wood, however, have entirely different shear strengths in the different directions. Wood tends to split or shear between its fibers (usually parallel to the beam axis). Since horizontal shear is rather critical for wood members, it is common in talking about wood formwork to use the term "horizontal shear."

The horizontal shearing stress in a rectangular wooden member can be calculated by the usual formula as follows:

$$f_v = \frac{VQ}{Ib} = \frac{(V)[b \times (h/2) \times (h/4)]}{(\frac{1}{12}bh^3)(b)} = \frac{3V}{2bh} = \frac{3V}{2A}$$

In this expression, V is equal to $w\ell/2$ for uniformly loaded simple spans. As in earlier chapters relating to reinforced concrete, it is permissible to calculate the shearing stress at a distance h from the face of the support. If h is given in inches and if w is the uniform load per foot, the external shear at a distance h from the support can be calculated as follows:

$$V = 0.5w\ell - \frac{h}{12}w$$

$$V = 0.5w\left(\ell - \frac{2h}{12}\right)$$

The sheathing, joists, stringers, wales, and so on for formwork are normally continuous. For uniformly loaded beams continuous over three or more spans, V is approximately equal to $0.6w\ell$. At a distance h from the support, it is *assumed* to equal the following value:

$$V = 0.6w\left(\ell - \frac{2h}{12}\right)$$

The allowable shear stresses for construction grades of lumber are quite low, and the designer is not really out of line if he or she increases the given allowable shear stresses by something more than 25% if the forms in question are to be used only once. Such increases are included in the allowable stresses given later in this chapter. It is not usually necessary to check the shearing stresses in sheathing where they are normally very low.

Deflections

Formwork must be designed to limit deflections to certain maximum values. If deflections are not controlled, the end results will be unsightly looking concrete, with bulges and perhaps cracks marring its appearance. The amounts of deflection permitted are dependent on the desired finish of the concrete as well as its location. For instance, where a rough finish is used, small deflections might not be very obvious, but where smooth surfaces are desired, small deflections are easily seen and are quite objectionable.

For some concrete forms, the deflection limitations of $\ell/240$ or $\ell/270$ are considered to be satisfactory, while for others, particularly for horizontal surfaces, a limitation of $\ell/360$ may be required. For some cases, particularly for architectural exposure, even tighter limitations may be required, perhaps $\ell/480$.

It should also be realized that perfectly straight beams and girders seem to the eye of a person below to sag downwards; consequently, just a little downward deflection seems to be very large and quite detrimental to the appearance of the structure. For this reason it is considered good practice to camber the forms of such members so they do not appear to be deflecting. A camber frequently used is about $\frac{1}{4}$ in. for each 10 ft of length.

For uniformly loaded equal-span beams continuous over three or more spans, the maximum centerline deflections can be approximately determined with the following expression:

$$\delta = \frac{w\ell^4}{145EI}$$

In using deflection expressions it is to be noted that the deflections that occur in the formwork are going to be permanent in the completed structure, and thus the modulus of elasticity E should not be increased even if the forms are to be used just one time. As a matter of fact, it may be necessary to reduce E rather than increase it for some parts of the formwork. When wood becomes wet, it becomes more flexible, with the result that larger deflections occur. You can see that wet formwork, particularly the sheathing, is the normal situation. As a result, it is considered wise to use a smaller E than normally permitted and a reduction factor of $\frac{10}{11}$ is commonly specified.

Three tables useful for both the analysis and design of formwork are presented in this section. The first of these, Table 19.1, gives the properties of several sizes of lumber that are commonly used for formwork. These properties include the dressed dimensions, cross-sectional areas, moments of inertia, and section moduli.

TABLE 19.1 PROPERTIES OF AMERICAN STANDARD BOARD, PLANK, DIMENSION, AND TIMBER SIZES COMMONLY USED FOR FORM CONSTRUCTION (Courtesy of the American Concrete Institute)

Nominal size in inches $b \times h$	American Standard size in inches $b \times h$ S4S^b 19% maximum moisture	Area of section, $A = bh$ (sq. in.)		Moment of inertia (in.^{4}) $I = \dfrac{bh^2}{12}$		Section modulus (in.^{3}) $S = \dfrac{bh^2}{6}$		Board feet per linear foot of piece
		Rough	S4S	Rough	S4S	Rough	S4S	
4×1	$3\frac{1}{2} \times \frac{3}{4}$	3.17	2.62	0.20	0.12	0.46	0.33	$\frac{1}{3}$
6×1	$5\frac{1}{2} \times \frac{3}{4}$	4.92	4.12	0.31	0.19	0.72	0.52	$\frac{1}{2}$
8×1	$7\frac{1}{4} \times \frac{3}{4}$	6.45	5.44	0.41	0.25	0.94	0.68	$\frac{2}{3}$
10×1	$9\frac{1}{4} \times \frac{3}{4}$	8.20	6.94	0.52	0.32	1.20	0.87	$\frac{5}{6}$
12×1	$11\frac{1}{4} \times \frac{3}{4}$	9.95	8.44	0.63	0.39	1.45	1.05	1
$4 \times 1\frac{1}{4}$	$3\frac{1}{2} \times 1$	4.08	3.50	0.43	0.29	0.76	0.58	$\frac{5}{12}$
$6 \times 1\frac{1}{4}$	$5\frac{1}{2} \times 1$	6.33	5.50	0.68	0.46	1.19	0.92	$\frac{5}{8}$
$8 \times 1\frac{1}{4}$	$7\frac{1}{4} \times 1$	8.30	7.25	0.87	0.60	1.56	1.21	$\frac{5}{6}$
$10 \times 1\frac{1}{4}$	$9\frac{1}{4} \times 1$	10.55	9.25	1.11	0.77	1.98	1.54	$1\frac{1}{24}$
$12 \times 1\frac{1}{4}$	$11\frac{1}{4} \times 1$	12.80	11.25	1.35	0.94	2.40	1.87	$1\frac{1}{4}$
$4 \times 1\frac{1}{2}$	$3\frac{1}{2} \times 1\frac{1}{4}$	4.98	4.37	0.78	0.57	1.14	0.91	$\frac{1}{2}$
$6 \times 1\frac{1}{2}$	$5\frac{1}{2} \times 1\frac{1}{4}$	7.73	6.87	1.22	0.89	1.77	1.43	$\frac{3}{4}$
$8 \times 1\frac{1}{2}$	$7\frac{1}{4} \times 1\frac{1}{4}$	10.14	9.06	1.60	1.18	2.32	1.89	1
$10 \times 1\frac{1}{2}$	$9\frac{1}{4} \times 1\frac{1}{4}$	12.89	11.56	2.03	1.50	2.95	2.41	$1\frac{1}{4}$
$12 \times 1\frac{1}{2}$	$11\frac{1}{4} \times 1\frac{1}{4}$	15.64	14.06	2.46	1.83	3.58	2.93	$1\frac{1}{2}$
4×2	$3\frac{1}{2} \times 1\frac{1}{2}$	5.89	5.25	1.30	0.98	1.60	1.31	$\frac{2}{3}$
6×2	$5\frac{1}{2} \times 1\frac{1}{2}$	9.14	8.25	2.01	1.55	2.48	2.06	1
8×2	$7\frac{1}{4} \times 1\frac{1}{2}$	11.98	10.87	2.64	2.04	3.25	2.72	$1\frac{1}{3}$
10×2	$9\frac{1}{4} \times 1\frac{1}{2}$	15.23	13.87	3.35	2.60	4.13	3.47	$1\frac{2}{3}$
12×2	$11\frac{1}{4} \times 1\frac{1}{2}$	18.48	16.87	4.07	3.16	5.01	4.21	2

Plywood is the standard material used in practice for floor and wall sheathing for formwork, and it is so used for the examples presented in this chapter. Almost all exterior plywoods manufactured with waterproof glue can be satisfactorily used, but the industry produces a particular type called *plyform* intended especially for formwork. It is this product that is referred to in the remainder of this chapter.

Plyform is manufactured by gluing together an odd number of sheets of wood with the grain of each sheet being perpendicular to the grain of the sheet adjoining it. The alternating of the grains of the wood sheets greatly reduces the shrinkage and warping of the resulting panels. Plyform can be obtained in thicknesses from $\frac{1}{8}$ to $1\frac{1}{8}$ in. Actually, however, many suppliers only carry the $\frac{5}{8}$- and $\frac{3}{4}$-in. sizes and the others may have to be specially ordered.

TABLE 19.1   (*continued*)

Nominal size in inches $b \times h$	American Standard size in inches $b \times h$ S4S^b 19% maximum moisture	Area of section, $A = bh$ (sq. in.) Rough	S4S	Moment of inertia (in.^{4}) $I = \dfrac{bh^2}{12}$ Rough	S4S	Section modulus (in.^{3}) $S = \dfrac{bh^2}{6}$ Rough	S4S	Board feet per linear foot of piece
2 × 4	1½ × 3½	5.89	5.25	6.45	5.36	3.56	3.06	⅔
2 × 6	1½ × 5½	9.14	8.25	24.10	20.80	8.57	7.56	1
2 × 8	1½ × 7¼	11.98	10.87	54.32	47.63	14.73	13.14	1⅓
2 × 10	1½ × 9¼	15.23	13.87	111.58	98.93	23.80	21.39	1⅔
2 × 12	1½ × 11¼	18.48	16.87	199.31	177.97	35.04	31.64	2
3 × 4	2½ × 3½	9.52	8.75	10.42	8.93	5.75	5.10	1
3 × 6	2½ × 5½	14.77	13.75	38.93	34.66	13.84	12.60	1½
3 × 8	2½ × 7¼	19.36	18.12	87.74	79.39	23.80	21.90	2
3 × 10	2½ × 9¼	24.61	23.12	180.24	164.89	38.45	35.65	2½
3 × 12	2½ × 11¼	29.86	28.12	321.96	296.63	56.61	52.73	3
4 × 4	3½ × 3½	13.14	12.25	14.39	12.50	7.94	7.15	1⅓
4 × 6	3½ × 5½	20.39	19.25	53.76	48.53	19.12	17.65	2
4 × 8	3½ × 7¼	26.73	25.38	121.17	111.15	32.86	30.66	2⅔
4 × 10	3½ × 9¼	33.98	32.38	248.91	230.84	53.10	49.91	3⅓
6 × 3	5½ × 2½	14.77	13.75	8.48	7.16	6.46	5.73	1½
6 × 4	5½ × 3½	20.39	19.25	22.33	19.65	12.32	11.23	2
6 × 6	5½ × 5½	31.64	30.25	83.43	76.26	29.66	27.73	3
6 × 8	5½ × 7½	42.89	41.25	207.81	193.36	54.51	51.56	4
8 × 8	7½ × 7½	58.14	56.25	281.69	263.67	73.89	70.31	5⅓

* Rough dry sizes are ⅛ in. larger, both dimensions.

The various types of wood used for manufacturing plywood (fir, spruce, larch, redwood, cedar, etc.) have different strengths and stiffnesses. Those types of woods having similar properties are assigned to a particular species group. To simplify design calculations the plywood industry has accounted for the different species and the effects of gluing the wood sheets together with alternating grains in developing the properties given in Table 19.2. As a result, the designer needs only to consider the allowable stresses for the plyform and the plyform properties given in the table.

It will be noted that section properties are given in Table 19.2 for stresses applied both parallel to the face grain and perpendicular to it. Should the face grain be placed parallel to the direction of bending, the section is stronger; it is weaker if placed in the opposite direction. This situation is shown in Figure 19.4.

TABLE 19.2 SECTION PROPERTIES AND ALLOWABLE STRESSES FOR CERTAIN SIZES OF PLYFORM
(Courtesy of the American Plywood Association)

Thickness (inches)	Approximate weight (psf)	Properties for stress applied parallel with face grain			Properties for stress applied perpendicular to face grain		
		Moment of inertia, I (in.^{4}/ft)	Effective section modulus, K_S (in.^{3}/ft)	Rolling shear constant, Ib/Q (in.^{2}/ft)	Moment of inertia, I (in.^{4}/ft)	Effective section modulus, K_S (in.^{3}/ft)	Rolling shear constant, Ib/Q (in.^{2}/ft)
Class I							
$\frac{1}{2}$	1.5	0.077	0.268	5.127	0.035	0.167	2.919
$\frac{5}{8}$	1.8	0.130	0.358	6.427	0.064	0.250	3.692
$\frac{3}{4}$	2.2	0.199	0.455	7.854	0.136	0.415	4.565
$\frac{7}{8}$	2.6	0.280	0.553	8.204	0.230	0.581	5.418
1	3.0	0.427	0.737	8.871	0.373	0.798	7.242
$1\frac{1}{8}$	3.3	0.554	0.849	9.872	0.530	0.986	8.566
Class II							
$\frac{1}{2}$	1.5	0.075	0.267	4.868	0.029	0.182	2.671
$\frac{5}{8}$	1.8	0.130	0.357	6.463	0.053	0.320	3.890
$\frac{3}{4}$	2.2	0.198	0.455	7.892	0.111	0.530	4.814
$\frac{7}{8}$	2.6	0.280	0.553	8.031	0.186	0.742	5.716
1	3.0	0.421	0.754	8.614	0.301	1.020	7.645
$1\frac{1}{8}$	3.3	0.566	0.869	9.571	0.429	1.260	9.032
Structural I							
$\frac{1}{2}$	1.5	0.078	0.271	5.166	0.042	0.229	3.076
$\frac{5}{8}$	1.8	0.131	0.361	6.526	0.077	0.343	3.887
$\frac{3}{4}$	2.2	0.202	0.464	7.926	0.162	0.570	4.812
$\frac{7}{8}$	2.6	0.288	0.569	7.539	0.275	0.798	6.242
1	3.0	0.479	0.827	7.978	0.445	1.098	7.639
$1\frac{1}{8}$	3.3	0.623	0.955	8.841	0.634	1.356	9.031

Note: All properties are adjusted to account for reduced effectiveness of plies with grain perpendicular to applied stress.

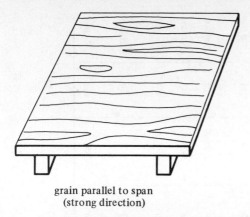

grain parallel to span
(strong direction)

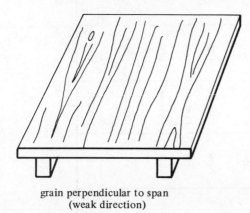

grain perpendicular to span
(weak direction)

Figure 19.4

In plyform the shearing stresses between the plies or along the glue lines are calculated with the usual expression $H = f_v = VQ/Ib$ and are referred to as the *rolling shear stresses*. For convenience in making the calculations, the values of Ib/Q are presented in the table for the different plyform sizes, and H can be simply determined for a particular case by dividing V by the appropriate Ib/Q value from the table.

The normal framing applications of plyform are for wet conditions, and adjustments in allowable stresses should be made for those conditions as well as for duration of load and other experience factors. As a result of these items, the following allowable stresses are normally given for plyform:

	Class I plyform	Class II plyform	Structural I plyform
Modulus of elasticity (psi)	1,650,000	1,430.000	1,650,000
Bending stress (psi)	1930	1330	1930
Rolling shear stress (psi)	80	72	102

TABLE 19.3 SUGGESTED WORKING STRESSES FOR DESIGN OF WOOD FORMWORK

Species and grade of form framing lumber	Class I formwork, single use or light construction					Class II formwork, multiple use or heavy construction				
	Allowable unit stress, psi					Allowable unit stress, psi				
	Extreme fiber bending	Compression ⊥ to grain	Compression ∥ to grain	Horizontal shear	Modulus of elasticity, psi	Extreme fiber bending	Compression ⊥ to grain	Compression ∥ to grain	Horizontal shear	Modulus of elasticity, psi
Douglas fir, coastal construction	1875	490	1500	180	1,760,000	1500	390	1200	145	1,760,000
Douglas fir inland, common structural	1815	475	1565	180	1,760,000	1450	380	1250	145	1,760,000
Hemlock, west coast construction	1875	455	1380	150	1,540,000	1500	365	1100	120	1,540,000
Larch, common structural	1815	490	1660	180	1,650,000	1450	390	1325	145	1,650,000
Pine, southern No. 1 SR	1875	490	1625–1875	220	1,760,000	1500	390	1300–1500	175	1,760,000
Redwood, heart structural	1625	400	1380	135	1,320,000	1300	320	1100	110	1,320,000
Spruce, eastern, 1450 f structural	1815	375	1310	160	1,320,000	1450	300	1050	130	1,320,000

TABLE 19.3 (continued) SHEATHING   wet or damp location assumed

Species and grade of form framing lumber	Class I formwork, single use or light construction					Class II formwork, multiple use or heavy construction				
	Allowable unit stress, psi					Allowable unit stress, psi				
	Extreme fiber bending	Compression ⊥ to grain	Compression ∥ to grain	Horizontal shear	Modulus of elasticity, psi	Extreme fiber bending	Compression ⊥ to grain	Compression ∥ to grain	Horizontal shear	Modulus of elasticity, psi
Plywood, Douglas fir, concrete form B-B Board sheathing, all species	2000 Use 100% of value for species, shown above	(Bearing on face) 435 Use 67% of value for species, shown above	1500 Use 90% of value for species, shown above	* Use 100% of value for species, shown above	1,600,000 Use ⅔ of value for species, shown above	1500 Use 100% of value for species, shown above	(Bearing on face) 325 Use 67% of value for species, shown above	1100 Use 90% of value for species, shown above	* Use 100% of value for species, shown above	1,600,000 Use ⅔ of value for species, shown above

Sources: Derived from "National Design Specification for Stress-Grade Lumber and Its Fastenings," 1960 edition, as amended, and from "Recommendations of the Douglas Fir Plywood Association."

* Shear is not a governing design consideration for plywood form panels except in the case of very short, heavily loaded spans. If 1-in. or 1⅛-in. panels are being used, check for rolling shear according to plywood manufacturer's suggestions.

Table 19.3 shows allowable bending stresses, allowable compression stresses perpendicular and parallel to the grain, allowable shearing stresses, and moduli of elasticity for several types of lumber often used in formwork.

Example 19.1 illustrates the calculations necessary to determine the bending and shear stresses and the maximum deflections that occur in the sheathing, joists, and stringers used as the formwork for a certain concrete slab and live loading. In an actual design problem, it may be necessary to consider several different arrangements of the joists, stringers, and so on and the different grades of lumber available with their different allowable stresses before a final design is selected.

■ EXAMPLE 19.1

The formwork for a 6-in-thick reinforced concrete floor slab of normal weight consists of $\frac{3}{4}$-in. class II plyform sheathing supported by 2- × 6-in. joists spaced 2 ft 0 in. on center, which in turn are supported by 2- × 8-in. stringers spaced 4 ft 0 in. on centers. The stringers are themselves supported by shores spaced 4 ft 0 in. on centers.

The joists and stringers are constructed with Douglas fir, coastal construction and are to be used many times. The allowable flexural stress in the fir members is 1875 psi, the allowable horizontal shear stress is 180 psi, and the modulus of elasticity is 1,760,000 psi. For the plyform the corresponding values are 1330 psi, 72 psi, and 1,430,000 psi, with the stress applied parallel with the face grain.

Using a live load of 150 psf, check the bending and shear stresses in the sheathing, joists, and stringers. Check deflections assuming that the maximum permissible deflection of each of these elements is $\ell/360$ for all loads.

SOLUTION

Sheathing Properties of 12-in.-wide piece of $\frac{3}{4}$ plyform from Table 19.2:

$$S = 0.455 \text{ in.}^3, \qquad I = 0.198 \text{ in.}^4, \qquad \frac{Ib}{Q} = 7.892 \text{ in.}^2$$

Loads (neglecting weight of forms):

$$\text{Concrete} = (\tfrac{6}{12})(150) = 75 \text{ psf}$$

$$LL = 150$$

$$\text{Total load} = \overline{225} \text{ psf}$$

Bending stresses (with reference made to Figure 19.5):

$$M = \frac{w\ell^2}{10} = \frac{(225)(2)^2}{10} = 90 \text{ ft-lb}$$

$$f_b = \frac{(12)(90)}{0.455} = 2374 \text{ psi} > 1330 \text{ psi} \qquad \underline{\text{no good}}$$

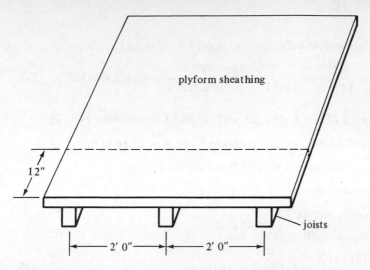

Figure 19.5

Shear stresses (not usually checked in sheathing):

$$V = 0.6w\left(\ell - \frac{2h}{12}\right) = (0.6)(225)\left(2 - \frac{2 \times \frac{3}{4}}{12}\right) = 253 \text{ lb}$$

$$f_v = \frac{VQ}{Ib} = \frac{V}{Ib/Q} = \frac{253}{7.892} = 32 \text{ psi} < 72 \text{ psi} \qquad \underline{\text{OK}}$$

Deflection:

$$\text{Permissible deflection} = (\tfrac{1}{360})(12 \times 2) = 0.067''$$

$$\text{Actual } \delta = \frac{w\ell^4}{145EI} = \frac{(\frac{225}{12})(12 \times 2)^4}{(145)(1.43 \times 10^6)(0.198)} = 0.152 > 0.067'' \quad \underline{\text{N.G.}}$$

Joists: Properties of 2×6 joist with finished dimensions $1\frac{1}{2} \times 5\frac{1}{2}$:

$$A = 8.250 \text{ in.}^2, \qquad I = 20.80 \text{ in.}^4, \qquad S = 7.56 \text{ in.}^3$$

$$\text{Load per foot} = (2)(225) = 450 \text{ lb/ft}$$

Bending and shear stresses:

$$M = \frac{w\ell^2}{10} = \frac{(450)(4)^2}{10} = 720 \text{ ft-lb}$$

$$f_b = \frac{(12)(720)}{7.56} = 1143 \text{ psi} < 1875 \text{ psi} \qquad \underline{\text{OK}}$$

$$V = 0.6w\left(\ell - \frac{2h}{12}\right) = (0.6)(450)\left(4 - \frac{2 \times 5.50}{12}\right) = 832.5 \text{ lb}$$

$$f_v = \frac{3V}{2A} = \frac{(3)(832.5)}{(2)(8.25)} = 151 \text{ psi} < 180 \text{ psi} \qquad \underline{\text{OK}}$$

Deflection:

$$\text{Permissible deflection} = \left(\tfrac{1}{360}\right)(12 \times 4) = 0.133''$$

$$\text{Actual } \delta = \frac{w\ell^4}{145EI} = \frac{\left(\tfrac{450}{12}\right)(12 \times 4)^4}{(145)(1.76 \times 10^6)(20.80)} = 0.0375 < 0.133'' \qquad \underline{\text{OK}}$$

Stringers   Properties of 2 × 8 stringer with finished dimensions $1\tfrac{1}{2} \times 7\tfrac{1}{4}$:

$$A = 10.87 \text{ in.}^2, \qquad I = 47.63 \text{ in.}^4, \qquad S = 13.14 \text{ in.}^3$$

$$\text{load/ft} = (4)(225) = 900 \text{ lb/ft}$$

Bending and shear stresses:

$$M = \frac{w\ell^2}{10} = \frac{(900)(4)^2}{10} = 1440 \text{ ft-lb}$$

$$f_b = \frac{(12)(1440)}{13.14} = 1315 \text{ psi} < 1875 \text{ psi} \qquad\qquad \underline{\text{OK}}$$

$$V = 0.6w\left(\ell - \frac{2h}{12}\right) = (0.6)(900)\left(4 - \frac{2 \times 7.25}{12}\right) = 1507 \text{ lb}$$

$$f_v = \frac{(3)(1507)}{(2)(10.87)} = 208 > 180 \text{ psi} \qquad\qquad \underline{\text{no good}}$$

Deflection:

$$\text{Permissible deflection} = \left(\tfrac{1}{360}\right)(12 \times 4) = 0.133''$$

$$\text{Actual } \delta = \frac{\left(\tfrac{900}{12}\right)(12 \times 4)^4}{(145)(1.76 \times 10^6)(47.63)} = 0.0328'' < 0.133'' \qquad \underline{\text{OK}} \ \blacksquare$$

19.10 DESIGN OF FORMWORK FOR FLOOR AND ROOF SLABS

From the same principles used for calculating the flexural and shear stresses and the deflections in the preceding section, it is possible to calculate maximum permissible spans for certain sizes of sheathing, joists, or stringers. As an illustration it is assumed that a rectangular section is used to support a total uniform load of w lb/ft over three or more continuous equal spans. The term ℓ used in the equations to follow is the span, center to center, of supports in inches, while f is the allowable flexural stress.

Moment

The bending moment can be calculated in inch-pounds and equated to the resisting moment also in inch-pounds. The resulting expression can be solved for ℓ, the maximum permissible span, in inches.

$$M = \frac{w\ell^2}{10} \text{ in foot-pounds} = \frac{w\ell^2}{120} \text{ in inch-pounds}$$

$$M_{res} = fS \text{ in inch-pounds}$$

Equating and solving for ℓ,

$$\frac{w\ell^2}{120} = fS$$

$$\ell = 10.95 \sqrt{\frac{fS}{w}}$$

Shear

In the same fashion, an expression can be written for the maximum permissible span from the standpoint of horizontal shear. In the expression to follow, H is the allowable shearing stress in pounds per square inch while V is the maximum external shear applied to the member at a distance h from the support previously assumed to equal $0.6w[\ell - (2h/12)]$.

$$H = \frac{3V}{2A} = \frac{(3)(0.6w)[\ell - (2h/12)]}{2bh}$$

The value of ℓ used in the preceding expression was given in feet; therefore, when the expression is solved for ℓ, it must be multiplied by 12 to give an answer in inches to coincide with the units of the moment and deflection expressions.

$$\ell = 12\left(\frac{Hbh}{0.9w} + \frac{2h}{12}\right)$$

Deflection

If the maximum centerline deflection is equated to the maximum permissible deflection, the resulting equation can be solved for ℓ. For this case the maximum permissible deflection is assumed to equal $\ell/360$.

$$\delta = \frac{w\ell^4}{145EI} = \frac{(w/12)(\ell^4)}{145EI} = \frac{\ell}{360}$$

$$\ell = 1.69 \sqrt[3]{\frac{EI}{w}}$$

■ EXAMPLE 19.2

It is desired to support a total uniform load of 200 psf for a floor slab with $\frac{3}{4}$-in. Class II plyform. If the forms are to be reused many times and the maximum permissible deflection is $\ell/360$, determine the maximum permissible span of the sheathing center to center, of the joists if $f = 1330$ psi and $E = 1,430,000$ psi. Neglect shear in the sheathing. Make similar calculations but consider shear values for 2- × 4-in. joists spaced 2 ft 0 in. on center if $f = 1500$ psi, $H = 180$ psi, and $E = 1,700,000$ psi.

SOLUTION

Sheathing   Assuming a 12-in. wide piece of plywood:

$$S = 0.455 \text{ in.}^3, \qquad I = 0.198 \text{ in.}^4$$

Moment:

$$\ell = 10.95 \sqrt{\frac{fS}{w}} = 10.95 \sqrt{\frac{(1330)(0.455)}{200}} = 19.05''$$

Deflection:

$$\ell = 1.69 \sqrt[3]{\frac{EI}{w}} = 1.69 \sqrt[3]{\frac{(1,430,000)(0.198)}{200}} = 18.98''$$

Joists: Properties of 2 × 4 joists with finished dimensions $1\frac{1}{2} \times 3\frac{1}{2}$:

$$A = 5.250 \text{ in.}^2, \qquad I = 5.36 \text{ in.}^4, \qquad S = 3.063 \text{ in.}^3$$

Load per foot if joists spaced 1'6" on center $= (1.5)(200) = 300$ lb/ft

Moment:

$$\ell = 10.95 \sqrt{\frac{(1500)(3.063)}{300}} = 42.85''$$

Shear:

$$\ell = (12)\left(\frac{Hbh}{0.9w} + \frac{2h}{12}\right) = (12)\left(\frac{180 \times 1.50 \times 3.50}{0.9 \times 300} + \frac{2 \times 3.50}{12}\right) = 49.00''$$

Deflection:

$$\ell = 1.69 \sqrt[3]{\frac{1,760,000 \times 5.36}{300}} = 53.34'' \qquad ■$$

For practical design work, many tables are available to simplify and expedite the design calculations. Table 19.4, which shows the maximum permissible spacings, center to center, of supports for joists, stringers, and other beams continuous over four or more supports for one particular grade of lumber, is one such example. From this table the maximum permissible spacing of the joists of Example 19.2 can be found to be 43 in., which corresponds with the 42.85 in. obtained in the example.

TABLE 19.4 SAFE SPACING (IN INCHES) OF SUPPORTS FOR JOISTS, STUDS (OR OTHER BEAM COMPONENTS OF FORMWORK), CONTINUOUS OVER THREE OR MORE SPANS (Courtesy of the American Concrete Institute)

f = 1875 psi   E = 1,700,000 psi   H = 225 psi

Uniform load, lb per lineal ft (equals uniform load on forms times spacing between joists or studs, ft)	2×4	2×6	2×8	2×10	2×12	3×4	3×6	3×8	3×10	4×2	4×4	4×6	4×8	6×2	6×4	6×6	6×8	8×2	8×8	10×2
100	76	111	137	164	190	90	126	156	187	43	98	138	169	50	110	154	195	55	210	60
200	59	92	115	138	160	72	106	131	157	34	80	116	142	40	92	130	164	44	177	47
300	48	75	99	125	145	62	96	118	142	30	70	105	129	35	81	117	148	38	160	41
400	41	65	86	110	133	54	84	110	132	27	63	97	120	32	74	109	138	35	149	38
500	37	58	77	98	119	48	75	99	125	24	57	89	113	29	69	103	130	32	141	35
600	33	52	69	88	107	44	69	91	116	22	52	81	107	28	65	98	124	30	134	33
700	29	46	61	78	95	40	64	84	107	20	48	75	99	26	60	94	120	29	129	31
800	27	42	55	70	86	38	59	78	100	19	45	70	93	24	56	88	116	28	125	30
900	24	38	51	65	79	36	56	74	94	18	42	66	87	23	53	83	112	26	121	29
1000	23	36	47	60	73	33	52	69	88	17	40	63	83	21	50	79	109	25	118	28
1100	21	33	44	56	68	31	48	64	82	16	38	60	79	20	48	75	103	24	115	27
1200	20	32	42	53	65	29	45	60	76	16	37	57	76	20	46	72	99	23	113	25
1300	19	30	40	50	61	27	43	56	72	15	35	55	73	19	44	69	94	22	110	24
1400	18	29	38	48	59	26	40	53	68	14	33	52	69	18	42	67	91	21	106	24
1500	17	27	36	46	56	24	38	51	65	13	31	49	65	18	41	64	88	20	103	23
1600	17	26	35	44	54	23	37	48	62	13	30	47	62	17	40	62	85	19	100	22
1700	16	26	34	43	52	22	35	46	59	12	29	45	59	16	38	61	83	19	97	21
1800	16	25	33	42	51	22	34	45	57	12	27	43	57	16	37	59	80	18	94	21

Nominal size of S4A lumber

TABLE 19.4 SAFE SPACING (IN INCHES) OF SUPPORTS FOR JOISTS, STUDS (OR OTHER BEAM COMPONENTS OF FORMWORK), CONTINUOUS OVER THREE OR MORE SPANS (Courtesy of the American Concrete Institute)

f = 1875 psi E = 1,700,000 psi H = 225 psi

Uniform load, lb per lineal ft (equals uniform load on forms times spacing between joists or studs, ft)	Nominal size of S4A lumber																			
	2×4	2×6	2×8	2×10	2×12	3×4	3×6	3×8	3×10	4×2	4×4	4×6	4×8	6×2	6×4	6×6	6×8	8×2	8×8	10×2
1900	15	24	32	40	49	21	33	43	55	11	26	41	55	16	36	57	78	18	91	20
2000	15	23	31	39	48	20	32	42	53	11	25	40	53	15	35	56	76	17	89	20
2100	14	23	30	38	47	19	31	40	51	10	24	38	51	15	34	54	74	17	87	19
2200	14	22	29	37	45	19	30	39	50	10	24	37	49	14	33	52	71	17	85	19
2300	14	22	29	37	44	18	29	38	49	10	23	36	48	14	32	50	69	16	83	18
2400	14	21	28	36	44	18	28	37	47	10	22	35	46	13	31	49	66	16	81	18
2500	13	21	27	35	43	17	27	36	46	9	22	34	45	13	30	47	63	16	80	18
2600	13	20	27	34	42	17	27	35	45	9	21	33	44	12	29	46	61	15	78	17
2700	13	20	27	34	41	17	26	35	44	9	20	32	43	12	28	45	60	15	77	17
2800	13	20	26	33	41	16	26	34	43	9	20	32	42	12	28	43	59	14	75	17
2900	12	19	26	32	40	16	25	33	42	8	20	31	41	11	27	42	58	14	73	16
3000	12	19	25	32	39	16	25	32	41	8	19	30	40	11	26	41	56	13	71	16
3200	12	19	25	31	38	15	24	31	40	8	18	29	38	11	25	39	54	13	68	16
3400	12	18	24	31	37	15	23	30	39	8	18	28	37	10	24	38	51	13	65	15
3600	11	18	24	30	37	14	22	30	38	7	17	27	36	10	23	36	49	12	62	15
3800	11	17	23	30	36	14	22	29	37	7	17	26	34	9	22	35	48	12	59	14
4000	11	17	23	29	35	14	21	28	36	7	16	25	33	9	21	34	46	11	57	13
4500	10	16	22	28	34	13	20	27	34	6	15	24	31	8	20	31	43	10	53	12
5000	10	16	21	27	33	12	19	25	32	6	14	22	30	8	18	29	40	9	49	11

Note: Span values above the solid line are governed by deflection. Values within dashed line box are spans governed by shear.

$\Delta_{max} = \ell/360$, but not to exceed $\frac{1}{4}$ in.

19.11 DESIGN OF SHORING

Wood shores are usually designed as simply supported columns using a modified form of the Euler equation. If the Euler equation is divided by a factor of safety of 3, and if r is replaced with 0.3 times d (the least lateral dimension of square or rectangular shores), as follows,

$$\frac{P}{A} = \frac{\pi^2 E}{3(\ell/0.3d)^2}$$

the so-called National Forest Products Association formula results:

$$\frac{P}{A} = \frac{0.3E}{(\ell/d)^2}$$

The maximum ℓ/d value normally specified is 50. If a round shore is being used, it may be replaced for calculation purposes with a square shore having the same cross-sectional area. Should a shore be braced at different points laterally so that it has different unsupported lengths along its different faces, it will be necessary to calculate the ℓ/d ratios in each direction and use the largest one to determine the allowable stress.

The allowable stress used may not be greater than the value obtained from this equation, nor greater than the allowable unit stress in compression parallel to the grain for the grade and type of lumber in question. If the formwork is to be used only once, the allowable stress in the column can reasonably be increased by 25%. Example 19.3 shows the calculation of the permissible column load for a particular shore. Tables are readily available for making these calculations for shores as they are for sheathing and beam members. For example, Table 19.5 gives the allowable column loads as determined here for a set of simple shores.

■ EXAMPLE 19.3

Are 4- × 6-in. shores (S4S) 10 ft long and 4 ft on centers satisfactory for supporting the floor system of Example 19.1? Assume the allowable compression stress parallel to the grain is 1000 psi and $E = 1.70 \times 10^6$ psi.

SOLUTION

Using 4 × 6 Shores (Dressed Dimensions $3\frac{1}{2} \times 5\frac{1}{2}$, $A = 19.250$ in.^{2})

Load applied to each store = (4)(4)(225) = 3600 lb

$$\frac{\ell}{d} = \frac{(12)(10)}{3.5} = 34.29 < 50 \qquad \underline{\text{OK}}$$

Allowable $\dfrac{P}{A} = \dfrac{(0.3)(1.70 \times 10^6)}{(34.29)^2} = 434$ psi < 1000 psi $\qquad \underline{\text{OK}}$

Allowable $P = (434)(19.250) = 8354 > 3600$ lb $\qquad \underline{\text{OK}}$ ■

TABLE 19.5   ALLOWABLE LOAD IN POUNDS ON SIMPLE WOOD SHORES,* FOR LUMBER OF THE INDICATED STRENGTH, BASED ON UNSUPPORTED LENGTH (Courtesy of the American Concrete Institute)

$c \parallel$ to grain $= 1{,}150$ psi $\quad E = 1{,}400{,}000$ psi $\quad \dfrac{\ell}{d_{max}} = 50 \quad \dfrac{P}{A_{max}} = \dfrac{0.30E}{(\ell/d)^2}$

Nominal lumber size, in.:	2 × 4		3 × 4		4 × 4		4 × 2		4 × 3		4 × 6		6 × 6	
Unsupported length, ft	R**	S4S**	R	S4S	R	S4S	R	S4S	R	S4S	R	S4S	R	S4S
4	2,800	2,200	10,900	10,000	15,100	15,000	6,800	6,000	10,900	10,100	23,400	22,100	36,400	34,800
5	1,800	1,400	7,700	6,400	15,100	15,000	6,800	6,000	10,900	10,100	23,400	22,100	36,400	34,800
6	1,300	1,000	5,300	4,400	14,400	12,200	6,300	5,200	10,100	8,700	21,700	19,100	36,400	34,800
7			3,900	3,300	10,300	8,900	4,600	3,800	7,400	6,400	15,900	14,000	36,400	34,800
8			3,000	2,500	7,870	6,800	3,500	2,900	5,700	4,900	12,200	10,700	36,400	34,800
9			2,400	2,000	6,200	5,400	2,800	2,300	4,500	3,900	9,700	8,500	36,200	32,900
10			1,900	1,600	5,000	4,400	2,300	1,900	3,600	3,100	7,800	6,900	29,200	26,700
11			—	—	4,200	3,600	1,900	1,500	3,000	2,600	6,500	5,700	24,100	22,100
12					3,500	3,000	1,600	1,300	2,500	2,200	5,400	4,800	20,300	18,500
13					3,000	2,600	1,300	1,100	2,200	1,800	4,600	4,100	17,300	15,800
14					2,600	2,200	1,200	1,000	1,900	1,600	4,000	3,500	14,900	13,600
15					2,200	—	1,000	—	1,600	—	3,500	—	13,000	11,900
16					—		—		—		—		11,400	10,400
17													10,100	9,200
18													9,000	8,200
19													8,100	7,100
20													7,300	6,700

(For the 4 × 2 and 4 × 3 columns: Bracing needed†)

* Calculated to nearest 100 lb.   ** R indicates rough lumber; S4S indicates lumber finished on all four sides.

† The dimension used in determining ℓ/d is that shown first in the size column. Where this is the larger dimension, the column must be braced in the other direction so that ℓ/d is equal to or less than that used in arriving at the loads shown. For 4 × 2's, bracing in the plane of the 2-in. dimension must be at intervals not greater than 0.4 times the unsupported length. For 4 × 3's, bracing in the plane of the 3-in. dimension must be at intervals not more than 0.7 times the unsupported length.

TABLE 19.5 (continued)

c ∥ to grain = 1,100 psi   E = 1,600,000 psi $\dfrac{\ell}{d_{max}} = 50$ $\dfrac{P}{A_{max}} = \dfrac{0.30E}{(\ell/d)^2}$

Nominal lumber size, in.:	2 × 4		3 × 4		4 × 4		4 × 2		4 × 3		4 × 6		6 × 6	
	R**	S4S**	R	S4S	R	S4S	R	S4S	R	S4S	R	S4S	R	S4S
Unsupported length, ft							Bracing needed†							
4	3,200	2,500	9,500	8,700	13,100	12,200	5,900	5,200	9,500	8,700	20,400	19,200	31,600	30,200
5	2,100	1,600	8,700	7,300	13,100	12,200	5,900	5,200	9,500	8,700	20,400	19,200	31,600	30,200
6	1,400	1,100	6,100	5,100	13,100	12,200	5,900	5,200	9,500	8,700	20,400	19,200	31,600	30,200
7	—	—	4,500	3,700	11,700	10,200	5,300	4,400	8,500	7,300	18,200	16,000	31,600	30,200
8			3,400	2,800	9,000	7,800	4,000	3,300	6,500	5,600	13,900	12,300	31,600	30,200
9			2,700	2,200	7,100	6,200	3,200	2,600	5,100	4,400	11,000	9,700	31,600	30,200
10			2,200	1,800	5,800	5,000	2,600	2,100	4,200	3,600	8,900	7,900	31,600	30,200
11			—	—	4,800	4,100	2,100	1,800	3,400	2,900	7,400	6,500	27,600	25,200
12					4,000	3,500	1,800	1,500	2,900	2,500	6,200	5,500	23,200	21,200
13					3,400	3,000	1,500	1,300	2,500	2,100	5,300	4,600	19,700	18,000
14					2,900	2,500	1,300	1,100	2,100	1,800	4,600	4,000	17,000	15,600
15					2,600	—	1,100	—	1,800	—	4,000	—	14,800	13,600
16					—		—		—		—		13,000	11,900
17													11,500	10,500
18													10,300	9,400
19													9,200	8,400
20													8,300	7,600

* Calculated to nearest 100 lb.   ** R indicates rough lumber; S4S indicates lumber finished on all four sides.

† The dimension used in determining ℓ/d is that shown first in the size column. Where this is the larger dimension, the column must be braced in the other direction so that ℓ/d is equal to or less than that used in arriving at the loads shown. For 4 × 2's, bracing in the plane of the 2-in. dimension must be at intervals not greater than 0.4 times the unsupported length. For 4 × 3's, bracing in the plane of the 3-in. dimension must be at intervals not more than 0.7 times the unsupported length.

TABLE 19.5 (continued)

c ∥ to grain = 1,100 psi E = 1,700,000 psi $\dfrac{\ell}{d_{max}} = 50$ $\dfrac{P}{A_{max}} = \dfrac{0.30E}{(\ell/d)^2}$

Nominal lumber size, in.:	2 × 4		3 × 4		4 × 4		4 × 2		4 × 3		4 × 6		6 × 6	
Unsupported length, ft	R**	S4S**	R	S4S	R	S4S	R	S4S	R	S4S	R	S4S	R	S4S
							Bracing needed†							
4	3,400	2,600	9,500	8,800	13,100	12,300	5,900	5,300	9,500	8,800	20,400	19,200	31,600	30,200
5	2,200	1,700	9,300	7,700	13,100	12,300	5,900	5,300	9,500	8,800	20,400	19,200	31,600	30,200
6	1,500	1,200	6,500	5,400	13,100	12,300	5,900	5,300	9,500	8,800	20,400	19,200	31,600	30,200
7	—	—	4,700	4,000	12,500	10,800	5,600	4,600	9,000	7,700	19,400	17,000	31,600	30,200
8			3,600	3,000	9,600	8,300	4,300	3,600	6,900	5,900	14,800	13,100	31,600	30,200
9			2,900	2,400	7,600	6,600	3,400	2,800	5,500	4,700	11,700	10,300	31,600	30,200
10			2,300	1,900	6,100	5,300	2,700	2,300	4,400	3,800	9,500	8,400	31,600	30,200
11			—	—	5,100	4,400	2,300	1,900	3,700	3,100	7,900	6,900	29,300	28,000
12					4,200	3,700	1,900	1,600	3,100	2,600	6,600	5,800	24,600	22,500
13					3,600	3,100	1,600	1,300	2,600	2,200	5,600	4,900	21,000	19,200
14					3,100	2,700	1,400	1,200	2,300	1,900	4,800	4,300	18,100	16,500
15					2,700	—	1,200	—	2,000	—	4,200	—	15,800	14,400
16					—						—		13,800	12,600
17													12,300	11,200
18													10,900	10,000
19													9,800	9,000
20													8,900	8,100

* Calculated to nearest 100 lb.   **R indicates rough lumber; S4S indicates lumber finished on all four sides.

† The dimension used in determining ℓ/d is that shown first in the size column. Where this is the larger dimension, the column must be braced in the other direction so that ℓ/d is equal to or less than that used in arriving at the loads shown. For 4 × 2's, bracing in the plane of the 2-in. dimension must be at intervals not greater than 0.4 times the unsupported length. For 4 × 3's, bracing in the plane of the 3-in. dimension must be at intervals not more than 0.7 times the unsupported length.

Wood shores may consist of different pieces of lumber that are nailed or bolted together to form larger built-up columns or shores. Such members are usually given lower allowable stresses than those available for solid-sawn or round columns. These built-up shores may be spaced columns consisting of two pieces of lumber with spacer blocks in between.

Spaced shores are very important for formwork because of economic factors. Logs can be efficiently sawed into 2 × material (2 × 4's, 2 × 6's, etc.). The cost of larger pieces, such as 4 ×, 6 ×, and so on, does not make such efficient use of the logs, and prices are appreciably higher. As a result, built-up or spaced shores are frequently used. The allowable stresses, maximum ℓ/d ratios, and other factors are somewhat different from single shores.

There are a number of adjustable shoring systems available. The simplest type is made by overlapping two wood members. The workers use a portable jacking tool to make vertical adjustments. Various hardware is available for joining the shores to the stringers with a minimum amount of nailing. Another type of adjustable shore consists of dimension lumber combined with a steel column section and a jacking device. Several manufacturers have available all-metal shores, called *jack shores*, which are adjustable for heights from 4 to 16 ft.[3]

19.12 BEARING STRESSES

The bearing stresses produced when one member rests upon another may be critical in the design of formwork. These stresses need to be carefully checked where joists rest on stringers, where stringers rest on shores, and where studs rest on wales as well as where form ties bear on the wales through brackets or washers. In each of the cases mentioned, the bearing forces applied to the horizontal timber members cause compression stresses perpendicular to the grain. The allowable stresses for compression perpendicular to the grain are much less than they are for compression parallel to the grain and, in fact, will often be less than the allowable compression stresses in shores as determined by the National Forest Products Association formula.

As a first illustration the bearing stresses in stringers resting on shores are considered. One particular point that should be noted is that the bearing area may well be less than the full cross-sectional area of the shore. For instance, in Examples 19.1 and 19.3 the 2- × 8-in. stringers are supported by 4- × 6-in. shores, as shown in Figure 19.6. Obviously, the bearing area does not equal the full cross-sectional area of the 4 × 6 ($3.50 \times 5.50 = 19.25$ in.^{2}) but rather equals the smaller crosshatched area ($1.50 \times 5.50 = 8.25$ in.^{2}).

The bearing stress can be calculated by dividing the total load applied to the shore (3600 lb from Example 19.3) by the hatched area.

[3] Hurd, M. K., 1973, *Formwork for Concrete* (Detroit: American Concrete Institute), pp. 68–69.

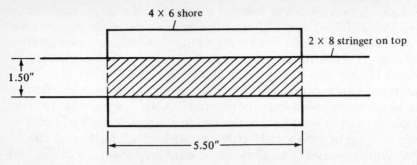

4 × 6 shore

2 × 8 stringer on top

1.50"

5.50"

Figure 19.6

$$\text{Bearing stress} = \frac{3600}{8.25} = 436 \text{ psi}$$

From Table 19.3 the allowable bearing stress perpendicular to the grain in the stringer is 490 psi for the Douglas fir, coastal construction lumber used in the examples of this chapter. The values given in this table for compression perpendicular to the grain are actually applicable to bearing spread over any length at the end of a member and for bearing at 6 in. or more in length at interior locations.

Should the bearing be located more than 3 in. from the end of a beam and be less than 6 in. long, the allowable compressive stresses perpendicular to the grain may be safely increased by multiplying them by the following factor, in which ℓ is the length of bearing measured along the grain. Should the bearing area be circular, as for a washer, the length of bearing is considered to equal the diameter of the circle.

$$\text{Multiplier} = \frac{\ell + \frac{3}{8}}{\ell}$$

If the shore is bearing at an interior point of the stringer in Figure 19.6, the allowable compression stress perpendicular to the grain equals

$$\left(\frac{\ell + \frac{3}{8}}{\ell}\right)(490) = \left(\frac{5.50 + 0.375}{5.50}\right)(490) = 523 \text{ psi} > 436 \text{ psi} \qquad \underline{\text{OK}}$$

If the calculated bearing stress exceeds the allowable stress, it is necessary to spread the load out, as by using larger members or by using a bearing plate or a hardwood cap on top of the shore.[4]

Another bearing stress situation is presented in Example 19.4, where the bearing stress caused in the wales of a wall form by the form ties is considered.

■ EXAMPLE 19.4

Form ties are subjected to an estimated force 5000 lb from the fresh concrete in a wall form. The load is transferred to double 2- × 4-in. wales through $3\frac{1}{2}$-in.

---

[4] Hurd, M. K., 1973, *Formwork for Concrete* (Detroit: American Concrete Institute), pp. 89–90.

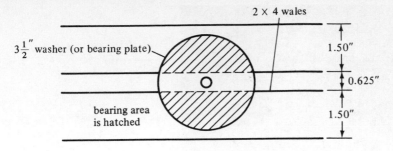

Figure 19.7

brackets or washers, as shown in Figure 19.7. Are the calculated bearing stresses within the permissible values if the wales are Douglas fir, coastal construction and if the bearing locations are at interior points in the wales?

SOLUTION

$$\text{Bearing area} = \frac{(\pi)(3.50)^2}{4} - (0.625)(3.50)$$

$$= 7.43 \text{ in.}^2$$

$$\text{Bearing stress} = \frac{5000}{7.43} = 673 \text{ psi}$$

$$\text{Allowable compression} \perp \text{to grain} = \left(\frac{\ell + \frac{3}{8}}{\ell}\right)(490)$$

$$= \left(\frac{3.50 + 0.375}{3.50}\right)(490)$$

$$= 543 \text{ psi} < 673 \text{ psi} \quad \underline{\text{no good}} \quad \blacksquare$$

Use larger washers or smaller spacing of ties because allowable bearing is not satisfactory.

19.13 DESIGN OF FORMWORK FOR WALLS

In Section 19.8 the lateral pressures on wall and column forms were discussed. These pressures increase from a minimum at the top to a maximum at the bottom of the semiliquid concrete. Occasionally in design this varying pressure is recognized, but most of the time it is practical to consider that the maximum pressure exists for the entire height. For convenience in construction, the sheathing, studs, wales, and ties are usually kept at the same size and spacings throughout the entire height of a wall, as are the wales and ties. In effect, therefore, uniform maximum pressures are assumed.

Column forms. (Courtesy of Burke Concrete Accessories, Inc.)

The sheathing and studs for a wall are designed as they are for the sheathing and joists for roof and floor slabs. In addition, the wales are designed as they are for the stringers for slabs, the spans being the center-to-center spacing of the ties.

If the wales are framed on both sides, the ties will extend through the formwork and be attached to the wales on each side, as was shown in Figure 19.2. The tensile force applied to each tie is then dependent on the pressure from the liquid concrete and on the horizontal and vertical spacing of the ties. If the spacing of ties for a certain wall is 2 ft vertically and 3 ft horizontally, and the calculated wall pressure is 800 psf, the total tensile force on each tie equals $2 \times 3 \times 800 = 4800$ lb, assuming a uniform pressure.

Should concrete be placed at 50°F at a rate of 4 ft/hr in a 12-ft-high wall, the lateral pressure for the semiliquid concrete could be calculated from the ACI expression as follows:

$$p = 150 + \frac{9000R}{T} = 150 + \frac{(9000)(4)}{50} = 870 \text{ psf}$$

The resulting pressure diagram for the wall would be as shown in Figure 19.8.

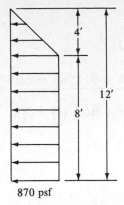

Figure 19.8

Assuming a uniform pressure of 870 psf, the spacing of the studs could be selected as was the spacing for the joists in the floor slab formwork. Example 19.5 shows the design procedure for the sheathing, studs, wales, and ties for a wall form.

■ **EXAMPLE 19.5**

Design the forms for a 12-ft high concrete wall for which the concrete is to be placed at a rate of 4 ft/hr at a temperature of 50°F. (Refer to Figure 19.8.) Use the following data:

1. Sheathing is to be $\frac{3}{4}$-in. Class I plyform, $f = 1930$ psi, and $E = 1,650,000$ psi.
2. Studs and wales are to consist of Douglas fir, coastal construction, $f = 1875$ psi, $H = 180$ psi, and $E = 1,760,000$ psi. Allowable compression perpendicular to the grain is 490 psi.
3. Ties can carry 5000 lb each, and the tie washers are the same size as the one shown in Figure 19.7.
4. Double wales are used to avoid drilling for ties.
5. Maximum deflection in any form component is $\frac{1}{360}$ of the span.

SOLUTION

Design of Sheathing Properties of 12-in.-wide piece of $\frac{3}{4}$ plyform:

$$S = 0.455 \text{ in.}^3, \qquad I = 0.198 \text{ in.}^4$$

$$\text{Load} = 870 \text{ lb/ft}$$

Moment:

$$\ell = 10.95 \sqrt{\frac{fS}{w}} = 10.95 \sqrt{\frac{(1930)(0.455)}{870}} = \underline{11.00''}$$

Deflection:

$$\ell = 1.62\sqrt[3]{\frac{EI}{w}} = 1.62\sqrt[3]{\frac{(1,650,000)(0.198)}{870}} = 11.69''$$

Use studs at 0'11" on center

Design of Studs (Spaced 0 ft 11 in. on Center)   Assuming 2×4 studs ($1\frac{1}{2} \times 3\frac{1}{2}$),

$$I = 5.36 \text{ in.}^4, \qquad S = 3.06 \text{ in.}^3$$

$$\text{Load} = (\tfrac{11}{12})(870) = 797.5 \text{ lb/ft}$$

Moment:

$$\ell = 10.95\sqrt{\frac{(1875)(3.06)}{797.5}} = 29.37''$$

Deflection:

$$\ell = 1.62\sqrt[3]{\frac{(1,760,000)(5.36)}{797.5}} = 36.91''$$

Shear:

$$\ell = 12\left(\frac{Hbh}{0.9w} + \frac{2h}{12}\right) = (12)\left(\frac{180 \times 1.50 \times 3.50}{0.9 \times 797.5} + \frac{2 \times 3.50}{12}\right)$$
$$= 22.80''$$

Use 1'10" spacing of wales on center

$$\text{Load on wales} = (\tfrac{22}{12})(870) = 1595 \text{ lb/ft}$$

$$\text{Maximum tie spacing} = \frac{5000}{1595} = 3.13'$$

Design of Wales   Assuming two 2×4 wales ($I = 2 \times 5.36 = 10.72$ in.^{4}, $S = 2 \times 3.06 = 6.12$ in.^{3})

Moment:

$$\ell = 10.95\sqrt{\frac{(1875)(6.12)}{1595}} = 29.37''$$

Deflection:

$$\ell = 1.62\sqrt[3]{\frac{(1,760,000)(6.12)}{1595}} = 30.62''$$

Shear:

$$\ell = (12)\left(\frac{180 \times 3.00 \times 3.50}{0.9 \times 1595} + \frac{2 \times 3.50}{12}\right) = 22.80''$$

Use ties at every other stud 1'10" on center

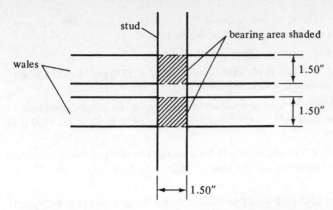

Figure 19.9

Check Bearing for Studs to Wales

$$\text{Maximum load transferred} = \frac{(22)(11)}{144}(870) = 1462 \text{ lb}$$

$$\text{Bearing area (see Figure 19.9)} = (2)(1.50)(1.50) = 4.50 \text{ in.}^2$$

$$\text{Bearing stress} = \frac{1462}{4.50} = 325 \text{ psi} < 490 \text{ psi} \qquad \underline{\text{OK}}$$

Checking Bearing from Ties to Wedges   Assume same dimensions as in Figure 19.7.

$$\text{Bearing load} = \frac{(22)(22)}{144}(870) = 2924 \text{ lb}$$

$$\text{Bearing stress} = \frac{2924}{7.43} = 394 \text{ psi} < 543 \text{ psi}$$

as determined in Example 19.4 \qquad \underline{\text{OK}} ■

PROBLEMS

19.1 Repeat Example 19.1 if the joists are spaced 1 ft 6 in. on centers, the stringers at 6 ft 0 in. on centers as are the shores, and the live load is 100 psf. (*Ans.* sheathing and joists ok but stringers unsatisfactory in both bending and shear)

19.2 Repeat Example 19.1 if $1\frac{1}{8}$ in. Class II plyform, 2 × 8 joists spaced 2 ft 0 in. on centers, and 2 × 8 stringers spaced 6 ft 0 in. on centers.

19.3 It is desired to support a total uniform load of 150 psf for a floor slab with $\frac{5}{8}$ in. Class I plyform. If the forms are to be used many times and the maximum permissible deflection is $\ell/360$ determine the maximum permissible span of the sheathing, center to center of the joists, if $f = 1930$ psi and $E = 1,650,000$ psi. Neglect shear in the sheathing. Make similar calculations but consider shear values for 2 × 4 joists (S4S) spaced 1 ft 6 in. on centers if $f = 1815$ psi, $H = 180$ psi and $E = 1,760,000$ psi. (*Ans.* sheathing 19.04″, joists 56.10″)

19.4 Repeat Prob. 19.3 if $\frac{3}{4}$-in. structural I plyform is used.

19.5 Repeat Prob. 19.3 if 1-in. Class II plyform is used with $f = 1330$ psi. Assume joists are spaced 2 ft 0 in. on center. (*Ans.* sheathing 28.17″, joists 47.11″)

19.6 Is a 4 × 4 in. shore (S4S) 10 ft long satisfactory to support an axial compression load of 4000 lbs if the allowable compression stress parallel to the grain is 1150 psi and $E = 1,400,000$ psi.

19.7 Repeat Prob. 19.6 if 6 × 6 in. shores made from rough lumber 14 ft long are used each to support an axial compression load of 16,000 lbs. (*Ans.* $P = 14,893$ lbs < 16,000 lbs N.G.)

19.8 Design the forms for a 15 ft-high concrete wall for which the concrete is to be placed at a rate of 5 ft/hr at a temperature of 70°F. Use the following data:

  1. Sheathing is to be $\frac{3}{4}$-in. Class II plyform, $f = 1330$ psi and $E = 1,430,000$ psi.

  2. Studs and wales are to consist of Douglas fir, coastal construction, $f = 2000$ psi, $H = 180$ psi and $E = 1,600,000$ psi. Allowable compression perpendicular to the grain is 435 psi.

  3. Ties can carry 5000 lbs each and the tie washers are the same size as those shown in Fig. 19.7.

  4. Double wales are used to avoid drilling for ties.

  5. Maximum deflection in any form component is $\frac{1}{360}$ of the span.

Appendix A

Tables and Graphs

TABLE A.1 VALUES OF MODULUS OF ELASTICITY FOR NORMAL-WEIGHT CONCRETE

Customary units		SI units	
f'_c (psi)	E_c (psi)	f'_c (MPa)	E_c (MPA)
3,000	3,160,000	20.7	21 760
3,500	3,410,000	24.1	23 503
4,000	3,640,000	27.6	25 130
4,500	3,870,000	31.0	26 650
5,000	4,070,000	34.5	28 090

Source Notes: Tables A.4, A.6, A.7, A.16, and A.17, as well as Graph 1, are reprinted from *Design of Concrete Structures* by Winter and Nilson. Copyright © 1972 by McGraw–Hill, Inc. and with permission of the McGraw–Hill Book Company. Graphs 2–13 reprinted from *Design Handbook, Volume 2, Columns* (SP-17A), 1978, with permission of American Concrete Institute.

Tabls A.3(a) and A.3(b) are reprinted from *Manual of Standard Practice*, 22nd ed., 1976, second printing. Concrete Reinforcing Steel Institute, Chicago, Ill.

Table A.5 is reprinted from *Reinforced Concrete Design*, 2nd ed., by Wang and Salmon (IEP, New York, 1973).

Table A.8 is reprinted from *Notes on ACI 318-71 Building Code Requirements with Design Applications*, 1972. Portland Cement Association, Skokie, Ill.

Tables A.18 through A.21 are reprinted from Commentary on Building Code Requirements For Reinforced Concrete (ACI 318-77) with permission of the American Concrete Institute.

TABLE A.2 DESIGNATIONS, AREAS, PERIMETERS, AND WEIGHTS OF STANDARD BARS

Bar no.	Customary units			SI units		
	Diameter (in.)	Cross-sectional area (in.^{2})	Unit weight (lb/ft)	Diameter (mm)	Cross-sectional area (mm^2)	Unit weight (kg/m)
3	0.375	0.11	0.376	9.52	71	0.560
4	0.500	0.20	0.668	12.70	129	0.994
5	0.625	0.31	1.043	15.88	200	1.552
6	0.750	0.44	1.502	19.05	284	2.235
7	0.875	0.60	2.044	22.22	387	3.042
8	1.000	0.79	2.670	25.40	510	3.973
9	1.128	1.00	3.400	28.65	645	5.060
10	1.270	1.27	4.303	32.26	819	6.404
11	1.410	1.56	5.313	35.81	1006	7.907
14	1.693	2.25	7.650	43.00	1452	11.384
18	2.257	4.00	13.600	57.33	2581	20.238

TABLE A.3(a) COMMON STOCK STYLES OF WELDED WIRE FABRIC

Style designation	Steel area (sq in. per ft)		Weight approx. (lb per 100 sq ft)
	Longit.	Transv.	
Rolls			
6 × 6—W1.4 × W1.4	0.03	0.03	21
6 × 6—W2 × W2	0.04	0.04	29
6 × 6—W2.9 × W2.9	0.06	0.06	42
6 × 6—W4 × W4	0.08	0.08	58
4 × 4—W1.4 × W1.4	0.04	0.04	31
4 × 4—W2 × W2	0.06	0.06	43
4 × 4—W2.9 × W2.9	0.09	0.09	62
4 × 4—W4 × W4	0.12	0.12	86
Sheets			
6 × 6—W2.9 × W2.9	0.06	0.06	42
6 × 6—W4 × W4	0.08	0.08	58
6 × 6—W5.5 × W5.5	0.11	0.11	80
4 × 4—W4 × W4	0.12	0.12	86

TABLE A.3(b)   SECTIONAL AREA AND WEIGHT OF WELDED WIRE FABRIC

Wire size number[a]		Nominal diameter (inches)	Nominal weight (lb/lin. ft)	Area in sq in. per ft of width for various spacings						
Smooth	Deformed			Center-to-center spacing						
				2"	3"	4"	6"	8"	10"	12"
W31	D31	0.628	1.054		1.24	0.93	0.62	0.465	0.372	0.31
W28	D28	0.597	0.952		1.12	0.84	0.56	0.42	0.336	0.28
W26	D26	0.575	0.934		1.04	0.78	0.52	0.39	0.312	0.26
W24	D24	0.553	0.816		0.96	0.72	0.48	0.36	0.288	0.24
W22	D22	0.529	0.748		0.88	0.66	0.44	0.33	0.264	0.22
W20	D20	0.505	0.680	1.20	0.80	0.60	0.40	0.30	0.24	0.20
W18	D18	0.479	0.612	1.08	0.72	0.54	0.36	0.27	0.216	0.18
W16	D16	0.451	0.544	0.96	0.64	0.48	0.32	0.24	0.192	0.16
W14	D14	0.422	0.476	0.84	0.56	0.42	0.28	0.21	0.168	0.14
W12	D12	0.391	0.408	0.72	0.48	0.36	0.24	0.18	0.144	0.12
W11	D11	0.374	0.374	0.66	0.44	0.33	0.22	0.165	0.132	0.11
W10	D10	0.357	0.340	0.60	0.40	0.30	0.20	0.15	0.12	0.10
W9.5		0.348	0.323	0.57	0.38	0.285	0.19	0.142	0.114	0.095
W9	D9	0.339	0.306	0.54	0.36	0.27	0.18	0.135	0.108	0.09
W8.5		0.329	0.289	0.51	0.34	0.255	0.17	0.127	0.102	0.085

TABLE A.3(b) (continued)

Wire size number[a] Smooth	Deformed	Nominal diameter (inches)	Nominal weight (lb/lin. ft)	Area in sq in. per ft of width for various spacings — Center-to-center spacing 2"	3"	4"	6"	8"	10"	12"
W8	D8	0.319	0.272	0.48	0.32	0.24	0.16	0.12	0.096	0.08
W7.5		0.309	0.255	0.45	0.30	0.225	0.15	0.112	0.09	0.075
W7	D7	0.299	0.238	0.42	0.28	0.21	0.14	0.105	0.084	0.07
W6.5		0.288	0.221	0.39	0.26	0.195	0.13	0.097	0.078	0.065
W6	D6	0.276	0.204	0.36	0.24	0.18	0.12	0.09	0.072	0.06
W5.5		0.265	0.187	0.33	0.22	0.165	0.11	0.082	0.066	0.055
W5	D5	0.252	0.170	0.30	0.20	0.15	0.10	0.075	0.06	0.05
W4.5		0.239	0.153	0.27	0.18	0.135	0.09	0.067	0.054	0.045
W4	D4	0.226	0.136	0.24	0.16	0.12	0.08	0.06	0.048	0.04
W3.5		0.211	0.119	0.21	0.14	0.105	0.07	0.052	0.042	0.035
W2.9		0.192	0.099	0.174	0.116	0.087	0.058	0.043	0.035	0.029
W2.5		0.178	0.085	0.15	0.10	0.075	0.05	0.037	0.03	0.025
W2		0.160	0.068	0.12	0.08	0.06	0.04	0.03	0.024	0.02
W1.4		0.134	0.048	0.084	0.056	0.042	0.028	0.021	0.017	0.014

Note: The above listing of smooth and deformed wire sizes represents wires normally selected to manufacture welded wire fabric styles to specific areas of reinforcement. Wire sizes and spacings other than those listed above may be produced provided the quantity required is sufficient to justify manufacture.

[a] The number following the prefix W or the prefix D identifies the cross-sectional area of the wire in hundredths of a square inch. The nominal diameter of a deformed wire is equivalent to the diameter of a smooth wire having the same weight per foot as the deformed wire.

TABLE A.4 AREAS OF GROUPS OF STANDARD BARS (in.^{2})

Bar no.	Number of bars								
	2	3	4	5	6	7	8	9	10
4	0.39	0.58	0.78	0.98	1.18	1.37	1.57	1.77	1.96
5	0.61	0.91	1.23	1.53	1.84	2.15	2.45	2.76	3.07
6	0.88	1.32	1.77	2.21	2.65	3.09	3.53	3.98	4.42
7	1.20	1.80	2.41	3.01	3.61	4.21	4.81	5.41	6.01
8	1.57	2.35	3.14	3.93	4.71	5.50	6.28	7.07	7.85
9	2.00	3.00	4.00	5.00	6.00	7.00	8.00	9.00	10.00
10	2.53	3.79	5.06	6.33	7.59	8.86	10.12	11.39	12.66
11	3.12	4.68	6.25	7.81	9.37	10.94	12.50	14.06	15.62
14	4.50	6.75	9.00	11.25	13.50	15.75	18.00	20.25	22.50
18	8.00	12.00	16.00	20.00	24.00	28.00	32.00	36.00	40.00

Bar no.	Number of bars								
	11	12	13	14	15	16	17	18	19	20
4	2.16	2.36	2.55	2.75	2.95	3.14	3.34	3.53	3.73	3.93
5	3.37	3.68	3.99	4.30	4.60	4.91	5.22	5.52	5.83	6.14
6	4.86	5.30	5.74	6.19	6.63	7.07	7.51	7.95	8.39	8.84
7	6.61	7.22	7.82	8.42	9.02	9.62	10.22	10.82	11.43	12.03
8	8.64	9.43	10.21	11.00	11.78	12.57	13.35	14.14	14.92	15.71
9	11.00	12.00	13.00	14.00	15.00	16.00	17.00	18.00	19.00	20.00
10	13.92	15.19	16.45	17.72	18.98	20.25	21.52	22.78	24.05	25.31
11	17.19	18.75	20.31	21.87	23.44	25.00	26.56	28.12	29.69	31.25
14	24.75	27.00	29.25	31.50	33.75	36.00	38.25	40.50	42.75	45.00
18	44.00	48.00	52.00	56.00	60.00	64.00	68.00	72.00	76.00	80.00

TABLE A.5 MINIMUM BEAM WIDTH (in.) ACCORDING TO THE 1977 ACI CODE[a,b]

Size of bars	Number of bars in single layer of reinforcement							Add for each added bar
	2	3	4	5	6	7	8	
#4	6.1	7.6	9.1	10.6	12.1	13.6	15.1	1.50
#5	6.3	7.9	9.6	11.2	12.8	14.4	16.1	1.63
#6	6.5	8.3	10.0	11.8	13.5	15.3	17.0	1.75
#7	6.7	8.6	10.5	12.4	14.2	16.1	18.0	1.88
#8	6.9	8.9	10.9	12.9	14.9	16.9	18.9	2.00
#9	7.3	9.5	11.8	14.0	16.3	18.6	20.8	2.26
#10	7.7	10.2	12.8	15.3	17.8	20.4	22.9	2.54
#11	8.0	10.8	13.7	16.5	19.3	22.1	24.9	2.82
#14	8.9	12.3	15.6	19.0	22.4	25.8	29.2	3.39
#18	10.5	15.0	19.5	24.0	28.6	33.1	37.6	4.51

Table shows minimum beam widths when stirrups are used.
For additional bars, add dimension in last column for each added bar.
For bars of different size, determine from table the beam width for smaller size bars and then add last column figure for each larger bar used.

[a] Assumes maximum aggregate size does not exceed three-fourths of the clear space between bars.

[b] Values given for smaller bars include minimum bend radius for stirrups.

TABLE A.6 AREAS OF BARS IN SLABS (in.²/ft)

Spacing (in.)	Bar no.								
	3	4	5	6	7	8	9	10	11
3	0.44	0.78	1.23	1.77	2.40	3.14	4.00	5.06	6.25
$3\frac{1}{2}$	0.38	0.67	1.05	1.51	2.06	2.69	3.43	4.34	5.36
4	0.33	0.59	0.92	1.32	1.80	2.36	3.00	3.80	4.68
$4\frac{1}{2}$	0.29	0.52	0.82	1.18	1.60	2.09	2.67	3.37	4.17
5	0.26	0.47	0.74	1.06	1.44	1.88	2.40	3.04	3.75
$5\frac{1}{2}$	0.24	0.43	0.67	0.96	1.31	1.71	2.18	2.76	3.41
6	0.22	0.39	0.61	0.88	1.20	1.57	2.00	2.53	3.12
$6\frac{1}{2}$	0.20	0.36	0.57	0.82	1.11	1.45	1.85	2.34	2.89
7	0.19	0.34	0.53	0.76	1.03	1.35	1.71	2.17	2.68
$7\frac{1}{2}$	0.18	0.31	0.49	0.71	0.96	1.26	1.60	2.02	2.50
8	0.17	0.29	0.46	0.66	0.90	1.18	1.50	1.89	2.34
9	0.15	0.26	0.41	0.59	0.80	1.05	1.33	1.69	2.08
10	0.13	0.24	0.37	0.53	0.72	0.94	1.20	1.52	1.87
12	0.11	0.20	0.31	0.44	0.60	0.78	1.00	1.27	1.56

TABLE A.7 DESIGN OF RECTANGULAR BEAMS AND SLABS

n and f_c'	f_s	f_c	k	j	ρ_s	K
	18,000	1000	0.357	0.881	0.0099	157
		1125	0.385	0.872	0.0120	189
10	20,000	1000	0.333	0.889	0.0083	148
(2500)		1125	0.360	0.880	0.0101	178
	24,000	1000	0.294	0.902	0.0061	133
		1125	0.319	0.894	0.0075	160
	18,000	1200	0.375	0.875	0.0125	197
		1350	0.403	0.866	0.0151	235
9	20,000	1200	0.351	0.883	0.0105	186
(3000)		1350	0.377	0.874	0.0128	223
	24,000	1200	0.310	0.897	0.0078	167
		1350	0.336	0.888	0.0095	201
	18,000	1600	0.416	0.861	0.0185	286
		1800	0.444	0.852	0.0222	341
8	20,000	1600	0.390	0.870	0.0156	272
(4000)		1800	0.419	0.860	0.0188	324
	24,000	1600	0.345	0.884	0.0116	246
		1800	0.375	0.875	0.0141	295
	18,000	2000	0.438	0.854	0.0243	374
		2250	0.467	0.844	0.0292	443
7	20,000	2000	0.412	0.863	0.0206	355
(5000)		2250	0.441	0.853	0.0248	423
	24,000	2000	0.368	0.877	0.0154	323
		2250	0.396	0.868	0.0186	387

Note: ρ_s = required percentage of steel.

TABLE A.8   BALANCED RATIO OF REINFORCEMENT ρ_b FOR RECTANGULAR SECTIONS WITH TENSION REINFORCEMENT ONLY

f_y	f'_c	2,500 psi (17.2 MPa) $\beta_1 = 0.85$	3,000 psi (20.7 MPa) $\beta_1 = 0.85$	4,000 psi (27.6 MPa) $\beta_1 = 0.85$	5,000 psi (34.5 MPa) $\beta_1 = 0.80$	6,000 psi (41.4 MPa) $\beta_1 = 0.75$
Grade 40	ρ_b	0.0309	0.0371	0.0495	0.0582	0.0655
40,000 psi	$0.75\rho_b$	0.0232	0.0278	0.0371	0.0437	0.0492
(275.8 MPa)	$0.50\rho_b$	0.0155	0.0186	0.0247	0.0291	0.0328
Grade 50	ρ_b	0.0229	0.0275	0.0367	0.0432	0.0486
50,000 psi	$0.75\rho_b$	0.0172	0.0206	0.0275	0.0324	0.0365
(344.8 MPa)	$0.50\rho_b$	0.0115	0.0138	0.0184	0.0216	0.0243
Grade 60	ρ_b	0.0178	0.0214	0.0285	0.0335	0.0377
60,000 psi	$0.75\rho_b$	0.0134	0.0161	0.0214	0.0252	0.0283
(413.7 MPa)	$0.50\rho_b$	0.0089	0.0107	0.0143	0.0168	0.0189
Grade 75	ρ_b	0.0129	0.0155	0.0207	0.0243	0.0274
75,000 psi	$0.75\rho_b$	0.0097	0.0116	0.0155	0.0182	0.0205
(517.1 MPa)	$0.50\rho_b$	0.0065	0.0078	0.0104	0.0122	0.0137

TABLE A.9 $f_y = 40,000$ psi (275.8 MPa); $f'_c = 3000$ psi (20.7 MPa)

ρ	$\dfrac{M_u}{\phi bd^2}$	ρ	$\dfrac{M_u}{\phi bd^2}$	ρ	$\dfrac{M_u}{\phi bd^2}$	ρ	$\dfrac{M_u}{\phi bd^2}$
ρ_{min} 0.0050	192.1	0.0081	303.4	0.0112	408.5	0.0143	507.6
0.0051	195.8	0.0082	306.8	0.0113	411.8	0.0144	510.7
0.0052	199.5	0.0083	310.3	0.0114	415.1	0.0145	513.8
0.0053	203.2	0.0084	313.8	0.0115	418.4	0.0146	516.9
0.0054	206.8	0.0085	317.3	0.0116	421.7	0.0147	520.0
0.0055	210.5	0.0086	320.7	0.0117	424.9	0.0148	523.1
0.0056	214.1	0.0087	324.2	0.0118	428.2	0.0149	526.1
0.0057	217.8	0.0088	327.6	0.0119	431.4	0.0150	529.2
0.0058	221.4	0.0089	331.1	0.0120	434.7	0.0151	532.2
0.0059	225.0	0.0090	334.5	0.0121	437.9	0.0152	535.3
0.0060	228.7	0.0091	337.9	0.0122	441.2	0.0153	538.3
0.0061	232.3	0.0092	341.4	0.0123	444.4	0.0154	541.4
0.0062	235.9	0.0093	344.8	0.0124	447.6	0.0155	544.4
0.0063	239.5	0.0094	348.2	0.0125	450.8	0.0156	547.4
0.0064	243.1	0.0095	351.6	0.0126	454.0	0.0157	550.4
0.0065	246.7	0.0096	355.0	0.0127	457.2	0.0158	553.4
0.0066	250.3	0.0097	358.4	0.0128	460.4	0.0159	556.4
0.0067	253.9	0.0098	361.8	0.0129	463.6	0.0160	559.4
0.0068	257.4	0.0099	365.2	0.0130	466.8	0.0161	562.4
0.0069	261.0	0.0100	368.5	0.0131	470.0	0.0162	565.4
0.0070	264.6	0.0101	371.9	0.0132	473.2	0.0163	568.4
0.0071	268.1	0.0102	375.3	0.0133	476.3	0.0164	571.4
0.0072	271.7	0.0103	378.6	0.0134	479.5	0.0165	574.3
0.0073	275.2	0.0104	382.0	0.0135	482.6	0.0166	577.3
0.0074	278.8	0.0105	385.3	0.0136	485.8	0.0167	580.2
0.0075	282.3	0.0106	388.6	0.0137	488.9	0.0168	583.2
0.0076	285.8	0.0107	392.0	0.0138	492.1	0.0169	586.1
0.0077	289.3	0.0108	395.3	0.0139	495.2	0.0170	589.1
0.0078	292.9	0.0109	398.6	0.0140	498.3	0.0171	592.0
0.0079	296.4	0.0110	401.9	0.0141	501.4	0.0172	594.9
0.0080	299.9	0.0111	405.2	0.0142	504.5	0.0173	597.8

TABLE A.9   (*continued*)

ρ	$\dfrac{M_u}{\phi bd^2}$	ρ	$\dfrac{M_u}{\phi bd^2}$	ρ	$\dfrac{M_u}{\phi bd^2}$		ρ	$\dfrac{M_u}{\phi bd^2}$
0.0174	600.7	0.0201	676.9	0.0228	748.4		0.0255	815.4
0.0175	603.6	0.0202	679.6	0.0229	751.0		0.0256	817.8
0.0176	606.5	0.0203	682.3	0.0230	753.5		0.0257	820.2
0.0177	609.4	0.0204	685.0	0.0231	756.1		0.0258	822.5
0.0178	612.3	0.0205	687.8	0.0232	758.6		0.0259	824.9
0.0179	615.2	0.0206	690.5	0.0233	761.2		0.0260	827.3
0.0180	618.0	0.0207	693.2	0.0234	763.7		0.0261	829.6
0.0181	620.9	0.0208	695.9	0.0235	766.2		0.0262	832.0
0.0182	623.8	0.0209	698.5	0.0236	768.7		0.0263	834.3
0.0183	626.6	0.0210	701.2	0.0237	771.2		0.0264	836.7
0.0184	629.5	0.0211	703.9	0.0238	773.7		0.0265	839.0
0.0185	632.3	0.0212	706.6	0.0239	776.2		0.0266	841.3
0.0186	635.1	0.0213	709.2	0.0240	778.7		0.0267	843.7
0.0187	638.0	0.0214	711.9	0.0241	781.2		0.0268	846.0
0.0188	640.8	0.0215	714.5	0.0242	783.7		0.0269	848.3
0.0189	643.6	0.0216	717.2	0.0243	786.2		0.0270	850.6
0.0190	646.4	0.0217	719.8	0.0244	788.6		0.0271	852.9
0.0191	649.2	0.0218	722.4	0.0245	791.1		0.0272	855.2
0.0192	652.0	0.0219	725.1	0.0246	793.6		0.0273	857.5
0.0193	654.8	0.0220	727.7	0.0247	796.0		0.0274	859.7
0.0194	657.6	0.0221	730.3	0.0248	798.5		0.0275	862.0
0.0195	660.3	0.0222	732.9	0.0249	800.9		0.0276	864.3
0.0196	663.1	0.0223	735.5	0.0250	803.3		0.0277	866.5
0.0197	665.9	0.0224	738.1	0.0251	805.7	ρ_{max}	0.0278	868.8
0.0198	668.6	0.0225	740.7	0.0252	808.2			
0.0199	671.4	0.0226	743.3	0.0253	810.6			
0.0200	674.1	0.0227	745.8	0.0254	813.0			

For SI units multiply $M_u/\phi bd^2$ by 0.006 895 for Tables A.9 through A.14.

TABLE A.10 $f_y = 40{,}000$ psi (275.8 MPa); $f'_c = 4000$ psi (27.6 MPa)

	ρ	$\dfrac{M_u}{\phi bd^2}$	ρ	$\dfrac{M_u}{\phi bd^2}$	ρ	$\dfrac{M_u}{\phi bd^2}$	ρ	$\dfrac{M_u}{\phi bd^2}$
ρ_{min}	0.0050	194.1	0.0091	344.5	0.0132	486.9	0.0173	621.4
	0.0051	197.9	0.0092	348.0	0.0133	490.2	0.0174	624.5
	0.0052	201.6	0.0093	351.6	0.0134	493.6	0.0175	627.7
	0.0053	205.4	0.0094	355.1	0.0135	497.0	0.0176	630.9
	0.0054	209.1	0.0095	358.7	0.0136	500.3	0.0177	634.1
	0.0055	212.9	0.0096	362.2	0.0137	503.7	0.0178	637.2
	0.0056	216.6	0.0097	365.8	0.0138	507.0	0.0179	640.4
	0.0057	220.3	0.0098	369.3	0.0139	510.4	0.0180	643.5
	0.0058	224.1	0.0099	372.9	0.0140	513.7	0.0181	646.7
	0.0059	227.8	0.0100	376.4	0.0141	517.1	0.0182	649.8
	0.0060	231.5	0.0101	379.9	0.0142	520.4	0.0183	653.0
	0.0061	235.2	0.0102	383.4	0.0143	523.7	0.0184	656.1
	0.0062	238.9	0.0103	387.0	0.0144	527.1	0.0185	659.2
	0.0063	242.6	0.0104	390.5	0.0145	530.4	0.0186	662.3
	0.0064	246.3	0.0105	394.0	0.0146	533.7	0.0187	665.5
	0.0065	250.0	0.0106	397.5	0.0147	537.0	0.0188	668.6
	0.0066	253.7	0.0107	401.0	0.0148	540.3	0.0189	671.7
	0.0067	257.4	0.0108	404.5	0.0149	543.6	0.0190	674.8
	0.0068	261.1	0.0109	408.0	0.0150	546.9	0.0191	677.9
	0.0069	264.8	0.0110	411.4	0.0151	550.2	0.0192	681.0
	0.0070	268.4	0.0111	414.9	0.0152	553.5	0.0193	684.1
	0.0071	272.1	0.0112	418.4	0.0153	556.7	0.0194	687.2
	0.0072	275.8	0.0113	421.9	0.0154	560.0	0.0195	690.3
	0.0073	279.4	0.0114	425.3	0.0155	563.3	0.0196	693.3
	0.0074	283.1	0.0115	428.8	0.0156	566.6	0.0197	696.4
	0.0075	286.7	0.0116	432.2	0.0157	569.8	0.0198	699.5
	0.0076	290.4	0.0117	435.7	0.0158	573.1	0.0199	702.5
	0.0077	294.0	0.0118	439.1	0.0159	576.3	0.0200	705.6
	0.0078	297.6	0.0119	442.6	0.0160	579.6	0.0201	708.6
	0.0079	301.3	0.0120	446.0	0.0161	582.8	0.0202	711.7
	0.0080	304.9	0.0121	449.4	0.0162	586.1	0.0203	714.7
	0.0081	308.5	0.0122	452.9	0.0163	589.3	0.0204	717.8
	0.0082	312.1	0.0123	456.3	0.0164	592.5	0.0205	720.8
	0.0083	315.7	0.0124	459.7	0.0165	595.7	0.0206	723.8
	0.0084	319.3	0.0125	463.1	0.0166	599.0	0.0207	726.9
	0.0085	322.9	0.0126	466.5	0.0167	602.2	0.0208	729.9
	0.0086	326.5	0.0127	469.9	0.0168	605.4	0.0209	732.9
	0.0087	330.1	0.0128	473.3	0.0169	608.6	0.0210	735.9
	0.0088	333.7	0.0129	476.7	0.0170	611.8	0.0211	738.9
	0.0089	337.3	0.0130	480.1	0.0171	615.0		
	0.0090	340.9	0.0131	483.5	0.0172	618.2		

TABLE A.10   (*continued*)

ρ	$\dfrac{M_u}{\phi bd^2}$	ρ	$\dfrac{M_u}{\phi bd^2}$	ρ	$\dfrac{M_u}{\phi bd^2}$	ρ	$\dfrac{M_u}{\phi bd^2}$
0.0212	741.9	0.0252	858.1	0.0292	966.8	0.0332	1067.9
0.0213	744.9	0.0253	860.9	0.0293	969.4	0.0333	1070.3
0.0214	747.9	0.0254	863.7	0.0294	972.0	0.0334	1072.7
0.0215	750.9	0.0255	866.5	0.0295	974.6	0.0335	1075.1
0.0216	753.9	0.0256	869.3	0.0296	977.2	0.0336	1077.6
0.0217	756.9	0.0257	872.1	0.0297	979.8	0.0337	1080.0
0.0218	759.8	0.0258	874.9	0.0298	982.4	0.0338	1082.4
0.0219	762.8	0.0259	877.7	0.0299	985.0	0.0339	1084.8
0.0220	765.8	0.0260	880.5	0.0300	987.6	0.0340	1087.2
0.0221	768.7	0.0261	883.2	0.0301	990.2	0.0341	1089.6
0.0222	771.7	0.0262	886.0	0.0302	992.7	0.0342	1091.9
0.0223	774.6	0.0263	888.7	0.0303	995.3	0.0343	1094.3
0.0224	777.6	0.0264	891.5	0.0304	997.9	0.0344	1096.7
0.0225	780.5	0.0265	894.3	0.0305	1000.4	0.0345	1099.1
0.0226	783.4	0.0266	897.0	0.0306	1003.0	0.0346	1101.5
0.0227	786.4	0.0267	899.7	0.0307	1005.6	0.0347	1103.8
0.0228	789.3	0.0268	902.5	0.0308	1008.1	0.0348	1106.2
0.0229	792.2	0.0269	905.2	0.0309	1010.7	0.0349	1108.5
0.0230	795.1	0.0270	907.9	0.0310	1013.2	0.0350	1110.9
0.0231	798.1	0.0271	910.7	0.0311	1015.7	0.0351	1113.2
0.0232	801.0	0.0272	913.4	0.0312	1018.3	0.0352	1115.6
0.0233	803.9	0.0273	916.1	0.0313	1020.8	0.0353	1117.9
0.0234	806.8	0.0274	918.8	0.0314	1023.3	0.0354	1120.2
0.0235	809.7	0.0275	921.5	0.0315	1025.8	0.0355	1122.6
0.0236	812.5	0.0276	924.2	0.0316	1028.3	0.0356	1124.9
0.0237	815.4	0.0277	926.9	0.0317	1030.8	0.0357	1127.2
0.0238	818.3	0.0278	929.6	0.0318	1033.3	0.0358	1129.5
0.0239	821.2	0.0279	932.3	0.0319	1035.8	0.0359	1131.8
0.0240	824.1	0.0280	935.0	0.0320	1038.3	0.0360	1134.1
0.0241	826.9	0.0281	937.6	0.0321	1040.8	0.0361	1136.4
0.0242	829.8	0.0282	940.3	0.0322	1043.3	0.0362	1138.7
0.0243	832.6	0.0283	943.0	0.0323	1045.8	0.0363	1141.0
0.0244	835.5	0.0284	945.6	0.0324	1048.2	0.0364	1143.3
0.0245	838.3	0.0285	948.3	0.0325	1050.7	0.0365	1145.6
0.0246	841.2	0.0286	950.9	0.0326	1053.2	0.0366	1147.8
0.0247	844.0	0.0287	953.6	0.0327	1055.6	0.0367	1150.1
0.0248	846.8	0.0288	956.2	0.0328	1058.1	0.0368	1152.4
0.0249	849.7	0.0289	958.9	0.0329	1060.5	0.0369	1154.6
0.0250	852.5	0.0290	961.5	0.0330	1063.0	0.0370	1156.9
0.0251	855.3	0.0291	964.1	0.0331	1065.4 ρ_{max}	0.0371	1159.2

TABLE A.11 $f_y = 50,000$ psi (344.8 MPa); $f'_c = 3000$ psi (20.7 MPa)

	ρ	$\dfrac{M_u}{\phi bd^2}$	ρ	$\dfrac{M_u}{\phi bd^2}$	ρ	$\dfrac{M_u}{\phi bd^2}$		ρ	$\dfrac{M_u}{\phi bd^2}$
ρ_{min}	0.0040	192.1	0.0082	376.9	0.0124	544.4		0.0166	694.5
	0.0041	196.7	0.0083	381.1	0.0125	548.2		0.0167	697.9
	0.0042	201.3	0.0084	385.3	0.0126	551.9		0.0168	701.2
	0.0043	205.9	0.0085	389.5	0.0127	555.7		0.0169	704.6
	0.0044	210.5	0.0086	393.6	0.0128	559.4		0.0170	707.9
	0.0045	215.0	0.0087	397.8	0.0129	563.2		0.0171	711.2
	0.0046	219.6	0.0088	401.9	0.0130	566.9		0.0172	714.5
	0.0047	224.1	0.0089	406.1	0.0131	570.6		0.0173	717.8
	0.0048	228.7	0.0090	410.2	0.0132	574.3		0.0174	721.1
	0.0049	233.2	0.0091	414.3	0.0133	578.0		0.0175	724.4
	0.0050	237.7	0.0092	418.4	0.0134	581.7		0.0176	727.7
	0.0051	242.2	0.0093	422.5	0.0135	585.4		0.0177	731.0
	0.0052	246.7	0.0094	426.6	0.0136	589.1		0.0178	734.2
	0.0053	251.2	0.0095	430.6	0.0137	592.7		0.0179	737.5
	0.0054	255.7	0.0096	434.7	0.0138	596.4		0.0180	740.7
	0.0055	260.1	0.0097	438.7	0.0139	600.0		0.0181	743.9
	0.0056	264.6	0.0098	442.8	0.0140	603.6		0.0182	747.1
	0.0057	269.0	0.0099	446.8	0.0141	607.2		0.0183	750.3
	0.0058	273.5	0.0100	450.8	0.0142	610.9		0.0184	753.5
	0.0059	277.9	0.0101	454.8	0.0143	614.5		0.0185	756.7
	0.0060	282.3	0.0102	458.8	0.0144	618.0		0.0186	759.9
	0.0061	286.7	0.0103	462.8	0.0145	621.6		0.0187	763.1
	0.0062	291.1	0.0104	466.8	0.0146	625.2		0.0188	766.2
	0.0063	295.5	0.0105	470.8	0.0147	628.7		0.0189	769.4
	0.0064	299.9	0.0106	474.8	0.0148	632.3		0.0190	772.5
	0.0065	304.2	0.0107	478.7	0.0149	635.8		0.0191	775.6
	0.0066	308.6	0.0108	482.6	0.0150	639.4		0.0192	778.7
	0.0067	312.9	0.0109	486.6	0.0151	642.9		0.0193	781.8
	0.0068	317.3	0.0110	490.5	0.0152	646.4		0.0194	784.9
	0.0069	321.6	0.0111	494.4	0.0153	649.9		0.0195	788.0
	0.0070	325.9	0.0112	498.3	0.0154	653.4		0.0196	791.1
	0.0071	330.2	0.0113	502.2	0.0155	656.9		0.0197	794.2
	0.0072	334.5	0.0114	506.1	0.0156	660.3		0.0198	797.2
	0.0073	338.8	0.0115	510.0	0.0157	663.8		0.0199	800.3
	0.0074	343.1	0.0116	513.8	0.0158	667.3		0.0200	803.3
	0.0075	347.3	0.0117	517.7	0.0159	670.7		0.0201	806.4
	0.0076	351.6	0.0118	521.5	0.0160	674.1		0.0202	809.4
	0.0077	355.8	0.0119	525.4	0.0161	677.5		0.0203	812.4
	0.0078	360.1	0.0120	529.2	0.0162	681.0		0.0204	815.4
	0.0079	364.3	0.0121	533.0	0.0163	684.4		0.0205	818.4
	0.0080	368.5	0.0122	536.8	0.0164	687.8	ρ_{max}	0.0206	821.3
	0.0081	372.7	0.0123	540.6	0.0165	691.1			

TABLE A.12 $f_y = 50,000$ psi (344.8 MPa); $f'_c = 4000$ psi (27.6 MPa)

	ρ	$\dfrac{M_u}{\phi bd^2}$	ρ	$\dfrac{M_u}{\phi bd^2}$	ρ	$\dfrac{M_u}{\phi bd^2}$	ρ	$\dfrac{M_u}{\phi bd^2}$
ρ_{min}	0.0040	194.1	0.0070	331.9	0.0100	463.1	0.0130	587.7
	0.0041	198.8	0.0071	336.4	0.0101	467.4	0.0131	591.7
	0.0042	203.5	0.0072	340.9	0.0102	471.6	0.0132	595.7
	0.0043	208.2	0.0073	345.3	0.0103	475.9	0.0133	599.8
	0.0044	212.9	0.0074	349.8	0.0104	480.1	0.0134	603.8
	0.0045	217.5	0.0075	354.3	0.0105	484.3	0.0135	607.8
	0.0046	222.2	0.0076	358.7	0.0106	488.6	0.0136	611.8
	0.0047	226.9	0.0077	363.1	0.0107	492.8	0.0137	615.8
	0.0048	231.5	0.0078	367.6	0.0108	497.0	0.0138	619.8
	0.0049	236.1	0.0079	372.0	0.0109	501.2	0.0139	623.7
	0.0050	240.8	0.0080	376.4	0.0110	505.4	0.0140	627.7
	0.0051	245.4	0.0081	380.8	0.0111	509.6	0.0141	631.7
	0.0052	250.0	0.0082	385.2	0.0112	513.7	0.0142	635.6
	0.0053	254.6	0.0083	389.6	0.0113	517.9	0.0143	639.6
	0.0054	259.2	0.0084	394.0	0.0114	522.1	0.0144	643.5
	0.0055	263.8	0.0085	398.4	0.0115	526.2	0.0145	647.5
	0.0056	268.4	0.0086	402.7	0.0116	530.4	0.0146	651.4
	0.0057	273.0	0.0087	407.1	0.0117	534.5	0.0147	655.3
	0.0058	277.6	0.0088	411.4	0.0118	538.6	0.0148	659.2
	0.0059	282.2	0.0089	415.8	0.0119	542.8	0.0149	663.1
	0.0060	286.7	0.0090	420.1	0.0120	546.9	0.0150	667.0
	0.0061	291.3	0.0091	424.5	0.0121	551.0	0.0151	670.9
	0.0062	295.8	0.0092	428.8	0.0122	555.1	0.0152	674.8
	0.0063	300.4	0.0093	433.1	0.0123	559.2	0.0153	678.7
	0.0064	304.9	0.0094	437.4	0.0124	563.3	0.0154	682.5
	0.0065	309.4	0.0095	441.7	0.0125	567.4	0.0155	686.4
	0.0066	313.9	0.0096	446.0	0.0126	571.4	0.0156	690.3
	0.0067	318.4	0.0097	450.3	0.0127	575.5	0.0157	694.1
	0.0068	322.9	0.0098	454.6	0.0128	579.6	0.0158	697.9
	0.0069	327.4	0.0099	458.9	0.0129	583.6	0.0159	701.8

TABLE A.12 $f_y = 50,000$ psi (344.8 MPa); $f'_c = 4000$ psi (27.6 MPa)

ρ	$\dfrac{M_u}{\phi bd^2}$	ρ	$\dfrac{M_u}{\phi bd^2}$	ρ	$\dfrac{M_u}{\phi bd^2}$		ρ	$\dfrac{M_u}{\phi bd^2}$
0.0160	705.6	0.0189	813.3	0.0218	914.7		0.0247	1010.0
0.0161	709.4	0.0190	816.9	0.0219	918.1		0.0248	1013.2
0.0162	713.2	0.0191	820.5	0.0220	921.5		0.0249	1016.4
0.0163	717.0	0.0192	824.1	0.0221	924.9		0.0250	1019.5
0.0164	720.8	0.0193	827.6	0.0222	928.3		0.0251	1022.7
0.0165	724.6	0.0194	831.2	0.0223	931.6		0.0252	1025.8
0.0166	728.4	0.0195	834.8	0.0224	935.0		0.0253	1029.0
0.0167	732.1	0.0196	838.3	0.0225	938.3		0.0254	1032.1
0.0168	735.9	0.0197	841.9	0.0226	941.6		0.0255	1035.2
0.0169	739.7	0.0198	845.4	0.0227	945.0		0.0256	1038.3
0.0170	743.4	0.0199	849.0	0.0228	948.3		0.0257	1041.4
0.0171	747.2	0.0200	852.5	0.0229	951.6		0.0258	1044.5
0.0172	750.9	0.0201	856.0	0.0230	954.9		0.0259	1047.6
0.0173	754.6	0.0202	859.5	0.0231	958.2		0.0260	1050.7
0.0174	758.3	0.0203	863.0	0.0232	961.5		0.0261	1053.8
0.0175	762.1	0.0204	866.5	0.0233	964.8		0.0262	1056.9
0.0176	765.8	0.0205	870.0	0.0234	968.1		0.0263	1059.9
0.0177	769.5	0.0206	873.5	0.0235	971.3		0.0264	1063.0
0.0178	773.2	0.0207	877.0	0.0236	974.6		0.0265	1066.0
0.0179	776.8	0.0208	880.5	0.0237	977.9		0.0266	1069.1
0.0180	780.5	0.0209	883.9	0.0238	981.1		0.0267	1072.1
0.0181	784.2	0.0210	887.4	0.0239	984.4		0.0268	1075.1
0.0182	787.8	0.0211	890.8	0.0240	987.6		0.0269	1078.2
0.0183	791.5	0.0212	894.3	0.0241	990.8		0.0270	1081.2
0.0184	795.1	0.0213	897.7	0.0242	994.0		0.0271	1084.2
0.0185	798.8	0.0214	901.1	0.0243	997.2		0.0272	1087.2
0.0186	802.4	0.0215	904.5	0.0244	1000.4		0.0273	1090.2
0.0187	806.0	0.0216	907.9	0.0245	1003.6		0.0274	1093.1
0.0188	809.7	0.0217	911.3	0.0246	1006.8	ρ_{max}	0.0275	1096.1

TABLE A.13 $f_y = 60{,}000$ psi (413.7 MPa); $f'_c = 3000$ psi (20.7 MPa)

ρ	$\dfrac{M_u}{\phi bd^2}$	ρ	$\dfrac{M_u}{\phi bd^2}$	ρ	$\dfrac{M_u}{\phi bd^2}$	ρ	$\dfrac{M_u}{\phi bd^2}$
ρ_{min} 0.0033	190.3	0.0066	365.2	0.0099	524.6	0.0131	664.5
0.0034	195.8	0.0067	370.2	0.0100	529.2	0.0132	668.6
0.0035	201.3	0.0068	375.3	0.0101	533.8	0.0133	672.8
0.0036	206.8	0.0069	380.3	0.0102	538.3	0.0134	676.9
0.0037	212.3	0.0070	385.3	0.0103	542.9	0.0135	681.0
0.0038	217.8	0.0071	390.3	0.0104	547.4	0.0136	685.0
0.0039	223.2	0.0072	395.3	0.0105	551.9	0.0137	689.1
0.0040	228.7	0.0073	400.3	0.0106	556.4	0.0138	693.2
0.0041	234.1	0.0074	405.2	0.0107	560.9	0.0139	697.2
0.0042	239.5	0.0075	410.2	0.0108	565.4	0.0140	701.2
0.0043	244.9	0.0076	415.1	0.0109	569.9	0.0141	705.2
0.0044	250.3	0.0077	420.0	0.0110	574.3	0.0142	709.2
0.0045	255.7	0.0078	424.9	0.0111	578.8	0.0143	713.2
0.0046	261.0	0.0079	429.8	0.0112	583.2	0.0144	717.2
0.0047	266.4	0.0080	434.7	0.0113	587.6	0.0145	721.1
0.0048	271.7	0.0081	439.5	0.0114	592.0	0.0146	725.1
0.0049	277.0	0.0082	444.4	0.0115	596.4	0.0147	729.0
0.0050	282.3	0.0083	449.2	0.0116	600.7	0.0148	732.9
0.0051	287.6	0.0084	454.0	0.0117	605.1	0.0149	736.8
0.0052	292.9	0.0085	458.8	0.0118	609.4	0.0150	740.7
0.0053	298.1	0.0086	463.6	0.0119	613.7	0.0151	744.6
0.0054	303.4	0.0087	468.4	0.0120	618.0	0.0152	748.4
0.0055	308.6	0.0088	473.2	0.0121	622.3	0.0153	752.3
0.0056	313.8	0.0089	477.9	0.0122	626.6	0.0154	756.1
0.0057	319.0	0.0090	482.6	0.0123	630.9	0.0155	759.9
0.0058	324.2	0.0091	487.4	0.0124	635.1	0.0156	763.7
0.0059	329.4	0.0092	492.1	0.0125	639.4	0.0157	767.5
0.0060	334.5	0.0093	496.8	0.0126	643.6	0.0158	771.2
0.0061	339.7	0.0094	501.4	0.0127	647.8	0.0159	775.0
0.0062	344.8	0.0095	506.1	0.0128	652.0	0.0160	778.7
0.0063	349.9	0.0096	510.7	0.0129	656.2	ρ_{max} 0.0161	782.5
0.0064	355.0	0.0097	515.4	0.0130	660.9		
0.0065	360.1	0.0098	520.0				

TABLE A.14 $f_y = 60,000$ psi (413.7 MPa); $f'_c = 4000$ psi (27.6 MPa)

	ρ	$\dfrac{M_u}{\phi bd^2}$	ρ	$\dfrac{M_u}{\phi bd^2}$	ρ	$\dfrac{M_u}{\phi bd^2}$		ρ	$\dfrac{M_u}{\phi bd^2}$
ρ_{min}	0.0033	192.2	0.0079	440.9	0.0125	667.0		0.0171	870.7
	0.0034	197.9	0.0080	446.0	0.0126	671.7		0.0172	874.9
	0.0035	203.5	0.0081	451.2	0.0127	676.3		0.0173	879.1
	0.0036	209.1	0.0082	456.3	0.0128	681.0		0.0174	883.2
	0.0037	214.7	0.0083	461.4	0.0129	685.6		0.0175	887.4
	0.0038	220.3	0.0084	466.5	0.0130	690.3		0.0176	891.5
	0.0039	225.9	0.0085	471.6	0.0131	694.9		0.0177	895.6
	0.0040	231.5	0.0086	476.7	0.0132	699.5		0.0178	899.7
	0.0041	237.1	0.0087	481.8	0.0133	704.1		0.0179	903.9
	0.0042	242.6	0.0088	486.9	0.0134	708.6		0.0180	907.9
	0.0043	248.2	0.0089	491.9	0.0135	713.2		0.0181	912.0
	0.0044	253.7	0.0090	497.0	0.0136	717.8		0.0182	916.1
	0.0045	259.2	0.0091	502.0	0.0137	722.3		0.0183	920.2
	0.0046	264.8	0.0092	507.1	0.0138	726.9		0.0184	924.2
	0.0047	270.3	0.0093	512.1	0.0139	731.4		0.0185	928.3
	0.0048	275.8	0.0094	517.1	0.0140	735.9		0.0186	932.3
	0.0049	281.2	0.0095	522.1	0.0141	740.4		0.0187	936.3
	0.0050	286.7	0.0096	527.1	0.0142	744.9		0.0188	940.3
	0.0051	292.2	0.0097	532.0	0.0143	749.4		0.0189	944.3
	0.0052	297.6	0.0098	537.0	0.0144	753.9		0.0190	948.3
	0.0053	303.1	0.0099	542.0	0.0145	758.3		0.0191	952.3
	0.0054	308.5	0.0100	546.9	0.0146	762.8		0.0192	956.2
	0.0055	313.9	0.0101	551.8	0.0147	767.2		0.0193	960.2
	0.0056	319.3	0.0102	556.7	0.0148	771.7		0.0194	964.1
	0.0057	324.7	0.0103	561.7	0.0149	776.1		0.0195	968.1
	0.0058	330.1	0.0104	566.6	0.0150	780.5		0.0196	972.0
	0.0059	335.5	0.0105	571.5	0.0151	784.9		0.0197	975.9
	0.0060	340.9	0.0106	576.3	0.0152	789.3		0.0198	979.8
	0.0061	346.2	0.0107	581.2	0.0153	793.7		0.0199	983.7
	0.0062	351.6	0.0108	586.1	0.0154	798.1		0.0200	987.6
	0.0063	356.9	0.0109	590.9	0.0155	802.4		0.0201	991.5
	0.0064	362.2	0.0110	595.7	0.0156	806.8		0.0202	995.3
	0.0065	367.6	0.0111	600.6	0.0157	811.1		0.0203	999.2
	0.0066	372.9	0.0112	.605.4	0.0158	815.4		0.0204	1003.0
	0.0067	378.2	0.0113	610.2	0.0159	819.7		0.0205	1006.8
	0.0068	383.4	0.0114	615.0	0.0160	824.1		0.0206	1010.7
	0.0069	388.7	0.0115	619.8	0.0161	828.3		0.0207	1014.5
	0.0070	394.0	0.0116	624.5	0.0162	832.6		0.0208	1018.3
	0.0071	399.2	0.0117	629.3	0.0163	836.9		0.0209	1022.0
	0.0072	404.5	0.0118	634.1	0.0164	841.2		0.0210	1025.8
	0.0073	409.7	0.0119	638.8	0.0165	845.4		0.0211	1029.6
	0.0074	414.9	0.0120	643.5	0.0166	849.7		0.0212	1033.3
	0.0075	420.1	0.0121	648.2	0.0167	853.9		0.0213	1037.1
	0.0076	425.3	0.0122	653.0	0.0168	858.1	ρ_{max}	0.0214	1040.8
	0.0077	430.5	0.0123	657.7	0.0169	862.3			
	0.0078	435.7	0.0124	662.3	0.0170	866.5			

TABLE A.15   BASIC TENSION DEVELOPMENT
LENGTHS ℓ_{db} (in.)

Bar no.	f_y (psi)	f'_c (psi) 3000	4000	5000	6000
2	40,000	1.43	1.24	1.11	1.01
	50,000	1.79	1.55	1.39	1.27
	60,000	2.15	1.86	1.66	1.52
3	40,000	3.21	2.78	2.49	2.27
	50,000	4.02	3.48	3.11	2.84
	60,000	4.82	4.17	3.73	3.41
4	40,000	5.73	4.96	4.43	4.05
	50,000	7.16	6.20	5.54	5.06
	60,000	8.59	7.44	6.65	6.07
5	40,000	8.97	7.77	6.95	6.34
	50,000	11.21	9.71	8.68	7.93
	60,000	13.45	11.65	10.42	9.51
6	40,000	12.91	11.18	10.00	9.13
	50,000	16.14	13.98	12.50	11.41
	60,000	19.37	16.77	15.00	13.69
7	40,000	17.56	15.20	13.60	12.41
	50,000	21.95	19.01	17.00	15.52
	60,000	26.33	22.81	20.40	18.62
8	40,000	22.93	19.86	17.76	16.21
	50,000	28.66	24.82	22.20	20.27
	60,000	34.40	29.79	26.64	24.32
9	40,000	29.21	25.30	22.63	20.66
	50,000	36.51	31.62	28.28	25.82
	60,000	43.82	37.95	33.94	30.98
10	40,000	37.01	32.05	28.67	26.17
	50,000	46.26	40.07	35.84	32.71
	60,000	55.52	48.08	43.00	39.26
11	40,000	45.60	39.49	35.32	32.24
	50,000	57.00	49.36	44.15	40.30
	60,000	68.40	59.24	52.98	48.37
14	40,000	62.08	53.76	48.08	43.89
	50,000	77.59	67.20	60.10	54.87
	60,000	93.11	80.64	72.12	65.84
18	40,000	91.29	79.06	70.71	64.55
	50,000	114.11	98.82	88.39	80.69
	60,000	136.93	118.59	106.07	96.82

Note: The ℓ_{db} values given here must be multiplied by the appropriate modification factors given in Table 6.1 of Chapter 6 of this book to obtain the ℓ'_{db} values. These values may not be less than $0.03 d_b f_y / \sqrt{f'_c}$ nor 12 in., whichever is larger. The ℓ_d values then equal the ℓ'_{db} values times any applicable modification factors from Table 6.2 of Chapter 6.

TABLE A.16 SIZE AND PITCH OF SPIRALS, ACI CODE

Diameter of column (in.)	Out to out of spiral (in.)	f'_c			
		2500	3000	4000	5000
$f_y = 40{,}000$:					
14, 15	11, 12	$\frac{3}{8}$–2	$\frac{3}{8}$–$1\frac{3}{4}$	$\frac{1}{2}$–$2\frac{1}{2}$	$\frac{1}{2}$–$1\frac{3}{4}$
16	13	$\frac{3}{8}$–2	$\frac{3}{8}$–$1\frac{3}{4}$	$\frac{1}{2}$–$2\frac{1}{2}$	$\frac{1}{2}$–2
17–19	14–16	$\frac{3}{8}$–$2\frac{1}{4}$	$\frac{3}{8}$–$1\frac{3}{4}$	$\frac{1}{2}$–$2\frac{1}{2}$	$\frac{1}{2}$–2
20–23	17–20	$\frac{3}{8}$–$2\frac{1}{4}$	$\frac{3}{8}$–$1\frac{3}{4}$	$\frac{1}{2}$–$2\frac{1}{2}$	$\frac{1}{2}$–2
24–30	21–27	$\frac{3}{8}$–$2\frac{1}{4}$	$\frac{3}{8}$–2	$\frac{1}{2}$–$2\frac{1}{2}$	$\frac{1}{2}$–2
$f_y = 60{,}000$:					
14, 15	11, 12	$\frac{1}{4}$–$1\frac{3}{4}$	$\frac{3}{8}$–$2\frac{3}{4}$	$\frac{3}{8}$–2	$\frac{1}{2}$–$2\frac{3}{4}$
16–23	13–20	$\frac{1}{4}$–$1\frac{3}{4}$	$\frac{3}{8}$–$2\frac{3}{4}$	$\frac{3}{8}$–2	$\frac{1}{2}$–3
24–29	21–26	$\frac{1}{4}$–$1\frac{3}{4}$	$\frac{3}{8}$–3	$\frac{3}{8}$–$2\frac{1}{4}$	$\frac{1}{2}$–3
30	27	$\frac{1}{4}$–$1\frac{3}{4}$	$\frac{3}{8}$–3	$\frac{3}{8}$–$2\frac{1}{4}$	$\frac{1}{2}$–$3\frac{1}{4}$

TABLE A.17   WEIGHTS, AREAS, AND MOMENTS OF INERTIA OF CIRCULAR
COLUMNS AND MOMENTS OF INERTIA OF COLUMN
VERTICALS ARRANGED IN A CIRCLE 5 in. LESS THAN THE
DIAMETER OF COLUMN

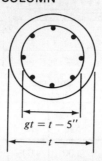

$gt = t - 5''$

t

Diameter of column h (in.)	Weight per foot (lb)	Area (in.^{2})	I (in.^{4})	A_s, where $\rho_g = 0.01$*	I_s (in.^{4})[†]
12	118	113	1,018	1.13	6.92
13	138	133	1,402	1.33	10.64
14	160	154	1,886	1.54	15.59
15	184	177	2,485	1.77	22.13
16	210	201	3,217	2.01	30.40
17	237	227	4,100	2.27	40.86
18	265	255	5,153	2.55	53.87
19	295	284	6,397	2.84	69.58
20	327	314	7,854	3.14	88.31
21	361	346	9,547	3.46	110.7
22	396	380	11,500	3.80	137.2
23	433	416	13,740	4.16	168.4
24	471	452	16,290	4.52	203.9
25	511	491	19,170	4.91	245.5
26	553	531	22,430	5.31	292.7
27	597	573	26,090	5.73	346.7
28	642	616	30,170	6.16	407.3
29	688	661	34,720	6.61	475.9
30	736	707	39,760	7.07	552.3

* For other values of ρ_g, multiply the value in the table by $100\rho_g$.

[†] The bars are assumed transformed into a thin-walled cylinder having the same sectional area as the bars. Then $I_s = A_s(\gamma t)^2/8$.

TABLE A.18 MOMENT DISTRIBUTION CONSTANTS FOR SLABS WITHOUT DROP PANELS*

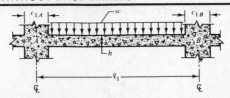

Column dimension		Uniform load FEM = Coef. $(w\ell_2\ell_1^2)$		Stiffness factor[†]		Carryover factor	
$\dfrac{c_{1A}}{\ell_1}$	$\dfrac{c_{1B}}{\ell_1}$	M_{AB}	M_{BA}	k_{AB}	k_{BA}	COF_{AB}	COF_{BA}
0.00	0.00	0.083	0.083	4.00	4.00	0.500	0.500
	0.05	0.083	0.084	4.01	4.04	0.504	0.500
	0.10	0.082	0.086	4.03	4.15	0.513	0.499
	0.15	0.081	0.089	4.07	4.32	0.528	0,498
	0.20	0.079	0.093	4.12	4.56	0.548	0.495
	0.25	0.077	0.097	4.18	4.88	0.573	0.491
	0.30	0.075	0.102	4.25	5.28	0.603	0.485
	0.35	0.073	0.107	4.33	5.78	0.638	0.478
0.05	0.05	0.084	0.084	4.05	4.05	0.503	0.503
	0.10	0.083	0.086	4.07	4.15	0.513	0.503
	0.15	0.081	0.089	4.11	4.33	0.528	0.501
	0.20	0.080	0.092	4.16	4.58	0.548	0.499
	0.25	0.078	0.096	4.22	4.89	0.573	0.494
	0.30	0.076	0.101	4.29	5.30	0.603	0.489
	0.35	0.074	0.107	4.37	5.80	0.638	0.481
0.10	0.10	0.085	0.085	4.18	4.18	0.513	0.513
	0.15	0.083	0.088	4.22	4.36	0.528	0.511
	0.20	0.082	0.091	4.27	4.61	0.548	0.508
	0.25	0.080	0.095	4.34	4.93	0.573	0.504
	0.30	0.078	0.100	4.41	5.34	0.602	0.498
	0.35	0.075	0.105	4.50	5.85	0.637	0.491
0.15	0.15	0.086	0.086	4.40	4.40	0.526	0.526
	0.20	0.084	0.090	4.46	4.65	0.546	0.523
	0.25	0.083	0.094	4.53	4.98	0.571	0.519
	0.30	0.080	0.099	4.61	5.40	0.601	0.513
	0.35	0.078	0.104	4.70	5.92	0.635	0.505
0.20	0.20	0.088	0.088	4.72	4.72	0.543	0.543
	0.25	0.086	0.092	4.79	5.05	0.568	0.539
	0.30	0.083	0.097	4.88	5.48	0.597	0.532
	0.35	0.081	0.102	4.99	6.01	0.632	0.524
0.25	0.25	0.090	0.090	5.14	5.14	0.563	0.563
	0.30	0.088	0.095	5.24	5.58	0.592	0.556
	0.35	0.085	0.100	5.36	6.12	0.626	0.548
0.30	0.30	0.092	0.092	5.69	5.69	0.585	0.585
	0.35	0.090	0.097	5.83	6.26	0.619	0.576
0.35	0.35	0.095	0.095	6.42	6.42	0.609	0.609

* Applicable when $c_1/\ell_1 = c_2/\ell_2$. For other relationships between these ratios, the constants will be slightly in error.

† Stiffness is $K_{AB} = k_{AB}E\dfrac{\ell_2 h^3}{12\ell_1}$ and $K_{BA} = k_{BA}E\dfrac{\ell_2 h^3}{12\ell_1}$.

TABLE A.19   MOMENT DISTRIBUTION CONSTANTS FOR SLABS WITH DROP PANELS*

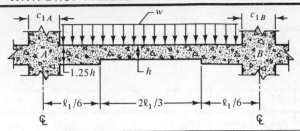

Column dimension		Uniform load FEM = Coef. $(w\ell_2\ell_1^2)$		Stiffness factor[†]		Carryover factor	
$\dfrac{c_{1A}}{\ell_1}$	$\dfrac{c_{1B}}{\ell_1}$	M_{AB}	M_{BA}	k_{AB}	k_{BA}	COF_{AB}	COF_{BA}
0.00	0.00	0.088	0.088	4.78	4.78	0.541	0.541
	0.05	0.087	0.089	4.80	4.82	0.545	0.541
	0.10	0.087	0.090	4.83	4.94	0.553	0.541
	0.15	0.085	0.093	4.87	5.12	0.567	0.540
	0.20	0.084	0.096	4.93	5.36	0.585	0.537
	0.25	0.082	0.100	5.00	5.68	0.606	0.534
	0.30	0.080	0.105	5.09	6.07	0.631	0.529
0.05	0.05	0.088	0.088	4.84	4.84	0.545	0.545
	0.10	0.087	0.090	4.87	4.95	0.553	0.544
	0.15	0.085	0.093	4.91	5.13	0.567	0.543
	0.20	0.084	0.096	4.97	5.38	0.584	0.541
	0.25	0.082	0.100	5.05	5.70	0.606	0.537
	0.30	0.080	0.104	5.13	6.09	0.632	0.532
0.10	0.10	0.089	0.089	4.98	4.98	0.553	0.553
	0.15	0.088	0.092	5.03	5.16	0.566	0.551
	0.20	0.086	0.094	5.09	5.42	0.584	0.549
	0.25	0.084	0.099	5.17	5.74	0.606	0.546
	0.30	0.082	0.103	5.26	6.13	0.631	0.541
0.15	0.15	0.090	0.090	5.22	5.22	0.565	0.565
	0.20	0.089	0.094	5.28	5.47	0.583	0.563
	0.25	0.087	0.097	5.37	5.80	0.604	0.559
	0.30	0.085	0.102	5.46	6.21	0.630	0.554
0.20	0.20	0.092	0.092	5.55	5.55	0.580	0.580
	0.25	0.090	0.096	5.64	5.88	0.602	0.577
	0.30	0.088	0.100	5.74	6.30	0.627	0.571
0.25	0.25	0.094	0.094	5.98	5.98	0.598	0.598
	0.30	0.091	0.098	6.10	6.41	0.622	0.593
0.30	0.30	0.095	0.095	6.54	6.54	0.617	0.617

[a] Applicable when $c_1/\ell_1 = c_2/\ell_2$. For other relationships between these ratios, the constants will be slightly in error.

[†] Stiffness occurs when $K_{AB} = k_{AB}E\dfrac{\ell_2 h^3}{12\ell_1}$ and $K_{BA} = k_{BA}E\dfrac{\ell_2 h^3}{12\ell_1}$.

TABLE A.20 MOMENT DISTRIBUTION CONSTANTS FOR SLAB-BEAM MEMBERS WITH COLUMN CAPITALS

FEM (uniform load w) $= Mw\ell_2(\ell_1)^2$

K (stiffness) $= kE\ell_2 h^3/12\ell_1$

Carryover factor $= C$

c_1/ℓ_1	c_1/ℓ_2	M	k	C	c_1/ℓ_1	c_2/ℓ_2	M	k	C
	0.00	0.083	4.000	0.500		0.00	0.083	4.000	0.500
	0.05	0.083	4.000	0.500		0.05	0.085	4.170	0.511
	0.10	0.083	4.000	0.500		0.10	0.086	4.346	0.522
	0.15	0.083	4.000	0.500		0.15	0.087	4.529	0.532
	0.20	0.083	4.000	0.500		0.20	0.088	4.717	0.543
0.00	0.25	0.083	4.000	0.500	0.20	0.25	0.089	4.910	0.554
	0.30	0.083	4.000	0.500		0.30	0.090	5.108	0.564
	0.35	0.083	4.000	0.500		0.35	0.091	5.308	0.574
	0.40	0.083	4.000	0.500		0.40	0.092	5.509	0.584
	0.45	0.083	4.000	0.500		0.45	0.093	5.710	0.593
	0.50	0.083	4.000	0.500		0.50	0.094	5.908	0.602
	0.00	0.083	4.000	0.500		0.00	0.083	4.000	0.500
	0.05	0.084	4.047	0.503		0.05	0.085	4.204	0.512
	0.10	0.084	4.093	0.507		0.10	0.086	4.420	0.525
	0.15	0.084	4.138	0.510		0.15	0.087	4.648	0.538
	0.20	0.085	4.181	0.513		0.20	0.089	4.887	0.550
0.05	0.25	0.085	4.222	0.516	0.25	0.25	0.090	5.138	0.563
	0.30	0.085	4.261	0.518		0.30	0.091	5.401	0.576
	0.35	0.086	4.299	0.521		0.35	0.093	5.672	0.588
	0.40	0.086	4.334	0.523		0.40	0.094	5.952	0.600
	0.45	0.086	4.368	0.526		0.45	0.095	6.238	0.612
	0.50	0.086	4.398	0.528		0.50	0.096	6.527	0.623
	0.00	0.083	4.000	0.500		0.00	0.083	4.000	0.500
	0.05	0.084	4.091	0.506		0.05	0.085	4.235	0.514
	0.10	0.085	4.182	0.513		0.10	0.086	4.488	0.527
	0.15	0.085	4.272	0.519		0.15	0.088	4.760	0.542
	0.20	0.086	4.362	0.524		0.20	0.089	5.050	0.556
0.10	0.25	0.087	4.449	0.530	0.30	0.25	0.091	5.361	0.571
	0.30	0.087	4.535	0.535		0.30	0.092	5.692	0.585
	0.35	0.088	4.618	0.540		0.35	0.094	6.044	0.600
	0.40	0.088	4.698	0.545		0.40	0.095	6.414	0.614
	0.45	0.089	4.774	0.550		0.45	0.096	6.802	0.628
	0.50	0.089	4.846	0.554		0.50	0.098	7.205	0.642
	0.00	0.083	4.000	0.500		0.00	0.083	4.000	0.500
	0.05	0.084	4.132	0.509		0.05	0.085	4.264	0.514
	0.10	0.085	4.267	0.517		0.10	0.087	4.551	0.529
	0.15	0.086	4.403	0.526		0.15	0.088	4.864	0.545
	0.20	0.087	4.541	0.534		0.20	0.090	5.204	0.560
0.15	0.25	0.088	4.680	0.543	0.35	0.25	0.091	5.575	0.576
	0.30	0.089	4.818	0.550		0.30	0.093	5.979	0.593
	0.35	0.090	4.955	0.558		0.35	0.095	6.416	0.609
	0.40	0.090	5.090	0.565		0.40	0.096	6.888	0.626
	0.45	0.091	5.222	0.572		0.45	0.098	7.395	0.642
	0.50	0.092	5.349	0.579		0.50	0.099	7.935	0.658

TABLE A.20   (*continued*)

c_1/ℓ_1	c_1/ℓ_2	M	k	C	c_1/ℓ_1	c_2/ℓ_2	M	k	C
	0.00	0.083	4.000	0.500		0.30	0.094	6.517	0.602
	0.05	0.085	4.289	0.515		0.35	0.096	7.136	0.621
	0.10	0.087	4.607	0.530		0.40	0.098	7.836	0.642
	0.15	0.088	4.959	0.546		0.45	0.100	8.625	0.662
	0.20	0.090	5.348	0.563		0.50	0.101	9.514	0.683
0.40	0.25	0.092	5.778	0.580		0.00	0.083	4.000	0.500
	0.30	0.094	6.255	0.598		0.05	0.085	4.331	0.515
	0.35	0.095	6.782	0.617		0.10	0.087	4.703	0.530
	0.40	0.097	7.365	0.635		0.15	0.088	5.123	0.547
	0.45	0.099	8.007	0.654		0.20	0.090	5.599	0.564
	0.50	0.100	8.710	0.672	0.50	0.25	0.092	6.141	0.583
	0.00	0.083	4.000	0.500		0.30	0.094	6.760	0.603
	0.05	0.085	4.311	0.515		0.35	0.096	7.470	0.624
	0.10	0.087	4.658	0.530		0.40	0.098	8.289	0.645
	0.15	0.088	5.046	0.547		0.45	0.100	9.234	0.667
	0.20	0.090	5.480	0.564		0.50	0.102	10.329	0.690
0.45	0.25	0.092	5.967	0.583					

TABLE A.21   MOMENT DISTRIBUTION CONSTANTS FOR
SLAB-BEAM MEMBERS WITH COLUMN
CAPITALS AND DROP PANELS

FEM (uniform load w) $= Mw\ell_2(\ell_1^2)$
K (stiffness) $= kEl_2h^3/12\ell_1$
Carryover factor $= C$

c_1/ℓ_1	c_2/ℓ_2	Constants for $h_2 = 1.25h_1$			Constants for $h_2 = 1.5h_2$		
		M	k	C	M	k	C
	0.00	0.088	4.795	0.542	0.093	5.837	0.589
	0.05	0.088	4.795	0.542	0.093	5.837	0.589
	0.10	0.088	4.795	0.542	0.093	5.837	0.589
0.00	0.15	0.088	4.795	0.542	0.093	5.837	0.589
	0.20	0.088	4.795	0.542	0.093	5.837	0.589
	0.25	0.088	4.795	0.542	0.093	5.837	0.589
	0.30	0.088	4.797	0.542	0.093	5.837	0.589
	0.00	0.088	4.795	0.542	0.093	5.837	0.589
	0.05	0.088	4.846	0.545	0.093	5.890	0.591
	0.10	0.089	4.896	0.548	0.093	5.942	0.594
0.05	0.15	0.089	4.944	0.551	0.093	5.993	0.596
	0.20	0.089	4.990	0.553	0.094	6.041	0.598
	0.25	0.089	5.035	0.556	0.094	6.087	0.600
	0.30	0.090	5.077	0.558	0.094	6.131	0.602

TABLE A.21   (*continued*)

c_1/ℓ_1	c_2/ℓ_2	Constants for $h_2 = 1.25h_1$			Constants for $h_2 = 1.5h_2$		
		M	k	C	M	k	C
	0.00	0.088	4.795	0.542	0.093	5.837	0.589
	0.05	0.088	4.894	0.548	0.093	5.940	0.593
	0.10	0.089	4.992	0.553	0.094	6.042	0.598
0.10	0.15	0.090	5.039	0.559	0.094	6.142	0.602
	0.20	0.090	5.184	0.564	0.094	6.240	0.607
	0.25	0.091	5.278	0.569	0.095	6.335	0.611
	0.30	0.091	5.368	0.573	0.095	6.427	0.615
	0.00	0.088	4.795	0.542	0.093	5.837	0.589
	0.05	0.089	4.938	0.550	0.093	5.986	0.595
	0.10	0.090	5.082	0.558	0.094	6.135	0.602
0.15	0.15	0.090	5.228	0.565	0.095	6.284	0.608
	0.20	0.091	5.374	0.573	0.095	6.432	0.614
	0.25	0.092	5.520	0.580	0.096	6.579	0.620
	0.30	0.092	5.665	0.587	0.096	6.723	0.626
	0.00	0.088	4.795	0.542	0.093	5.837	0.589
	0.05	0.089	4.978	0.552	0.093	6.027	0.597
	0.10	0.090	5.167	0.562	0.094	6.221	0.605
0.20	0.15	0.091	5.361	0.571	0.095	6.418	0.613
	0.20	0.092	5.558	0.581	0.096	6.616	0.621
	0.25	0.093	5.760	0.590	0.096	6.816	0.628
	0.30	0.094	5.962	0.590	0.097	7.015	0.635
	0.00	0.088	4.795	0.542	0.093	5.837	0.589
	0.05	0.089	5.015	0.553	0.094	6.065	0.598
	0.10	0.090	5.245	0.565	0.094	6.300	0.608
0.25	0.15	0.091	5.485	0.576	0.095	6.543	0.617
	0.20	0.092	5.735	0.587	0.096	6.790	0.626
	0.25	0.094	5.994	0.598	0.097	7.043	0.635
	0.30	0.095	6.261	0.600	0.098	7.298	0.644
	0.00	0.088	4.795	0.542	0.093	5.837	0.589
	0.05	0.089	5.048	0.554	0.094	6.099	0.599
	0.10	0.090	5.317	0.567	0.095	6.372	0.610
0.30	0.15	0.092	5.601	0.580	0.096	6.657	0.620
	0.20	0.093	5.902	0.593	0.097	6.953	0.631
	0.25	0.094	6.219	0.605	0.098	7.258	0.641
	0.30	0.095	6.550	0.618	0.099	7.571	0.651

TABLE A.22   STIFFNESS FACTORS AND CARRYOVER FACTORS FOR COLUMNS

a/b	ℓ_u/ℓ_n	0.95	0.90	0.85	0.80	0.75
0.20	k_{AB}	4.32	4.70	5.33	5.65	6.27
	C_{AB}	0.57	0.64	0.71	0.80	0.89
0.40	k_{AB}	4.40	4.89	5.45	6.15	7.00
	C_{AB}	0.56	0.61	0.68	0.74	0.81
0.60	k_{AB}	4.46	5.02	5.70	6.54	7.58
	C_{AB}	0.55	0.60	0.65	0.70	0.76
0.80	k_{AB}	4.51	5.14	5.90	6.85	8.05
	C_{AB}	0.54	0.58	0.63	0.67	0.72
1.00	k_{AB}	4.55	5.23	6.06	7.11	8.44
	C_{AB}	0.54	0.57	0.61	0.65	0.68
1.20	k_{AB}	4.58	5.30	6.20	7.32	8.77
	C_{AB}	0.53	0.57	0.60	0.63	0.66
1.40	k_{AB}	4.61	5.36	6.31	7.51	9.05
	C_{AB}	0.53	0.56	0.59	0.61	0.64
1.60	k_{AB}	4.63	5.42	6.41	7.66	9.29
	C_{AB}	0.53	0.55	0.58	0.60	0.62
1.80	k_{AB}	4.65	5.46	6.49	7.80	9.50
	C_{AB}	0.53	0.55	0.57	0.59	0.60
2.00	k_{AB}	4.67	5.51	6.56	7.92	9.68
	C_{AB}	0.52	0.54	0.56	0.58	0.59

Notes: 1. Values computed by column analogy method.

2. $k_c = (k_{AB} \text{ from table})\left(\dfrac{EI_0}{\ell_n}\right)$.

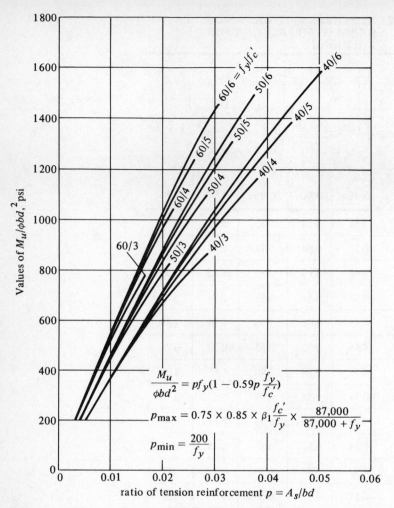

$$\frac{M_u}{\phi bd^2} = pf_y(1 - 0.59p\frac{f_y}{f_c'})$$

$$p_{max} = 0.75 \times 0.85 \times \beta_1\frac{f_c'}{f_y} \times \frac{87,000}{87,000 + f_y}$$

$$p_{min} = \frac{200}{f_y}$$

Graph 1 Moment capacity of rectangular sections.

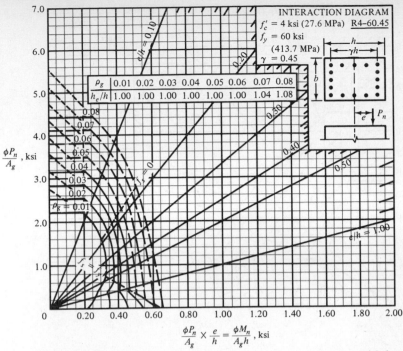

Graph 2 Column interaction diagram for rectangular tied column with bars on all four faces.

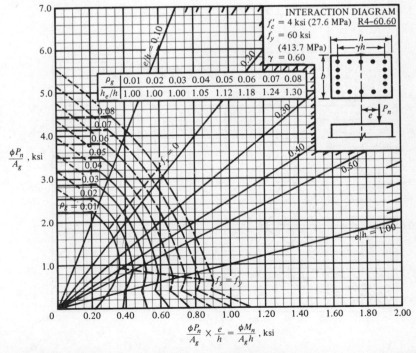

Graph 3 Column interaction diagram for rectangular tied column with bars on all four faces.

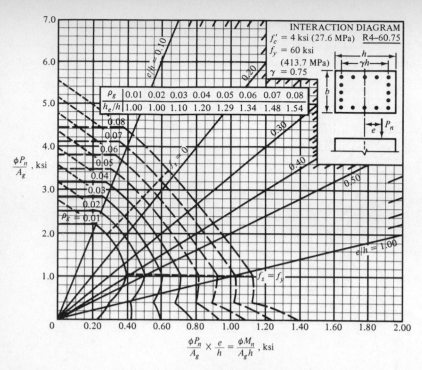

Graph 4 Column interaction diagram for rectangular tied column with bars on all four faces.

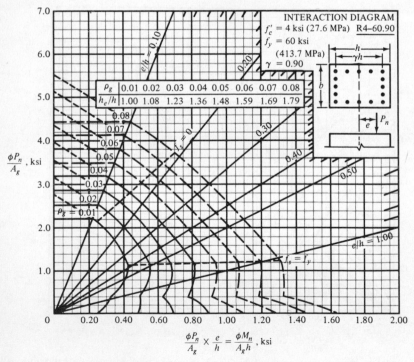

Graph 5 Column interaction diagram for rectangular tied column with bars on all four faces.

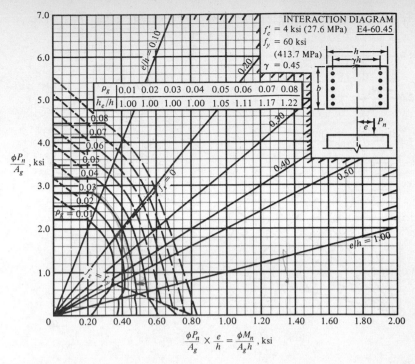

Graph 6 Column interaction diagram for rectangular tied column with bars on end faces.

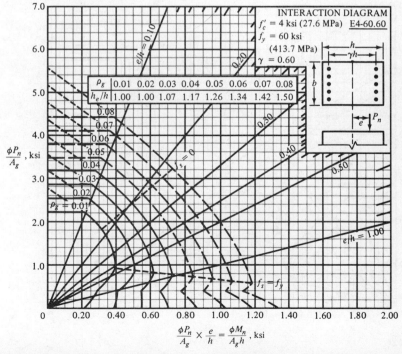

Graph 7 Column interaction diagram for rectangular tied column with bars on end faces.

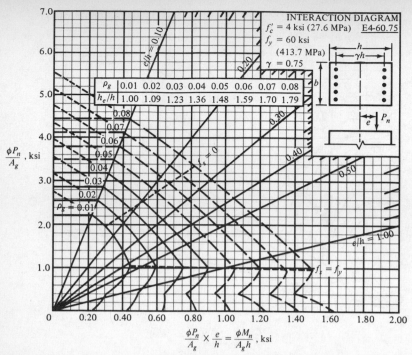

Graph 8 Column interaction diagram for rectangular tied column with bars on end faces.

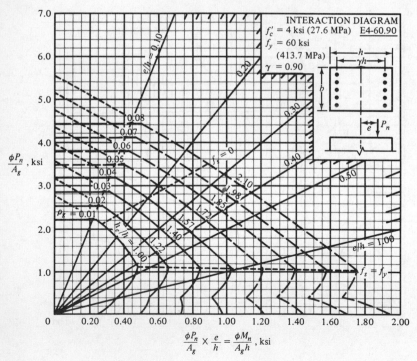

Graph 9 Column interaction diagram for rectangular tied column with bars on end faces.

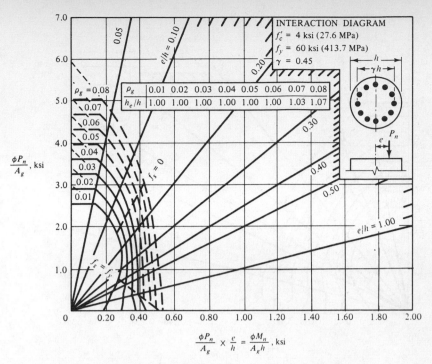

Graph 10 Column interaction diagram for circular spiral columns.

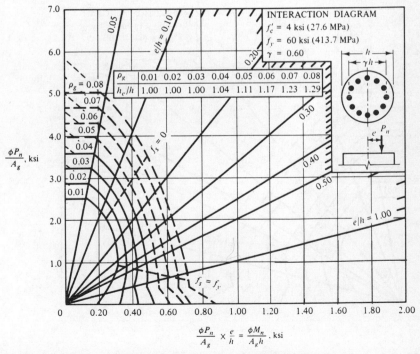

Graph 11 Column interaction diagram for circular spiral columns.

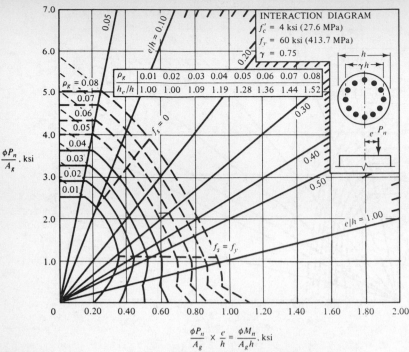

Graph 12 Column interaction diagram for circular spiral columns.

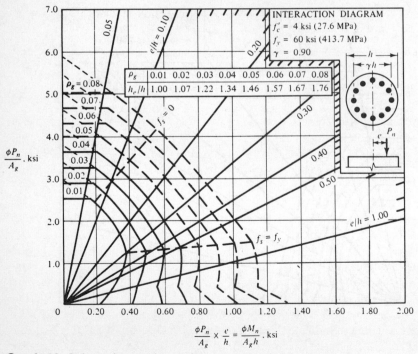

Graph 13 Column interaction diagram for circular spiral columns.

Appendix B

Alternate Design Method

B.1 INTRODUCTION

As an alternate to the strength design method, the ACI Code in its Appendix A permits the proportioning of reinforced concrete structures with an alternate method. This method is an allowable or working stress method in which a margin of safety is provided by permitting calculated flexure stresses to reach only a certain percentage of the ultimate strength of the concrete or of the yield strength of the reinforcing. These percentages are sufficiently small so that an approximately linear relation exists between the stress and strain in the concrete as well as in the reinforcing.

The Code permits the design of reinforced concrete members using service loads and the working-stress design method, except that the method is now referred to as the *alternate design method*. Appendix A of the Code presents in some detail the allowable stresses and other conditions to be used for the alternate design method. The placing of this information in an appendix may be thought by some to indicate that it is not an official part of the Code. For this reason, Section 8.1.2 of the Code states that nonprestressed reinforced concrete members can be designed using service or working loads in accordance with the provisions of Appendix A of the Code. This clearly makes it an official and legal part of the Code.

From Appendix A the maximum permissible concrete compressive stress in the extreme fiber of a member is $0.45f'_c$. The maximum allowable tensile stress in the reinforcing is 20,000 psi for grades 40 and 50 steels and 24,000 psi for grade 60 steels and steels with higher yield strengths. These values are summarized in Appendix Table A.7. For one-way slabs with clear spans of 12 ft or less and reinforced with #3 bars or with welded wire fabric (with diameters not larger than $\frac{3}{8}$ in.) the permissible stress is increased to $0.5f_y$ but may not exceed 30,000 psi.

Should reference be made to the AASHTO or AREA bridge specifications for reinforced concrete, it will be found that those groups are more conservative than the ACI, and for obvious reasons. Bridge structures are, in general, subjected to more severe weather conditions than are buildings because of their more exposed locations; in addition, they must support more violent loadings from trucks or trains. The allowable stresses given in the 1989 AASHTO specification are $0.40f'_c$ for concrete and 20,000 psi for grade 40 steel and 24,000 psi for grade 60 steel.

B.2 DERIVATION OF WSD DESIGN FORMULAS

This section presents the derivation of the formulas needed for the design of rectangular beams with tensile steel only. Reference is made to Figure B.1 for the beam cross section and the terms that will be used in the derivation. The steel is once again transformed into an equivalent area of fictitious concrete that can resist tension. This area is nA_s.

In the alternate design method the most economical design possible is referred to as *balanced design*. A beam designed by this method will, under full service load, have its extreme fibers in compression stressed to their maximum permissible value f_c and its reinforcing bars stressed to their maximum permissible value f_s. Balanced design is the situation assumed for the beam and stress diagram shown in Figure B.1.

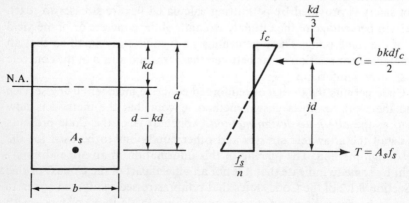

Figure B.1 Balanced design.

In the alternate design method the usual objective is to have a balanced design because it will often represent the most economical proportions. (Should a beam be designed in which the tensile steel reaches its maximum allowable tensile stress before the concrete fibers reach their maximum allowable compression stress, the beam is said to be *underreinforced.*)

The design formulas are derived on the basis of a consideration of the internal couple consisting of the two forces C and T. Once again, the total compression C equals the compression area bkd times the average compression stress $f_c/2$ and T equals $A_s f_s$. The sum of the horizontal forces in a beam in equilibrium is obviously zero, and thus $C = T$. The internal moment can be written as Cjd or Tjd, and these are equated to the external moment, M and the resulting expressions are solved for the beam dimensions and the steel area required.

As a straight-line variation of stress is assumed from f_c to f_s/n, the following ratio can be written and from it the design value of k obtained:

$$\frac{kd}{d} = \frac{f_c}{f_c + (f_s/n)}$$

$$k = \frac{f_c}{f_c + (f_s/n)} \tag{B.1}$$

In a similar manner j is determined:

$$jd = d - \frac{kd}{3}$$

$$j = 1 - \frac{k}{3} \tag{B.2}$$

Now, using the internal couples,

$$M = Cjd$$

$$M = \frac{bkdf_c}{2} jd$$

$$bd^2 = \frac{2M}{f_c kj} \tag{B.3}$$

$$M = Tjd$$

$$M = A_s f_s jd$$

$$A_s = \frac{M}{f_s jd} \tag{B.4}$$

These equations (B.1 through B.4) were derived for rectangular sections and they do not apply to sections where the compression area is not rectangular or to sections with compression reinforcing.

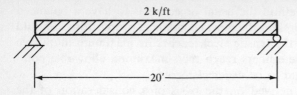

Figure B.2

■ EXAMPLE B.1

Design the beam shown in Figure B.2 for moment only. Compute stresses in the resulting section by using the transformed area method of Chapter 2; $f'_c = 3000$ psi, $f_c = 1350$ psi, $f_s = 20,000$ psi, and $n = 9$.

SOLUTION

Assume Bean Weight = 350 lb/ft

$$M = \frac{(2.35)(20)^2}{8} = 117.5 \text{ ft-k}$$

$$k = \frac{f_c}{f_c + (f_s/n)} = \frac{1350}{1350 + (20,000/9)} = 0.378$$

$$j = 1 - \frac{k}{3} = 1 - \frac{0.378}{3} = 0.874$$

$$
bd^2 = \frac{2M}{f_c kj} = \frac{(2)(12)(117,500)}{(1350)(0.378)(0.874)} = 6323
\qquad
\begin{array}{cc}
b & d \\
12 \times 22.95 \\
14 \times 21.25 \\
16 \times 19.88
\end{array}
$$

Minimum edge distance $= 1.50 + 0.375 + \dfrac{1.27}{2} = 2.51''$ (say $2\frac{1}{2}''$)

Try 14 × 24 Beam (d = 21.5 in.)

$$\text{Beam weight} = \frac{(14)(24)}{144}(150) = 350 \text{ lb/ft} \qquad \underline{\text{OK}}$$

$$A_s = Mf_s jd = \frac{(12)(117,500)}{(20,000)(0.874)(21.5)} = 3.75 \text{ in.}^2$$

Use 3 #10 bars (3.79 in.²) (Table A.4 of Appendix A).

Discussion   An effective depth of 21.25 in. was required, but to simplify the dimensions a d of 21.50 in. was selected (which when added to an edge distance of $2\frac{1}{2}$ in. gives a total depth of 24 in.). In addition, the 3 #10 bars selected have

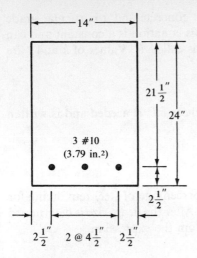

Figure B.3

a little more area than was required. As a result of these changes, the design is no longer a perfectly balanced one, but for all practical purposes it is assumed to be. A sketch of the beam is shown in Figure B.3.

Review of Design by Transformed Area

$$(14x)\left(\frac{x}{2}\right) = (9)(3.79)(21.5 - x)$$

Solving quadratic we get $x = 8.08$ in.

$$I_x = (\tfrac{1}{3})(14)(8.08)^3 + (9)(3.79)(13.42)^2 = 8605 \text{ in.}^4$$

$$f_c = \frac{(12)(117,500)(8.08)}{8605} = 1324 \text{ psi} < 1350 \text{ psi} \qquad \underline{\text{OK}}$$

$$f_s = (9)\frac{(12)(117,500)(13.42)}{8605} = 19,790 \text{ psi} < 20,000 \text{ psi} \qquad \underline{\text{OK}} \quad \blacksquare$$

Example B.2 illustrates the design of another simple beam using the alternate design method. The beam is unsymmetrically loaded and care must be taken in calculating the maximum moment. For such cases it is usually desirable to draw the shear and moment diagrams.

In the previous example the proportions of the beam were selected by use of the following expression:

$$bd^2 = \left|\frac{2}{f_c kj}\frac{M}{}\right|$$

It will be noted that for a particular grade of concrete and a particular grade of steel, the reciprocal of the circled part of this equation is a constant and can be taken from Appendix Table A.7, where it is called K. Values of k and j are also provided in the same table.

$$K = \tfrac{1}{2} f_c k j$$

This value is used in the following example where bd^2 is needed and is written as

$$bd^2 = \frac{M}{K}$$

Table A.7 also provides values of ρ, the percentage of steel reinforcing for the different grades of concrete and steel given. After the beam size is determined from bd^2, the steel area may be determined from the expression

$$A_s = \frac{M}{f_s j d}$$

or it may be determined by substituting into $A_s = \rho b d$, where ρ is taken from Table A.7. The values obtained from these two expressions will vary a little from each other unless the calculated values of b and d (not the round-off values, say, for d) are used in both calculations.

■ EXAMPLE B.2

Design the beam shown in Figure B.4, assuming $f'_c = 3000$ psi, $f_c = 1350$ psi, $f_s = 20,000$ psi, and $n = 9$.

SOLUTION

$K = 223$ and $j = 0.874$ from Appendix Table A.7. Assume beam weight = 800 lb/ft. Drawing shear and moment diagrams (Figure B.5),

$$bd^2 = \frac{M}{K} = \frac{(12)(450,000)}{223} = 24,215 \begin{cases} \begin{array}{cc} b & d \\ 16 \times 38.9 \\ 18 \times 36.7 \\ 20 \times 34.8 \end{array} \end{cases}$$

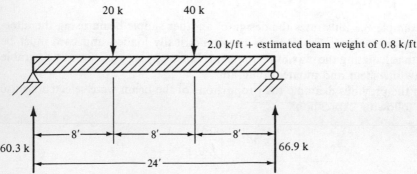

20 k   40 k

2.0 k/ft + estimated beam weight of 0.8 k/ft

60.3 k |← 8' →|← 8' →|← 8' →| 66.9 k

|← 24' →|

Figure B.4

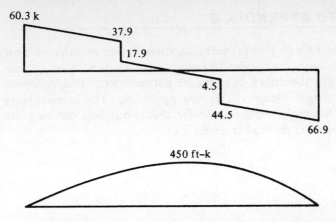

Figure B.5

Try 20 × 38 (d = 35.5)

$$\text{Beam weight} = \frac{(20)(38)}{144}(150) = 792 \text{ lb/ft} < 800 \text{ lb/ft} \quad \underline{\text{OK}}$$

$$\text{Minimum edge distance} = 1.50 + 0.375 + \frac{1.27}{2} = 2.51'' \qquad (\text{say } 2\tfrac{1}{2}'')$$

$$A_s = \frac{M}{f_s jd} = \frac{(12)(450,000)}{(20,000)(0.874)(35.5)} = 8.70 \text{ in.}^2$$

Use 7 #10 bars (8.86 in.²).

A sketch of the beam is shown in Figure B.6. ∎

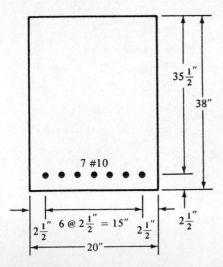

Figure B.6

CONCLUSION TO APPENDIX B

Appendix A of the ACI Code also provides alternate design requirements for shear and torsion, for members subject to compression loads with or without bending, and so on. For the design of members not included, the appropriate provisions of the Strength Design Code are applicable. The serviceability requirements of the Strength Design Code for deflections and cracking are applicable whichever design method is used.

Appendix C

Overreinforced Beams

C.1 INTRODUCTION

The Code (10.3.3) does not permit the design of overreinforced beams. Occasionally, however, you may be reviewing an existing structure and encounter an overreinforced beam. The member's nominal resisting moment may be computed as described in the next few pages and multiplied by an approximate ϕ factor to obtain its permissible strength. Rather than using a value of 0.9 as for underreinforced beams, a value of 0.7 is commonly used to account for the brittle nature of overreinforced beams.

For an overreinforced beam the strain in the steel when the concrete crushes will obviously be less than its yield strain. This situation is shown in Figure C.1.

If we can in some way obtain the value of c, we can easily determine ε_s by proportions and than compute f_s and $a = \beta_1 c$. The nominal resisting moment will then be

$$M_n = A_s f_s \left(d - \frac{a}{2} \right)$$

It is possible to determine c by a rather tedious trial-and-error procedure. We could assume a value of c (obviously greater than c_{bal}), compute ε_s, f_s, and a, and finally determine $T = A_s f_s$ and $C = 0.85 f'_c\, ab$. If T and C are equal, we will have the correct c; if not, we can try again, and so on.

A direct method for computing c is much to be preferred. Such a method is presented in the 11th edition of *Design of Concrete Structures* by A. H. Nilson and the late G. Winter (1991, McGraw–Hill, pages 92–93).

681

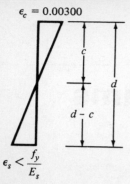

Figure C.1 Strain diagram for an overreinforced beam.

An expression for the strain in the steel, ε_s is written with reference to Figure C.1.

$$\frac{\varepsilon_c}{c} = \frac{\varepsilon_s}{d - c}$$

$$\varepsilon_s = \frac{\varepsilon_c}{c}(d - c)$$

The usual equation $C = T$ is written

$$0.85f'_c ab = A_s f_s$$

Replacing A_s with ρbd and f_s with $\varepsilon_s E$

$$0.85f'_c ab = \rho bd\varepsilon_s E_s$$

and then ε_s with $(\varepsilon_c/c)(d - c)$ where $\varepsilon_c = 0.003$

$$0.85f'_c ab = \rho bd(\varepsilon_c/c)(d - c)E_s$$

and replacing a with $\beta_1 c$

$$0.85f'_c \beta_1 cb = \rho bd(\varepsilon_c/c)(d - c)E_s$$

and then letting $k_u = c/d$ and $m = E_s\varepsilon_c/0.85\beta_1 f'_c$, we get a quadratic equation

$$k_u^2 + m\rho k_u = m\rho = 0$$

Solving for k_u, we obtain

$$k_u = \sqrt{m\rho + \left(\frac{m\rho}{2}\right)^2} - \frac{m\rho}{2}$$

Once we have k_u we can find $c = k_u d$. Example C.1. presents the analysis of an overreinforced beam using this procedure.

■ EXAMPLE C.1

Compute M_u for the beam of Figure C.2 if $f_y = 60,000$ psi and $f'_c = 3000$ psi.

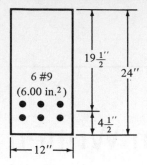

Figure C.2

SOLUTION

$$\rho = \frac{A_s}{bd} = \frac{6.00}{(12)(19.5)} = 0.0256 > \rho_{max} \text{ of } 0.0161$$

∴ It's overreinforced

$$m = \frac{E_s \varepsilon_c}{0.85 \beta_1 c_c'} = \frac{(29,000)(0.003)}{(0.85)(0.85)(3.0)} = 40.138$$

$$m\rho = (40.138)(0.0256) = 1.0275$$

$$k_u = \sqrt{m\rho + \left(\frac{m\rho}{2}\right)^2} - \frac{m\rho}{2} = \sqrt{1.0275 + \left(\frac{1.0275}{2}\right)^2} - \frac{1.0275}{2}$$

$$= 0.6227$$

$$c = k_u d = (0.6227)(19.50) = 12.14 \text{ in.}$$

$$a = \beta_1 c = (0.85)(12.14) = 10.32 \text{ in.}$$

$$\varepsilon_s = \varepsilon_c \frac{d - c}{c} = (0.003)\left(\frac{19.50 - 12.14}{12.14}\right) = 0.00182$$

$$f_s = \varepsilon_s E_s = (0.00182)(29 \times 10^3) = 52.78 \text{ ksi}$$

$$M_n = A_s f_s \left(d - \frac{a}{2}\right)$$

$$= (6.00)(52.78)\left(19.50 - \frac{10.32}{2}\right) = 4541.2 \text{ in.-k}$$

$$= 378.4 \text{ ft-k}$$

$$M_u = \phi M_n = (0.70)(378.4) = \underline{264.9 \text{ ft-k}}$$

∎

Appendix D

Concrete
A Computer Program Written by Nadim M. Aziz*

D.1 INTRODUCTION

CONCRETE is an interactive program consisting of 12 modules for analysis/
design of reinforced concrete beams, columns and footings. The modules are:

1. Flexural Analysis of Rectangular Beams
2. Design of Singly Reinforced Beams with Given ρ
3. Design of Singly Reinforced Beams with Given Dimensions
4. Analysis of T Beams
5. Design of T Beams
6. Analysis of Doubly Reinforced Rectangular Beams
7. Design of Doubly Reinforced Rectangular Beams
8. Design of Shear Reinforcing
9. Analysis or Design of Short Square or Rectangular Columns
10. Design of Wall Footings
11. Design of Square/Rectangular Footings for Square/Rectangular Columns
12. Design of Rectangular Beams for Shear and Torsion

The program is written in Quick Basic 7 and runs on IBM microcomputers
and compatibles. The program requires 150K bytes of memory to run.

D.2 MOVEMENT AND PROGRAM CONTROL

A module is selected from the main menu, and when execution of a module
is ended, control returns to the main menu for selection of other options. Once

---

* Associate Professor of Civil Engineering & Engineering Graphics at Clemson University.

a module is selected, the program displays a set of lines that define the input requests for the module. The screen of each program module is divided into two areas: the upper portion is for input to the program while the bottom is for output and other user input that which is output dependent. A highlighted area is an area in which the user should input values. When the user enters the last value on the upper portion of the screen, the program will automatically run and will show the results. In some program modules, the cursor returns to the upper portion of the screen indicating that the program is done. Here the user may elect to change an input value and run the program again. In order to do so, this time the user can enter the changes and press the F5 key to rerun the program with the new data. Moreover, the user may instead elect to print the results by pressing the F10 key or return to the main menu by pressing the ESC key. In some program modules, the user is requested to enter a selection of reinforcing bars from a displayed list or simply enter a value that is needed at an intermediate point in the program. An input cell is displayed in the bottom portion of the screen and the user should enter the required value and press the ENTER key.

A help line is always displayed at the bottom of the screen. On-line help is available for each module by pressing the F2 key. These help messages refer to sections in the textbook and in the ACI Code and help the user in understanding the method of solution. Users are encouraged to enter the data correctly the first time around. However, the programs can be used to study the effect on the design of changing one or more input values. The program, however, should not be expected to produce reasonable results for unrealistic input.

Additional help on the special keys used for movement around the program is obtained by pressing the F1 key. The following is a summary of the special keys and their functions:

F1	Help on special keys
F2	Technical help on individual program modules
F5	Rerun module
F10	Print results
ESC	Return/exit
ENTER	Accept input and move to next input request
Up Arrow	Accept input and move backward to previous input request
Down Arrow	Accept input and move forward to the next input request
Right Arrow	Move right inside an input cell
Left Arrow	Move left inside an input cell
Del	Delete the number at cursor location
Backspace	Delete the previous number inside an input cell
Page Up	Move to the previous page on the help screen
Page Down	Move to the next page on the help screen

Glossary

Aggregate interlock Shear or friction resistance by the concrete on opposite sides of a crack in a reinforced concrete member (obviously larger with narrower cracks).

Balanced failure condition The simultaneous occurrence of the crushing of the compression concrete on one side of a member and the yielding of the tensile steel on the other side.

Bond stresses The shear-type stresses produced on the surfaces of reinforcing bars as the concrete tries to slip on those bars.

Camber The construction of a member bent or arched in one direction so that it won't look so bad when the service loads bend it in the opposite direction.

Capacity reduction factors Factors that take into account the uncertainties of material strengths, approximations in analysis, and variations in dimensions and workmanship. They are multiplied by the nominal or theoretical strengths of members to obtain their permissible strengths.

Cast in place concrete Concrete cast at the building site in its final position.

Column capital A flaring or enlarging of a column underneath a reinforced concrete slab.

Composite column A concrete column which is reinforced longitudinally with structural steel shapes.

Concrete A mixture of sand, gravel, crushed rock, or other aggregates held together in a rocklike mass with a paste of cement and water.

Cover A protective layer of concrete over reinforcing bars to protect them from fire and corrosion.

Cracking moment The bending moment in a member when the concrete tensile stress equals the modulus of rupture and cracks begin to occur.

Creep or plastic flow When a concrete member is subjected to sustained compression loads, it will continue to shorten for years. The shortening which occurs after the initial or instantaneous shortening is called *creep* or *plastic flow* and is caused by the squeezing of water from the pores of the concrete.

Dead load Loads of constant magnitude that remain in one position. Examples: weights of walls, floors, roofs, plumbing, fixtures, structural frames, and so on.

Development length The length of a reinforcing bar needed to anchor or develop its stress at a critical section.

Doubly reinforced beam Concrete beams which have both tensile and compression reinforcing.

Drop panels A thickening of a reinforced concrete slab around a column.

Effective depth The distance from the compression face of a flexural member to the center of gravity of the tensile reinforcing.

Factored load A load that has been multiplied by a load factor, thus providing a safety factor.

Flat plate Solid concrete floor or roof slabs of uniform depths that transfer loads directly to supporting columns without the aid of beams or capitals or drop panels.

Flat slab Reinforced concrete slab with capitals and/or drop panels.

Formwork The mold in which semiliquid concrete is placed.

Grade 40 (60) steel Steel with a minimum yield stress of 40,000 psi (60,000 psi).

Honeycomb Areas of concrete where there is segregation of the coarse aggregate or rock pockets where the aggregate is not surrounded with mortar. It is caused by the improper handling and placing of the concrete.

Inflection point   A point in a flexural member where the bending moment is zero and where the moment is changing from one sign to the other.

Influence line   A diagram whose ordinates show the magnitude and character of some function of a structure (shear, moment, etc.) as a load of unity moves across the structure.

Interaction curve   A diagram showing the interaction or relationship between two functions of a member, usually axial column load and bending.

L Beam   A T beam at the edge of a reinforced concrete slab.

Lightweight concrete   Concrete where lightweight aggregate (such as zonolite, expanded shales, sawdust, etc.) is used to replace the coarse and/or fine aggregate.

Limit state   A condition at which a structure or some part of that structure ceases to perform its intended function.

Live loads   Loads that change position and magnitude. They move or are moved. Examples: trucks, people, wind, rain, earthquakes, temperature changes, and so on.

Load factor   A factor generally larger than one which is multiplied by a service or working load to provide a factor of safety.

Long columns   *See* Slender columns.

Microcrack   A crack too fine to be seen with the naked eye.

Modulus of elasticity   The ratio of stress to strain in elastic materials. The higher its value, the smaller the deformations in a member.

Modulus of rupture   The flexural tensile strength of concrete.

Monolithic concrete   Concrete cast in one piece or in different operations but with proper construction joints.

Nominal strength   The theoretical ultimate strength of a member such as M_n (nominal moment), V_n (nominal shear), and so on.

One-way slab   A slab designed to bend in one direction.

Overreinforced members   Members for which the tensile steel will not yield (nor will cracks and deflections appreciably change) before failure, which will be sudden and without warning due to crushing of the compression concrete.

P-delta moments   *See* Secondary moments.

Plain concrete   Concrete with no reinforcing whatsoever.

Plastic centroid of column   The location of the resultant force produced by the steel and the concrete.

Plastic deformation   Permanent deformation occurring in a member after its yield stress is reached.

Plastic flow   *See* Creep or plastic flow.

Poisson's ratio   The lateral expansion of a member divided by its longitudinal shortening when the member is subjected to compression forces. (Average value for concrete is about 0.16.)

Posttensioned concrete   Prestressed concrete for which the steel is tensioned after the concrete has hardened.

Precast concrete   Concrete cast at a location away from its final position. It may be cast at the building site near its final position but usually is done back at a concrete yard.

Prestressing   The imposition of internal stresses into a structure that are of an opposite character to those that will be caused by the service or working loads.

Pretensioned   Prestressed concrete for which the steel is tensioned before the concrete is placed.

Primary moments   Computed moments in a structure which do not account for structure deformations.

Ready-mixed concrete   Concrete that is mixed at a concrete plant and then is transported to the construction site.

Reinforced concrete   A combination of concrete and steel reinforcing wherein the steel provides the tensile strength lacking in the concrete. (The steel reinforcing can also be used to resist compressive forces.)

Secondary moments   Moments caused in a structure by its deformations under load. As a column bends laterally, a moment is caused equal to the axial load times the lateral deformation. It is called a *secondary* or *P-delta* moment.

Serviceability   Pertains to the performance of structures under normal service loads and is concerned with such items as deflections, vibrations, cracking, and slipping.

Service loads   The actual loads that are assumed to be applied to a structure when it is in service (also called *working loads*).

Shearheads   Cross-shaped elements such as steel channels and I beams placed in reinforced concrete slabs above columns to increase their shear strength.

Shores   The temporary members (probably wood or metal) that are used for vertical support for formwork into which fresh concrete is placed.

Short columns   Columns with such small slenderness ratios that secondary moments are negligible.

Slender columns (or long columns)   Columns with sufficiently large slenderness ratios that secondary moments appreciably weaken them (to the ACI an appreciable reduction in strength in columns is more than 5%).

Spalling   The breaking off or flaking off of a concrete surface.

Spirals   Closely spaced wires or bars wrapped in a continuous spiral around the longitudinal bars of a member to hold them in position.

Spiral column   A column that has a helical spiral made from bars or heavy wire wrapped continuously around its longitudinal reinforcing bars.

Split-cylinder test   A test used to estimate the tensile strength of concrete.

Stirrups   Vertical reinforcement added to reinforced concrete beams to increase their shear capacity.

Strength design   A method of design where the estimated dead and live loads are multiplied by certain load or safety factors. The resulting so-called *factored loads* are used to proportion the members.

T beam   A reinforced concrete beam that incorporates a portion of the slab which it supports.

Tendons   Wires, strands, cables, or bars used to prestress concrete.

Tied column   A column with a series of closed steel ties wrapped around its longitudinal bars to hold them in place.

Ties   Individual pieces of wires or bars wrapped at intervals around the longitudinal bars of a member to hold them in position.

Top bars   Horizontal reinforcing bars that have at least 12 in. of fresh concrete placed beneath them.

Transformed area   The cross-sectional area of one material theoretically changed into an equivalent area of another material by multiplying it by the ratio of the moduli of elasticity of the two materials. For illustration, an area of steel is changed to an equivalent area of concrete, expressed as $A_{steel} \dfrac{E_{steel}}{E_{concrete}}$.

Two-way slabs   Floor or roof slabs supported by columns arranged so that the slabs can bend in two directions.

Underreinforced member   A member that is designed so that the tensile steel will begin to yield (resulting in appreciable deflections and large visible cracks) while the compression concrete is still understressed. Thus a warning is provided before failure occurs.

Web reinforcement   Shear reinforcement in flexural members.

Working loads   The actual loads that are assumed to be applied to a structure when it is in service (also called *service loads*).

Working stress design   A method of design where the members of a structure are so proportioned that the estimated dead and live loads do not cause elastically computed stresses to exceed certain specified values. Method is also referred to as *allowable stress design, elastic design,* or *service load design.*

Index